EXPERIMENTS IN BIOLOGY

A Manual to Accompany BIOLOGY: *A Journey Into Life*

by Karen Arms and Pamela S. Camp

Illustrations by Monica Howland and Fine Line Illustrations, Inc.

CAROLYN EBERHARD
Cornell University

SAUNDERS COLLEGE PUBLISHING
Philadelphia Fort Worth Chicago
San Francisco Montreal Toronto
London Sydney Tokyo

Text Typeface: 10/12 Meridien
Compositor: Bi-Comp, Inc.
Acquisitions Editor: Edward Murphy
Developmental Editor: Lynne Gery
Art Director: Carol C. Bleistine
Text Designer: Dorothy Chattin
Cover Designer: Lawrence R. Didona
Text Artwork: Monica Howland and Fine Line Illustrations, Inc.
Production Manager: Merry Post

EXPERIMENTS IN BIOLOGY

ISBN 0-03-012842-0

Printed in the United States of America.

9 032 9876543

Preface

This manual is designed to be used in the laboratory section accompanying a general biology course. The order of topics is similar to that in the accompanying textbook, *Biology: A Journey into Life*, by Karen Arms and Pamela S. Camp. However, the descriptive material in this laboratory manual is sufficiently comprehensive and detailed that it can be used with other texts.

Since the content and form of previous editions have successfully met the needs of many instructors, this edition is a revision rather than a drastically altered manual. A section on Mendelian genetics in corn and a new exercise on DNA have been added. The exercise on behavior in schooling fish now appears as a project. There are many new or redrawn figures, and a special insert of color plates illustrates some material that does not lend itself well to black and white. A short reference list for students appears at the end of each topic in the laboratory manual, and references are included in the instructor's manual. There are four open-ended projects at the end of the manual for courses in which students are asked to independently carry out work that is less structured. Directions for writing a laboratory report and for statistical analysis to complement the projects are given in appendices. A description of the metric system, now including a short set of problems and answers, is given in the final appendix.

To accommodate the varying requirements of instructors teaching biology laboratory, 31 topics have been included—more than would ordinarily be covered in a one- or two-semester course. Among these topics is introductory material important to both majors and nonmajors taking laboratory: seven topics on the diversity of living organisms, seven topics emphasizing mammalian structure and function (including four involving the fetal pig), and two topics dealing with plant biology. Other topics in this manual, such as those on enzyme action, genetics, muscle physiology, and animal behavior, help students learn how to gather and analyze data and to design experiments. We believe it is necessary that a general biology course provide a thorough introduction to the basic biology of plants and animals because many students, including biology majors, do not pursue studies in these areas. Moreover, those who do take more advanced courses often find that the instructor assumes they are familiar with material that in fact was not emphasized or was eliminated altogether from their introductory biology courses.

Each topic begins with a formal *Objectives* section that describes what knowledge or skills the student should be able to demonstrate after completing the exercise. *Text References* to the appropriate sections of the Arms and Camp text are given, and *Key Topics* then state the subjects and concepts relevant to each exercise. The exercises contain questions concerning information recall, concepts, and analysis of laboratory material to stimulate student discussion during the laboratory period. Worksheets are provided at the end of many topics for students to use to record data and observations and draw conclusions for particular experiments. These sheets can then be collected by the instructor for grading at the end of the laboratory period. An index is provided for the convenience of both the student and the instructor in using the manual.

The exercises present techniques such as pipeting, chromatography, and the use of the spectrophotometer and kymograph. We have tried to minimize the use of complicated procedures and elaborate equipment wherever possible so that each experiment is feasible in some form at most colleges and universities. The major recurrent budget items for these exercises are preserved fetal pigs and live material. Although we encourage the use of live material, we have often provided for the use of preserved material as well in the event that live material is unavailable or too expensive.

Four projects for independent study are given at the end of the manual, in which groups of students are asked to design and/or conduct and analyze experiments with schooling fish, plant hormones, plant nutrition, and pond microecosystems. The projects emphasize input from the students rather than the use of elaborate equipment, and they have been used successfully with students who had varying levels of preparation. These projects are very different from one another, so at least one should be appropriate for any given course.

The separate *Instructor's Manual*, which may be obtained from the publisher free of charge, suggests how many topics may be combined, shortened, or expanded. It also provides lists of material and equipment needed for these exercises, sources for specific supplies, directions for preparing solutions and handling live material, and an estimate of the cost of each exercise. A list of reference books, films, and film loops is also included. Requests for the *Instructor's Manual* should be addressed to: Order Fulfillment Department, Holt Rinehart & Winston, Inc., Orlando, FL 32887-0454.

ACKNOWLEDGMENTS

The foundation for these laboratory exercises is based upon a long association with colleagues in Introductory Biology at Cornell University: Antonie W. Blackler, the late William T. Keeton, Melvin Kreithen, Carol Hardy McFadden, Neil Campbell, William McFarland, P. Richard Ecklund, and Robert Turgeon. We are indebted to all of our colleagues at Cornell University and at Ithaca College for their support, suggestions and contributions to the manual large and small: Renee Alexander, Robert Anderson, Jeanne Appling, Alan Bennett, Everette Busbee, Judith Byrnes, Donna Hill Canney, Nancy Clum, Gary Creason, Lin Davidson, Jan Factor, Gail Ferstandig, Sue Giffen, Jon Glase, Larry Greenberg, Monica Howland, Barry Ingber, Leanne Ketcham, Debra Kirchhof-Glazier, Marie Knowlton, Joshua Korman, Martha Lackey, Carolyn Louie, Martha Lyon, Carol Mapes, Michael Meredith, Elizabeth Mort, Sharman O'Neill, Donald Rhodes, Helen Roland, William Schrader, Buffy Silverman, Paul Sisco, Ann Snook, Daniel Solomon, Adrian Srb, George Swenson, Craig Tendler, Eileen Whalen, Richard Wodzinski, Nancy Wolf, and Hal Yingling.

We are extremely grateful to Monica Howland for researching and preparing many new and revised figures at short notice and long distance, as well as critiquing many of the exercises.

Philip Snider from the University of Houston suggested and provided helpful comments on the corn genetics exercise, and Nancy Clum, Cornell University, reviewed the comparative vertebrate anatomy and evolution exercise. Robert Anderson from Ithaca College gave important pointers on data handling and statistics.

We also wish to acknowledge the following individuals who offered instructive comments on the manual manuscript at some point: Betty D. Allamong, Ball State University; Robert Amy, Southwestern at Memphis; Paul Davidson, Rollins College; John Erickson, Western Washington University; Cesar Fermin, Florida Institute of Technology; Eileen Gregory, Rollins College; Terry William Hill, Southwestern at Memphis; Russell C. Hollingsworth, Tarrant County Junior College, Northeast; Pius Horner, San Bernardino Valley College; Virginia Keville, Salem State College; Stephen Klein, Schenendehowa Central High School; William Leonard, University of Nebraska; Karen Loftin, University of Houston; Carmita Love, Community College of Philadelphia; W. Wallace Martin, Randolph Macon College; Karel Rogers, Adams State College; Roger Sawyer, University of South Carolina; Allen Schroeder, Gettysburg College; Robert Smith, San Bernardino Valley College; Gail Stetten, Johns Hopkins Hospital; Joseph Wood, University of Missouri; and Sherry Zachary, student at San Bernardino Valley College.

We are grateful to Sara Jane Hymes for typing the instructor's manual manuscript, and to Saunders' Biology Editor Ed Murphy and the editorial staff, especially Developmental Editor Lynne Gery and Project/Copy Editor Robin Bonner for bringing the manual to fruition across the considerable barrier of the Atlantic Ocean.

Finally we wish to express our gratitude to the literary executor of the late Sir Ronald Fisher, F.R.S., Dr. Frank Yates, F.R.S., and Longman Group, Ltd. in London for permission to reprint material from their book, *Statistical Tables for Biological, Agricultural, and Medical Research* (6th edition, 1974).

Ithaca, New York C.E.
November 1987

Contents

To the Student

OBJECTIVES

At the beginning of each topic in this manual is a list of specific Objectives describing the subject areas and skills that you should master upon completion of the topic. In addition, there are other, broader objectives that you should try to achieve in the course of using this manual. We hope that you will:

1. Interact with other students and your instructor by asking questions about the laboratory material, by comparing your findings, and, when possible, by presenting your results in class discussions at the end of the laboratory session.
2. Acquire knowledge that will provide a foundation for further scientific study.
3. Apply what you have learned in this course to your daily life, and try to make more informed decisions about such matters as nutrition, medicine, natural resources, endangered species, and environmental hazards.

SAFETY

Experiments in biology laboratories sometimes involve working with potentially dangerous chemicals and expensive equipment that must be used properly. To ensure your personal safety, follow these simple rules:

1. Do not eat in the laboratory.

2. Do not drink in the laboratory.

3. Never smoke in the laboratory.

4. Protect your face and eyes by wearing safety glasses and contact lenses. Tie back long loose hair.

5. Roll up long sleeves when using equipment or open flames.

6. Keep open flames away from flammable chemicals or materials and *yourself!*

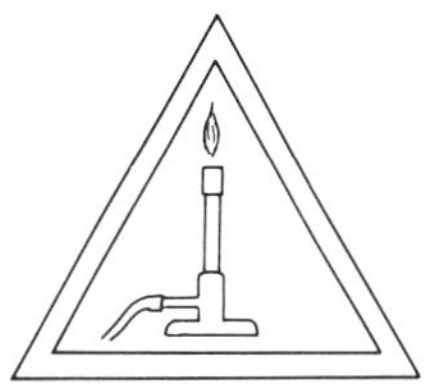

7. Report any injury to yourself or another student, no matter how minor, to the instructor.

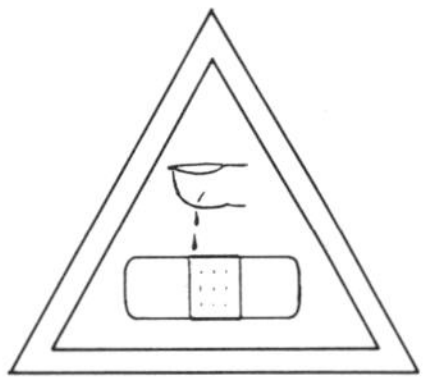

8. Report any spills to your instructor, and clean up the spills immediately.

9. Determine the shortest exit route from the building, and be prepared to leave the room immediately in case of a fire.

PREPARATION

1. Read the assigned topic before the laboratory session. Otherwise your time will be spent less productively than it should be. Pay particular attention to the ***Objectives,*** which explain what you should get out of each laboratory.
2. Use your textbook to complement each lab by reading about the ***Key Topics,*** listed before the introduction to each topic, while the laboratory work is still fresh in your mind. If this material has not yet been discussed in lecture, read the relevant text sections *before* the laboratory session.

METRIC SYSTEM AND SCIENTIFIC NOTATION

If you are not familiar with the **metric system,** carefully read over Appendix D at the end of this manual. Since you will be expected to use the metric system in this course, make sure you understand the problem set given there.

Some measurements in biology have very small or very large units, so it is awkward to write them out as decimals. Instead, we use **scientific notation** in which the decimal point is placed after the first digit and the number is expressed as a multiple of the power of 10. For example, 506 kilograms is expressed in scientific notation as $\mathbf{5.06 \times 10^2}$ **kg,** and $\frac{1}{200}$ (0.005) of a second is expressed as $\mathbf{5 \times 10^{-3}}$ **sec.**

Scientific notation makes it easy to use the metric system. In our second example, 5×10^{-3} sec equals 5 msec because 10^{-3} sec equals 1 msec (or 1 second equals 10^3 msec). The formal calculation in which 10^{-3} cancels out would be: 5×10^{-3} sec $\times$ 1 msec/10^{-3} sec = 5 msec. Always use scientific notation for values with an order of magnitude of 10^{-3} or less and 10^3 or greater.

Introductory biology laboratory may be one of your few opportunities to explore firsthand many of the topics covered in this manual. We hope that you will take advantage of this time to interact with your classmates and your instructor as much as possible so that this experience will be both rewarding and enjoyable.

C.E.

TOPIC 1

Biological Chemistry

OBJECTIVES

When you have completed this topic, you should be able to:

1. Define the term functional group and give the functional group or groups characteristic of an acid, an amino acid, a carbohydrate, a fatty acid, and an alcohol.
2. Describe how to carry out a simple laboratory test for a monosaccharide, a polysaccharide, a protein, and a fat.
3. Describe how a protein can be broken down into its building blocks, amino acids.
4. Show by thin-layer chromatography that different amino acids have different chemical properties.
5. Explain how chromatography works to separate chemically different substances.

TEXT REFERENCES Chapter 2; Chapter 3

KEY TOPICS

atoms	chemical reactions	water
bonds	oxidation	carbon chemistry
compounds	reduction	biological molecules
polarity		

INTRODUCTION

An element is a particular type of atom that is different from the atoms of all the other elements. The elements most abundant in living organisms are:

Carbon	= C	Nitrogen	= N
Hydrogen	= H	Sulfur	= S
Oxygen	= O	Phosphorus	= P

These elements can be combined in enormous variety to make up the chemical molecules on which life is based.

In most biological molecules, the atoms are held together by **covalent bonds** in which electrons are shared between adjacent atoms. Ionic bonds, hydrogen bonds, and disulfide bonds are also important. If you are not sure how each of these bonds is formed, refer back to your text and review them.

Fortunately for us, covalent bonds are not very easily broken down under ordinary conditions. Some small molecules are especially stable and are used as the building blocks or **monomers** of very large molecules called **polymers.** A polymer often consists of many similar monomers strung together in a long chain. It is easier to break down a polymer into its monomers than to break apart the monomers themselves.

Biological molecules can be categorized on the basis of similar parts called **functional groups:**

Carboxyl Hydroxyl Amino

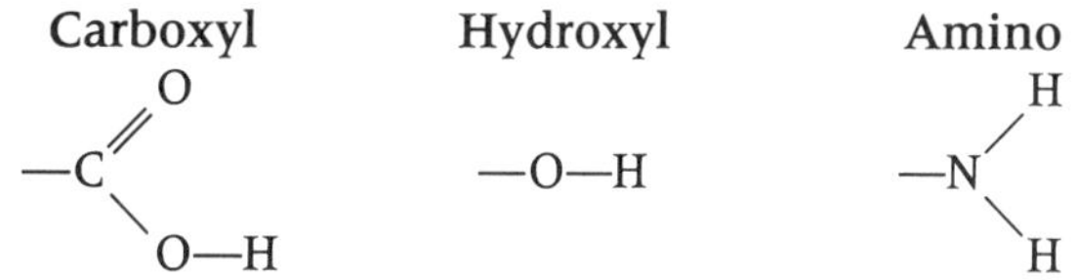

Aldehyde

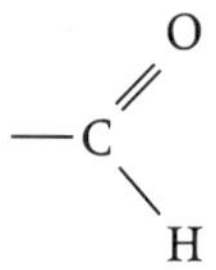

Ketone

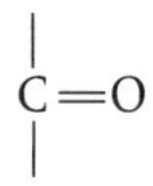

Sulfhydryl

—S—H

Hydrocarbon

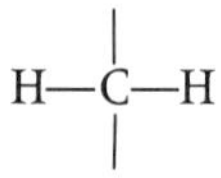

For example, any molecule that has the carboxyl group 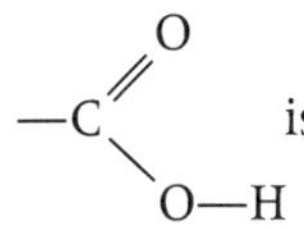 is an acid. (Remember that one line represents one covalent bond and two lines represent a double bond.) The term **fatty acid** refers to the group of acids that have the carboxyl group and several 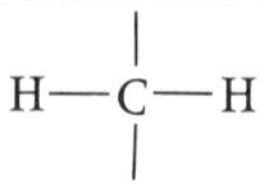 groups besides. An **amino acid** molecule has an amino group and a carboxyl group. A carbohydrate such as a monosaccharide or simple sugar has an aldehyde or ketone group and some hydroxyl groups. (Aldehyde and ketone groups are specific examples of carbonyl compounds that always contain 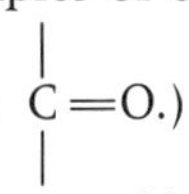.)

Amino acid molecules may have additional functional groups (sulfhydryl, additional carboxyl groups, additional hydroxyl groups, and so on), but they all share the functional groups characteristic of an amino acid.

Functional groups can be divided into two groups, polar and nonpolar, depending on whether they contain **polar covalent bonds** or are **ions.** Polar or charge-bearing groups tend to make the molecule **hydrophilic** (= water loving). Groups with **nonpolar covalent bonds** such as C—C and C—H tend to make the molecule **hydrophobic** (= water fearing).

Polar Functional Groups

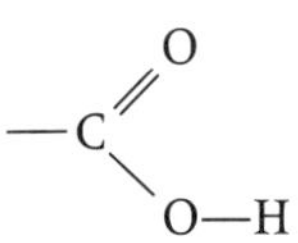 carboxyl group

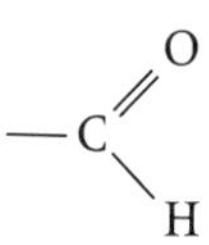 aldehyde group

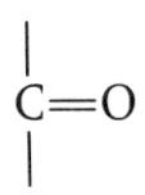 ketone group

—O—H hydroxyl group

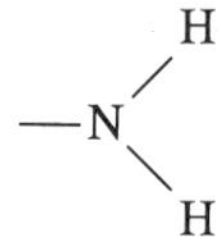 amino group

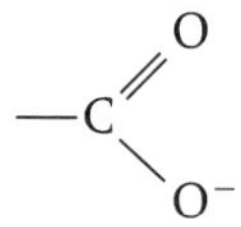 anion (negative charge)

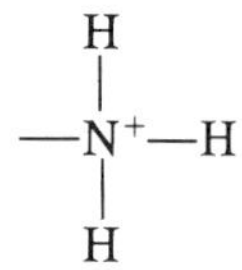 cation (positive charge)

Nonpolar Functional Groups

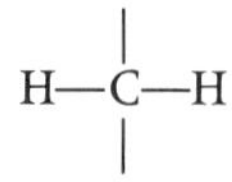 hydrocarbon group

—S—H sulfhydryl group

Learn the names and structures of these functional groups in order to understand the chemistry of today's lab.

Polar functional groups are hydrophilic because they interact with each other and with water, another polar substance; they do not interact significantly with nonpolar hydrophobic functional groups or with nonpolar solvents.

For example, oil does not mix with vinegar, a solution of acetic acid and water.

What kind of functional group must oil therefore contain? ______________

IDENTIFICATION OF BIOLOGICAL MOLECULES

Molecules of a certain class have similar chemical properties because they have the same functional groups. A chemical test that is sensitive to these groups can be used to identify molecules that are in that class. Practice the following tests by using them to see what various foods contain. In each test, you will be including a substance that does not react in the test to serve as a **negative control.** A substance known to give a reaction in the test would be a **positive control.**

Please follow the directions carefully, so that your test results are unambiguous. Record and analyze your results in Table 1–2, at the end of this topic.

If you are using pipets in this lab, use 2 mL of solution per cm, and omit marking the tubes.

Carbohydrates

______________ ______________

These molecules contain the very polar **aldehyde** and **hydroxyl** functional groups. Label them correctly.

Certain mono- and disaccharides can be detected because of their aldehyde groups. Some polysaccharides can be detected because of their specific three-dimensional structure. Carbohydrates, which are water soluble or hydrophilic substances, are not detected by their hydroxyl groups.

Given the structure of water, can you explain why not?

____

____

Benedict's Test

This **reducing sugar** test depends on a free aldehyde group:

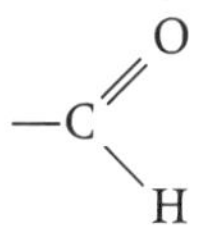

Monosaccharides and some disaccharides have a free aldehyde group and consequently will react to give a positive **Benedict's test.** Such a sugar acts as a **reducing agent** and is called a **reducing sugar.** The reaction goes like this:

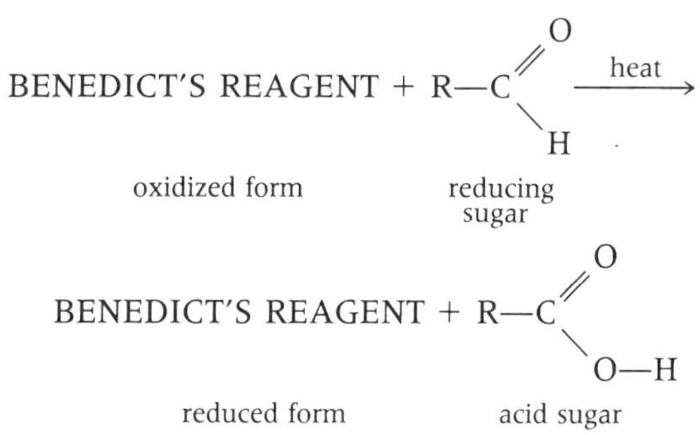

This is a typical **oxidation-reduction reaction** in which oxidation of the sugar occurs simultaneously with reduction of the Benedict's reagent. Remember that:

REDUCTION means removal of oxygen or addition of hydrogen. **OXIDATION means addition of oxygen or removal of hydrogen.**

In this case the sugar obviously gained an oxygen during the reaction, and therefore the sugar underwent oxidation.

Some disaccharides (e.g. sucrose) and all polysaccharides (e.g. starch and cellulose) are not reducing sugars because they do not have free aldehyde groups.

Will ordinary table sugar give a positive Benedict's test? ______________

You'll use the Benedict's test to compare the reducing sugar content of onion juice with that of potato juice. Enter your results on Table 1–2, at the end of this topic.

- Mark 3 test tubes at 1 cm and 3 cm from the bottom.
- Add onion juice up to the 1-cm mark of tube #1.
- Add potato juice to the 1-cm mark of tube #2.
- Add water to the 1-cm mark of tube #3, the control.
- Add Benedict's Reagent to the 3-cm marks of all 3 tubes and mix.
- Heat the tubes for 3 minutes in a boiling water bath.
- Remove the tubes using the tongs provided, and record their colors.

A change from clear blue to red orange indicates an abundance of reducing sugars. A change to green indicates the presence of a smaller amount of reducing sugar.

Which vegetable, potato or onion, has more reducing sugar? ______________

What was the purpose of tube #3?

____

Iodine Test

Another test, the **Iodine test,** can be used to distinguish **starch** from mono-, di-, and other polysaccharides. Starch is a polymer of glucose in which the chains are coiled up in a particular way so that they can interact with the iodine molecules to give a distinctive blue-black color. Other polymers, even those of glucose, lack the precise coiled structure of starch and do not give the dark color.

Compare the starch content of onions and potatoes, and enter your results on Table 1–2, at the end of this topic.

- Mark 3 tubes at 1 cm.
- Fill one to the mark with onion juice, another with potato juice, and the third with water.
- Add 3 drops of iodine solution to each tube and mix.

A positive test will give a blue-black color. A yellow or brown result is negative.

Which vegetable, onion or potato, stores more starch?

____

Sudan III Test

A fat is a hydrophobic molecule consisting of a glycerol molecule joined to three fatty acids. The important functional groups involved in a fat are the following:

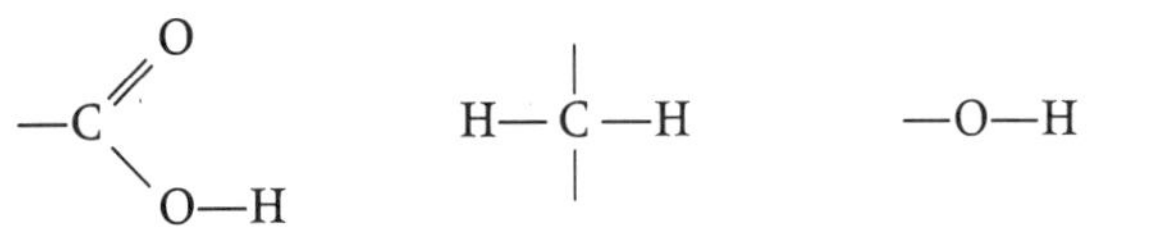

______________ ______________ ______________

The hydroxyl groups of the glycerol (label the hydroxyl functional group above) react with the carboxyl groups of the fatty acids (label the carboxyl group) in a condensation reaction, so these functional groups are not available in the fat itself for a test reaction. Instead, the **Sudan III test** depends on the detection of the hydrocarbon groups remaining. The colored dye, Sudan III, and the hydrocarbon groups are **nonpolar** and stick tightly together in their **polar** surroundings. This is called **hydrophobic interaction** and is thus the basis for this rather inelegant test.

Since fats are not very soluble in water, extracts of four foods have been made using 95% ethanol (ethanol is the alcohol found in alcoholic beverages such as beer and wine.) A fifth tube contains only water and ethanol to be used for the control. Test each of the tubes for the presence of fat as follows, and enter your results on Table 1–2, at the end of this topic.

- Using a lead *pencil,* mark a filter paper disc with a "W" (water), "F" (flour), "K" (cream), "C" (coconut), and an "M" (margarine) so that the letters are fairly equally spaced on the filter paper. Draw a small circle next to each letter as shown in Figure 1–1. Add your initials somewhere.
- *Using the dedicated Pasteur pipet,* add a small drop from each tube to the appropriate circled spot on the filter paper. Return each pipet to its proper solution when you finish with it.
- Allow the paper to dry *completely.* Use the blow dryer if necessary.
- Soak the paper for 3 minutes in the Sudan III solution.
- Use forceps to remove the paper from the stain.
- Rinse the paper in the water bath for one minute.
- Examine the intensity of orange staining of the 5 spots.

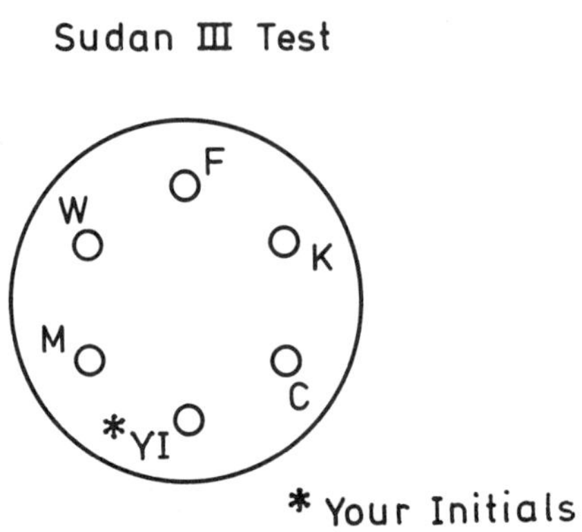

Figure 1–1. Filter paper ready for the Sudan III test.

Note: The filter paper itself will pick up some of the dye, so the excess must be soaked off in order for you to see the spots. Rinse some more if no spots are apparent. You should see at least one spot!

- Rate the four foods as 0 = no color over the filter paper background, + = faint orange color, and ++ = definite orange color, and record the results.

What was the purpose of the "water" spot?

____

Which foods contain detectable amounts of fat?

____

Which had the highest fat content? ______________

Biuret Test

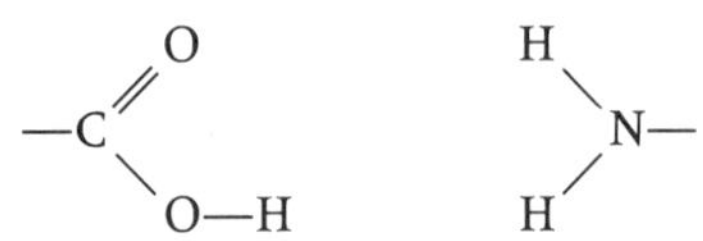

carboxyl group amino group

Proteins are polymers of amino acids in which the carboxyl group of one amino acid residue joins head-to-tail with the amino group of its neighbor in a **peptide bond.** In this reaction a molecule of water is formed. Compare the functional groups above with the peptide bond shown below and circle the atoms lost to form the water molecule. In a peptide bond, the bound amino group is nevertheless sufficiently reactive to change the **Biuret reagent** from blue to violet. Thus the basis of the test is a subtle interaction between the **copper ions** in the Biuret reagent and the amino groups in the peptide bonds. This test is specific for the **amino group** found in the peptide bonds of proteins:

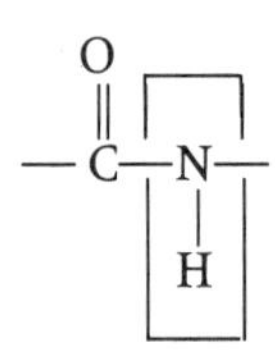

You'll observe how the blue biuret reagent reacts in the presence of protein by comparing the amount of protein in egg white with the amount in chicken soup. The solutions contain a serving of egg white or of chicken broth in equivalent volumes. Enter your results on Table 1–2, at the end of this topic.

- Mark 3 test tubes at 3 cm and 5 cm from the bottom.
- Add egg white solution to the 3-cm mark of the first tube, chicken soup solution to the 3-cm mark of the second tube, and water to the 3-cm mark of the third tube.
- Add sodium hydroxide (NaOH) to the 5-cm mark of each tube and **mix.** Be careful not to get NaOH on yourself.
- Add 5 drops of 1% copper sulfate ($CuSO_4$), the Biuret test reagent, to each tube and **mix.**
- Hold the tubes against a white background, note the color of each, and record the results.

A color change from blue to violet occurs when copper ions react with the amide bonds, and such a change indicates the presence of proteins.

Which food contains more protein? ______________

Testing Unknowns

Your instructor will provide each of you with an unknown food. The letter gives you the group of possible unknowns that you might have:

A	B	C
soy flour	corn starch	ground coffee
glucose	honey	instant coffee
powdered milk	table sugar	
enriched flour	egg white	

D	E
Vermont Maid syrup	glucose
maple syrup	table salt
	potato starch
	gelatin

The compositions of these foods are given in Table 1–1. Low levels of components will probably not show up in these tests.

- Record the letter and number of the unknown, e.g. A–3, in Table 1–3, at the end of this topic.
- Determine the results you would *expect* for each possible food in each of the tests you have used in

TABLE 1–1. Composition of Foods*

Grams per 100 grams edible portion (or %)
AMOUNTS IN PARENTHESES NOT DETECTED IN TESTS

	Water	Calories	% Protein	% Fats	% Carbohydrate
Corn Starch	12	362	(.3)	(trace)	87.6 starch
Egg White	87.6	51	10.9	(trace)	(0.8)
Egg Yolk	51.1	348	16.0	30.6	(0.6)
Enriched flour (white)	12	364	(10.5)	(1.0)	76.1 starch
Gelatin (dry)	13.0	335	85.6	(.1)	(0)
Glucose	9	335	(0)	(0)	91
Ground Coffee	4.1	301	(12.5)	15.4	28.5
Honey	17.2	304	(0.3)	(0)	82.3 glucose + fructose
Instant Coffee	2.6	129	(0)	(0)	35.0
Maple Syrup (estimated)	21.5	299	(0)	(0)	(77) only sucrose
Potato Starch	7.6	351	(8.0)	(0.8)	79.9 starch
Powdered Skim Milk	4	352	35.8	(0.7)	51.6 lactose
Soy Flour (high fat)	8	380	41.2	12.1	33.3
Table Salt	.2	0	(0)	(0)	(0)
Table Sugar	.5	385	(0)	(0)	(99.5)
Vermont Maid Syrup (estimated)	25	286	(0)	(0)	74 glucose + fructose

* Watt, Bernice K., and Annabel L. Merrill. "Composition of Foods. Raw. Processed. Prepared." Agricultural Handbook No. 8. Revised December 1963. Superintendent of Documents, U.S. Govt. Printing Office. Washington, D.C.

Table 1–3. For group A, that would be soy flour (–,+,+,+), glucose (+,–,–,–), etc.

- **USING PART OF THE UNKNOWN** for each test, place a small sample in a clean test tube and then test it for reducing sugar, starch, fat, and protein with the tests you've just used.

Do not add reagents to the entire original unknown.

Test only a small portion of solid unknowns and dissolve them in distilled water when appropriate, as for the Benedict's test. The starch test may also be done in a test tube. Be sure to **mix all components** thoroughly so that you don't get a false negative reaction.

- As needed, include a negative and positive control.
- Record the results in Table 1–3.
- **SAVE ALL TUBES SO THAT YOU MAY CHECK YOUR RESULTS WITH YOUR LAB INSTRUCTOR.**
- For each test, decide which, if any, substance on your list of possible unknowns was ruled out, and enter it in the table. For example, a positive test for reducing sugar would rule out corn starch (starch must give a negative test). A negative test for reducing sugar would rule out honey (glucose in honey must give a positive test.) You may ask your lab instructor about the composition of any of the foods.
- Identify your unknown, and enter the name of the food in the space provided.
- Ask your lab instructor to check your conclusion with the master key before you leave.
- If necessary, discuss your results with your instructor and repeat any doubtful tests.

AMINO ACID CHROMATOGRAPHY

If you touch a pen containing washable black ink to a piece of tissue, the ink will spread out on the paper and will separate into red, green, and blue bands. As the liquid in which the colored pigments are dissolved is absorbed by the cellulose fibers of the tissue, the pigments can't keep up: Some will move almost as fast as the solvent, but some will move more slowly. This is because each pigment has a tendency to stick or **adsorb** to the cellulose fibers, and those that stick more strongly will be slowed down the most. As a result, each substance will have its characteristic rate of movement, and the pigments will separate from each other. This is the essential principle of **chromatography.**

Chromatography has been adapted to separate all kinds of mixtures into their individual components. The substances do not have to be colored, so long as they can be identified somehow when the chromatography is finished. To make the chromatography reproducible, the mixture is applied in a tiny spot to the **matrix,** or stationary phase, which might be cellulose, silica gel, alumina, or some other inert substance. One end of the matrix is then dipped into a **solvent,** which it will **absorb.** As the solvent rises up the matrix by capillary action, it dissolves the substances in the mixture, and allows them to migrate along the matrix. Different substances will migrate different distances. The solvent is carefully formulated so that the substances to be separated are not completely soluble in it, or they would move as rapidly as the solvent itself. Nor must they be completely insoluble or they would not migrate at all. As they move they tend to adsorb or stick to the matrix to different degrees, so adsorption also helps to separate them. When the solvent has almost reached the edge of the matrix, the chromatography is stopped, and the finished **chromatogram** is dried after noting the exact position of the **solvent front,** the leading edge of the area wet by the solvent. Finally the individual substances are identified, and the distance moved by each one from the **origin** is carefully measured. This distance is compared with the distance travelled by the solvent itself: The ratio of the two is the R_f and is always constant for a given substance in a particular solvent-matrix system:

$$R_f = \frac{\text{distance travelled by substance (cm)}}{\text{distance travelled by solvent (cm)}}$$

When a protein is broken down by hydrolysis into its monomer amino acids, the result is a mixture of 20 or more individual amino acids.

$$\text{protein} + H_2O \xrightarrow{\text{hydrolysis}} 20 \text{ amino acids}$$

How are proteins broken down during digestion in the body?

____

____

____

The problem is to determine that these substances are indeed amino acids and to identify the ones present. Although each amino acid has its characteristic R_f, when 20 amino acids are involved it isn't practical to try to separate them all in a single chromatography. Instead you will try to separate a mixture of two or three at a time to see how chromatography works. The matrix you will be using is silica gel and the solvent contains butanol, acetic acid, and water. You will be given pure amino acid solutions to use as **standards** from which to calculate their correct R_f values. Then you will also chromatograph unknown solutions, calculate the R_f values of their components, and identify them by comparison with the standard R_f's.

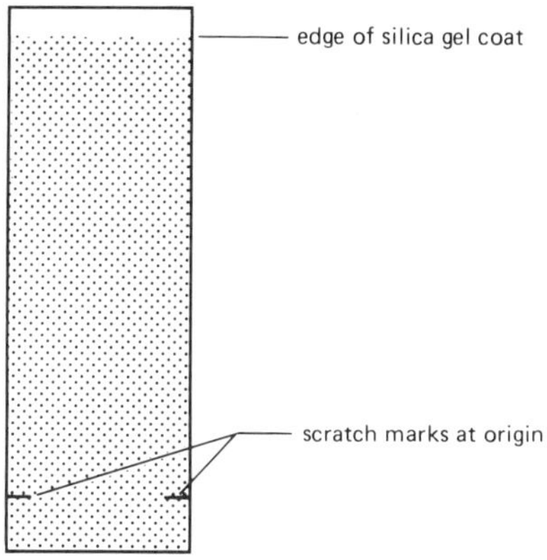

Figure 1–2. Chromatography slide ready to spot.

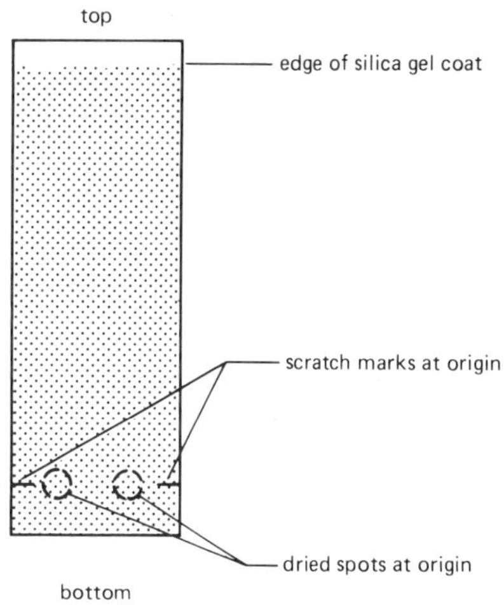

Figure 1–3. Chromatography slide after spotting samples.

You will need three slides to run your chromatograms and will chromatograph two solutions on each slide.

- To prepare a slide for thin-layer chromatography, first stir the slurry of silica gel with the glass rod to mix it well.
- Then use a pair of forceps to dip the clean slide into the slurry to within 5 mm of the top, gently move it back and forth to make sure it is evenly coated, and carefully withdraw it, holding it in a vertical position as you do so.
- Let the slide dry in air, choose the most smoothly coated side, and wipe the silica gel off the *other* side with a paper towel. Touch the slide as little as possible so that you do not contaminate it with finger grease.
- Use a teasing needle to mark the origin: Make a small scratch on each edge of the silica gel layer about 1 cm from the bottom so that your slide looks like the one shown in Figure 1–2.
- Using a separate capillary pipet for each solution as you spot, apply two standard amino acid solutions between the origin marks: Touch the gel very quickly with only the end of the pipet or your spot will be too large. After the first spot dries, apply a second drop beside it. The two amino acid spots should be at least 1 cm apart, and the spots must not overlap. (See Figure 1–3.)
- Let the spots dry thoroughly in air.
- While your first slide is drying, prepare your second and third slides in the same way: Spot two more standard amino acid solutions and two different mixtures of unknown amino acids, unknown mixtures #1 and #2. Your instructor may ask you to substitute unknown amino acids for some of the standards.
- When you are ready to chromatograph, use forceps to set the slide *with the amino acid spots at the bottom* into a Coplin jar containing the solvent. The level of solvent should be low enough so that your spots are above, not in, the solvent at the start. (Your spots will dissolve in the solvent rather than migrate if they are in the solvent.)
- Let the solvent move up the gel until it is within 1 cm of the top of the silica gel layer. This takes about 30 min. (See Figure 1–4.)

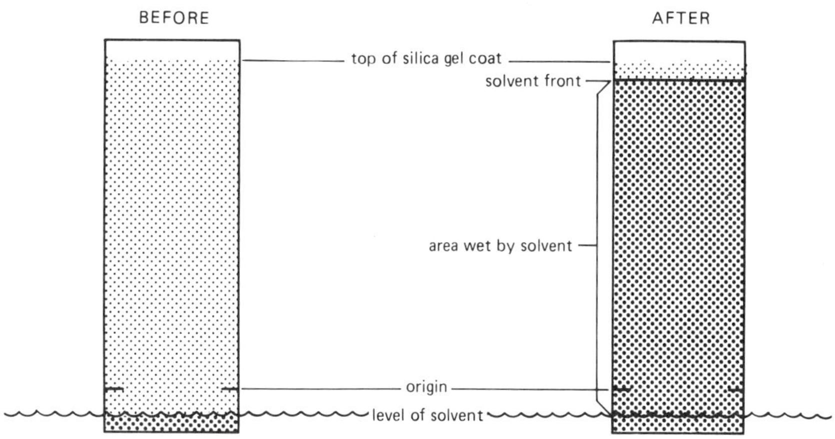

Figure 1–4. Running a chromatogram. Stop the chromatography before the solvent reaches the top of the silica gel.

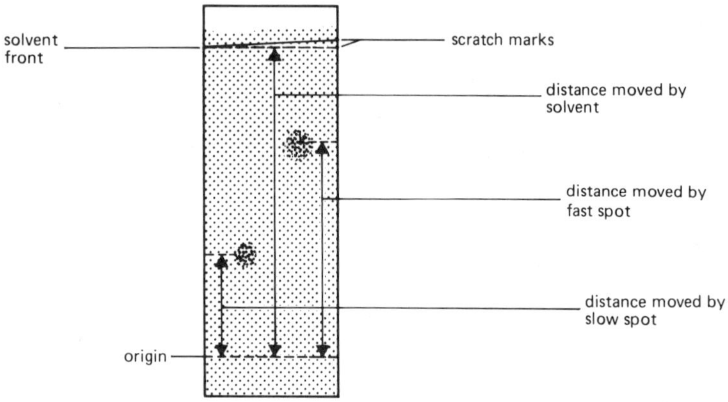

Figure 1–5. Measuring a chromatogram. Measure the distance from the origin to the center of each spot.

- Remove the slide with forceps and immediately make a scratch mark on each edge of the silica gel to show the position of the solvent, the **solvent front.**
- Allow nothing to touch the silica gel while you let it dry in air.
- Use forceps to transfer the slide to the box provided for ninhydrin spraying. Ninhydrin reacts with amino acid spots to produce identifiable visible spots.

Be very careful when using toxic ninhydrin spray and work in a fume hood if at all possible.

- Spray the slide as demonstrated by your instructor.
- Allow the slides to dry until colored spots appear.

Did your amino acids migrate away from the origin?

____

- For each spot, record the color in Table 1–4 (at the end of this topic), and measure the distance travelled. The exact distance is the distance from the center of the original spot at the origin to the center of the colored spot.
- Measure the exact distance travelled by the solvent in each case, so that you can calculate the R_f values for each amino acid. (See Figure 1–5.)
- Calculate the R_f value for each spot.
- Finally you can now identify the amino acids in your unknown amino acid mixtures by comparing the color and R_f of each spot with the color and R_f values for the standard amino acids. Since you determined the R_f values for only four amino acids, you may have to obtain color and R_f information from other members of the class to identify some of your unknown spots.

Return all material and solutions and clean up your work area before leaving the laboratory. Hand in your results according to the instructions given.

REFERENCES AND FURTHER READING

Baker, Jeffrey J.W., and Garland E. Allen. *Matter, Energy, and Life: An Introduction to Chemical Concepts,* 4th ed. Reading, MA: Addison-Wesley, 1981.

Barker, R. *Organic Chemistry of Biological Compounds.* Englewood Cliffs, NJ: Prentice-Hall, 1971.

Berlow, P.P., D.J. Burton, and J.I. Routh. *Introduction to the Chemistry of Life.* Philadelphia: Saunders College Publishing, 1982.

Calvin, M., and W.A. Pryor. *Organic Chemistry of Life: Readings from Scientific American.* San Francisco: W.H. Freeman, 1973.

Hill, J.W. *Chemistry for Changing Times,* 3rd ed. Minneapolis: Burgess, 1980.

Lehninger, A.L. *Principles of Biochemistry.* New York: Worth, 1982.

White, E.H. *Chemical Background for the Biological Sciences,* 2nd ed. Englewood Cliffs, NJ: Prentice-Hall, 1970.

Journal Articles

Block, Eric. "The chemistry of garlic and onions." *Scientific American,* 114, March 1985.

Grosser, A.E. "The culinary alchemy of eggs." *American Scientist* 71: 129, 1983.

Scrimshaw, Nevin S., and Lance Taylor. "Food." *Scientific American,* 74, September 1980.

Sharon, Nathan. "Carbohydrates." *Scientific American,* 80, November 1980.

Student Name ______________________ **Date** ____________

TABLE 1–2. Results From Chemical Tests

Record the results from each of the chemical tests.

Benedict's Test: Green to Red Is Positive

Tube	Substance	Color	Reducing Sugar Present?
1	onion juice		
2	potato juice		
3	water		

Iodine Test: Blue to Black Is Positive

Tube	Substance	Color	Starch Present?
1	onion juice		
2	potato juice		
3	water		

Sudan III Test: Orange Spot Is Positive

Code	Substance	Color of Spot	Fat Present?
W	water		
F	flour		
K	cream		
C	coconut		
M	margarine		

Biuret Test: Violet Is Positive

Tube	Substance	Color	Protein Present?
1	egg white		
2	chicken soup		
3	water		

TABLE 1–3. Unknown Results

Unknown Group and Number: ______________________________

Expected results for each possible food in each test (+ or −):

Possible Unknowns in My Unknown Group				
Component				
reducing sugar				
starch				
fat				
protein				

Actual results of tests (+ or −):

Test	Test Result (+ or −)	Food(s) Ruled Out
reducing sugar		
starch		
fat		
protein		

Unknown was: ______________________________ (Check against master key)

Student Name ______________________ **Date** ______________

TABLE 1–4. Results From Amino Acid Chromatography

Standard Amino Acids	Color	Distance Moved by Amino Acid	Distance Moved by Solvent	R_f
Alanine				
Arginine				
Asparagine				
Aspartic acid				
Cysteine				
Glutamic acid				
Glutamine				
Glycine				
Histidine				
Isoleucine				
Leucine				
Lysine				
Methionine				
Phenylalanine				
Proline				
Serine				
Threonine				
Tryptophan				
Tyrosine				
Valine				
Unknown Amino Acids				
#1 =				
#2 =				
#3 =				
#4 =				
Unknown Mixture #1				
Unknown Mixture #2				

What amino acids did unknown mixture #1 contain? ______________________

What amino acids did unknown mixture #2 contain? ______________________

TOPIC 2

Enzyme Action

OBJECTIVES

When you have completed this topic, you should be able to:

1. Give the class of macromolecules to which peroxidase belongs and the monomers of which it is composed.
2. Name the substrates and products of the peroxidase catalyzed reaction.
3. Explain the role of guaiacol in this experiment.
4. Define enzyme, activation energy, active site, pH, and denaturation.
5. Distinguish between oxidation/reduction, activation energy/catalysis, substrate/product, modulator/prosthetic group, and hydrogen peroxide/peroxidase.
6. Define the term optimum with respect to peroxidase activity.
7. Describe how temperature, pH, enzyme concentration, and substrate concentration affect the reaction rate.
8. Explain why peroxidase is a necessary enzyme for all aerobic or oxygen-utilizing cells.

TEXT REFERENCES Chapter 3

KEY TOPICS

chemical reactions	proteins	enzyme-substrate interactions
catalyst	enzymes	pH

INTRODUCTION

A catalyst speeds up a chemical reaction by lowering the **activation energy** required. (See Text Figure 6–4.) Enzymes are **biological catalysts** that carry out the thousands of chemical reactions that occur in living cells. They are generally large proteins made up of several hundred amino acids. Sometimes an enzyme incorporates an important nonprotein component into its structure. If this component is attached to the enzyme's protein covalently, it is a **prosthetic group.** If it is held more loosely by other types of bonds, it is a **cofactor.** Examples of cofactors are inorganic **ions** and organic **coenzymes.** You are already familiar with many coenzymes that are required in the human diet and are known as vitamins.

In an enzyme-catalyzed reaction, the substance to be acted upon, or **substrate,** binds to the **active site,** or business end, of the enzyme. The enzyme and substrate are held together in an **enzyme-substrate complex** by hydrophobic interactions, hydrogen bonds, and ionic bonds. (See Text Figure 3–23.)

The enzyme then converts the substrate to the reaction products in a process that often requires several chemical steps and may involve covalent bonds. Finally, the products are released into solution and the enzyme is ready to form another enzyme-substrate complex. As is true of any catalyst, the enzyme is not used up as it carries out the reaction but is recycled over and over. One enzyme molecule can carry out thousands of reaction cycles every minute.

Each enzyme is specific for a certain reaction because its amino acid sequence is unique and causes it to have a unique three-dimensional structure. (See Text Figures 3–19 and 3–20.) The active site also has a specific shape so that only one or a few of the thousands of compounds present in the cell can interact with it. If there is a prosthetic group on the enzyme, it will form part of the active site. Any substance that blocks or changes the shape of the active site will interfere with the activity and efficiency of the enzyme.

If these changes are large enough, the enzyme can no longer act at all, and is said to be **denatured.** There are several factors that are especially important in determining the enzyme's shape, and these are closely regulated both in the living organism and in laboratory experiments to give the **optimum** or most efficient enzyme activity:

1. **Salt concentration.** If the salt concentration is very low or zero, the charged amino acid side chains of the enzyme molecules will stick together. The enzyme will denature and form an inactive precipitate. If, on the other hand, the salt concentration is very high, normal interaction of charged groups will be blocked, new interactions will occur, and again the enzyme will precipitate. An intermediate salt concentration such as that of blood (0.9%) or cytoplasm is the optimum for most enzymes.
2. **pH.** pH is a logarithmic scale that measures the acidity or H^+ concentration in a solution. The scale runs from 0 to 14 with 0 being highest in acidity and 14 lowest. When the pH is in the range of 0–7, a solution is said to be acidic; if the pH is around 7, the solution is neutral; and if the pH is in the range of 7–14, the solution is basic. Amino acid side chains contain groups such as 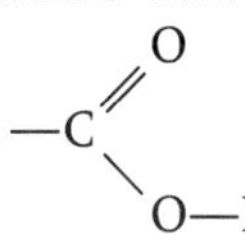 and 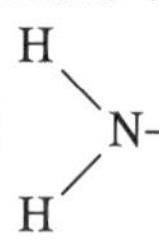 that readily gain or lose H^+ ions. As the pH is lowered, an enzyme will tend to gain H^+ ions, and eventually enough side chains will be affected so that the enzyme's shape is disrupted. Likewise, as the pH is raised, the enzyme will lose H^+ ions and eventually lose its active shape. Many enzymes have an optimum in the neutral pH range and are denatured at either extremely high or low pH. Some enzymes, such as those that act in the human stomach where the pH is very low, will have an appropriately low pH optimum. (See Text Figure 3–25.) A **buffer** is a compound that acts like a sponge to pick up any extra H^+ or OH^- ions so that a constant pH is maintained. The buffer molecules (B) must be in two forms to do this:

$$B + H^+ \rightarrow BH^+$$

$$BH^+ + OH^- \rightarrow B + H_2O$$

   Different buffers are designed to keep the pH at various pH levels.
3. **Temperature.** All chemical reactions speed up as the temperature is raised. As the temperature increases, more of the reacting molecules have enough kinetic energy to undergo the reaction. Since enzymes are catalysts for chemical reactions, enzyme reactions also tend to go faster with increasing temperature. However, if the temperature of an enzyme-catalyzed reaction is raised still further, a **temperature optimum** is reached: Above this point the kinetic energy of the enzyme and water molecules is so great that the structure of the enzyme molecules starts to be disrupted. The positive effect of speeding up the reaction is now more than offset by the negative effect of denaturing more and more enzyme molecules. Many proteins are denature by temperatures around 40–50°C, but some are still active at 70–80°C, and a few even withstand being boiled. (See Text Figure 3–26.)
4. **Modulator molecules.** Many small molecules other than the substrate may interact with an enzyme. If such a molecule increases the rate of the reaction it is an **activator,** and if it decreases the reaction rate it is an **inhibitor.** The cell can use these modulators to regulate how fast the enzyme acts. Any substance that tends to unfold the enzyme, such as an organic solvent or detergent, will also act as an inhibitor. Some inhibitors act by reducing the —S—S— bridges that stabilize the enzyme's structure. Many inhibitors act by reacting with side chains in or near the active site to change or block it. Others may damage or remove the prosthetic group. Many well-known poisons such as potassium cyanide and curare are enzyme inhibitors that interfere with the active site of a critical enzyme.

TURNIP PEROXIDASE

In this experiment you will study the enzyme **peroxidase** from turnips. Peroxidases are widely distributed in plant and animal cells and catalyze the oxidation of organic compounds by hydrogen peroxide as follows:

$$\underset{\text{reduced form}}{\text{H}-\boxed{\text{R}-\text{O}}-\text{H}} + \text{H}_2\text{O}_2 \xrightarrow{\text{peroxidase}} \underset{\text{oxidized form}}{\boxed{\text{R}=\text{O}}} + 2\text{H}_2\text{O}$$

What are the substrates in this reaction?

___

What are the products?

___

Any cell using molecular oxygen in its metabolism will produce small amounts of H_2O_2 as a highly toxic byproduct. A peroxide has a very reactive —O—O— structure, so it is critical that it be quickly removed by enzymes such as peroxidase before it can do damage to the cell.

In order to follow the reaction as it proceeds, you

will be using a substrate, the reduced, colorless form of a substance called guaiacol. It is an organic chemical produced by the guaiac tree of Central America and it corresponds to H—R—O—H in the equation. In the reaction, guaiacol will donate hydrogens and thereby become oxidized. It has a brown color in its oxidized form and you will be able to measure the amount of oxidized guaiacol produced by determining the intensity of the color in the spectrophotometer.

4 guaiacol + 4 hydrogen peroxide $\xrightarrow{\text{peroxidase}}$
4 H—R—O—H + 4 H—O—O—H
colorless

tetraguaiacol + water
R—O—O—R + 8 H_2O
| |
R—O—O—R
brown

In this reaction, H_2O_2 is reduced to water by giving up an atom of oxygen as guaiacol gives up 8 of its hydrogens. The oxygen and hydrogen combine to form more water. The decomposition of H_2O_2 is thus a good example of an oxidation-reduction reaction.

THE SPECTROPHOTOMETER

A colored solution such as oxidized guaiacol solution appears that way because some of the light entering the solution is absorbed by the colored substance. A clear solution will allow almost all of the light to pass through. The amount of **absorbance** can be determined by using a spectrophotometer, which measures quantitatively what fraction of the light passes through a given solution, and indicates on the absorbance scale the amount of light absorbed compared to that absorbed by a clear solution. The darker the solution, the greater its absorbance.

Inside the machine there is a light that shines through a filter (which can be adjusted to control the color, or wavelength, of light), then through the sample and onto a light-sensitive phototube. The phototube produces an electric current proportional to the amount of light striking it. The absorbance meter measures how much light has been blocked by the sample and thereby prevented from striking the phototube. A clear tube of water or other solvent is the **blank** and has zero absorbance. A solution that contains a small amount of a colored substance might show an absorbance of 0.1, a solution with a moderate amount might show an absorbance of 0.4, and so forth. In fact, in the lower portion of the absorbance scale, the amount of substance in solution is directly proportional to the absorbance reading so that a graph of absorbance versus concentration will give a straight line. This very useful relationship is known as **Beer's law.**

When you are ready to read a sample, use Figure 2–1 to follow these steps:

1. Turn on the instrument with the power switch knob. Allow 5 min for warm-up time.
2. Adjust to the desired wavelength (in this case use 470 nm).
3. With the sample chamber *empty* and the cover closed, use the power switch knob to set the meter needle to read **infinity absorbance.** (When the chamber is empty, a shutter blocks all light from the phototube.)

Always read from the absorbance scale, not the transmission scale.

4. Fill a spectrophotometer tube or cuvette halfway with distilled water (or other solvent when water is not the solvent in your experiment) to serve as the blank. Wipe it free of moisture or fingerprints with a lint-free tissue (Kimwipe), and insert it into the sample holder. Line up the etched mark with the raised line on the front of the sample holder, and close the cover.
5. With the right-hand knob, adjust the meter to read **zero absorbance.** Remove the tube, empty it, and shake it as dry as possible. This precaution is essential for accuracy.
6. Fill and insert the sample cuvette.
7. Read the absorbance directly from the meter. Rinse the cuvette with clean water and shake it as dry as possible.
8. It is necessary to reset the machine to infinity absorbance and zero absorbance before each set of

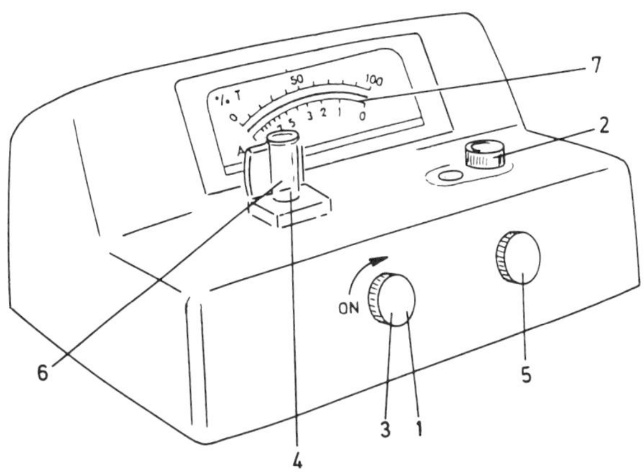

Figure 2–1. Spectronic 20 spectrophotometer. The numbered controls correspond to steps in the operating instructions (see text).

readings because the settings drift a bit. Whenever the wavelength is changed, the infinity and zero absorbance also must be reset.

EXPERIMENTAL PROCEDURE

Preparation of the Turnip Extract

- Weigh out 2 g of turnip (the peeled, inner portion). More may be required, depending on the condition of the turnip.
- Place it in a blender with 200 mL of distilled water.
- Blend it thoroughly at the high setting for about 1 min.

This suspension is the **turnip extract,** and contains the enzyme peroxidase. The activity of the turnip extract will vary from day to day, depending on the size and age of the turnip and the extent of blending. You should adjust the turnip suspension to be more dilute or more concentrated so that your absorbance for the **base line** is in the range 0.1–0.2.

Kinetics of the Peroxidase Reaction

Pipets will be used to measure accurately the solutions used in this experiment. Your instructor will demonstrate how to use a pipet correctly.

Be sure to use a different pipet for each solution so that the reagents are not ruined by cross-contamination.

- Label the pipets with tape or a marking pencil so each one can be reused with the proper solution.
- Obtain two spectrophotometer tubes and label them C (control) and R (reaction).
- Obtain three test tubes and label them #1, #2, and #3. #1 will contain a control reaction without H_2O_2. The contents of #2 (substrate) and #3 (enzyme) will be mixed to start the reaction.
- Set up the three tubes as follows, and make a record of these additions in Table 2–1 (Run 1, base line) at the end of this topic.

  Tube #1 (control tube without H_2O_2): Add 0.1 mL of guaiacol, 1.0 mL of turnip extract, and 8.9 mL of distilled water; mix well.

  Tube #2 (substrate): Add 0.1 mL of guaiacol, 0.2 mL of 0.1% H_2O_2, and 4.7 mL of distilled water.

  Tube #3 (enzyme): Add 1.0 mL of turnip extract and 4.0 mL of distilled water.
- Adjust the Spectronic 20 to zero absorbance using tube C filled with solution from tube #1 and following steps 1–5 in the instructions for its operation.

You have now set up the instrument so that any difference in the meter reading with a change in sample will reflect a difference in oxidized guaiacol concentration.

- Obtain a stopwatch, wind it up, and be sure that you understand how to use it correctly.
- Prepare the sample: have ready spectrophotometer tube R, tissue, and tubes #2 and #3 filled with the solutions given above.

You will have 30 sec to mix the contents of tubes #2 and #3, pour the contents into the clean spectrophotometer tube, wipe the tube, and take your first reading.

- When you are completely ready, mix the contents of tubes #2 and #3, pour the contents back and forth two times, and then pour quickly into tube R; start the stopwatch when the tubes are mixed (t = 0 when the tubes are mixed.)
- Wipe the outside of the tube and place it in the spectrophotometer.
- Take your first reading 30 sec or as soon as possible after the tubes were mixed (t = 30 sec).
- Continue to read the absorbance every 30 sec for 3 min.

Record the readings in Table 2–2, and graph the absorbance versus time in Figure 2–2 (at end of topic).

This curve will represent the **base line** of enzyme activity with which the enzyme activity under varying conditions will be compared. In the following experiments you will vary one condition at a time and compare the results with the base line. Be sure you have already recorded the setup for your base line experiment under run 1 in Table 2–1. Check your graph with your instructor before proceeding with the rest of the experiment.

Since this exercise contains too many projects to be performed by one pair of students, the instructor will specify how the remaining work will be divided up. Use the experimental setup space in Table 2–1 to record your experiment number and what you put into your tubes; record your data in Table 2–2, and graph your results. An extra page for graphing is included, in case you need more space.

1. Effect of Enzyme Concentration

What happens when you use:

(a) TWICE THE AMOUNT OF ENZYME

Tube #1. Add 0.1 mL of guaiacol, 2.0 mL of turnip extract, and 7.9 mL of distilled water.

Tube #2. Add 0.1 mL of guaiacol, 0.2 mL of 0.1% H_2O_2, and 4.7 mL of distilled water.

Tube #3. Add 2.0 mL of turnip extract and 3.0 mL of distilled water.

(b) HALF THE AMOUNT OF ENZYME

Tube #1. Add 0.1 mL of guaiacol, 0.5 mL of turnip extract, and 9.4 mL of distilled water.

Tube #2. Add 0.1 mL of guaiacol, 0.2 mL of 0.1% H_2O_2, and 4.7 mL of distilled water.

Tube #3. Add 0.5 mL of turnip extract and 4.5 mL of distilled water.

Notice that tube #1 always contains 10 mL and tubes #2 and #3 together total 10 mL. Why is this important?

____

____

- Repeat the procedure for preparing the sample, read it, record the data, and graph your results.

How does changing the concentration of enzyme affect the rate of the reaction?

____

2. Effect of Varying the Substrate Concentration

- Design and carry out an experiment to determine whether varying the concentration of the substrate H_2O_2 affects the rate of reaction. If substrate is present in excess, there would be little or no effect.

Conclusion:

____

____

What would happen if the concentration of this substrate were increased further and further?

____

3. Effect of Temperature

What is the effect of temperature on enzyme activity?

- Set up an experiment to show the difference in activity found between 0° and 60°C. Use the conditions of your base line experiment, but run the reaction in the water baths at the different temperatures that are available to you, such as 4°, 15°, room temperature (about 22°), 30°, 37°, and 55°C. Record the exact temperature for each run.
- Place tubes #2 and #3 in the water bath to equilibrate for 5 min before mixing.
- Read the absorbance immediately after mixing; then return tube R to the water bath and wait.
- Read the absorbance again at 3 min.
- Graph your two data points for each temperature.
- Next, set up another run but heat tube #3 in a boiling water bath at 100° for 10 min, and test its activity as in Run 1.

Is the boiled enzyme active? ____________

Explain what happened during boiling.

____

____

____

____

____

At what temperature in the range 0–100° do you see maximal activity? ____________

This temperature is called the enzyme's temperature ____________.

4. Effect of pH

- Measure the effect of pH by using the base line conditions but substituting buffers at pH values between 4 and 9 (4.0, 5.5, 7, and 8.5) for distilled water in tube #3. Compare each pH rate with the others. (You cannot compare the base line and buffer results, because buffer itself will change the rate.)

At which pH is the enzyme most effective?

____

Why is this important biologically?

____

____

____

5. Effect of Inhibitors

Hydroxylamine is a small molecule whose structure is similar enough to that of H_2O_2 that it attaches to the iron atom that is part of the peroxidase enzyme. Its inhibition is **reversible** and **competitive.** To test how this substance affects enzyme activity, in tube #3 substitute 3.0 mL of distilled water and 1.0 mL of 10% hydroxylamine. Mix this tube and wait 1 min. Then measure the activity of the peroxidase as usual.

Is the peroxidase still active? ____________

How could you restore activity in the presence of this inhibitor?

____

____

Explain at the molecular level the action of hydroxylamine on the enzyme's activity:

____

____

____

____

____

When you are finished with your work, wash and thoroughly rinse your glassware, return the reagents, and discard any debris left from your experiment. Rinse thoroughly but **DO NOT BRUSH** your spectrophotometer tubes.

REFERENCES AND FURTHER READING

Dixon, M., and E.C. Webb. *Enzymes,* 3rd ed. New York: Academic, 1979.

Fersht, Alan. *Enzyme Structure and Mechanism,* 2nd ed. San Francisco: W.H. Freeman, 1984.

Lehninger, A.L. *Principles of Biochemistry.* New York: Worth, 1982.

Stryer, L. *Biochemistry,* 2nd ed. San Francisco: W.H. Freeman, 1981.

Journal Articles

DeDuve, Christian. "Microbodies in the living cell." *Scientific American,* 74, May 1983.

Koshland, D.E., Jr. "Protein shape and biological control." *Scientific American,* 52, October 1973.

Student Name ______________________________ **Date** ________________

TABLE 2–1. Experimental Design for Project # ______ List the contents of each tube.

	Tube #1	Tube #2	Tube #3
Run 1 (Base line)			
Run 2			
Run 3			
Run 4			
Run 5			
Run 6			
Run 7			

TABLE 2–2. Results From Turnip Peroxidase Experiment

Read absorbance every 30 sec. Read at 30 sec and 180 sec for the temperature experiment.

Time (sec)	Run 1 (Base line)					
30						
60						
90						
120						
150						
180						

Time (sec)						
30						
60						
90						
120						
150						
180						

Student Name ______________________ **Date** ____________

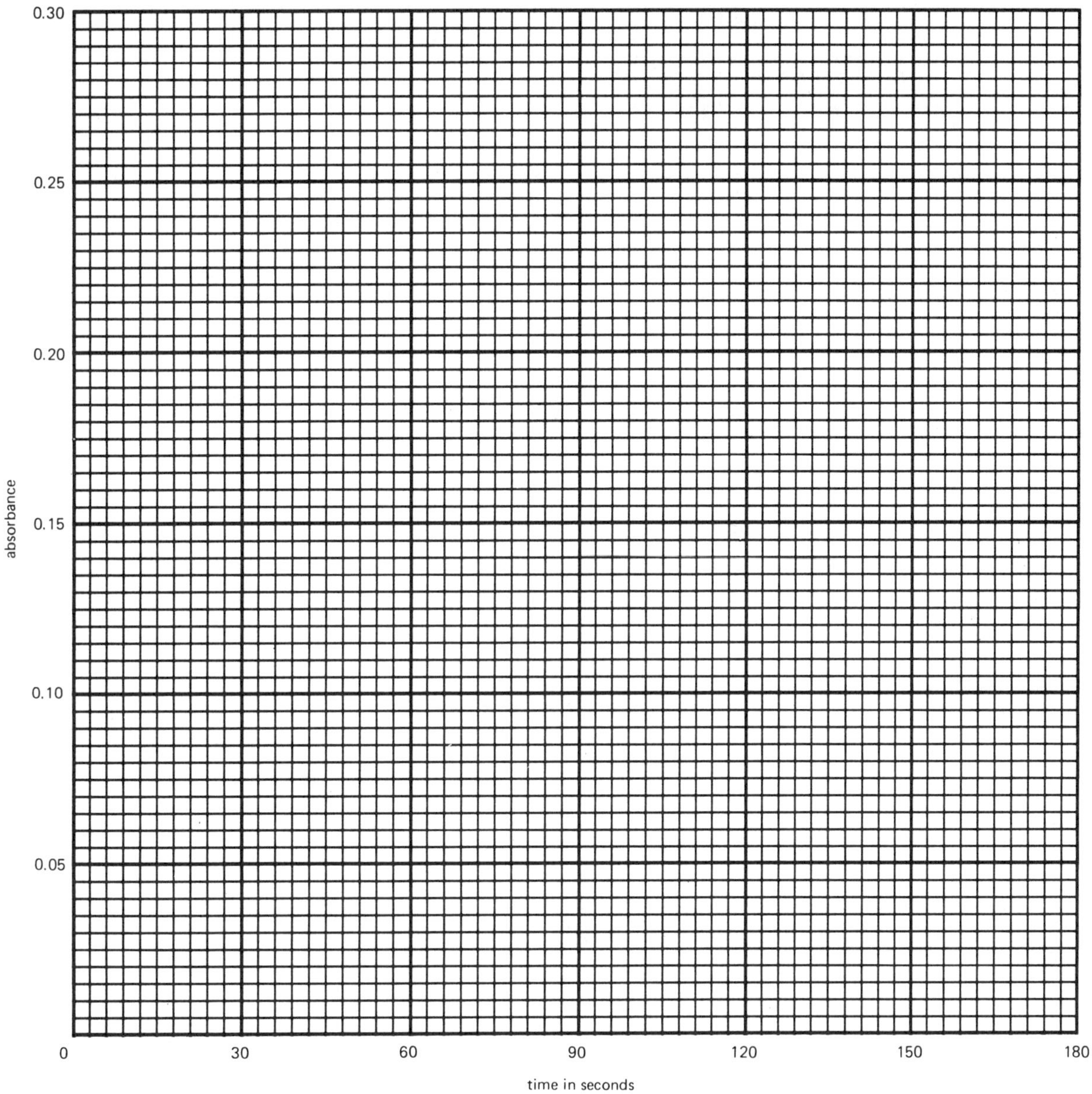

Figure 2–2. Kinetics of turnip peroxidase activity.

Results of experiment # ____________

Conclusions:

____

____

____

____

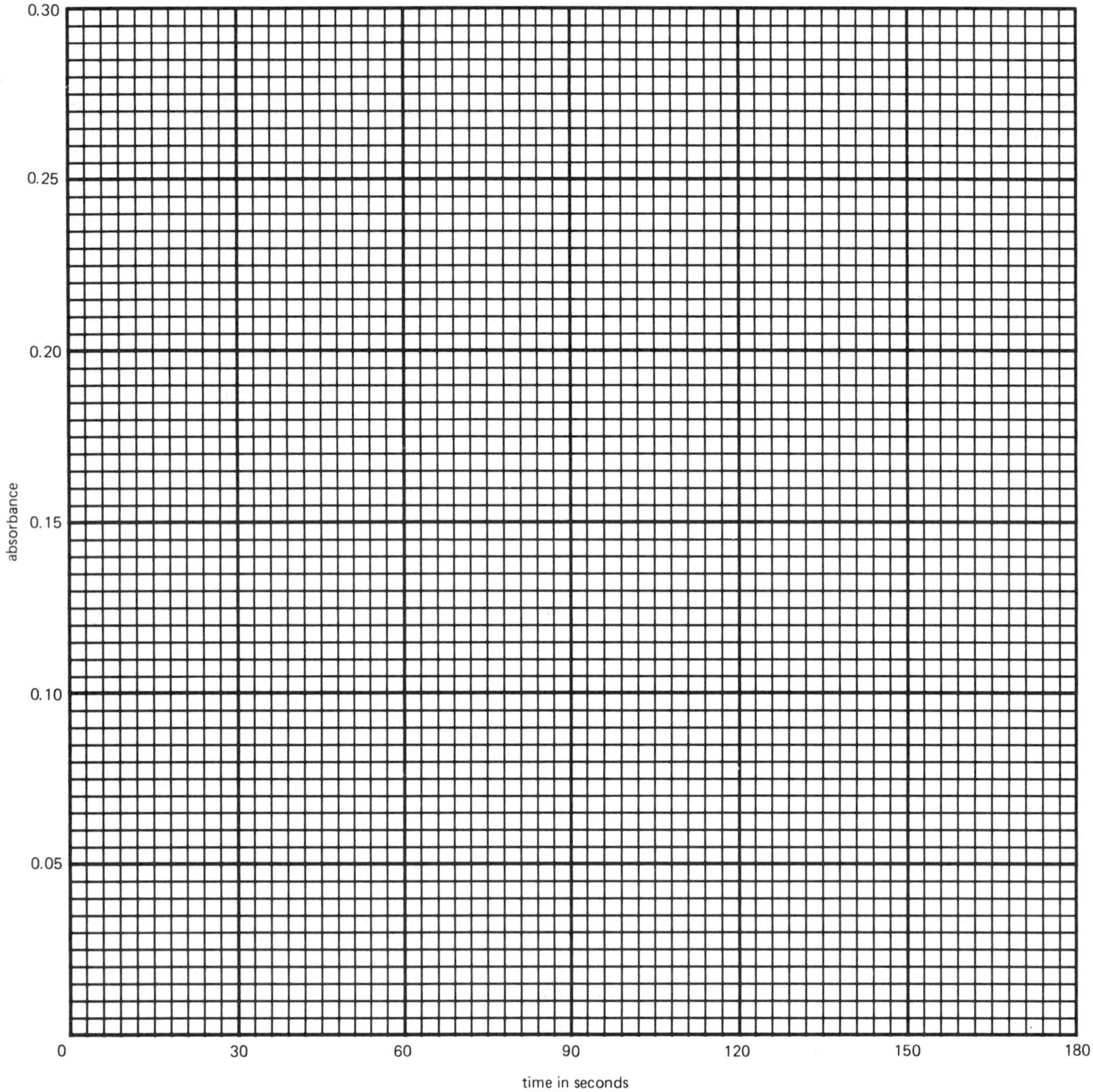

Results of experiment # ____________

Conclusions:

____

____

____

____

TOPIC 3

Cells and the Microscope

OBJECTIVES When you have completed this topic, you should be able to:

1. Carry a microscope correctly and place it correctly on your lab desk.
2. Explain the function of each part of the microscopes as shown in Figure 3–2 and 3–4.
3. State the advantages and disadvantages of a standard compound microscope and a dissecting microscope.
4. Prepare a wet mount.
5. Clean the microscope lenses correctly using lens paper.
6. Prepare the microscope for examining a ready-made (prepared or permanent) slide.
7. Locate the object of interest and focus on it under low and high power.
8. View a slide under oil immersion, and remove the oil from the slide and lens afterward.
9. Determine the magnification of the microscope.
10. Measure a tiny object by viewing it in the microscope field at a known magnification.
11. Store the microscopes properly.

TEXT REFERENCES Chapter 5

KEY TOPICS

metric units
light microscopes
specimen preparation

INTRODUCTION

The purpose of a microscope is to see inside organisms and cells—to see what is invisible to the naked eye. The eye can be aided with a **simple microscope,** which is nothing more than a magnifying glass with one lens, or with a **compound microscope,** which has two lenses at opposite ends of a tube. The **ocular lens** is the one nearest the eye, and the **objective lens** is nearest the object or specimen. Both microscopes you will be using today are compound and quite a bit more sophisticated than the exquisite simple microscopes with which Anthony van Leeuwenhoek made his discovery of the microbes. "Microbes" are organisms too small to be seen with the naked eye. They may be a single cell or may contain many cells. Plants, animals, fungi, protists, and bacteria all include organisms that are microscopic in size. Some examples are shown in Figure 3–1.

How small are microbes? They are small, but actually many can be detected with the naked eye if you know where to look. As a point of reference, consider the **millimeter** (1 mm = $\frac{1}{1000}$ of a meter), which is the smallest division on your metric ruler, about [-] long. (The metric system is explained in Appendix D.) Microbes are less than a millimeter in size and are usually measured in **micrometers** (1 μm = $\frac{1}{1000}$ of a mm). They could range in size from 1 to 1000 μm. You should be able to see the organism with your naked eye if it were about 100 μm or larger, but you wouldn't be able to see much

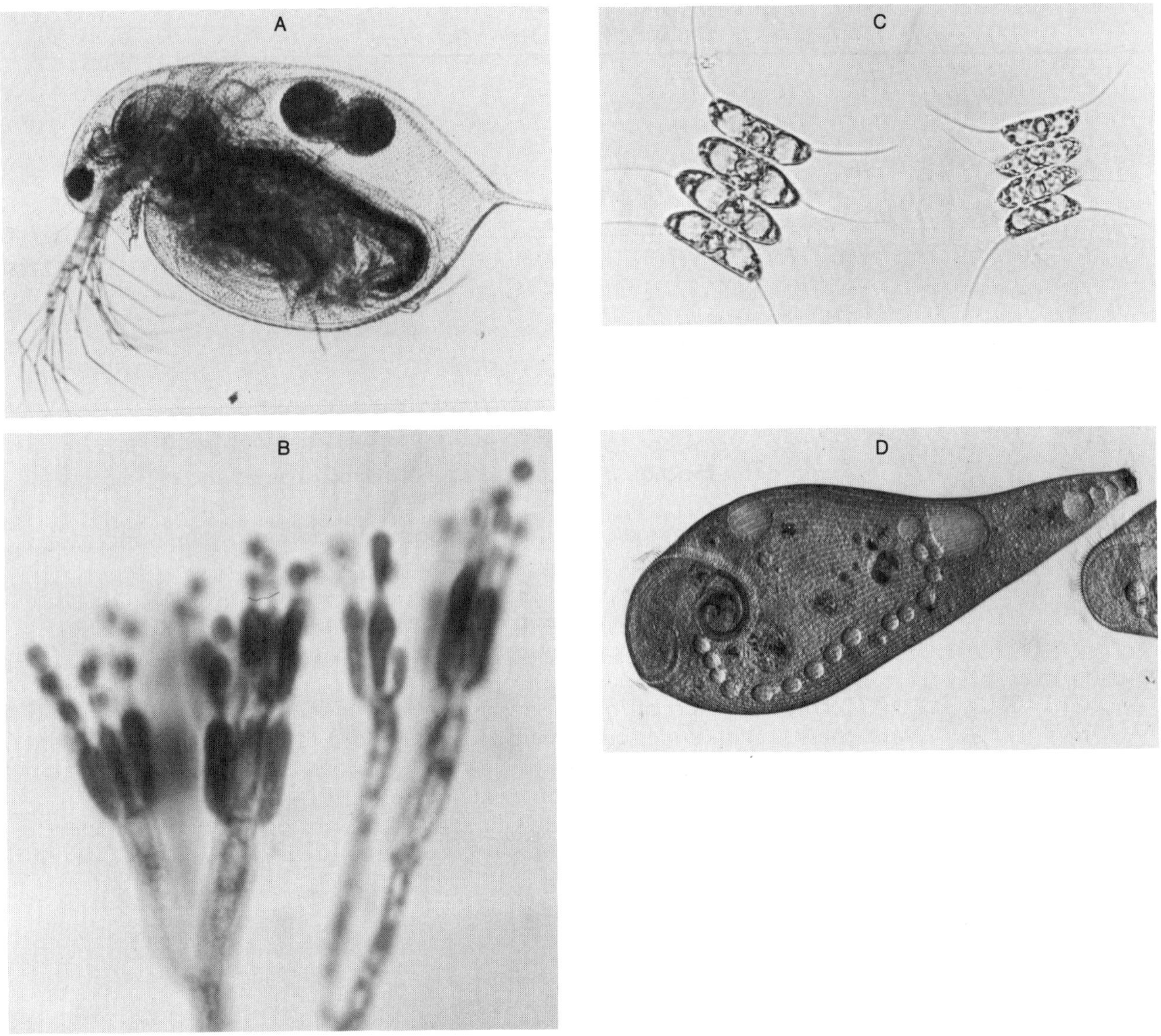

Figure 3–1. Microbes. (A) *Daphnia,* a tiny animal that is a freshwater crustacean and may protect its young in a brood pouch. (B) *Penicillium,* a fungus, which provides us with penicillin. (C) A green alga, *Scenedesmus,* which belongs to the plants. (D) A giant animal-like protist, *Stentor,* which is a single cell. (Photos by Carolina Biological Supply Company)

detail. Plant and animal cells are typically 5 to 40 μm in size, so they are not visible unless magnified under the microscope.

Magnification is a measure of how big an object looks to your eye compared to "life size." Life-size images are specified as 1. Magnification is usually written by a number followed by "×" which means "times life size." For example, 10× means 10 times life size. A simple lens like a magnifying glass magnifies 3× to 10×, a dissecting microscope about 10× to 30× and a standard light microscope about 40× to 440×.

HOW TO USE THE DISSECTING MICROSCOPE

In today's lab we will start out with the dissecting microscope which, at its least powerful setting, magnifies things about as much as a strong magnifying glass. If your dissecting microscope looks like the one in Figure 3–2, it is a binocular microscope and is really two microscopes with two oculars so that objects can be viewed with binocular vision in three dimensions. The **field of view,** the area you see when you look into the microscope, is large, and the **depth of field** that is in focus at the same time is also large compared with that of the standard microscope. The light source can be directed down onto the object in addition to or instead of up through the object, so that you can view objects too thick to transmit light. The magnification may be varied continuously if your microscope has a "zoom" knob.

A microscope is a precision instrument and easily damaged. You should not attempt to use it until you have grasped the function of each vital part.

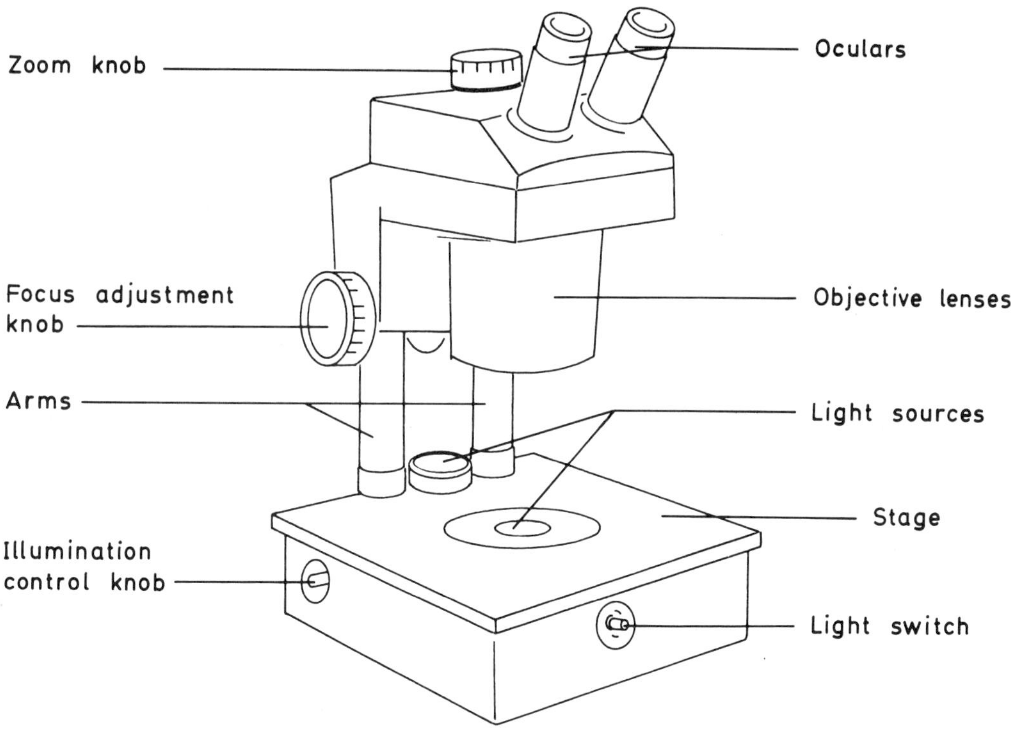

Figure 3–2. Dissecting microscope. Your microscope may be similar to the one shown in this diagram. Your instructor may be able to provide a diagram that represents your microscope exactly.

■ Carrying the dissecting microscope by its arm, place it on the table, and locate the following parts using Figure 3–2 as your guide.

Base with Built-in Illuminator

Light Switch: Turns on the illuminator. Push the switch on the front of the base to turn on the light. If your microscope has a separate light source, turn it on and aim it down onto the stage.

Illumination Control Knob: Rotates to provide reflection or transmission or, in some microscopes, to combine transmission and reflection.

1. **Transmission:** Light travels up through specimen.
2. **Reflection:** Light is directed down onto the speciman from which it is reflected up into the microscope.
3. **Transmission-Reflection:** Combines (1) and (2). Rotate the knob to get the best possible illumination for a particular sample.

Oculars: Move the oculars together or apart until you feel comfortable looking into the microscope. The two fields of view should overlap completely so that you see a single circle of light.

"Zoom" Knob: Controls magnification. It is located next to the oculars on the body of the microscope. Turn it until the indicator mark is on 4× or the lowest magnification. Some microscopes have revolving objectives much like a standard microscope.

Stage: Place a small ruler on the stage.

Focus Knob: A single knob that you should turn until the ruler is in sharp focus.

How big is the field under the lowest possible magnification? ____________ mm ____________ μm

Turn the zoom knob to the highest magnification. How big is the field? ____________ mm ____________ μm

Now look at your finger or some other handy object under low power. Is it right side up or upside down?

____

Move your finger to the right; which way does the image move? ____________

Your instructor will provide objects or cultures of biological interest for you to view in the dissecting microscope such as:

insect	*Volvox*
leaf	*Planaria*
fungus	*Hydra*

Always place the specimen on a slide or glass dish before viewing.

Never put specimens directly on the microscope stage.

If the organisms are in liquid culture, make a wet mount as described in the section "Preparing a Wet

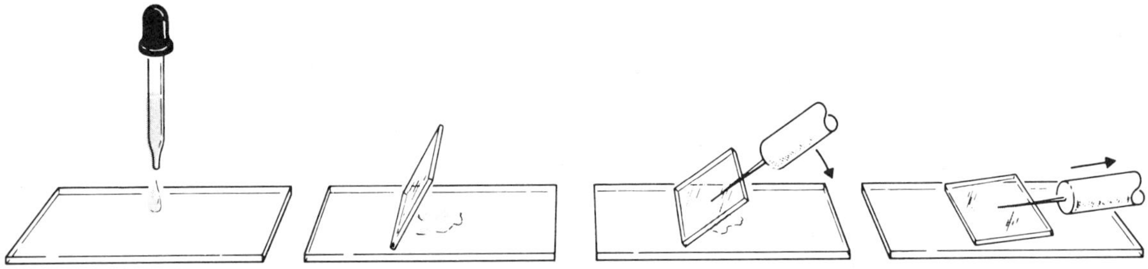

Figure 3–3. Preparing a wet mount. Mount your living specimen or culture as shown to avoid trapping air bubbles under the coverslip.

Mount," or place a few drops in a depression slide or small dish.

When you are finished, remove the sample, turn off the light, clean the lenses with lens paper (see below, "Cleaning the Lenses"), cover the dissecting microscope, and store it in its box or cabinet.

You can see that this microscope is extremely useful when a small amount of magnification is needed, or when you need to manipulate or dissect things too small to be easily seen with the naked eye. One task that is greatly facilitated under a dissecting microscope is removing scratches from a stereo record with a fine needle: The grooves that have been crushed can be easily straightened out again so that the needle no longer skips. Similar microscopes are used in human microsurgery.

Preparing a Wet Mount

Wet mounts are used to study fresh, living material. They can be used only for a little while because they will soon dry out, but they are useful for observing qualities such as color, movement, or behavior that cannot be observed on dead, stained material. Follow the steps shown in Figure 3–3 to prepare your wet mount.

- Obtain a slide, coverslip, and teasing needle from your instructor.
- Place a drop of the culture on your slide.
- Touch the coverslip to one edge of the drop, and gently lower it with the teasing needle, as shown.

If you have been careful, the slide will not have any bubbles. If not, you will see them as circles of various sizes with very dark edges when you look at your sample.

Use lower or higher magnification to view your organisms, depending on their size.

HOW TO USE THE STANDARD COMPOUND MICROSCOPE

- Study Figure 3–4 as you locate each of the following parts:

Arm: Supports the body tube and is the part that you can grasp to carry the microscope.

- Pick up your microscope by its arm, keeping it upright, and supporting it underneath with your free hand. Set it gently on your lab desk.

Base: Gives the microscope a firm, steady support.

Ocular Lens: Magnifies ten times (10×). This lens is often unattached, and thus it may fall out unless the microscope is kept upright.

Objective Lens: Magnifies the object by the factor marked on the particular lens. Low power (10×) gives the smallest image, high power (sometimes called high dry) gives a large image (40×), and oil immersion gives the largest image (100×). Sometimes a very low power scan objective (4×) replaces the oil immersion lens. Objective lenses are always used in the following order: low, high, oil immersion.

Nosepiece: The revolving part to which objectives are attached. It must be firmly clicked into position when the objective is changed. Rough treatment can cause it to snap off.

Body Tube: Joins the nosepiece to the ocular lens.

Stage: Supports the slide that is held onto it by **stage clips,** and has a hole so that light can shine up through the specimen. Always center the specimen over this hole.

Coarse Adjustment: Moves the body tube or stage up and down, depending on the design of the microscope, to approximately the right position so that the specimen is in focus. This knob is used only with the **low power** or **scan** (focus finder) **objective.**

Fine Adjustment: Moves the body tube (or stage) up and down to precisely the right position so that the specimen is perfectly in focus. Use it to achieve fine focus with the low power objective and for *all focusing* with the high power and oil immersion objectives.

Light Source: Usually a small electric light beneath the stage that is controlled by a push-button light switch. Sometimes a mirror is used to reflect light from another source into the microscope.

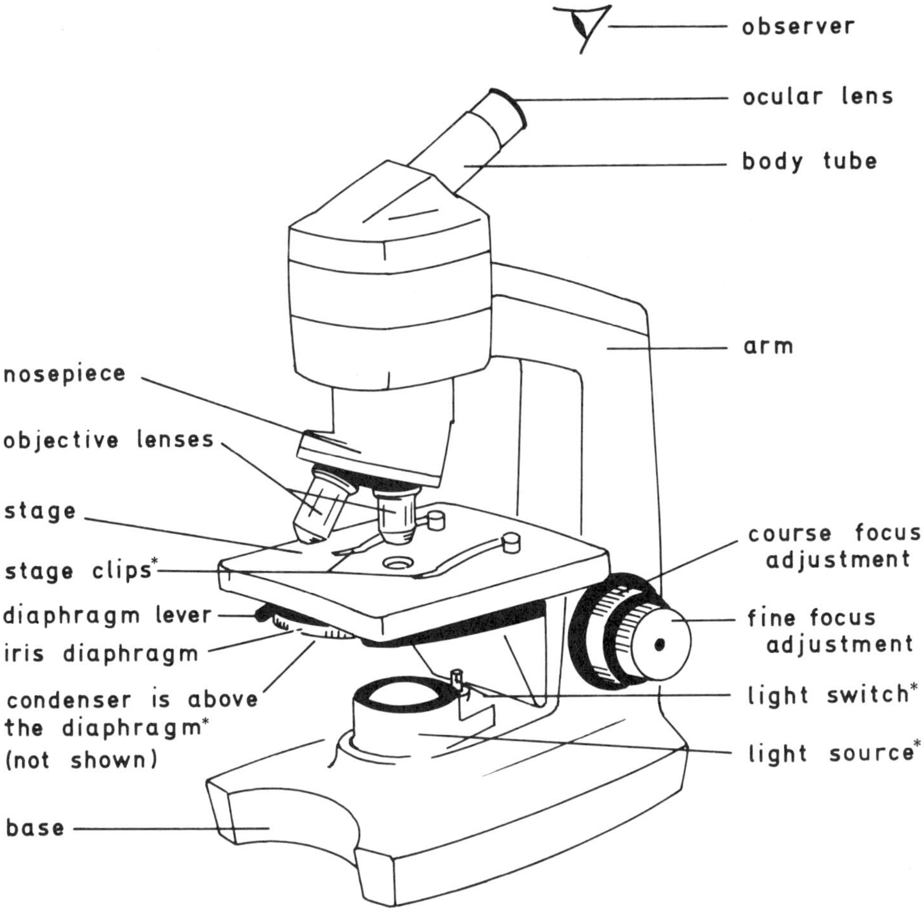

Figure 3–4. Standard compound microscope. Your microscope should resemble the one shown in this diagram. Some of the parts that may not be present on your microscope are marked with an asterisk (*).

Iris Diaphragm: Regulates how much light and lamp heat go through the specimen. It is controlled by a lever that is moved back and forth.

Condenser: A lens (substage condenser or light director), located above the diaphragm, which concentrates the light before it passes through the specimen. Its position is controlled by a knob on microscopes in which it is adjustable.

Observer: The microscope is useless when the observer *looks* but does not see.

Cleaning the Lenses

Clean lenses are necessary for a clear view of your specimen.

Nothing should touch the microscope lenses except the special lens paper and appropriate solvent.

Use a clean piece of lens paper to wipe the ocular and objective lenses. Sometimes you may need to moisten the lens paper with alcohol or xylene to remove grease or oil, which will blur the image. Follow the directions of your instructor regarding the use of solvent.

If there's dirt in your microscope scene,
Use only lens paper to clean,
If you do it just right,
Your field will be bright,
If not, you can try some xylene.

Viewing a Prepared Slide

Develop familiarity with your microscope by looking at a prepared slide of a letter ("e," "k," and so on) or some other object.

- First click the lowest power objective into position.
- Then use the coarse adjustment to raise the ocular (or lower the stage) as far as possible.
- Place the slide (oriented so that you can read the letter "e" properly) on the stage and position the letter over the hole in the stage.
- Now look through the microscope and adjust the diaphragm lever (and condenser knob) so that enough light is coming through to make the field bright.

In which direction do you move the lever to increase the amount of light? ____________

- View the objective lens *from the side* and turn the coarse adjustment until the lens *almost* touches the slide.

Did you have to turn the knob toward you or away from you? ____________

- Look into the microscope and slowly turn the coarse adjustment in the opposite direction so that the objective lens is moving *away* from the slide.
- Continue turning until the letter "e" comes into focus.

Remember that the coarse adjustment is used only on low power, so you should not have to use it again with this slide.

- Finish focusing with the fine adjustment knob.

Does the object appear normal or upside down?

____

- Move the slide a tiny bit away from you while observing the object in the microscope.

Does it move away from you or toward you?

____

- Move the slide back and forth and note that the inverted image always moves in the direction opposite to the direction in which the slide is moved.

When you change objectives to increase magnification, the **field of view** will be *smaller* as the magnification increases. As you take a closer look under greater magnification, you can see less and less at a time: First the quarter, then Washington's face, then his left eyeball. If the eyeball is not exactly in the center of the field at low magnification, you won't see it when you get to higher magnification; you might be looking at the nose instead.

As the field of view becomes smaller, the **depth of field** also *decreases*. The thickness of the part of the specimen in focus is larger under low power, so it is easier to focus using the coarse focus. Only a thin layer is in focus when you use high magnification, so fine focus adjustment is critical. By focusing up and down you can view different parts of a single cell or different cell layers. The depth of field also decreases when you close down the iris diaphragm to reduce the illumination.

- Center the image in the middle of the field of view.
- Move the high power lens into position, making sure that it clicks in correctly; you may have to focus a little with the fine focus adjustment knob but do *not* touch the coarse focus adjustment; you may also have to increase the amount of light in order to have a clear view.

If you do not see anything in the field, you probably did not have the image properly centered. Return to low power, center the image, and then switch to high power. Only the central portion of the low power field is visible under high power. Ask your instructor for help if you are still having trouble.

- When you have finished viewing the letter and are ready to go on, return to low power, turn the coarse adjustment knob to move the objective away from the slide, and return the slide to the front desk.

Using Oil Immersion

Learn to use oil immersion by viewing one of the designated slides. In this technique the resolution of the image is increased by filling the space between the oil immersion lens and the slide with oil. It works because oil has a much higher index of refraction than air and contributes to the magnification of the image.

Oil is used only with the oil immersion lens. Do not use oil with any of the other lenses.

- With your low power lens in position, locate the object and center it in the field.
- Switch to high power and refocus.
- Swing the high power lens away from the slide, and place a drop of immersion oil in the center of the coverslip and over the specimen.
- Swing the oil immersion lens into position and refocus using the fine focus only. Note the improved resolution of detail obtained by using immersion oil.
- When you are finished viewing, swing the immersion lens away, but do not put any other lens into position. Raise the body tube with the coarse adjustment and remove the slide. First wipe most of the oil off with lens paper, then use a piece of lens paper wet with xylene to wipe the rest of the oil off the lens and the slide. Finally, wipe the lens and slide with dry lens paper.
- Click the low power lens into position and you are ready to go on.

Computing Magnification

Magnification is only part of the problem in viewing tiny objects. If something is greatly magnified but its image is fuzzy, we can't make out its detail. **Resolution** measures how clearly you can see details in the microscope, and is usually given as the distance between two objects that can just barely be resolved. It depends in part on the quality of the lenses used. Resolution increases with magnification up to a theoretical limit: The maximum resolution in an ordinary light microscope is $\frac{1}{2}\lambda$ where λ is the wavelength of light used. If $\lambda = 500$ nm, as in green light, the resolution is $\frac{1}{2} \times 500$ nm $= 250$ nm (0.25 μm).* The use of immersion oil will increase the resolving power somewhat to about 0.18 μm. Your naked eye, on the other hand, has a resolution of only about 100 μm.

The ocular lens on your microscope gives 10× magnification of the image made by the objective lens. The objective lens magnifies the object 4× or more, so total magnification is the magnification of the first lens *times* the magnification of the second lens. The magnification of the objective lens may be 4×, 10×, 40×, or 100×. Fill in the magnifications of the objective lenses of your microscope and the total magnification:

Ocular	Objective	Total Magnification
10×	____×	____×
10×	____×	____×
10×	____×	____×
10×	____×	____×

Measuring Objects in the Microscope

If the field of view is a certain size under the lowest power or magnification, it will be correspondingly *smaller* as the power is increased. If the magnification is known, the relative sizes of the two fields can be calculated.

■ Place a ruler on your stage and measure the size of the field under the lowest magnification:

field = ____ mm (a)

lowest total magnification = ____ × (b)

■ Switch to another magnification:

total magnification = ____ × (d)

■ Calculate the field size of the second magnification where:

(a) (b) (c) (d)
field 1 × magnification 1 = field 2 × magnification 2

$$\text{field 2} = c = \frac{a \times b}{d} = \text{\_\_\_\_ mm}$$

■ Calculate the approximate field sizes in μm for your microscope:

Objective	Total Magnification	Field size in μm
scan	____ ×	____ μm
low power	____ ×	____ μm
high power	____ ×	____ μm
oil immersion	____ ×	____ μm

Once the size of the field in micrometers is known, it is possible to estimate the size of an object by comparing its size to the size of the field. For example, consider the organisms in Figure 3–5. Suppose that you are viewing them under low power where the size of the field is 1500μm. Their length is about 1.5 divisions out of 10 total, which equals 0.15 or 15%. Their estimated size would thus be:

$$0.15 \times 1500\ \mu\text{m} = 230\ \mu\text{m}$$

■ Obtain a prepared slide containing small organisms.

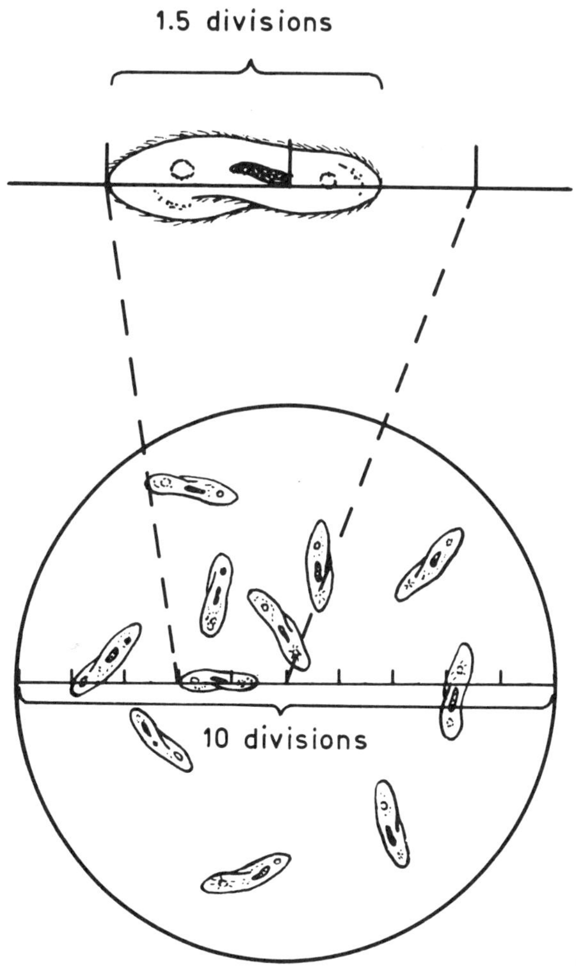

Figure 3–5. Estimating size. This organism is about 0.15 or 15% the size of the total field. Multiply the actual size of the field in μm by 0.15 to estimate the organism's size (see text).

* Refer to the description of metric units in Appendix D if these terms are unfamiliar to you.

- View the slide under low power and determine whether higher magnification is needed to determine the size of the organisms.
- Switch to higher powers if necessary.
- Note the size of the organism relative to the diameter of the field, and record the objective lens used, the magnification, and the size of the field in the chart below.
- Calculate the approximate size of the organism in micrometers.

Organism measured	
Objective lens	
Total magnification	
Size of field (μm)	
Relative size of organism	
Size of organism (μm)	

Studying Biological Organisms

Your instructor will provide cultures of living organisms or prepared slides for you to study in the microscope. always view the organisms under low power first, then increase the magnification as necessary. You may be asked to sketch and estimate the size of some of the following organisms:

yeast	*Paramecium caudatum*	*Amoeba*
Volvox	*Paramecium bursaria*	*Euplotes*
Spirogyra	*Chlamydomonas*	*Stentor*

- Make a wet mount of a culture of your organism.
- Locate your organism under low power and then switch to high power.
- Make a sketch of what you see and record the total magnification:

Total magnification = ____________ ×

Estimated size = ____________ μm

Identify the organism if you can, or describe it:

____

____

____

- When you are finished, discard the coverslip and clean the slide.

Storing the Microscope

- Return to the low power objective.
- Raise the body tube or lower the stage with the coarse adjustment so the lenses can't strike the stage accidentally.
- Clean all lenses with lens paper and (if you have not already done so) clean the oil immersion lens with xylene.
- Turn off the light.
- Cover the microscope, if there is a cover for it.
- Pick up the microscope by its arm, supporting it under the base, and return it to its storage box or cabinet.

When you are finished with today's lab, return all cultures and supplies to the front desk and clean off your workspace for the next student.

REFERENCES AND FURTHER READING

Corrington, Julian D. *Getting Acquainted with the Microscope.* Rochester: Bausch and Lomb, 1969.

Headstrom, Richard. *Adventures with a Hand Lens.* New York: Dover, 1976.

———, *Adventures with a Microscope.* New York: Dover, 1977.

Leeuwenhoek, Anthony van. *Anthony van Leeuwenhoek and his Little Animals.* New York: Dover, 1960.

Journal Articles

Binnig, Gerd, and Heinrich Rohrer. "The scanning tunneling microscope." *Scientific American,* 40, August 1985.

Fein, Jack M. "Microvascular surgery for stroke." *Scientific American,* 58, April 1978 (#1385).

Quate, Calvin F. "The acoustic microscope." *Scientific American,* 62, October 1979 (#3061).

TOPIC 4

Cells and Their Membranes

OBJECTIVES

When you have completed this topic, you should be able to:

1. Define solute, solvent, and concentration or electrochemical gradient.
2. Define the following processes and identify the characteristics that distinguish them from one another: diffusion, osmosis, and active transport.
3. Give three factors that affect the rate of diffusion and state whether they increase or decrease the rate.
4. Define and use correctly the following terms: osmotic potential, osmotic pressure, permeable, hypertonic, isotonic, hypotonic.
5. Describe what happens when an animal or plant cell is put into distilled water; then describe what happens when they are put into a concentrated salt solution. Use the following terms correctly in your answer: turgor, plasmolysis, lysis, wall pressure, hypertonic, hypotonic, osmosis.
6. State the function of the contractile vacuole in protozoans like *Paramecium* and *Amoeba.*
7. State the function of the central vacuole and tonoplast in plant cells.
8. Given any two solutions of different osmotic potentials that are separated by a selectively permeable membrane, state which solution is hypertonic and in which direction there will be a net flow of water.

TEXT REFERENCES Chapter 4

KEY TOPICS

diffusion, plasma membrane, permeability, osmosis, active transport

INTRODUCTION

In today's laboratory you will study two ways in which small molecules can move into and out of cells: diffusion and active transport. In both cases the molecules move *through* the plasma membrane. The organization of the plasma or cell membrane is shown in Figure 4–1. It is **selectively permeable** in that some substances may pass through it and others may not. Molecules, depending on their size and functional groups, may pass between the membrane components, through channels surrounded by protein, or they may be escorted across the lipid bilayer by specific transport proteins called **carriers.** The plasma membrane is thus not only a physical container for the cell but is an important regulator of the cell's contents.

Large molecules and particles may enter or leave the cell only if they are enclosed in sacs called vesicles which are formed by the membrane itself: Solid aggregates enter by **endocytosis** and liquids by **pinocytosis.** These processes are fundamentally different from diffusion and active transport because the material does not pass through the membrane as it enters the cell.

What kinds of substances must enter a typical human cell?

____

____

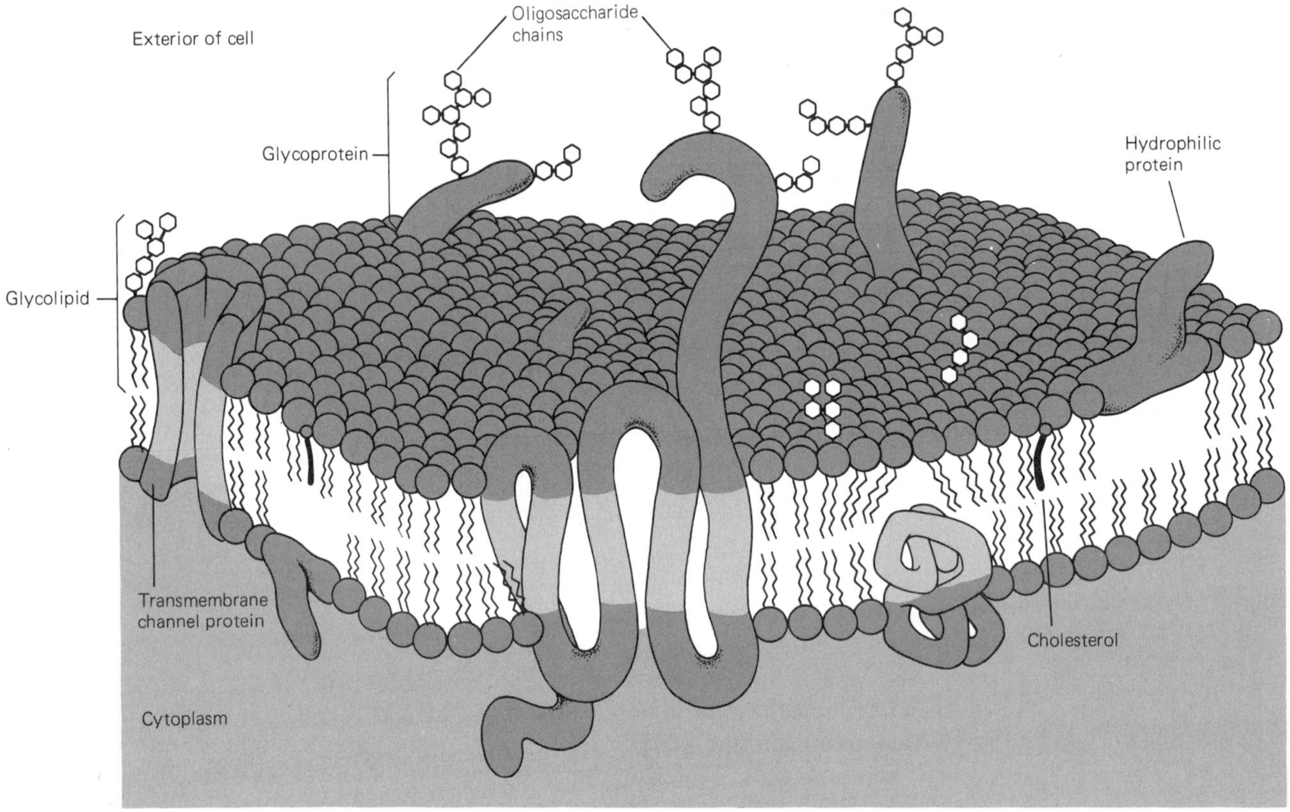

Figure 4–1. Fluid mosaic model of a plasma membrane. The lipid bilayer has a nonpolar interior and a polar exterior. Proteins and glycoproteins may move about in the lipid bilayer. Small molecules, if they are sufficiently nonpolar, may cross the membrane by diffusion through the lipid bilayer or they may cross through a protein channel. Some membrane proteins act as carriers to shuttle small molecules across.

What kinds of substances must move out of such a cell?

____

____

Do hydrophilic or hydrophobic substances pass more readily through a cell membrane?

____

If you put a high concentration of a solid substance, the **solute,** into a liquid, the **solvent,** the randomly moving molecules of the dissolving solute will spread throughout the liquid. This spontaneous or "downhill" process is called **diffusion** and it does not require any input of energy. (See Text Figure 4–2.) As they spread out, the molecules form a **concentration** or **electrochemical gradient** with the highest concentration near the dissolving solute and the lowest concentration farthest away. The diffusing molecules will tend to move *down the concentration gradient* from higher to lower concentration until they reach equilibrium. Then they will be spread out fairly evenly and there will be no further *net* movement. The free energy of the solution will be *lower* than it was before and the entropy (disorder) will be *higher* after diffusion has occurred.

Diffusion takes place rapidly in a gas, but more slowly in a liquid or solid. You are experiencing both types of diffusion when you smell odors. Diffusion in air to your nose is followed by diffusion in a liquid, as a chemical odor reaches your nasal membranes, where it must dissolve before exciting the sensory cells.

When diffusion occurs across a plasma membrane, solutes may pass into or out of the cell in the process called **dialysis.** Each solute will move from the side where its concentration is higher to the side with a lower concentration, because the direction of movement is always down the concentration gradient. Its movement will be independent of the movement of the other solutes.

In the special case of diffusion in which the substance diffusing across the membrane is water, the process is called **osmosis.** During osmosis water will flow from the side with the most water, and hence the least negative (higher) osmotic potential and lowest solute concentration, to the side with the lower water concentration, most negative (lower) osmotic potential, and higher solute concentration.

Pure water has an osmotic potential of zero and any solution will have a negative osmotic potential of less than zero.

Osmotic potential is thus a measure of the tendency of a solution to gain water compared with pure water in an ideal system: The more negative the potential, the more water tends to enter the solution. The negative of osmotic potential is **osmotic pressure,** the hydrostatic force needed to *prevent* pure water from entering a solution in an ideal system. As a point of reference, remember that strong salt solutions will have a strong tendency to gain water from dilute solutions:

SALT SUCKS

Any two solutions can be compared by their tendency to gain water. The solution that has a stronger tendency to gain water is said to be **hypertonic** *with respect to* another solution that gains water to a lesser extent. The solution that has a lower tendency to gain water is **hypotonic** *with respect to* the first solution. Pure water is the most hypotonic solution possible, since it can never gain water. Two solutions with exactly the same tendency to gain water are **isotonic** *with respect to* each other.

Pure water has an osmotic pressure of zero and is the most hypotonic solution possible.

When the solutions on either side of a selectively permeable membrane differ in concentration, there will be a net flow of water by osmosis across the membrane (all membranes are permeable to water). During osmosis, water acts like any small molecule undergoing diffusion, so osmosis can be considered to be a special case of diffusion: Water will always flow from the region of *higher water concentration* to the region of *lower water concentration.* Pure water has the highest possible water concentration: 100%. The more solute a solution contains, the less water it will have.

Cells cannot control the process of osmosis directly since cell membranes are freely permeable to water, and since osmosis is a spontaneous downhill process requiring no energy input. Cells can and must control their water content, however, in order to keep their solutes at the correct concentrations required for life processes to function. They can do this by removing excess water with special structures such as the contractile vacuole of protozoans, or they can expend energy to increase their solute concentration, permitting water to enter the cell by osmosis.

Many substances are brought into the cell or removed from the cell *against* their chemical gradients in the process of **active transport.** Since this is an uphill process, a biologically active membrane that can directly or indirectly utilize energy from cell respiration is required. The process of active transport is highly specific in that each particular substance to be carried across the membrane must be transported by a specific **carrier** molecule. The carriers are membrane proteins and their transport function must always be coupled with a source of energy. If there is no cell respiration, or if the integrity of the membrane is destroyed, active transport cannot occur.

DIFFUSION

Diffusion of a Liquid in a Liquid

Since liquids are highly sensitive to convection currents, the success of the experiment depends on not allowing your setup to be disturbed or jiggled once the experiment is under way.

- Fill a small beaker with water and set it in a protected place. Wait 10 min.
- Add a drop of red dye to the surface of the solution as gently as possible. Try not to disturb the surface, but hold the dropper as close to it as you can.
- Watch the dye spread through the solution from time to time during the lab period.

At the end of the period answer the questions given below:

Is the **concentration** of the colored part of the solution greater at the beginning or end of the experiment?

____

Is the **free energy** of the solution greater at the beginning or at the end? ______________

Is the **entropy** of the solution greater at the beginning or at the end? ______________

What difference would you have seen if you had originally set up the beaker in the refrigerator?

____

____

Diffusion in a Solid

The solid you will use is agar, which is clear like jello so you can see what is happening. The substances that will be diffusing are solids that react with each other to give colored compounds. The position of the colored bands shows where the two substances first

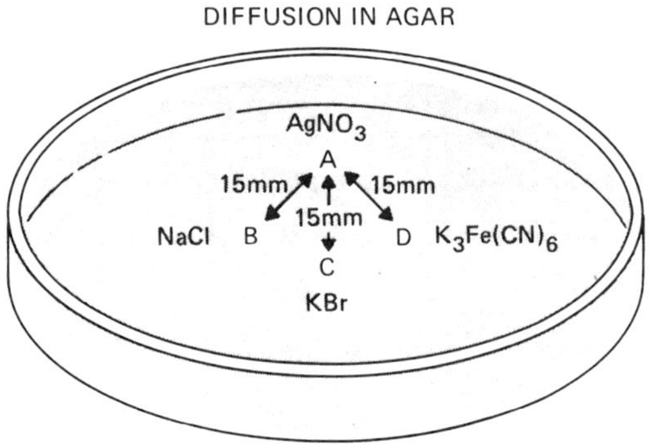

Figure 4–2. Diffusion in a solid. A Petri dish of agar has four holes filled with the following solutions: A contains 1 M $AgNO_3$; B contains 1 M NaCl; C contains 1 M KBr; and D contains 1 M $K_3Fe(CN)_6$. Run the experiment in the dark.

come in contact. The band will be closer to the substance that diffuses more slowly. The setup for this experiment is shown in Figure 4–2.

- Take a Petri dish of 1.5% agar and turn it upside down. Make a dot near one edge and label it A. Use a ruler to measure 15 mm from A and make a second dot B. Makes dots C and D also exactly 15 mm from A and as evenly spaced as you can.
- Use a cork borer to make holes in the agar at A, B, C, and D. Work carefully and slowly so you don't damage the agar.
- Use a Pasteur pipet or dropper to add these solutions to the corresponding holes. Be careful *not to overfill the holes* and try to fill them all to about the same level.

A:	1 M $AgNO_3$	silver nitrate
B:	1 M NaCl	sodium chloride
C:	1 M KBr	potassium bromide
D:	1 M $K_3Fe(CN)_6$	potassium ferricyanide

- Place the Petri dish in the dark and check it from time to time until you can see colored bands forming where the solutions meet solution A.

Ag^+ will react with negative ions from the other solutions to give colored substances. Record what colors appear in Table 4–1 at the end of this topic.

- Measure the distance between each hole and the colored band, and record it in the table. Then calculate the ratio of the distance each ion moved from holes B, C, and D versus the distance moved by the Ag^+ ion: Cl^- distance/Ag^+ distance, and so on. Record these ratios in the table. Since time is constant, the ratio of distances traveled is equal to the ratio of rates.

$$\frac{D_1}{D_2} = \frac{R_1}{R_2}$$

Rank the relative rates of the Cl^-, Br^-, and $Fe(CN)_6^{\equiv}$ ion as slow, faster, or fastest.

As you might expect, the larger an ion is, the slower it will diffuse through the agar. Given that the molecular weights (MW) of the ions in holes B, C, and D can be ranked $Cl^- < Br^- < Fe(CN)_6^{\equiv}$, which ion should move fastest?

____

Which should move slowest? ______________

Did your results agree with your expectations?

____

Calculating Diffusion Rates

The diffusion rate is inversely related to the square root of the molecular weight, so the ratio of rates will be

$$\frac{\text{rate}_1 = \dfrac{1}{\sqrt{MW_1}}}{\text{rate}_2 = \dfrac{1}{\sqrt{MW_2}}} \quad \text{or} \quad \frac{R_1}{R_2} = \sqrt{\frac{MW_2}{MW_1}}$$

- Given the molecular weights of the ions, calculate the expected ratio of their diffusion rates, and enter your results in Table 4–1.
- Enter your observed ratios next to your calculated ratios for comparison.

OSMOSIS

Osmosis in a Model Cell

You will use the following system to become familiar with osmosis: Two solutions will be separated by a **selectively permeable membrane** that is **permeable** to water, glucose, and small molecules, but **impermeable** to larger molecules. Water will diffuse from one to the other in the process of **osmosis,** and small molecules will diffuse across the membrane in the process of **dialysis.** This setup is shown in Figure 4-3.

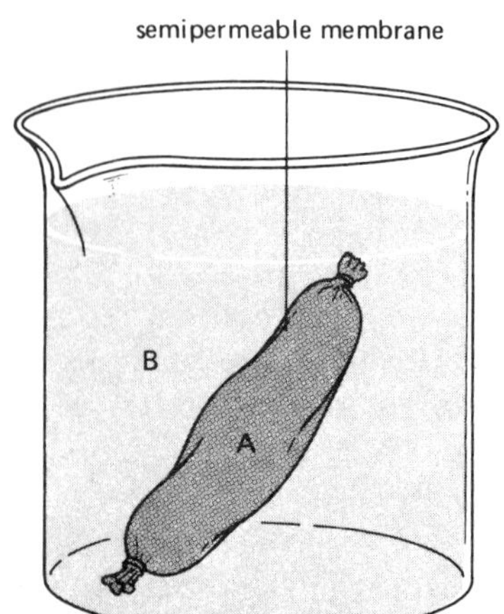

Figure 4–3. Model cell. A selectively permeable membrane filled with solution A represents the cell. It is placed in solution B to start the experiment.

- Obtain a short piece of dialysis tubing (12–15 cm), and soften it by soaking it in distilled water.
- Fold one end over and tie it tightly with a piece of thin string or strong thread.
- Fill the bag with solution A, a liquid "meal" containing:

  25% glucose
  0.5% egg albumin
  1% starch

- Fold over the top, squeeze out all the air, and tie tightly. The bag should be limp.
- Place the bag in a small beaker and add enough of solution B (distilled water containing a very small amount of iodine) to cover it. Let it stand for an hour or so while you go on.

At the end of your experiment answer the questions in Table 4–2, remembering that the bag is permeable to glucose, iodine, and water but not to starch or protein.

Osmosis in a Living Cell

To watch osmosis across a living membrane, see what happens when a plant cell comes in contact with a hypertonic solution. Note that the plant is kept in an aquarium that contains fresh water that is hypotonic to the plant cell. The cell swells and becomes **turgid** but is prevented from undergoing **lysis** (bursting) because its rigid cellulose walls exert **wall pressure** on the cell contents. Many animal cells have no such protection against lysis and can exist only in an isotonic solution.

- Make a wet mount of an *Elodea* leaf and focus on it under high power.
- Add a drop or two of 10% NaCl solution to the edge of the coverslip.
- Touch a piece of tissue to the opposite side of the coverslip to pull the solution through and observe.

Figure 4–4 shows the cell *before;* sketch the cell *after* NaCl in the space provided.

The cell is now (turgid, flaccid) ______________.

The cell has undergone (lysis/plasmolysis/death) ______________.

- Now add a few drops of distilled water to the edge of the coverslip and pull it through with tissue while watching the cell.

Is the process you have observed reversible? ______________

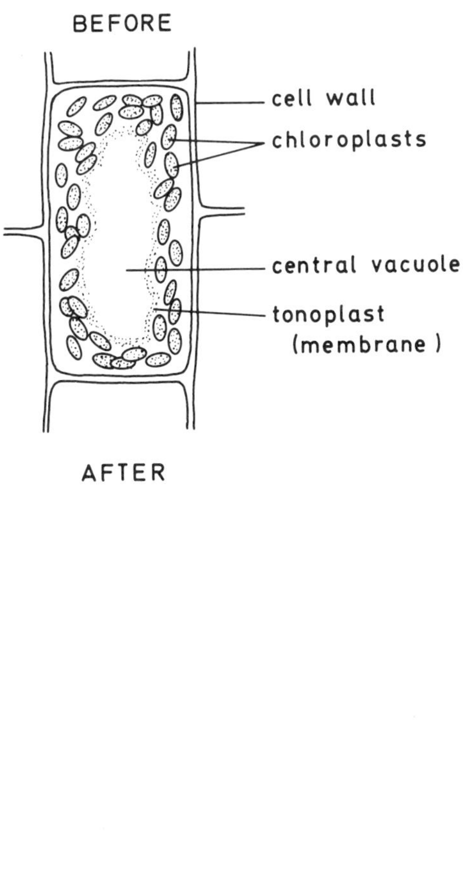

Figure 4–4. *Elodea* cell. Sketch the appearance of the cell after exposure to NaCl solution.

ACTIVE TRANSPORT

Work in pairs to study active transport in yeast, a unicellular fungus. The substance to be transported is neutral red, a dye that is red in a slightly acidic solution and yellow in a basic solution. Sodium carbonate (Na_2CO_3) will be used to make the dye solution basic initially. Conditions inside living cells are relatively acidic.

The experimental plan is shown in Figure 4–5. The

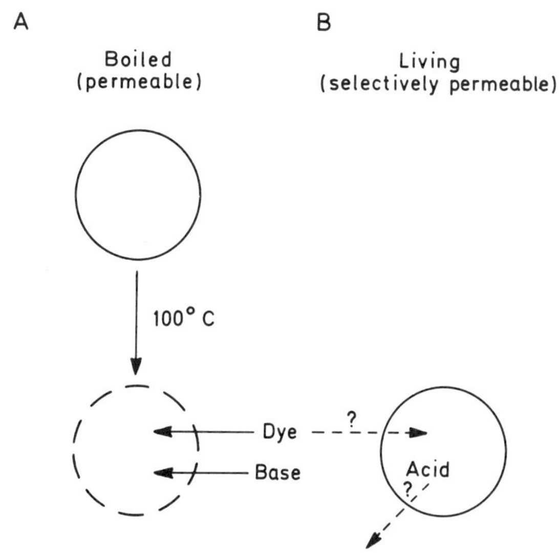

Figure 4–5. Active transport experiment. Nonliving (A) and living (B) yeast cells are compared.

cells in Flask A will be boiled, so their membranes are permeable to all molecules. Those in Flask B are living and their membranes may be permeable or impermeable to dye—this is the question you are going to investigate. Record your results in Table 4–3, at the end of this topic.

- Obtain two 1-g samples of baker's yeast.
- In a 250-mL Erlenmeyer flask (Flask A) place 1 g of yeast, 25 mL of 0.75% Na_2CO_3 (sodium carbonate), and mix well. Boil this flask gently for 2 min and cool. The cells are now dead.
- Add 25 mL of 0.02% neutral red. Record the color of the suspension.
- In a second flask (Flask B) place 25 mL of 0.75% Na_2CO_3 and 25 mL of 0.02% neutral red. Record the color of the solution.
- Add 1 g of yeast cells, swirl the flask, and watch for a color difference compared to Flask A. Record the final color of each flask.

Either the cells in Flask B put out something to change the solution or the dye entered the cells and was changed in the cytoplasm. To test this compare cells in Flasks A and B:

- Filter a portion of the contents of Flask A into a test tube. Record the color of the yeast cells and the solution.
- Filter a portion of the contents of Flask B into a test tube. Record the color of the cells and the solution.

Think about your observations, then answer the questions in Table 4–3 about Flask B.

- Add an equal volume of 0.75% acetic acid to the remaining contents of Flask A to acidify the suspension. Record the color of the suspension in Flask A.
- Filter a portion of the suspension. Record the color of the cells and complete Table 4–3.

When you are finished, wash the glassware that you have used and clean up the lab desk for the next student.

REFERENCES AND FURTHER READING

DeRobertis, E.D.P., and E.M.F. DeRobertis. *Cell and Molecular Biology,* 7th ed. Philadelphia: W.B. Saunders, 1980.

Giese, A.C. *Cell Physiology,* 5th ed. Philadelphia: W.B. Saunders, 1979.

Lehninger, A.L. *Principles of Biochemistry.* New York: Worth, 1982.

Stryer, Lubert. *Biochemistry.* 2nd ed. San Francisco: W.H. Freeman, 1981.

Journal Articles

Bretscher, Mark S. "The molecules of the cell membrane." *Scientific American,* 86, October 1985.

Dautry-Varsat, Alice, and Harvey F. Lodish. "How receptors bring proteins and particles into cells." *Scientific American,* 52, May 1984.

Lodish, Harvey F., and James E. Rothman. "The assembly of cell membranes." *Scientific American,* 48, January 1979 (#1415).

Rubenstein, Edward. "Diseases caused by impaired communication among cells." *Scientific American,* 102, March 1980 (#1463).

Singer, S.J., and G. Nicholson. "The fluid mosaic model of the structure of cell membranes." *Science* 175: 720, 1972.

Staehelin, L. Andrew, and Barbara E. Hull. "Junctions between living cells." *Scientific American,* 140, May 1978 (#1388).

Unwin, Nigel, and Richard Henderson. "The structure of proteins in biological membranes." *Scientific American,* 78, February 1984.

Student Name ______________________________ **Date** ______________

TABLE 4–1. Results from the Diffusion Experiment

Band	Reaction	Color
AB	$Ag^+ \rightarrow Cl^- = AgCl$	
AC	$Ag^+ \rightarrow Br^- = AgBr$	
AD	$Ag^+ \rightarrow Fe(CN)_6^{\equiv} = Ag_3Fe(CN)_6$	

Ion	Measurement	Distance (mm)	Ratio $D_1/D_2 + R_1/R_2$	Relative Rate
Cl^- Ag^+	B → AB band A → AB band	$D_1 =$ $D_2 =$	$Cl^-/Ag^+ =$	Cl^- moved ______
Br^- Ag^+	C → AC band A → AC band	$D_1 =$ $D_2 =$	$Br^-/Ag^+ =$	Br^- moved ______
$Fe(CN)_6^{\equiv}$ Ag^+	D → AD band A → AD band	$D_1 =$ $D_2 =$	$Fe(CN)_6^{\equiv}/Ag^+ =$	$Fe(CN)_6^{\equiv}$ moved ______

MW_1	MW_2	MW_2/MW_1	$R_1/R_2 + \sqrt{MW_2/MW_1}$	Observed Ratio R_1/R_2
$Cl^- = 35$	$Ag^+ = 108$			$Cl^-/Ag^+ =$
$Br^- = 80$	$Ag^+ = 108$			$Br^-/Ag^+ =$
$Fe(CN)_6^{\equiv} = 212$	$Ag^+ = 108$			$Fe(CN)_6^{\equiv}/Ag^+ =$

TABLE 4–2. Results from the Osmosis Experiment

Which solution was hypertonic (A or B)? ______

Which solution gained water (A or B)? ______

Which solution had the higher osmotic potential (A or B)? ______

Which substance inside the membrane reacted with the iodine? ______

Which substance(s) moved out through the membrane?

______

What test that you learned in Topic 1 could you use to test for the presence of glucose outside?

______

If you swallowed some of solution A, which substance(s) could be absorbed directly by the cells of your gut, assuming that they had the same permeability as the dialysis tubing?

______

Which substance(s) would have to be further digested?

______

If your dialysis bag was flaccid or limp at the beginning of the experiment, at the end was it very flaccid, flaccid, turgid, or very turgid?

______

Which of the following statements best describes the situation at equilibrium if you let the system stand for a long time? ______

1. No molecules move across the membrane.
2. All molecules cross the membrane equally often in either direction.
3. Molecules to which the membrane is permeable cross equally often in either direction.
4. Only water molecules cross the membrane equally often in either direction.

TABLE 4–3. Results from the Active Transport Experiment

	Flask A	Flask B
Initial color of suspension		
Final color of suspension		
Color of cells		
Color of solution		
Is dye inside cells?	XXXXX	
Did Na_2CO_3 enter cells?	XXXXX	
Is the membrane permeable?	XXXXX	
Color of acidified suspension		XXXXX
Color of cells		XXXXX
Did neutral red enter cells?		XXXXX
Did acetic acid enter cells?		XXXXX
What term describes a membrane that is permeable to some substances but not to others? __________ Are dead cells permeable or impermeable to acetic acid? __________		

TOPIC 5

Cell Structure and Function

OBJECTIVES

When you have completed this topic, you should be able to:

1. Describe the structures that you can expect to see in a typical plant cell with the light microscope.
2. Describe the structures that you can expect to see in a typical animal cell with the light microscope.
3. Describe how a bacteria cell looks under the light microscope.
4. Explain how a prokaryotic cell such as a bacterium basically differs from a typical eukaryotic cell.
5. List structures found in unicellular organisms that are an adaptation to their free-living life, and are not usually found in the cells of multicellular organisms.
6. List the advantages and disadvantages to an individual cell of being part of a multicellular organism.
7. Describe some specialized adaptations found in cells in multicellular organisms and in tissues of higher organisms.

TEXT REFERENCES Chapter 5; Chapter 20 (A,G,H,N,O); Chapter 21 (C).

KEY TOPICS

light microscopy	prokaryotic cells	animal tissues
staining	protists	plant tissues
eukaryotic cell structure	green algae	

INTRODUCTION

In today's laboratory you will study biology from the viewpoint of the individual cell. Although the cell is considered to be the building block of all organisms, because it can carry on all activities associated with living things, cells differ enormously in shape, size, and capability. The purpose of your observations should be, first, to review the basic organization of cells and, second, to study how the cell has become adapted to a variety of biological roles through the process of evolution.

No one cell can be considered "typical," so you will be studying several examples. Every cell (with some highly specialized exceptions) has **DNA** in a **nuclear area** or in a **nucleus, cytoplasm,** and a **plasma membrane.** It also has many organelles, vital to its survival, which are too tiny to be seen in the light microscope and consequently cannot be studied here. All cells contain **ribosomes,** the organelles that carry out protein synthesis.

There are two major types of cells. **Prokaryotic cells** are small (0.5 to 5 μm) and less complex, usually exist as unicellular organisms, and have limited capabilities compared with **eukaryotic cells.** Higher organisms, namely plants, animals, and fungi, are made up of highly integrated aggregations of large (10 to 50 μm), specialized eukaryotic cells. Some sophisticated organisms, known as protists, consist of a single eukaryotic cell. These cells are large (5 to 200 μm or more), have a membrane-bound nucleus, and have membrane-bound organelles.

All cells, including those that develop into humans, come into existence from preexisting cells. Thus life is a continuum extending backwards in time to the very first cell. Because the organization of the DNA in eukaryotes is very complex, cell division in these cells is fundamentally different from that in prokaryotes. The most common type of cell division in eukaryotes is called **mitosis** and will be studied in Topic 10. Prokaryotes, which have only a nuclear area rather than a nucleus, divide by a simpler process of **binary fission.** This difference in cell division is another important feature by which eukaryotic and prokaryotic cells are distinguished from one another.

A "TYPICAL" CELL

No one cell will show all the features that a cell might have, so you will instead look at three ordinary cells to see what they are like. The first two have a true nucleus with a nuclear envelope and other characteristics not found in prokaryotes.

Onion Epidermis (Eukaryotic Plant Cell)

- Break or cut off with a razor blade a piece of a single layer of onion from within the onion (the outermost layer may contain only dead cells). See Figure 5–1.
- Return the onion to its storage container.
- Snap the piece in half and then use forceps to peel off a bit of tissuelike transparent epidermis from the inner layer as shown in Figure 5–1.
- Mount the epidermis in tap water on a slide so that it is flat and not doubled over on itself. Add a coverslip.
- View the slide under low power.

What is the shape of the cells?

____

What is the thick layer that surrounds each of the cells?

____

What does the clear part of each cell contain?

____

If you are looking at a confusing mix of overlapping cell parts, you probably do not have a good piece of epidermis. Ask your instructor for help. Some cells will not have a **central vacuole,** if they are young. Also, the **plasma membrane,** which lines the inner surface of the cell wall, is too thin to be seen in the light microscope, but you can see exactly where it must be.

- Raise the coverslip and add a drop of 45% acetocarmine dye; this will stain the **cytoplasm** and **nucleus** of your cells pink or red.
- Distinguish between the vacuoles and the cytoplasm. The cytoplasm is within the cell membrane and is "living," whereas the contents of the vacuoles are often inert storage materials.
- Locate the nucleus, which now should be red. It is a true, eukaryotic nucleus and is surrounded by a nuclear envelope.

What kinds of molecules are located in the nucleus?

____

____

What is the function of the nucleus?

____

- Make a sketch of the onion cell in Figure 5–5 at the end of this topic to show the parts that you saw. Indicate the size of a typical cell. (Make this simple calculation by comparing the size of the cell to the known diameter of your microscope field.)

Sketches for this lab should be made in the spaces provided at the end of the exercise. Your instructor may ask that they be handed in when you leave.

- When you are finished with each specimen, discard the coverslip and clean the slide for the next specimen. Your instructor will tell you whether you should discard, rather than wash, a slide on which you used stain.

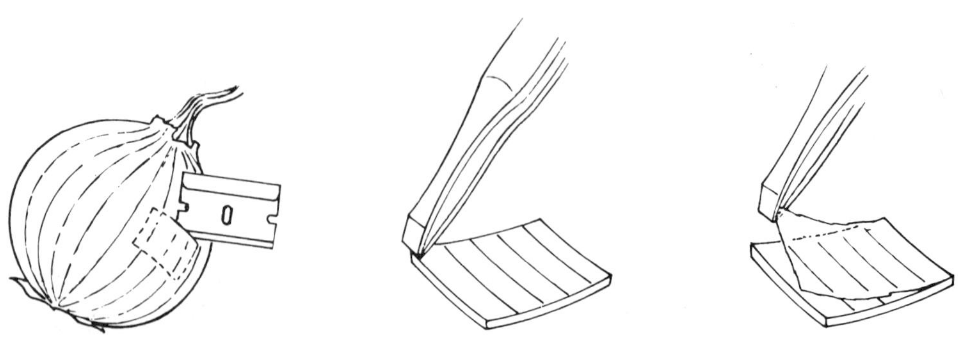

Figure 5–1. Onion epidermis. The desired tissue is extremely thin and transparent.

Human Cheek Epithelium (Eukaryotic Animal Cell)

- Use a clean toothpick to scrape the inside of your cheek gently.
- Stir the scrapings into a drop of tap water and add a drop of methylene blue dye to stain the cheek cells. Add a coverslip.
- View your cells under low power and locate some that are spread out rather than bunched up in a big clump.

The cells you are looking for are pale blue and have a very dark blue nucleus. You may see some that are folded over so that you can see how flat they are. (See Text Figure 4–1.)

- Switch to high power and take a closer look at one of the cells.

What shape is the cell?

____

Do you see a vacuole or cell wall? ____________

Again, you cannot see the plasma membrane but can assume it surrounds the light blue cytoplasm.

- Look closely at the nucleus and locate the nucleolus within it (try another cell if necessary).

What is the function of the nucleolus?

____

How does this cell type compare in size to the onion cell? ____________

This cell is highly specialized and has a protective function. As a result, it cannot carry out certain functions.

What are some of these functions?

____

____

____

____

Bacterium (Prokaryotic Cell)

- Use a toothpick to place a tiny dab of yogurt on your slide. Add a small drop of tap water and stir the toothpick around to spread out the cells. Add a coverslip.
- Focus on the slide under low power and then switch to high power.

The swarms of rod-shaped organisms that you see are cells of yogurt bacterium *Lactobacillus.* These bacteria are adapted to live on milk sugar (lactose) and are used by human beings to convert ordinary milk into yogurt, which is acidic and keeps much longer than milk. Their nutrition is thus by **absorption.** Bacteria can use an enormous variety of food sources and many can carry out **photosynthesis** to make their own food.

- Add a drop of immersion oil to your slide and view the bacteria under oil immersion if your microscope has an oil immersion lens.

Can you see any structure within the cells?

____

Bacteria have tough cell walls but do not have any of the membrane-bound organelles that are found in eukaryotic cells. Their DNA is found in a **nuclear area.**

What organelle, not visible in the light microscope, is found in both eukaryotic and prokaryotic cells?

____

If time permits, look at some of your own oral bacteria:

- Use a clean toothpick to scrape some of the tartar from between your teeth.
- Smear out the scum on a slide and pass the slide over the small flame of an alcohol lamp or Bunsen burner to fix the cells onto the slide.
- Add a drop of crystal violet stain and wait 1 min.
- Wash off the excess dye with a gentle stream of water and blot off the excess water with a tissue, being careful not to touch the spot where the bacteria are.
- If you have an oil immersion lens, let the slide dry and then add a drop of oil and view under oil immersion.
- If you do not have oil immersion, add a drop of water and coverslip and view under high power.

The bacteria are tiny and darkly stained. They may be rod-shaped, round, or spiral-shaped, and some may stick together in clumps. Choose some different types, and draw what you see in the space provided in Figure 5–5 at the end of this topic.

- If a demonstration slide is available, view the three main shapes of bacteria using oil immersion.

What shapes can you see? ____________

____

____

SPECIALIZED PROTISTS

Protists are unicellular organisms not clearly related to a group of multicellular organisms.

Amoeba

This organism is a single cell with an irregular shape and many internal organelles. Cytoplasmic streaming allows it to move by changing shape so it actively seeks prey and can **ingest** food by trapping it within arms called **pseudopods.** (See Text Figure 4–12). *Amoeba* digests the food within its membrane-bound **food vacuoles.** It has a **contractile vacuole** to take up excess water and eject it to the exterior so that the *Amoeba* doesn't swell up and burst. The **nucleus** controls the activity of the cell and is duplicated when the cell divides by simply balling up and splitting in half during mitosis.

- Use a dropping pipet to take a sample from the bottom of the *Amoeba* culture. Try to get your sample near some debris.

***Amoeba* is sometimes large enough to see with the naked eye. If not, you may use a dissecting microscope when locating an *Amoeba* and transferring it to your slide.**

- Place the drop on your slide and view it under low power to make sure that you have at least one *Amoeba;* if not, try again using a dissecting microscope to make sure that your dropper sucks one up.

Small, fast-moving organisms are food, not *Amoebae*!

- Add a coverslip and view the *Amoeba* under high power.

A healthy *Amoeba* puts on quite an active performance, extending pseudopods in all directions, creeping along the slide and perhaps even encircling some of the food organisms in the culture fluid. This organism works in slow motion, so you may have to watch carefully to see exactly what your *Amoeba* is doing.

Keep in mind that the *Amoeba* is very sensitive to heat from the microscope lamp, so you should turn off the lamp frequently to let it cool, and always turn off the lamp when you are not looking through the microscope.

What is the approximate diameter of your *Amoeba?*

_______________ μm

Note that the inner part of the cytoplasm is doing most of the streaming. This is the **sol** part of the cytoplasm whereas the more rigid outer part is called the **gel.** Cytoplasm can be interconverted from gel to sol. A gel-to-sol transition occurs when actin molecules of the **microfilaments** (not visible in the light microscope) disassemble and it is controlled by the concentration of Ca^{++} in the cell. Cytoplasmic streaming can then begin in the sol, which causes new pseudopods to form and results in **amoeboid movement.**

- Locate the contractile vacuole that periodically ejects its contents to the outside of the cell.

About how long does each cycle take?

____

Some of the white blood cells in your own body are very similar to this *Amoeba.* The circulate between the cells of your tissues and engulf harmful bacteria and cell debris that would otherwise accumulate.

- If possible, return your *Amoeba* to the supply dish before cleaning your slide for the next event.

Paramecium

This organism is also a single cell with some of the same features of the *Amoeba;* it moves, ingests particles of food, digests the food in vacuoles, removes water with a contractile vacuole, and reproduces by dividing in half (see Color Plate 2). Its life style in doing these things is so different, however, that it is hard to see how *Amoeba* and *Paramecium* are similar at all. The *Paramecium* is in constant motion, thanks to its **cilia,** which cover the whole body and beat in a coordinated sequence to propel the organism along. (See Text Figure 5–16.) The body has a definite shape which allows it to move efficiently, and there is an **oral groove** that tends to accumulate possible food items for efficient ingestion. Stinging organelles called **trichocysts** on the surface of the *Paramecium* are used to trap and immobilize its prey. Cytoplasmic streaming helps circulate vacuoles around the cytoplasm, and wastes tend to be eliminated from old food vacuoles at the same site on the membrane.

- Place a small drop of the *Paramecium* culture on a clean slide and add a drop of methyl cellulose (a viscous solution such as Protoslo that slows the speed of the organisms). Stir the drop with a toothpick and add a coverslip.
- View the slide under low power so that you can see what *Paramecium* looks like, and then switch to high power.

Approximately how large is your specimen?

_______________ μm

- Make an outline drawing of *Paramecium* in the space provided at the end of this topic, showing the oral groove and labeling the anterior end.

Does a *Paramecium* always swim in the same direction? _______________

- Locate the large nucleus, which is the macronucleus. It controls growth, metabolism, and asexual reproduction. Sexual reproduction is under the control of micronuclei.
- Locate the active contractile vacuole(s). Observe

the cycle through which it slowly fills with excess water and then suddenly contracts to expel water. (See Text Figure 4–6.)

Yeast cells are a good food source for *Paramecium*. When they are heated and stained with Congo red dye, they can be easily seen even after they have been ingested and are enclosed in food vacuoles.

- Add a small drop of *Paramecium*, a drop of yeast cells stained with Congo red dye, and a drop of Protoslo to a clean slide and stir with a toothpick. Add a coverslip and wait a few minutes for the yeast to be taken up.
- Locate a *Paramecium* under high power.

Where are the yeast cells found inside the cell?

____

____

____

What color are the yeast cells when they are inside the *Paramecium?* ______________

What factor affected the color of the dye?

____

When you are finished, clean your slide for the next preparation.

Peridinium

This organism has a life style fundamentally different from that of *Amoeba* and *Paramecium* in that it is **photosynthetic** and can make its own food. Because it is an **autotroph** ("self feeder") it does not need any mechanism for ingestion. *Peridinium* lacks a cell wall; instead, it has **cellulose plates** just under the plasma membrane which give the cell form and protection

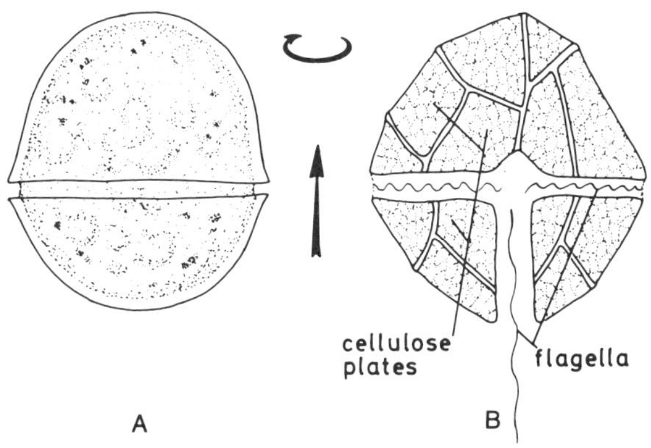

Figure 5–2. ***Peridinium.*** This single, complex cell moves by means of two flagella and can carry out photosynthesis. It belongs to the unique group of plantlike protists called dinoflagellates. (A) *Peridinium cinctum* as it appears in the light microscope (shown here about 800×). (B) A diagram showing its two flagella and cellulose armor plates that can be clearly seen only in the scanning electron microscope (shown here about 800×).

(see Figure 5–2). It moves by means of two **flagella** located in grooves between the cellulose plates. These appendages are characteristic of the group **dinoflagellates,** to which *Peridinium* belongs.

- If a living culture is available, make a wet mount of a culture of *Peridinium* and observe under low power.
- Note the spinning movement of the cells as they swim along.

What color are these cells? ______________

Dinoflagellates do have green chlorophyll, but it is masked by photosynthetic pigments of other colors.

- Make another wet mount, stirring in a drop of Protoslo before adding the coverslip, or obtain a prepared slide of *Peridinium*, and observe under high power.
- Reduce the illumination by closing the diaphragm and try to see the flagella. The structure of a flagellum is basically like that of the cilium in *Paramecium* except that the flagellum is longer.

How big is *Peridinium* in μm? ______________ μm

In spite of their small size, and in contrast to animal-like protists such as *Amoeba* and *Paramecium*, dinoflagellates are extremely important in ecology. They form a substantial part of the **plankton,** which is the collection of tiny organisms found in enormous numbers in lakes and oceans. Although one generally associates photosynthesis with green plants, planktonic autotrophs, or phytoplankton, carry out photosynthesis on a large scale in the oceans and may actually have a greater combined photosynthetic output than plants. Dinoflagellates play a further important role in our economy because some of them store energy in the form of oil. Masses of ancient dinoflagellates have been preserved to form our underground oil deposits.

Can you suggest a reason why some dinoflagellates store energy as oil rather than as starch, for example?

____

SPECIALIZED ANIMAL CELLS

Intestinal Cells

In more complex multicellular organisms, the cells are actually much simpler than some of the cells that you have already seen; the complexity is created by *combinations* of cells rather than the cells themselves. Each individual cell tends to lose some functions, such as motility or protection, and to specialize in other functions, such as support, secretion, or contraction. A **tissue** is a group of similar cells specialized for a particular function. In animals, there are four main types: **epithelium** for protection; **connective** tissue for support, storage, and other functions; **muscle** for contraction; and **nervous** tissue for the conduction of nervous impulses and coordination. An **organ** is a structure, such as the intestine, that contains several types of tissue organized to interact for a general function, in this case digestion of food.

- Take a prepared slide of the cross section of intestine and view it under low power.
- Note the hollow inner space, which is the **lumen.**

There are five or more layers of different tissues surrounding the lumen. They are organized in sequence from the inner side into the **mucosa, submucosa, muscle,** and the **peritoneum.**

- Study the cell types in each layer in turn, using Figure 5–3 as your guide.
- Make a sketch of each cell type in the space provided in Figure 5–6 at the end of this topic.

Columnar Epithelial Cells: These cells make up the inner layer of the mucosa. They follow its highly folded inner surface and secrete mucus, enzymes, and fluid into the lumen of the gut. Their specialty is protection and secretion.

Goblet Cells: These have large cuplike spaces facing the lumen of the gut and are filled with the mucus, which they secrete. These cells are highly specialized for secretion.

Connective Tissue Cells: These cells make up the bulk of the structure of the submucosa. The dense connective tissue cells secrete collagen fibers that support the other cells and the many blood vessels and lymph vessels that pass through this layer. Since blood is a fluid connective tissue itself, the blood cells within the vessels are another type of connective tissue cell specialized for oxygen transport or defense against harmful organisms. There may also be some adipose cells specialized for the storage of fat droplets.

Smooth Muscle Cells: These cells are specialized for contraction. You can see the darkly stained cell nuclei. The muscle cells are in two layers: The cells of the inner layer (closer to the lumen) have a spindle shape. They are oriented in rings so that they have been cut in longitudinal section. The cells in the outer layer look like circles. They are also spindle-shaped, but are oriented along the length of the intestine and so have been cut in cross section. The two layers of muscle cells counteract each other: If the inner layer contracts, the section of gut will become long and thin, but if the outer layer contracts, the section will become short and fat. Contraction is involuntary and is under the control of neurons contained in nervous tissue.

Squamous Epithelial Cells: The peritoneum is the outermost layer of cells and is only one cell layer thick. The cells are flattened and serve to protect the outer surface of the intestine. The nuclei may be seen as small elevations. The cells lining the blood vessels are also squamous epithelium.

What type of squamous epithelial cells did you see earlier in this exercise?

____

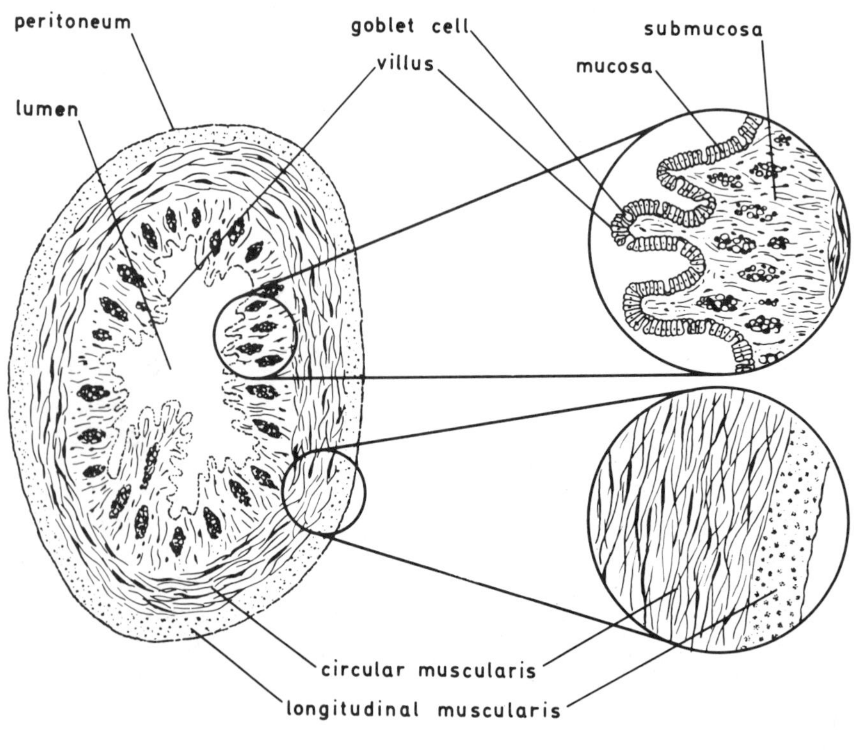

Figure 5–3. Intestine in cross section. The lumen is the hollow interior of the intestine where you would find undigested food. The projections into the lumen are the villi, which increase the area for absorption.

SPECIALIZED PLANT CELLS

All plants are **autotrophs** and contain **chloroplasts,** organelles specialized for synthesizing food through photosynthesis (see Color Plate 1). Most plants are multicellular, but there are a few exceptions among the green algae. This group includes some single-celled organisms such as *Eremosphaera* that have many features found in more complex plants, particularly chloroplasts containing the same types of chlorophyll (See Figure 5–4). For this reason the green algae, including many single-celled forms, are considered to be plants.

Eremosphaera

- Make a wet mount of *Eremosphaera* and observe under low power.

Is this cell motile? ______________

How big is the typical cell? ______________ μm

- Switch to high power and locate the large nucleus in the center of the cell.

Where are the chloroplasts located? ______________

Do you see empty spaces that could be vacuoles?

____

- Sketch this organism in Figure 5–5 at the end of this topic.

As an autotroph, *Eremosphaera* is nutritionally almost independent of its surroundings except that it must take in raw materials.

What important environmental factor is absolutely required for its survival? ______________

How is *Eremosphaera* adapted to obtain this factor?

____

____

Figure 5–4. *Eremosphaera.* This tiny plant is a single cell and has many chloroplasts located in the strands of cytoplasm that radiate out from the center of the cell. It is one of the largest of the green algae, the simplest of the plants.

Leaf Cells

In a multicellular plant, the leaf is usually the chief organ of photosynthesis (see Color Plate 5). Within the leaf the cells are organized into three tissues:

1. **surface tissue**—epidermis
2. **fundamental tissue**—parenchyma (mesophyll)
3. **vascular tissue**—xylem and phloem

Each tissue has its own characteristic types of cells specialized to carry out certain functions. Surface tissue includes **epidermis** cells specialized to protect the leaf. Fundamental tissue includes **parenchyma** cells specialized for manufacture or storage of materials. In the leaf they are called **mesophyll** ("middle leaf") cells and are specialized for photosynthesis and gas exchange. Vascular tissue consists of **xylem** and **phloem** tissue for transport.

- Take a prepared slide of the cross section of a leaf and refer to Figure 5–7 at the end of this topic as you study the types of cells that it contains. Start with low power.

Epidermis: Cells of the epidermis are closely packed and form a single protective layer on both the upper and lower surfaces of the leaf. They secrete a waxy **cuticle** that waterproofs the leaf surface and helps protect it.

- Note the openings, called **stomata,** in the lower leaf epidermis. The two small cells on either side of each opening are the **guard cells,** specialized epidermal cells that control the passage of gases into and out of the leaf.

Mesophyll: These cells make up the bulk of the inner part of the leaf.

- Find the densely packed cells of the **palisade mesophyll** below the upper epidermis. These cells are specialized to carry out photosynthesis very efficiently: They contain many chloroplasts, and their shape lets them trap much of the light striking the leaf.
- Locate the **spongy mesophyll** beneath the palisade layer. Here the cells are loosely packed and irregular in shape so that gases can be freely exchanged between the cells and air within the leaf.
- Notice how gases entering at the guard cells can travel in the air spaces to the palisade layer.
- Now examine the mesophyll cells under high power to reemphasize the characteristics typical of plant cells.
- Note the thick, tough **cell wall** that lies just outside the plasma membrane. It is made of fibers of **cellulose,** the most abundant organic molecule on earth. (See Text Figure 3–6.) Cellulose is harvested and used by us for a variety of purposes such as paper, wood for construction, and cotton clothing.

In the plant, cellulose gives structure to the cells and to the plant as a whole.

- Note that the mesophyll cells generally have large internal spaces called **central vacuoles** surrounded by the membrane called the tonoplast. These are characteristic of mature plant cells but are not found in animal cells. The vacuole contains water, salts, and organic substances.
- Finally note the **chloroplasts** in the cytoplasm of the cell. These organelles are the ultimate source of energy-rich molecules for the plant and for herbivores ("plant eaters") such as ourselves.

Xylem: Vascular tissue occurs as bundles of xylem and phloem and transports material into and out of the leaf. Xylem is one type of vascular tissue that is so specialized to carry out its function of transport that it is dead, having lost its nucleus and cytoplasm.

- Use low power to locate oval or round **vascular bundles** in your leaf. They may be stained red.
- Switch to high power to locate the xylem vessels. They are the large, empty-looking structures within the vascular bundles. (Any cell that has a nucleus is *not* xylem.)
- Find the spiral thickenings in the walls of xylem vessels. These are easy to see where the vessels run parallel to the cut through the leaf and usually stain red.

In the stem of woody plants, many xylem vessels are bunched together to make up the woody part. The cell walls are made of cellulose and are greatly thickened to give support to the stems of these large plants and to conduct water and minerals over great distances without collapsing.

Phloem: Phloem tissue is also specialized for transport, but has only living cells and controls the substances that travel in them. Phloem conducts food, usually in the form of the disaccharide **sucrose** from the leaves to the stem and roots for storage, and then from storage back to the next crop of leaves.

- Locate the phloem cells alongside the xylem vessels; they are smaller, more lightly stained, and may lack a nucleus.
- Label the important cell structures in Figure 5–7 at the end of this topic.
- Use arrows to indicate on the diagram how external gases reach the palisade mesophyll and how sucrose synthesized there reaches the phloem.

When you are finished with your observations, return all materials and store your microscopes properly.

REVIEW

When you have time, review what you have learned about cells by completing Table 5–1 at the end of this topic.

REFERENCES AND FURTHER READING

Berns, M.W. *Cells,* 2nd ed. Philadelphia: Saunders College Publishing, 1983.

Brock, Thomas D. et al. *Biology of Microorganisms,* 4th ed. Englewood Cliffs, NJ: Prentice-Hall, 1984.

DeRobertis, E.D.P., and E.M.F. DeRobertis. *Cell and Molecular Biology,* 7th ed. Philadelphia: W.B. Saunders, 1979.

Fawcett, D.W. *An Atlas of Fine Structure.* Philadelphia: W.B. Saunders, 1966.

Fawcett, D.W. *The Cell,* 2nd ed. Philadelphia: W.B. Saunders, 1981.

Giese, A.C. *Cell Physiology,* 5th ed. Philadelphia: W.B. Saunders, 1979.

Jahn, T.L., and F. Jahn. *How to Know the Protozoa,* 2nd ed. Dubuque, IA: W.C. Brown, 1979.

Kessel, Richard G., and Randy H. Kardon. *Tissues and Organs: A Text-Atlas of Scanning Electron Microscopy.* San Francisco: Freeman, 1979.

Leeson, C.R., T.S. Leeson, and A.A. Paparo. *Textbook of Histology,* 5th ed. Philadelphia: W.B. Saunders, 1985. (Especially for the four primary animal tissues.)

Nester, E.W., C.E. Evans, M.E. Lidstrom, N.N. Pearsall, and M.T. Nester. *Microbiology,* 3rd ed. Philadelphia: Saunders College Publishing, 1983.

Prescott, G.W. *How to Know the Freshwater Algae,* 3rd ed. Dubuque, IA: W.C. Brown, 1978.

Weiss, L., and R.O. Greep. *Histology,* 4th ed. New York: McGraw-Hill, 1977. (Especially for the digestive system.)

Journal Articles

Albrecht-Buehler, Guenter. "The tracks of moving cells." *Scientific American,* 68, April 1978 (#1386).

Dustin, Pierre. "Microtubules." *Scientific American,* 66, August 1980 (#1477).

Lake, James A. "The ribosome." *Scientific American,* 84, August 1981.

Lazarides, Elias, and Jean Paul Revel. "The molecular basis of cell movement." *Scientific American,* 100, May 1979 (#1427).

Margulis, Lynn. "The origin of plant and animal cells." *American Scientist* 59:230, 1971.

Porter, Keith R., and Jonathan B. Tucker. "The ground substance of the living cell." *Scientific American,* 56, March 1981.

Sloboda, Roger D. "The role of microtubules in cell structure and cell division." *American Scientist* 68:290, 1980.

Student Name ______________________________ **Date** ______________

Onion Skin Cell __________ μm

Bacteria __________ μm

Paramecium __________ μm

Eremosphaera __________ μm

Figure 5–5. Sketches of cell types.

Columnar Epithelial Cells

Goblet Cells

Connective Tissue Cells

Smooth Muscle Cells

Squamous Epithelial Cells

Figure 5–6. Sketches of cells in intestinal tissue.

Student Name ______________________________ **Date** ______________

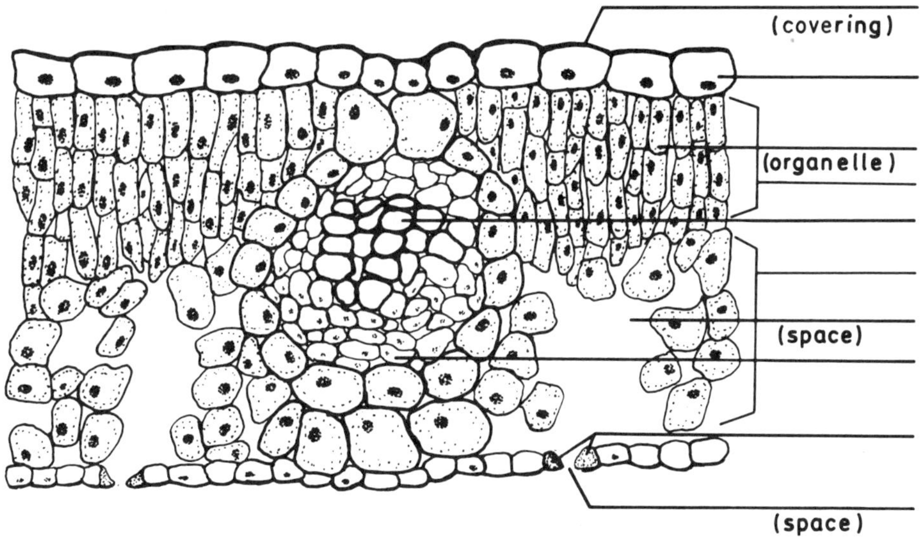

Figure 5–7. A leaf in cross section. Label the important cell types or structures.

TABLE 5–1. Summary and Review of Cells.

A = usually or always; M = sometimes; N = never; S = small (0.5–5μm); L = large (10–200μm)

Structure or Characteristic	Plant	Animal	Protist	Bacteria
photosynthetic				
a single cell				
size range of cells (μm)				
single cells motile				
plasma membrane present				
nucleus (organelle) present				
cell wall present				
chloroplasts present				
ribosomes present				

TOPIC 6

Cell Respiration and Fermentation

OBJECTIVES

When you have completed this topic, you should be able to:

1. Distinguish anaerobic from aerobic cellular respiration in terms of ATP, oxygen, and chemiosmosis.
2. Define fermentation and give its starting materials and end products.
3. Measure CO_2 production by fermenting yeast.
4. List the starting materials and end products of each of the following processes: glycolysis, citric acid cycle, electron transport system.
5. Describe an experiment to measure the respiration rate of a living plant or animal.
6. State the importance of the electron transport system and the role of O_2 in it.
7. Describe an experiment that will measure the reaction of oxygen with an enzyme of the electron transport system.
8. Describe an experiment that will show the effect of an inhibitor on the electron transport system.

TEXT REFERENCES Chapter 6; Chapter 7

KEY TOPICS

energy	coenzymes	electron transport system
ATP	glycolysis	chemiosmosis
oxidation reduction reactions	citric acid cycle	fermentation
	oxidative phosphorylation	

INTRODUCTION

Heterotrophs ("other feeder") obtain energy by oxidizing organic **food molecules** and **coupling** these exergonic oxidation reactions to the synthesis of ATP, an endergonic process. The ATP is then used to carry out metabolic reactions necessary to maintain the organisms's physical integrity, and to support all its other activities. (See Text Figure 6–7.)

Some organisms are able to exist in the absence of molecular oxygen, and may even be harmed by the presence of oxygen. They still carry out oxidation reactions but do not use molecular oxygen itself.

Define **oxidation:**

____

____

In any case, the cytoplasm of all cells contains the enzymes needed in the ancient, central pathway of **glycolysis,** in which ATP is synthesized during the oxidation of energy-rich glucose to pyruvic acid.

Anaerobic respiration in the absence of oxygen is called **fermentation.** It begins with glucose, and utilizes the enzymes of glycolysis to oxidize it to pyruvic acid in order to make ATP. It then uses pyruvic acid as a sink for leftover hydrogen, producing ethanol and CO_2, or a mixture of organic acids and other compounds. There is a net gain of only two molecules of ATP for each molecule of glucose oxidized in fermentation. These ATP molecules are formed by **substrate level phosphorylation** during glycolysis.

The process of fermentation, in yeast for example, may be summarized as:

$$\underset{\text{glucose}}{C_6H_{12}O_6} \xrightarrow{\text{fermentation}} \underset{\text{ethanol}}{2\ C_2H_5OH} + \underset{\text{carbon dioxide}}{2CO_2} + \underset{\textbf{ATP and heat}}{\textbf{energy}}$$

Most organisms use molecular oxygen in the process of **cell respiration.** Pyruvic acid, produced by the reactions of glycolysis, loses a molecule of CO_2, then enters another set of reactions in the mitochondrion, the citric acid cycle (tricarboxylic acid cycle) in which the carbon skeleton is completely oxidized to CO_2. (See Text Figure 7–7.) Certain **coenzymes,** NAD^+ and FAD, are reduced by the reactions of glycolysis and the citric acid cycle and act as hydrogen carriers: NADH and $FADH_2$ transfer their hydrogen atoms to a third set of proteins called the **electron transport system** or **electron transport chain.** The flow of electrons along this chain of protein carriers is **coupled** to the synthesis of ATP. The very last enzyme in the chain, **cytochrome oxidase,** reacts directly with molecular oxygen to produce water. This step is the point where the oxygen that organisms take in is used and as a result forms **metabolic water.** The purpose of the oxygen is to return the cytochrome oxidase to an oxidized form so that it can pick up more electrons from its neighboring electron carrier, cytochrome c. Aerobic respiration of a molecule of glucose gives a maximum net yield of 38 molecules of ATP.

This is the overall process of cell respiration in summary:

$$\underset{\text{glucose}}{C_6H_{12}O_6} + \underset{\text{oxygen}}{6\ O_2} \xrightarrow{\text{respiration}} \underset{\text{carbon dioxide}}{6\ CO_2} + \underset{\text{metabolic water}}{6\ H_2O} + \underset{\textbf{ATP and heat}}{\textbf{energy}}$$

Electron flow from the hydrogen carriers to oxygen generates ATP in the process of **oxidative phosphorylation:** These redox reactions drive the phosphorylation of ADP. The coupling of electron flow in cellular respiration, as well as in photosynthesis, is explained by the **chemiosmotic hypothesis.** In cellular respiration, a proton gradient is set up as electron flow pumps H^+ ions into the compartment between the inner and outer mitochondrial membranes. This electrochemical gradient represents stored energy. Proton flow back through special channels in the inner membrane, from the higher to lower H^+ concentration, provides the energy for ATP synthesis. ATP is made by **ATP synthetases,** enzymes associated with these channels, and energy is stored in its chemical bonds. Of the maximum of 38 ATPs produced by the complete aerobic respiration of one glucose molecule, 34 come from chemiosmosis.

FERMENTATION IN YEAST

Yeasts are eukaryotic fungi that are commercially very important. They are necessary for the production of beer, wine, bread, and industrial chemicals. In this experiment you will study the production of CO_2 during the fermentation of various carbohydrates by yeast cells under anaerobic conditions.

What molecule is split in the reaction that produces the CO_2?

____

When carbohydrate is fermented by yeast, only one third of the carbons are released in the form of CO_2. The rest of the carbon is released as molecules of

_______________.

Why is it advantageous for an organism to be able to carry on fermentation?

____

____

Study the experiment shown in Figure 6–1 for measuring fermentation under anaerobic conditions. Work with a partner.

- Obtain or start an actively fermenting yeast culture (yeast dissolved in a molasses solution).
- Obtain a test tube rack; four clean, dry test tubes; four graduated pipets; four Pasteur pipets; and pieces of Parafilm or clay.
- Label the tubes #1–#4,
- Fill the tubes as follows:

Tube #1: 1 mL of sucrose + 1 mL of yeast suspension.
Tube #2: 1 mL of galactose + 1 mL of yeast suspension.
Tube #3: 1 mL of molasses + 1 mL of yeast suspension.
Tube #4: 1 mL of water + 1 mL of yeast suspension.

- Mix the solution in tube #1, then draw it up in a graduated pipet, and place your finger over the top of the pipet.
- Seal the opposite end of the pipet with a piece of Parafilm or clay.

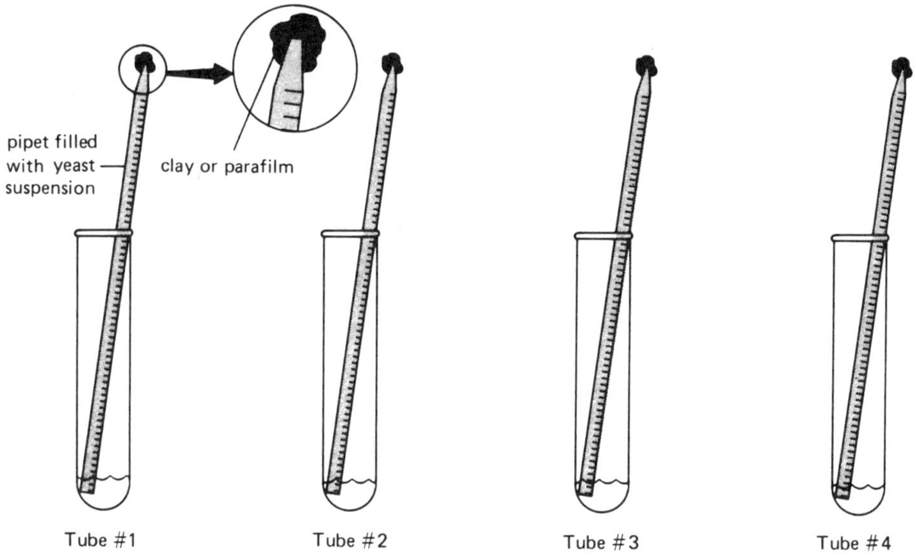

Figure 6–1. Fermentation experiment. Fermentation by yeast within the pipet produces CO_2. The pipet must be completely filled with yeast suspension at the beginning of the experiment.

- Using the Pasteur pipet, completely fill the graduated pipet to overflowing with solution from tube #1.
- Place the pipet *upside down* in tube #1, and set the tube back in the rack.
- Repeat for the other test tubes.

As gas forms during the fermentation, it will rise in the pipet and collect at the plugged end. Note the exact level of the gas in each pipet every minute for about 20 min, and record your readings in Table 6–1 at the end of this topic.

Did the type of carbohydrate being fermented affect the rate of gas formation? ______________

Which carbohydrate supported the fastest rate of fermentation? ______________

Your instructor may ask you to vary this basic experiment by using different concentrations of carbohydrate or yeast, or by adding an inhibitor such as ethanol.

RESPIRATION IN PLANTS AND ANIMALS: CO_2 PRODUCTION

In aerobic respiration the carbon skeleton of glucose is completely broken down to CO_2. The CO_2 must be quickly removed via the transport and gas exchange systems, so that it does not build up and make the body fluids so acidic that the organism cannot function.

In which reactions is CO_2 released during respiration?

____

____

____

Why does CO_2 make tissues acidic?

____

____

____

____

In this experiment you will measure the rate of CO_2 production in several organisms. This rate is an indicator of the respiration rate, or metabolic rate. A diagram of the experiment is shown in Figure 6–2. During the experiment a certain amount of CO_2 will be put out into the beakers, making the water more acidic. At the end of the experiment you will **titrate** each solution by adding a base, NaOH, until a certain, fixed pH is reached. When that pH is reached, an indicator that you will add before starting to titrate turns color at its **end point.** The more CO_2 in a solution, the more NaOH will be required to get to the end point.

- Obtain five 150-mL beakers, and label them #1–#5.
- Set up the beakers as follows using **adjusted aquarium water** or a source of water specified by your instructor.

	Adjusted Water	Organism
Beaker #1	100 mL	1 fish
Beaker #2	25 mL	2 snails
Beaker #3	100 mL	5 cm of *Elodea*
Beaker #4	100 mL	5 cm of *Elodea*
Beaker #5	100 mL	None

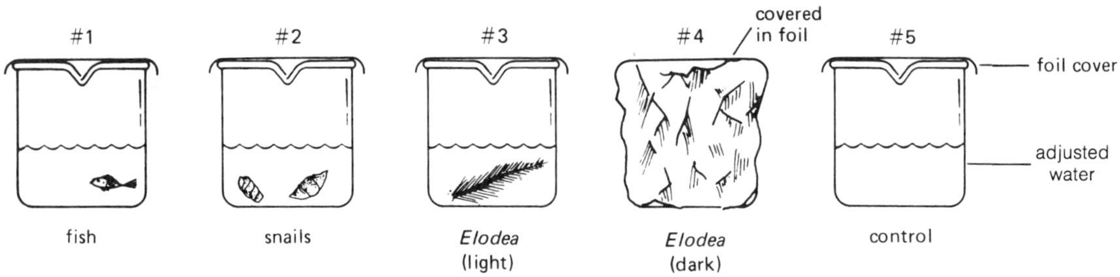

Figure 6–2. Respiration experiment. These aquatic organisms produce CO_2. Each 150-mL beaker contains 100 mL or 25 mL of adjusted water and its top is covered with aluminum foil. Beaker #4 is completely covered with foil to exclude light.

- Cap beakers #1, #2, #3, and #5 with aluminum foil and note the time.
- Completely cover beaker #4 with aluminum foil so that the *Elodea* is kept dark, and do not open this beaker until the end of the experiment.
- At the end of 30 min return the living organisms to the proper containers on the instructor's desk.
- Obtain a 125-mL flask, a 25-mL graduated cylinder, a 1-mL graduated pipet* (or a buret), a dropping bottle of phenolphthalein (1% stock, indicator), and a bottle of NaOH (0.0025M).
- Add 25 mL from the contents of beaker #1 to the flask and place it on a white background.
- Add 4 drops of phenolphthalein solution and swirl to mix thoroughly.
- Fill your pipet (or buret) with NaOH exactly to the top mark.
- Add NaOH drop by drop to the flask, swirling after each drop, until the solution definitely turns color. **DO NOT DISCARD THE REMAINING NaOH.**
- Wait 10 seconds. If there is still a faint, but definite color, you have reached the end point. If the color disappears, you are very near the end point and should cautiously continue adding drops of NaOH.
- Read off the level of NaOH remaining and record in Table 6–2 (at the end of this topic) the **volume of NaOH used** from the pipet (or buret). If "0" is at the top of your pipet, read the volume directly; if "0" is at the bottom, subtract the reading from 1.0 mL (or the full volume).
- If you use up the NaOH in the pipet without reaching the end point, record what you have used, refill the pipet, and continue adding drops. Several mL may be required, but add the drops slowly.
- Repeat for beakers #2, #3, #4, and #5. These titrations will use less NaOH than #1.
- Multiply your values by 4 for the experiments where the total volume was 100 mL (beaker #1, #3, #4, and #5).

Before calculating the CO_2 production of your organisms, you will need to know their approximate volumes. Your instructor may provide you with this information. If not, use either of the following methods to determine their volumes. Record the volumes in Table 6–2 at the end of this topic.

The most practical way to measure the volume of a small organism (less than 5 mL) is to weigh the organism and assume that its density is 1.0 g/mL. Use this method if a balance is available.

- Add 50 mL of water to a 100-mL beaker. A larger amount may be needed for a large fish.
- Weigh the beaker and water and record this **tare** weight.
- Add the organism to the beaker; reweigh, and record the **total** weight.
- Return the organism to the supply dish or aquarium.
- Subtract the tare weight from the total weight to determine the organism's weight in grams. If 1 g = 1 mL, then its weight is the same as its volume in mL.

If a balance is not available, use this method:

- Add 50 mL of water and the organism to a 100-mL beaker.
- Make a horizontal mark on the beaker exactly at the bottom of the meniscus (lowest point of the "surface" layer). Memorize exactly where the water surface is in relation to the line.
- Remove the organism and return it to the supply dish or aquarium.
- Fill a 5-mL pipet exactly to the mark with water.
- Add water from the pipet to the beaker until the water level returns exactly to your mark.
- Record the volume of water that you added. This is the same as the volume of the organism.

In the case of the snail, much of the volume is nonliving shell and does not metabolize.

- If an empty snail shell is available, measure its volume and subtract it from the snail's total volume to determine more accurately the volume of the living snail.

Use the following equation to calculate the CO_2 production in μM/mL/hr.

* If you will be using an ungraduated pipet to add NaOH, count the number of drops used for each titration and convert this number to mL by dividing by 20.

$$\frac{[\text{mL NaOH (exptl.)} - \text{mL NaOH (control)} \times 2.5\ \mu\text{M NaOH/mL}]}{\text{volume of organism (mL)} \times \text{time (hr)}}$$

Which organism had the highest metabolic rate?

____

Which one had the lowest rate? ______________

What is the significance of a negative metabolic rate (a *decrease* in CO_2)?

____

____

Explain your results with *Elodea* in the dark versus in the light:

____

____

Which *Elodea* experiment was a better measure of the metabolic rate? ______________

Why?

____

____

____

THE ELECTRON TRANSPORT CHAIN: CYTOCHROME OXIDASE

Most of the ATP in a cell is made in the electron transport system by **oxidative phosphorylation.** The last carrier, **cytochrome oxidase,** picks up electrons from its neighbor, cytochrome c, and donates them to molecular oxygen:

$$2e^- + 2\,H^+ + \tfrac{1}{2}\,O_2 \xrightarrow{\text{cytochrome oxidase}} H_2O$$

The water produced in this way is called **metabolic water** because it is formed by the organisms's metabolism and not taken in from its environment.

In today's experiment you will study cytochrome oxidase from beans. This carrier will be in the form of a crude extract from bean seedlings. An added reagent (Nadi reagent) will provide electrons for cytochrome c. If cytochrome oxidase is active, it will pass the electrons on to oxygen and permit a continuous flow of electrons from the Nadi reagent to oxygen. The oxidized form of Nadi reagent will accumulate over time and turn the solution blue. This is diagrammed in Figure 6–3.

If oxygen is removed from the electron transport system, all of the component carriers will remain in their reduced forms, and electron flow will cease. No ATP can be synthesized by oxidative phosphorylation and the organism must rely on anaerobic respiration only. Many organisms cannot survive without a functioning electron transport system.

Cytochrome oxidase, like all of the cytochromes, carries a **heme** prosthetic group. (See Text Figure 24–12.) Any molecule that interacts with the heme group will poison the cytochrome chain and have the same effect as cutting off oxygen. Potassium cyanide and sodium azide are lethal poisons that act in this way.

Where in the cell are the carriers of the electron transport chain located?

____

Record your results and analysis of this experiment in Table 6–3 at the end of this topic.

- Obtain four test tubes and label them #1–#4.
- Add the following reagents to the tubes:

	Bean Extract (Cytochrome Oxidase)	Sucrose	Sodium Azide
Tube #1	—	4 mL	—
Tube #2	2 mL	2 mL	—
Tube #3	2 mL	1.8 mL	5 drops (0.2 mL)
Tube #4	2 mL boiled*	2 mL	—

* Use boiled enzyme if it is available; otherwise, add 2 mL bean extract to the tube, and then hold it in a boiling water bath for 10 min and cool before adding the other reagents.

- Add 5 drops (0.2 mL) of Nadi reagent A to each tube. Mix.
- Add 5 drops (0.2 mL) of Nadi reagent B to each tube. Mix.
- Observe the tubes every 10 min for 30 min, and note the color changes.
- Explain your results in Table 6–3 at the end of this topic.

Can the electron flow in this experiment be used for ATP synthesis? ______________

Explain your answer:

____

____

Sodium azide is a potent inhibitor of cytochrome oxidase, and will block the reaction of this enzyme with molecular oxygen.

Did inhibition of cytochrome oxidase block the reaction of cytochrome *c* with the Nadi reagent?

____

If the electron transport system were blocked at an early step, for example to the left of cytochrome b, but

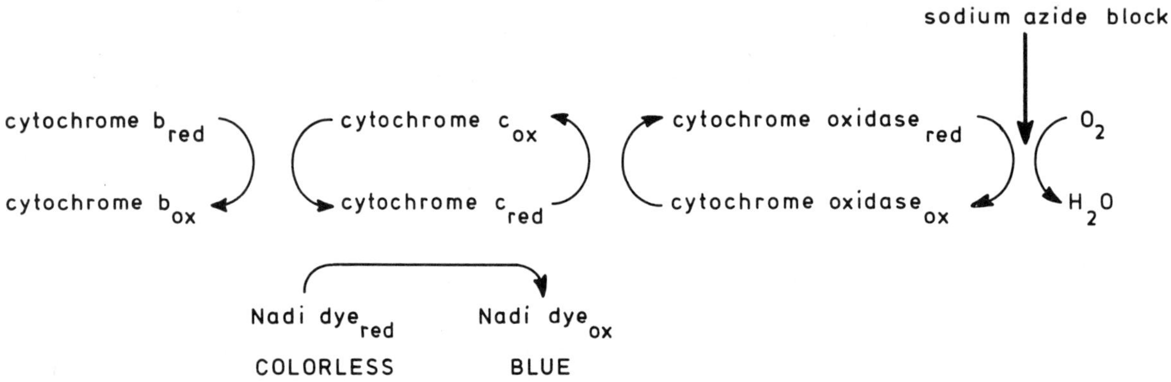

Figure 6–3. Cytochrome oxidase experiment. Colorless Nadi dye$_{red}$ donates electrons to reduce cytochrome c. Cytochrome c in turn passes the electrons to cytochrome oxidase and then to oxygen. Blue Nadi dye$_{ox}$ accumulates as long as electrons flow.

oxygen and active cytochrome oxidase were present, would the reaction of the Nadi reagent continue in the absence of electron flow from the other carriers? Explain your answer.

____

____

____

____

____

When you are finished with your work, return all animals and materials, and wash your glassware.

REFERENCES AND FURTHER READING

Baker, Jeffrey J.W., and Garland E. Allen. *Matter, Energy and Life: An Introduction to Chemical Concepts,* 4th ed. Reading, MA: Addison-Wesley, 1981.

Berlow, P.P., D.J. Burton, and J.I. Routh. *Introduction to the Chemistry of Life.* Philadelphia: Saunders College Publishing, 1982.

Lehninger, A.L. *Biochemistry,* 2nd ed. New York: Worth, 1975.

Lehninger, A.L. *Bioenergetics: The Molecular Basis of Biological Energy Transformations,* 2nd ed. Menlo Park, CA: Benjamin/Cummings, 1971.

Lehninger, A.L. *Principles of Biochemistry.* New York: Worth, 1982.

Rawn, J. David. *Biochemistry.* New York: Harper and Row, 1983.

Stryer, Lubert. *Biochemistry,* 2nd ed. San Francisco: W.H. Freeman, 1981.

Journal Articles

Dickerson, Richard E. "Cytochrome c and the evolution of energy metabolism." *Scientific American,* 136, March 1980 (#1464).

Hinkle, Peter C., and Richard E. McCarty. "How cells make ATP." *Scientific American,* 104, March 1978 (#1383).

Kosikowski, Frank V. "Cheese." *Scientific American,* 88, May 1985.

Revelle, Roger. "Carbon dioxide and the world climate." *Scientific American,* 35, August 1982.

Waggoner, Paul E. "Agriculture and carbon dioxide." *American Scientist* 72(2): 179, 1984.

Webb, A. Dinsmoor. "The science of making wine." *American Scientist* 72(4): 360, July 1984.

Student Name ______________________________ **Date** ____________________

In the following table, note the exact level of the gas in the pipet above the yeast suspension every minute.

TABLE 6–1. Fermentation in Yeast

Time (min)	Tube #1	Tube #2	Tube #3	Tube #4
0				
1				
2				
3				
4				
5				
6				
7				
8				
9				
10				
11				
12				
13				
14				
15				
16				
17				
18				
19				
20				
Total CO_2 in mL/hr:				
Substrates				Control

Write the net equation for the fermentation of the most rapidly fermented substrate:

______________________ → ______________________

Where are the enzymes of fermentation located in the yeast cell?

____

TABLE 6–2. CO_2 Production in Plants and Animals

Record the number of drops or mL (20 drops = 1 mL) needed to change color. Calculate CO_2 production in each case.

Beaker	Organism	NaOH		Volume of Organism (mL)	$\frac{\mu M\ CO_2^*}{mL/hr}$
		drops	mL		
#1	Fish				
#2	Snail				
#3	*Elodea* (light)				
#4	*Elodea* (dark)				
#5	None (control)				

$$^* \frac{[\text{mL NaOH (exptl.)} - \text{mL NaOH (control)}] \times 2.5\ \mu M\ \text{NaOH/mL}}{\text{volume of organism (mL)} \times \text{time (hr)}}$$

TABLE 6–3. Activity of Cytochrome Oxidase

Record the color of each tube and indicate whether a reaction occurred.

Time (min)	Tube #1	Tube #2	Tube #3	Tube #4
0				
10				
20				
30				
Reaction				

Explain your results for each tube in which no reaction or only a slight reaction occurred.

Tube #1: ______________________________

____

Tube #2: ______________________________

____

Tube #3: ______________________________

____

Tube #4: ______________________________

____

TOPIC 7

Photosynthesis

OBJECTIVES

When you have completed this topic, you should be able to:

1. Identify the pigments involved in photosynthesis by paper chromatography.
2. Give the function of chlorophyll and the accessory pigments.
3. Determine the wavelengths most useful for photosynthesis by determining the absorption spectrum of a pigment extract.
4. Distinguish between the electron transport and carbon fixation reactions of photosynthesis.
5. Demonstrate the effect of changing the level of light on electron transport as measured by O_2 production.
6. Demonstrate the effect of changing the level of light on electron transport by measuring reduction of a dye in the Hill reaction.
7. Explain why stimulating carbon fixation affects electron transport as measured by O_2 production.
8. Describe an experiment designed to measure the extent of carbon fixation as a function of light.

TEXT REFERENCES Chapter 8

KEY TOPICS

chloroplasts	photochemical reactions	chemiosmosis
electromagnetic radiation	photosystems I and II	photophosphorylation
photosynthetic pigments	electron transport	carbon fixation

INTRODUCTION

The overall reaction of photosynthesis can be understood as two closely linked processes. Light driven **electron transport** produces ATP and NADPH, which feed into the assembly line of the Calvin cycle where enzymes carry out **carbon fixation** by converting CO_2 to glucose. A diagram showing how these processes are linked is shown in Figure 7–1. Reactions on the left side take place in the **thylakoid** membranes of the chloroplast; those on the right occur in intervening fluid regions of **stroma.**

The first step in photosynthesis is **light absorption** by pigments. This drives the flow of electrons from water to $NADP^+$ along a chain of carriers. The flow also sets up a H^+ electrochemical gradient that is used for **chemiosmotic ATP synthesis.** This entire process, by which light energy has been converted to chemical energy in the form of NADPH and ATP, can be summarized as **non-cyclic photophosphorylation.** Electrons flow in a **non-cyclic** fashion from water to $NADP^+$, which is reduced to NADPH. This electron transport depends directly on light **(photo-)** and permits the coupled synthesis of ATP from ADP **(-phosphorylation)** through chemiosmosis. Sometimes ATP alone is made by cyclic photophosphorylation, but this process is much harder to study because O_2 and NADPH are not produced. (See Text Figure 8–9.)

In the final part of photosynthesis, **carbon fixation,** the energy of NADPH and ATP is used to reduce **CO_2** to **glucose** in a series of reactions known as the **Calvin cycle.** These reactions are light-independent in that they can occur in the absence of

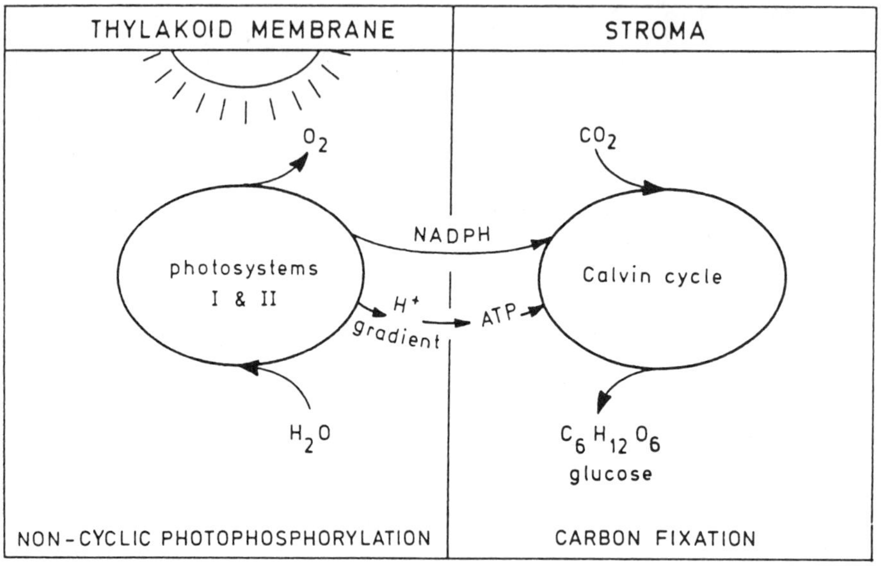

Figure 7–1. Photosynthesis. ATP synthesis by chemiosmosis depends on a hydrogen ion electrochemical gradient set up as a result of electron transport. The Calvin cycle is tightly coupled to electron flow and quickly stops if light is not available.

light as long as NADPH and ATP are provided. In intact photosynthetic systems, however, light is an absolute necessity, since sufficient NADPH and ATP cannot be provided without it. (See Text Figure 8–10.)

Write out the balanced equation for the synthesis of 1 glucose from 6 carbon dioxide and 12 water molecules.

In your equation, circle those oxygen atoms in the starting material(s) that will give rise to the waste product O_2.

PHOTOSYNTHETIC PIGMENTS: CHROMATOGRAPHY

If you have never carried out chromatography, turn back to Topic 1 and read the description in the section "AMINO ACID CHROMATOGRAPHY." You need to understand the following terms: **adsorption, matrix, chromatogram, origin, solvent front,** and $\mathbf{R_f}$.

In this experiment the matrix will be the cellulose of Whatman #1 filter paper, and the solvent is petroleum ether : acetone (9 : 1). Although substances do not have to be colored to be separated by chromatography, it is a little more dramatic when you can actually see the different substances as they pull apart to form distinct spots. You will begin by applying a green mixture of **PIGMENT** containing **chlorophylls** and **accessory pigments** to the paper, and will obtain a separation of the mixture into several distinct components at the end of the chromatography. The separation depends on the **size, solubility,** and **affinity** for paper or adsorption of the various pigment molecules.

Preparing the Chromatogram

A correctly prepared chromatogram is shown in Figure 7–2.

- Obtain a rectangle of filter paper (about 10 × 15 cm), and lay it on a clean surface, *touching the paper as little as possible* and only at the edges.
- Use a pencil to mark a straight line 1.5 cm from

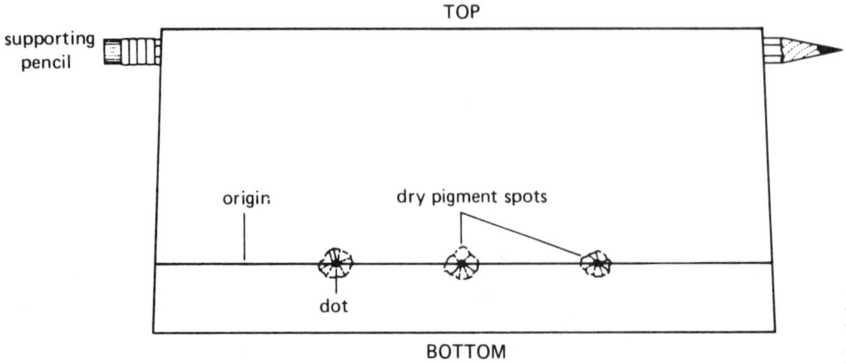

Figure 7–2. Chromatography of photosynthetic pigments. A paper chromatogram after spotting three samples.

and parallel to the long edge of the paper; this is the origin.

- Make three pencil dots on the origin that are at least 3 cm apart.
- Place a support (such as a pencil) under the top edge of the paper so that the origin doesn't touch the surface.
- Use a capillary pipet to spot **PIGMENT** onto the dots by touching the pipet to the paper as briefly as possible so that your spots will be quite small.

Do not spot chloroplasts or chlorophyll extract onto the paper. Use only pigment for this experiment. Small spots are better than big spots.

- Use forceps to hold the paper with three wet spots in front of the blower until the spots are dry, or blow on them yourself. Touch the paper with your fingers as little as possible.
- Again make a small spot on each dot with the capillary pipet, and dry.
- Repeat until you have made a minimum of five applications on each dot.
- Dry the chromatogram thoroughly before starting your chromatography.

Running the Chromatogram

- Form your dry chromatogram into a cylinder, as shown in Figure 7–3, and staple it at the top and bottom so that the edges are close together but *do not touch or overlap.*
- Obtain a chromatography jar and cover. Add enough solvent to the jar to just cover the bottom (about 20 mL). Replace the cover on the jar, and close the solvent stock bottle.
- When the solvent has had a chance to evaporate and saturate the air in the chromatography jar, pick up your chromatogram with forceps, and set it gently in the center of the jar. It should stand up straight and *not touch the sides of the jar.*
- Cover the jar and leave it undisturbed during the chromatography.
- Check the jar every few minutes, and note the progress of the solvent as it wets the paper and moves toward the top.

Do not let the solvent reach the top of the paper.

- After about 20 min, the solvent will be close to the top of the paper, and you must immediately stop the chromatography. Use the forceps to remove the chromatogram. Replace the cover on the jar.
- *Immediately* mark the **solvent front** (edge of the part wet by the solvent) with a pencil. You will not be able to calculate any R_f values if you fail to do this before the rapidly drying solvent dries.
- Dry your chromatogram with the blower, or hang it up with clothespins for a few minutes before analyzing it.

Analyzing the Chromatogram

The number of spots that you observe will depend on the exact condition of the extract that you used. These are the major pigments in the extract:

Chlorophyll *a*: a blue-green pigment
Chlorophyll *b*: a yellow-green pigment
Carotenes: yellow-orange accessory pigments
Xanthophylls: yellow accessory pigments

Pigment molecules in the faster spots will be more soluble in the solvent; they will be smaller and will have less affinity for the paper.

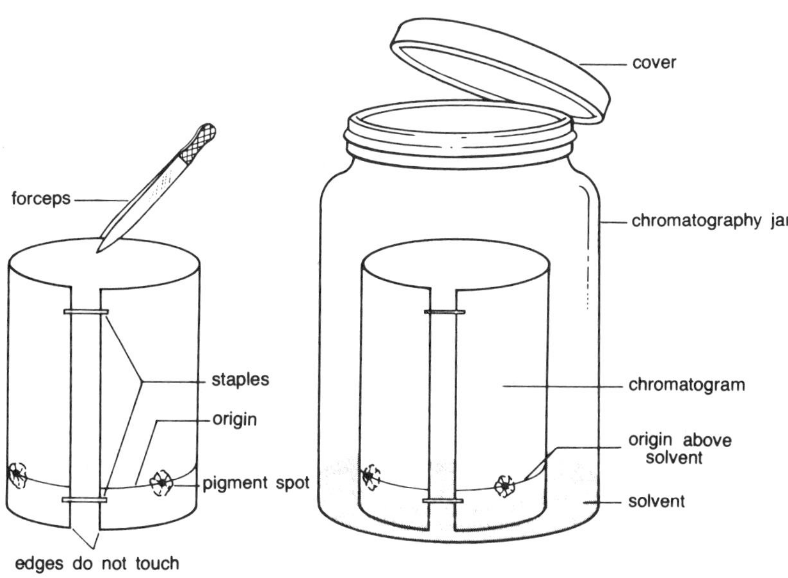

Figure 7–3. Chromatography of photosynthetic pigments. A paper chromatogram ready to run has been stapled so that its edges do not touch. The dry chromatogram should be put into the jar so that it does not touch the sides, and the origin should be above the level of the solvent. Keep the jar covered.

- Outline each visible spot with a pencil, and note its color on the chromatogram.
- Make a dot with your pencil in the center of each spot. If the spot is irregular or band-shaped, use your judgment as to where the approximate center might be if it were a round spot.
- Take your chromatogram into a dark room, or put it into a darkbox, and view it under long-wavelength ultraviolet light.

Never look directly at ultraviolet light. It can damage your eyes.

- Note the bright pink **fluorescence** of chlorophyll.
- Use the difference in fluorescence to help distinguish the chlorophylls and any other fluorescent compounds present. Outline the spots with pencil.
- Note whether there are any dark areas that contain compounds that absorb UV light, but do not show fluorescence. Outline them with pencil.
- Finally, carefully measure the distance from the origin to the solvent front, and record it in Table 7–2 at the end of this topic.
- Measure and record the distance from the origin to the dot in the center of each spot observed in visible or UV light.
- Calculate the R_f for each spot, and describe its appearance under visible and under UV light.
- If you can, identify the major pigments, and record your R_f values for them in the table.
- Tape the chromatogram into the space provided.

These pigments are located in the thylakoid membranes of the chloroplasts where they normally would absorb light, become chemically excited, and pass electrons to the electron carriers of photosystems I and II. When they are free, however, as they are on the chromatography paper, there are no molecules to accept the electrons, and the UV light-excited molecules release their energy as photons of red light during fluorescence. Only the molecules of certain compounds are able to give up the energy as light. Those of other compounds lose it in the form of heat.

PHOTOSYNTHETIC PIGMENTS: ABSORPTION SPECTRUM

Only the light that is absorbed by chloroplasts can be used in photosynthesis. As you have already seen, the pigments of a green plant look green to the eye because they permit green light to pass through, but absorb the red and blue light. The particular wavelengths of light that are **absorbed** by a certain substance form a pattern called its **absorption spectrum.** The spectrum is determined by illuminating a solution of the substance with each wavelength of light in turn, and measuring the absorption in each case. Using visible light, you will determine the visible spectrum for a mixture of pigments extracted from chloroplasts. The spectrum can then be plotted as a graph of absorbance versus the wavelength or color of light used.

In this experiment you will have to *be careful not to spill the solvent* on or into the spectrophotometer because it will dissolve the plastic.

- Turn on the Spectronic 20 spectrophotometer to warm up for a few minutes while you turn back to Topic 2, and review the procedure for using this instrument described in the section "THE SPECTROPHOTOMETER."
- Fill a spectrophotometer cuvette halfway with 80% acetone to use as your blank.
- Fill a second tube halfway with diluted **CHLOROPHYLL EXTRACT.**
- Set the wavelength at 400 nm, and adjust the needle to ∞ absorbance with the left knob, referring to Figure 2–1, if necessary.
- Insert your blank, and adjust the needle to zero absorbance with the right-hand knob.
- Insert the tube containing chlorophyll, and take your first reading. It should be in the vicinity of 0.8. If it is not, ask your instructor for help. Record the absorbance in Table 7–3 at the end of this topic.

The zero and ∞ absorbance will change each time you adjust the wavelength, so you will have to reset the spectrophotometer with and without the blank for each of your readings.

- Change the wavelength to 405 nm, reset the spectrophotometer, and take your second reading; enter the absorbance in Table 7–3.
- Continue your readings every 5 nm until the readings have become quite low. Then switch to every 10 or 20 nm.
- When the absorbance starts to rise, switch back to taking readings every 5 nm, and continue until you reach 700 nm.

The purpose of taking the spectrum is to get the shape as accurately as possible, especially around the regions of greatest absorbance (low 400s and high 600s). When the absorbance rises and then falls again, the peak is called the **absorption maximum,** and the wavelength at which it occurs is used as a characteristic for identifying a compound. If you are not sure where at least two maxima are in your spectrum, go back to the regions of high absorbance and take some more readings.

- Plot your data on the grid provided in Table 7–3, and label each absorbance maximum with its corresponding wavelength.

PHOTOSYNTHESIS IN *ELODEA*

In an intact plant both electron transport and carbon fixation will go on as long as the plant is illuminated, and it is possible to study the effect of light intensity and the concentration of CO_2 on the rate of photosynthesis. A simple way to measure photosynthetic rate is to observe the rate at which oxygen is produced as a result of splitting water. An actively photosynthesizing plant will produce an abundance of oxygen bubbles.

- Obtain a healthy green sprig of *Elodea,* and measure back about 10 cm from the tip.
- Wipe a sharp razor blade with alcohol, let it dry, and make a diagonal cut across the stem of the plant as gently as you can. Save the tip.
- Put the tip into a test tube upside down so that the cut end of the stem is about 3 cm from the top of the tube.
- Immerse the whole tube in the aquarium, or fill it with tap water until it is completely full and the cut stem is well under the surface of the water.
- Place the tube 50 cm from the floodlamp, making sure that there is a container of water between the lamp and *Elodea* to act as a heat sink.

Your setup should look like that shown in Figure 7–4.

- Wait a few minutes for the plant to become adjusted. Bubbles should appear at the cut end of the stem. If there are hardly any bubbles after 15 min, ask your instructor for help.
- When there are two or more bubbles per minute, begin counting, and record the bubbles produced in a 3-min interval in Table 7–4 at the end of this topic.
- Repeat for another 3-min interval. Calculate the average bubbles per minute for the light intensity at 50 cm.
- Move the tube to 25 cm, and allow it to adjust. Repeat your measurements and record the results.
- Now move the plant to 75 cm. Allow it to readjust, and repeat your measurements.

Did you see a different rate of oxygen evolution when the light intensity was increased or decreased?

____

- Replace the water in the tube with 0.5% $NaHCO_3$ (sodium bicarbonate, which will act as a source of CO_2 for the plant).
- Place the plant back at 25 cm, and measure the rate of bubble release for two 3-min intervals.

Does the initial reaction of CO_2 in carbon fixation catalyzed by RuBP carboxylase require light?

____

Is oxygen produced as a result of carbon fixation or of electron transport?

____

Can electron transport continue to produce NADPH if there is no CO_2? Why or why not?

____

____

Can the Calvin cycle enzymes continue to fix CO_2 when the light is turned off? Why or why not?

____

____

____

____

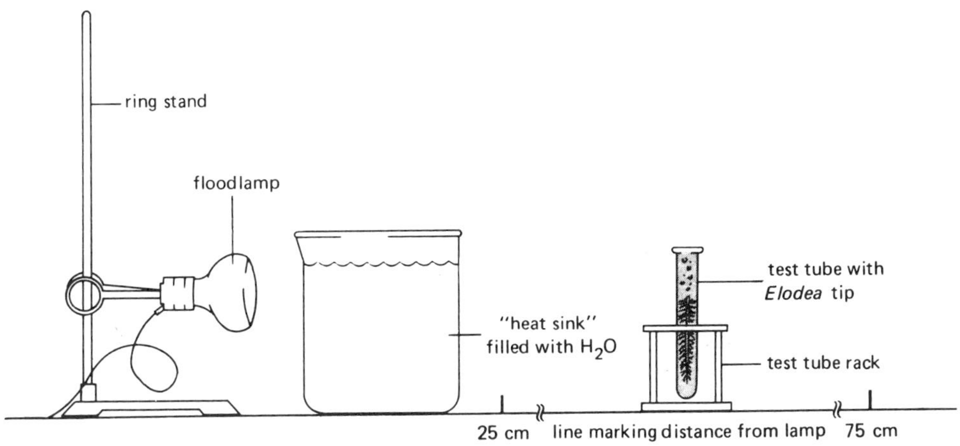

Figure 7–4. Experimental setup to measure oxygen production. *Elodea* can be exposed to different light intensities by moving it toward or away from the light source. The rate of oxygen evolution at different light intensities can be measured by counting the rate of bubble production. The "heat sink" must be used so that the water in the tube doesn't heat up and kill the plant.

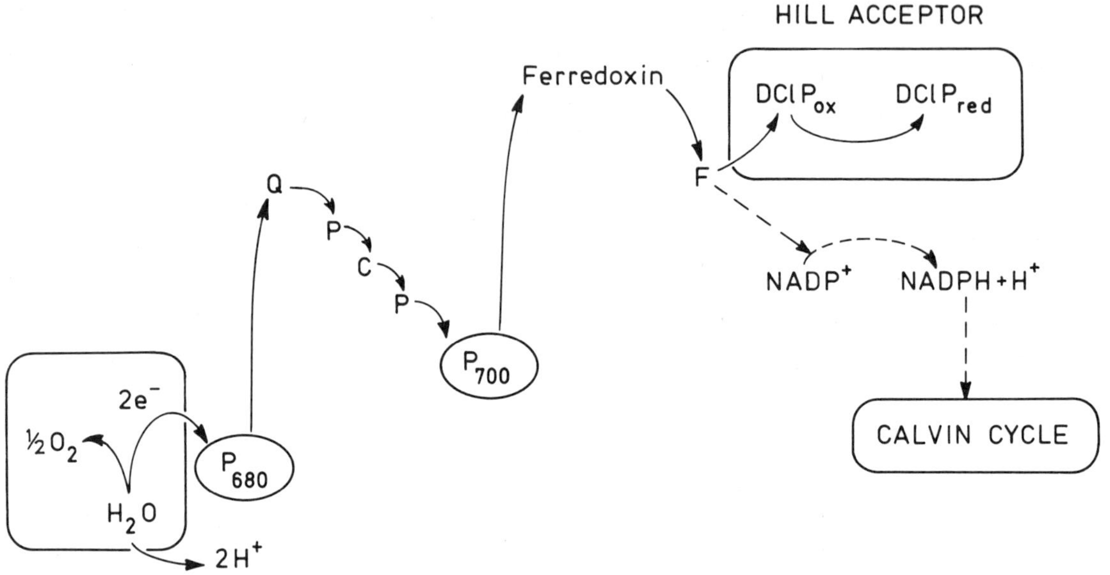

Figure 7–5. The Hill reaction. In this experiment, an electron acceptor, DCIP, is provided to pick up electrons from Photosystem I. This allows non-cyclic electron transport to continue in the absence of the Calvin cycle. DCIP is observed to turn blue as it is reduced.

THE HILL REACTION

In 1937 Robin Hill showed that chloroplasts continue to carry out electron flow and oxygen production in the absence of CO_2 as long as they are provided with an electron acceptor. (See Figure 7–5.)

What is the fate of the electrons when CO_2 is present?

____

Hill used a dye that loses color when it becomes reduced, so that the rate of photosynthesis could be measured by observing the rate at which the dye changed color. In this experiment you will use a blue dye nicknamed DCIP (2,6-dichlorophenolindolphenol), which is blue when oxidized, and becomes colorless after it has been reduced:

$$\underset{\text{blue}}{DCIP_{ox}} + 2e^- + 2H^+ \rightarrow \underset{\text{colorless}}{DCIP \cdot H_{2\,red}}$$

The experimental tube will contain "living" chloroplasts, buffer, and the blue dye, and will be held in the light.

What change will you look for in the experimental tube?

____

What should the control tube contain?

____

Should you hold it in the light? ______________

The chloroplasts for this experiment are diluted in sucrose buffer so there are 100 μg of chlorophyll per milliliter, and they should be kept cold until you are ready to use them. Be sure to use the suspension labeled **CHLOROPLASTS.**

- Set the wavelength of the spectrophotometer at 620 nm.
- Obtain three spectrophotometer tubes and label them #1, #2, and #3,
- Add the reagents in the order given (left to right) in Table 7–1.
- Cover the tubes with Parafilm or stoppers and mix the ingredients by inverting the tube several times.
- Immediately cover the tubes with aluminum foil, and keep them covered except when they are being measured.
- Adjust the spectrophotometer to ∞ absorbance, insert tube #1, and adjust the absorbance to 0.
- Insert tube #2, and record its absorbance in Table 7–5 at the end of this topic. Cover the tube and wait 3 min.

TABLE 7–1. Components For the Hill Reaction

	Chloroplasts	Cold Buffer	Distilled Water	DCIP
Tube #1: blank	0.5 mL	3.0 mL	1.5 mL	—
Tube #2: reaction	0.5 mL	3.0 mL	0.5 mL	1.0 mL
Tube #3: control*	—	3.0 mL	1.0 mL	1.0 mL

* A tube identical to tube #2 but held in the dark would be an equally good control but would use up the chloroplasts.

- Again read the absorbance of tube #2.

Did any change in the dye occur in the dark control?

____

- Choose one light intensity (1000, 1500, or 2000 foot-candles), or take the one assigned to you, and hold your tube at the appropriate position for exactly 30 sec. Cover the tube.
- Read the absorbance; cover the tube again. Record the value in Table 7–5.
- Hold the tube in the light for another 30-sec period, and again read the absorbance.
- Continue until you have 12 readings.
- Since the experiment is meaningless without a light control, read the absorbance of tube #3 (no chloroplasts).
- Hold the tube for 3 min at the same light intensity that you used in the experiment.
- Read the absorbance again.
- Graph your results in Figure 7–6 (at the end of this topic) as percent DCIP reduced as a function of seconds exposed to the light. Label the curve with the light intensity that you used.

Join forces with students who used the other light intensities available and pool your results for comparison. Work together to answer the questions on the reverse of your graph.

CARBON FIXATION

Although carbon fixation can occur in the absence of light, it will quickly slow down and stop because the ATP and NADPH will soon be used up. A leaf or part of the plant that is not illuminated will obtain food from the other parts of the plant, but will not be able to carry on photosynthesis itself. Roots, tree trunks, and the other nongreen parts of plants must routinely obtain carbohydrate from the green photosynthetic parts. In this experiment you will study the effect of depriving part of a green leaf of light. The extent of carbon fixation will be measured as the amount of the storage carbohydrate, starch, that is formed.

Why can't starch be transported to all of the leaf cells? (Think back to the osmosis experiments.)

____

- Fold a piece of black paper in half, and cut out two copies of a small square, rectangle, or some other interesting shape.
- Stick a label on one of the pieces, and write your initials and the date on it.
- Use a paper clip to fasten the black shapes to the top and bottom surfaces of a leaf of a plant such as *Coleus* so that they are matched up exactly.
- Place the plant in sunlight until the next laboratory period.

Wait one week.

- Remove the leaf that was partially covered and take off the black paper.
- Drop the leaf into a beaker of boiling 95% alcohol in the hood, and boil it until it turns white.
- Remove the leaf, and place it in a finger bowl.
- Flood the leaf with iodine solution, and let it absorb the iodine for a few minutes.
- Rinse the leaf with tap water, and observe which parts stain darkly.

What happens when the leaf is boiled in alcohol?

____

____

What part of the leaf stained dark? ______________

Why does carbon fixation occur only in the light?

____

____

If you tested the leaf for glucose, would you expect to find any in the unstained part of the leaf? ________

When you finish your work, discard any debris, and clean your glassware thoroughly.

REFERENCES AND FURTHER READING

Baker, Jeffrey J.W., and Garland E. Allen. *Matter, Energy and Life: An Introduction to Chemical Concepts,* 4th ed. Reading, MA: Addison-Wesley, 1981.

Galston, A.W., P.J. Davies, and R.L. Satter. *The Life of the Green Plant,* 3rd ed. Englewood Cliffs, N.J.: Prentice-Hall, 1980.

Lehninger, A.L. *Biochemistry,* 2nd ed. New York: Worth, 1975.

Lehninger, A.L. *Principles of Biochemistry.* New York: Worth, 1982.

Rawn, J. David. *Biochemistry.* New York: Harper and Row, 1983.

Stryer, Lubert. *Biochemistry,* 2nd ed. San Francisco: W.H. Freeman, 1981.

Journal Articles

Björkman, Olle, and Joseph Barry. "High-efficiency photosynthesis." *Scientific American,* 80, October 1973 (#1281).

Govindjee, and R. Govindjee. "The primary events of photosynthesis." *Scientific American,* 68, December 1974 (#1310).

Miller, Kenneth R. "The photosynthetic membrane." *Scientific American,* 102, October 1979 (#1448).

Stein, William H., and Stanford Moore. "Chromatography." *Scientific American,* 35, March 1951 (#81).

Stoeckenius, Walther. "The purple membrane of salt-loving bacteria." *Scientific American,* 38, June 1976 (#1340).

Waggoner, Paul E. "Agriculture and carbon dioxide." *American Scientist* 72(2): 179, 1984.

Woodwell, George M. "The carbon dioxide question." *Scientific American,* 34, January 1978 (#1376).

Student Name ______________________ **Date** ______________

Extra space is provided on these data sheets in case some observations are repeated.

TABLE 7–2. Data From Chromatography

Spot	cm moved	R_f (cm/cm Solvent)	Color (Visible)	Color (UV)	Identification
Solvent front		———	———	———	———
Fastest spot					
Slowest spot					

Chromatogram:

R_f values: carotene (yellow-orange) = ____________
xanthophyll (yellow) = ____________
chlorophyll *a* (blue-green) = ____________
chlorophyll *b* (yellow-green) = ____________

TABLE 7–3. Absorption Spectrum of Photosynthetic Pigments

DATA

Wavelength	Absorbance	Wavelength	Absorbance	Wavelength	Absorbance
400 nm					

SPECTRUM

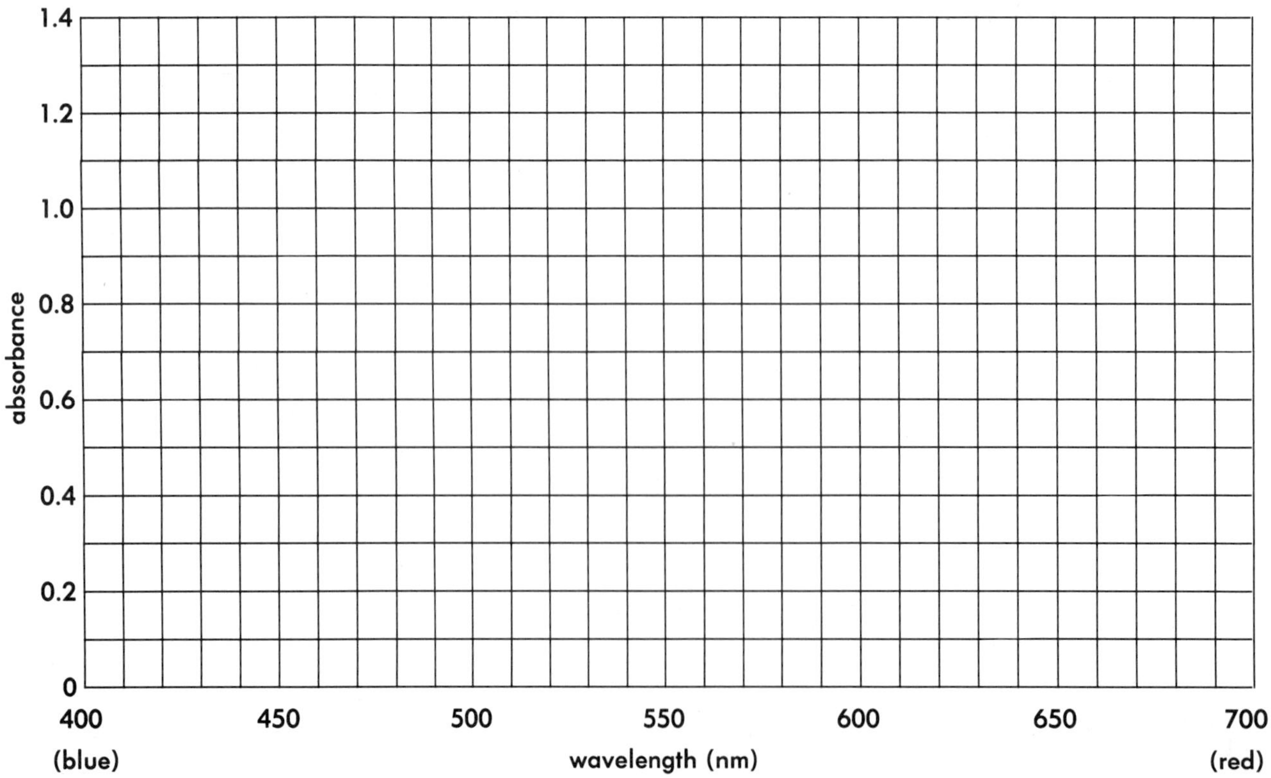

Student Name ______________________ **Date** ______________

TABLE 7–4. Data on Oxygen Evolution by *Elodea*

TAP WATER

Distance	Bubbles/3 min	Avg. Bubbles/min

$NaHCO_3$

Distance	Bubbles/3 min	Avg. Bubbles/min

SUMMARY

Treatment	Distance	Avg. Bubbles/min
None	75 cm	
None	50 cm	
$NaHCO_3$	25 cm	
None	25 cm	

What effect did decreasing light (increasing distance) have on the rate of photosynthesis?

____

What effect did increasing CO_2 ($NaHCO_3$) have on the rate of photosynthesis?

____

TABLE 7–5. Data From the Hill Reaction

1. The average initial absorbance of tube #2 was ______________.
2. Calculate the change from this value for each reading by subtracting the **new** absorbance from the **initial** absorbance.
3. Calculate the % DCIP reduced = (change × 100)/initial.

TUBE #2 REACTION (extra space provided)

Seconds of Light	(1) Absorbance	(2) Change in Absorbance	(3) % DCIP Reduced
0 (initial)		0	0
0 (3 min dark)			
30			
60			

TUBE #3 CONTROL

Seconds of Light	(1) Absorbance	(2) Change in Absorbance	(3) % DCIP Reduced
0		0	0
180			

Student Name ______________________________ **Date** ________________

HILL REACTION AT A LIGHT INTENSITY OF ____________ FOOT-CANDLES

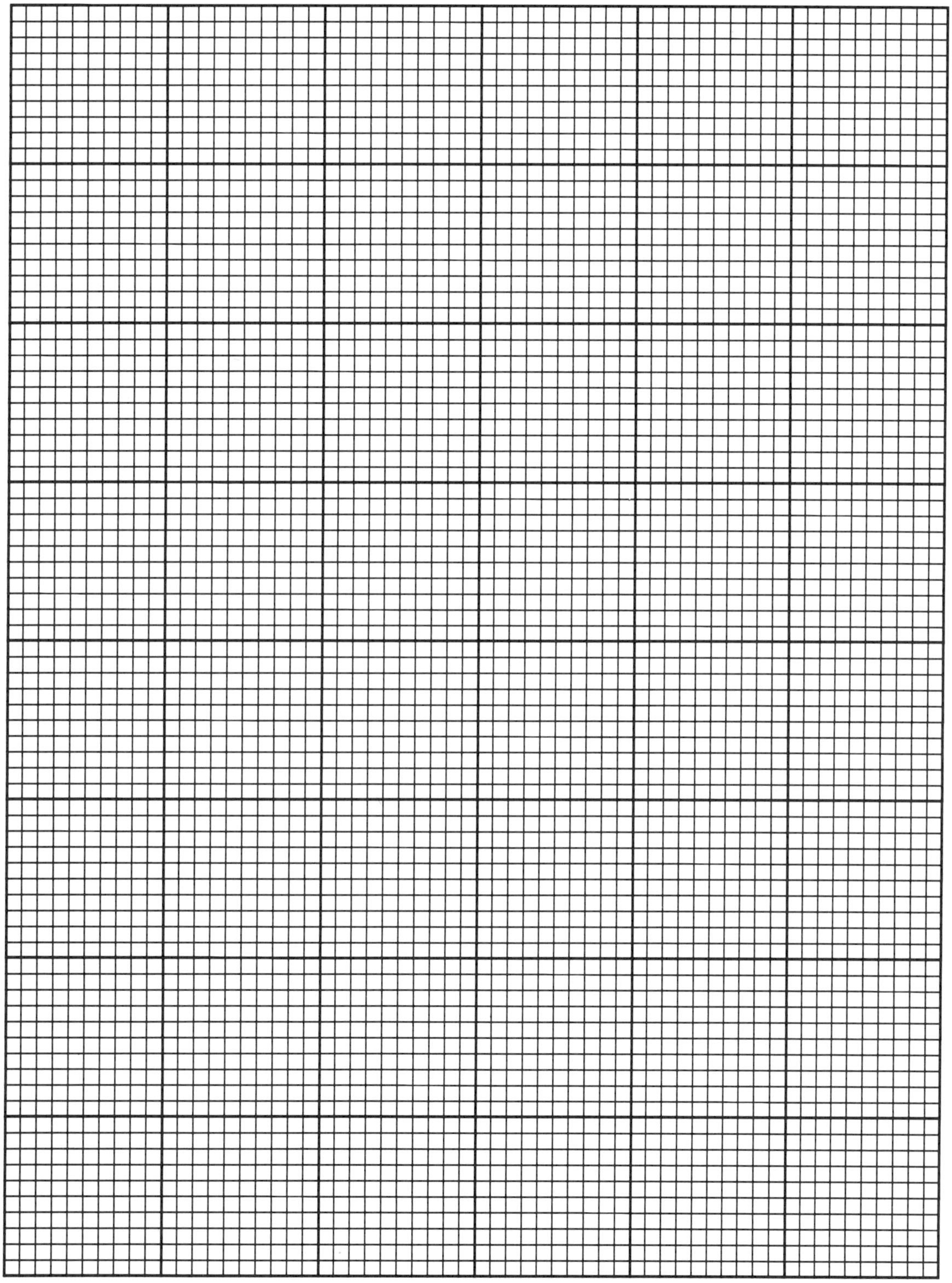

Figure 7–6. Hill reaction results. Plot the percent DCIP reduced as a function of seconds exposed to light. Be sure to label both axes of your graph. Answer the questions on the next page.

Compare your curve with those of other class members who used different light intensities. Your instructor may ask you to draw additional curves on your own graph.

Did the rate of dye reduction depend on the intensity of the light? ____________

Did the lowest light intensity used still show more dye reduction than seen in the control? ____________

If the light intensity could be increased even more without heating up the chloroplasts, would the rate of photosynthesis keep increasing indefinitely? Explain your answer.

____

____

TOPIC 8

DNA, RNA, and Protein Synthesis

OBJECTIVES

When you have completed this topic, you should be able to:

1. Describe the structure of a deoxyribonucleotide.
2. Explain how nucleotides are arranged in a molecule of DNA.
3. Describe the role of the bases and base pairing in the structure, function, and replication of DNA.
4. Explain the structure of giant salivary gland chromosomes in flies and cite evidence that these chromosomes are active in RNA transcription.
5. Define regulation, induction, and repression.
6. Name the genes of the lactose operon and their products, if any.
7. Explain the effect of lactose on the genes of the lactose operon.

TEXT REFERENCES Chapter 9; Chapter 10.

KEY TOPICS

DNA structure	giant salivary gland chromosomes
deoxyribonucleotides	gene regulation
base pairing	lactose operon

INTRODUCTION

DNA is the macromolecule that contains the blueprints for life. In prokaryotes it is a small circle with a single copy in each cell. In eukaryotes each chromosome contains a very long DNA molecule associated with RNA and protein. The **genes** for all of the cell's proteins, including the enzymes that make the other cell components, are encoded in the structure of DNA. A prokaryotic chromosome may contain about 2,500 genes and there are many thousands of genes, 30,000 for humans, in the nucleus of a eukaryotic cell. The structure of DNA is usually constant from generation to generation; any change is a **mutation,** and most such changes in DNA result in altered or deranged cell function. It seems to be highly adaptive for an organism to keep most of its genes identical from generation to generation. The structure of DNA, which inherently provides a mechanism for its exact replication, allows almost perfect conservation of our genes.

The building blocks of DNA are **deoxyribonucleotides,** each containing one **deoxyribose** 5 carbon sugar, to which is attached a **phosphate** group and a nitrogenous **base.** The phosphate group joins to the sugar of the adjacent nucleotide in a **condensation reaction;** hundreds of nucleotides are joined end-to-end in this way to make up the backbone of the DNA molecule in which sugar and phosphate groups alternate. A base extends from each sugar. Two such backbones coil around each other in a **double helix,** with the backbones on the outside of the helix and the bases on the inside. (See Text Figures 9–7 and 9–8.) Each base pairs with its opposite number and is held precisely in position by **hydrogen bonds** between appropriate functional groups, as shown in Figure 8–1.

The key to DNA's structure, function, and repli-

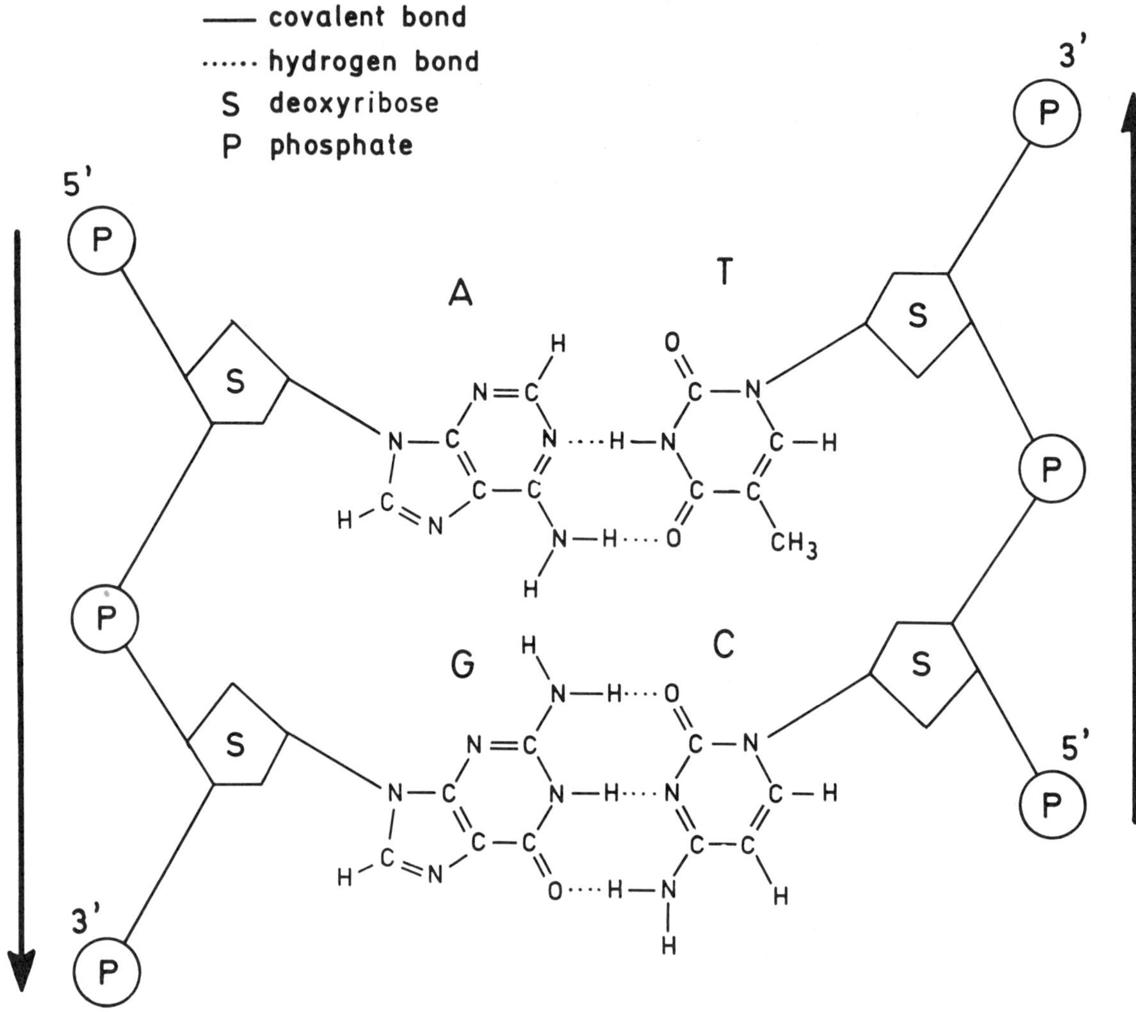

Figure 8–1. A dinucleotide section of DNA. Here are two base pairs in each of which the bases are held together by hydrogen bonds. Each base is part of a nucleotide composed of one deoxyribose sugar, one phosphate, and one base. The two strands of DNA run in opposite directions and so are antiparallel.

cation is the arrangement of the bases. This is the secret of life. The bases used are the double-ringed **purines, adenine** and **guanine,** and the single-ringed **pyrimidines, thymine** and **cytosine.** The distance between the backbones is just wide enough to accommodate a purine and a pyrimidine, but not two purines. There cannot be two pyrimidines either, because they would be too short to bridge the gap. A second constraint is that their respective functional groups permit adenine and thymine to hydrogen bond and likewise guanine and cytosine. The other combinations cannot form the requisite hydrogen bonds and are not normally found. The permitted combinations are called **complementary base pairs: AT, TA, GC,** and **CG.**

The next constraint on the structure of DNA is the order of the base pairs along the length of the molecule. This **sequence** is the code that is interpreted during protein synthesis to make the specific proteins characteristic of an organism. The fixed sequence of the bases allows millions of amino acids to be specified in exactly the right order to make the thousands of proteins needed by every cell.

Finally, the pairing of the bases provides a mechanism for **self-replication** of DNA. The information in one backbone of the helix is sufficient to specify the structure of the entire double helix and replication is **semi-conservative** in that each newly synthesized DNA double helix contains one "old" strand and one "new." During replication, the enzyme DNA helicase opens up the helix at the replication fork so that each strand can act as a **template** for binding the complementary bases of the new strands (See Text Figures 9–9 and 9–10). DNA polymerase enzymatically joins the incoming bases into the new complementary strands. Because the structure of DNA is likely to suffer some alteration during the lifetime of a cell, and because replication is not 100% perfect, normal cells contain enzyme systems to proofread replicated DNA and to repair some types of damage.

Certain cells of fly larvae are very large and necessarily have many times the amount of nuclear DNA normal for the organism. In the salivary glands of *Drosophila* larvae, for example, each cell contains 1024 times as much DNA as other cells and it is highly organized into **giant chromosomes.** All of the identical DNA strands in each polytene (polytene = many-stranded) chromosome are bundled together lengthwise and show a characteristic pattern of coiling: Parts of the chromosome are more condensed and other parts are more fluffy and expanded. The dense regions are all in register and likewise the less dense regions. When the cells are treated with a stain that binds most strongly to DNA, the most dense regions of each chromosome will stain darker and other regions will stain less well. The result is a pattern of dark bands and light interbands that is characteristic for a particular chromosome. Each band is thought to represent one gene.

Most genes are under some form of **regulation,** so that they can be switched on or off according to the prevailing conditions. **Gene expression** means that the gene is "on" and its DNA is being used to make RNA and usually protein. When a gene is active or "on," it must be in an uncondensed form so that the helix can be opened up and made accessible to RNA polymerase. This enzyme uses one of the DNA strands as a template during **transcription** or synthesis of mRNA. The mRNA will in turn be used to make protein during **translation** after it has travelled into the cytoplasm. Visible evidence of regulation in which only certain genes are turned on can be seen in giant chromosomes when they form **puffs** or **Balbiani rings** at different locations along the chromosomes during larval development (see Text Figure 10–15). It has been shown, by radioactive labeling of newly synthesized RNA, that active mRNA synthesis goes on in the puffs, so they represent genes that are turned on. The fine structure of an active puff is shown in Figure 8–2.

In bacteria there is genetic and biochemical evidence of regulation as well. If the colon bacterium *Escherichia coli* is given two sugars to metabolize for energy, it does an interesting thing: It uses up one completely before starting to use the second. This phenomenon is called **diauxie** ("double eating") and led to the idea that the second sugar was not used right away *because the enzymes needed for its metabolism were not present.*

If the sugars are glucose and lactose, for example, glucose will be metabolized first, appropriately enough since the enzymes for glycolysis are always present in all cells. Lactose, on the other hand, is a disaccharide and must be split to glucose and galactose before it can enter the glycolytic pathway. The presence of an abundance of lactose will usually cause the derepression or **induction** of the synthesis of the required enzymes within a few minutes. In this process, **repressed** genes that have previously been inactive in transcription are switched on. They begin making mRNA that is immediately translated into proteins needed for lactose metabolism. These processes in active genes of *E. coli* are shown in Text Figure 10–6.

The system of lactose induction is quite sophisticated in that if glucose is present along with lactose, it acts to block induction of lactose digesting enzymes while glucose itself is being used up. When the glucose is finally finished, induction proceeds, and the bacteria continue to grow on their new carbon source.

There is an obvious selective advantage in having a mechanism to regulate or switch genes on or off: If there is no immediate need for certain genes, a lot of energy can be saved and used for other purposes if these unnecessary enzymes are not made. There must be a quick, specific mechanism for their synthesis, however, as soon as a need arises. In *E. coli,* the enzymes for lactose metabolism can be immediately switched on whenever its host mammal drinks milk. At other times the enzymes are not needed and are not made.

The **lactose operon** is an inducible operon. It consists of a regulator gene, here a **repressor** gene, the **structural** genes under its control and the **operator** switch by which the repressor protein turns them on or off. The **inducer** that switches the operon on is allolactose, present only in cells exposed to lactose.

■ Identify the genes of the lactose operon in Figure 8–3:

I codes for the repressor protein, the regulator for this operon.
O, the operator, is the 21-base segment of DNA to which the repressor protein binds.
P, the promoter, is the DNA-sequence where RNA polymerase binds and begins transcription.
Z, a structural gene coding for β-galactosidase, the enzyme that hydrolyzes lactose into glucose and galactose.
Y, a structural gene coding for permease, which transports lactose into the cell.
A, a structural gene coding for transacetylase.

Only genes I, Z, Y, and A are transcribed and code

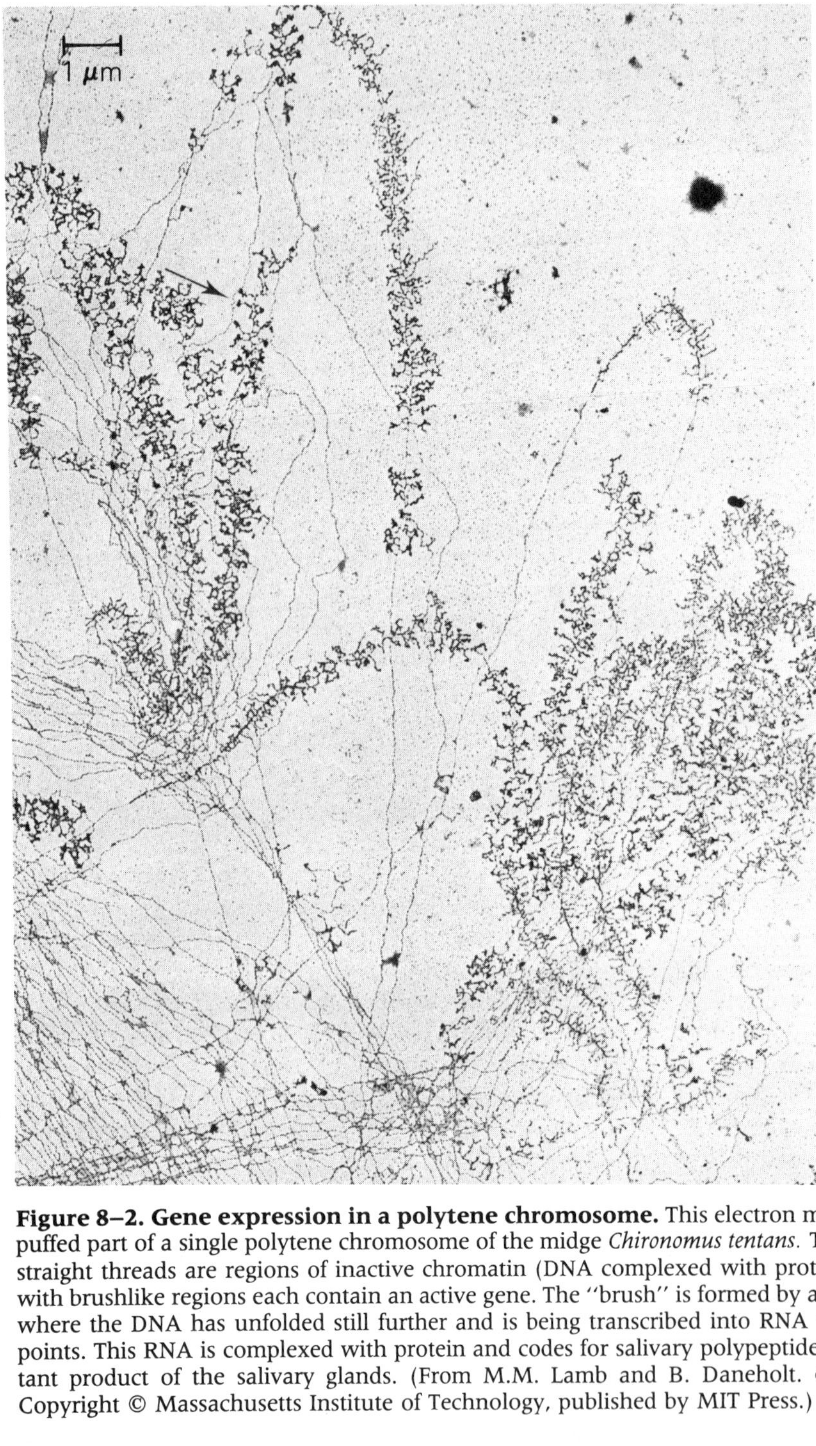

Figure 8–2. Gene expression in a polytene chromosome. This electron micrograph shows a puffed part of a single polytene chromosome of the midge *Chironomus tentans.* The parallel packed straight threads are regions of inactive chromatin (DNA complexed with protein) and the loops with brushlike regions each contain an active gene. The "brush" is formed by a loop of chromatin where the DNA has unfolded still further and is being transcribed into RNA "bristles" at many points. This RNA is complexed with protein and codes for salivary polypeptides, the most important product of the salivary glands. (From M.M. Lamb and B. Daneholt. *Cell* 17:838, 1979. Copyright © Massachusetts Institute of Technology, published by MIT Press.)

for proteins. O and P are gene sequences important in regulation but they do not code for proteins. Gene I is always transcribed and translated into repressor protein. There are only about 10 molecules of repressor per cell.

■ Study the following situations shown in Figure 8–3:

A. Repression. When *E. coli* is in the typical situation where there is no lactose present, repressor binds to the operator. This binding is extremely specific for the lactose operator and extremely tight. The RNA polymerase cannot bind to the promoter while the repressor is bound to the operator, because the P and O sequences partly overlap. There is no transcription or translation of structural protein. The operon is OFF.

B. Induction. When lactose arrives, allolactose is formed, binds to the repressor, and converts it to an alternate, inactive form. Inactive

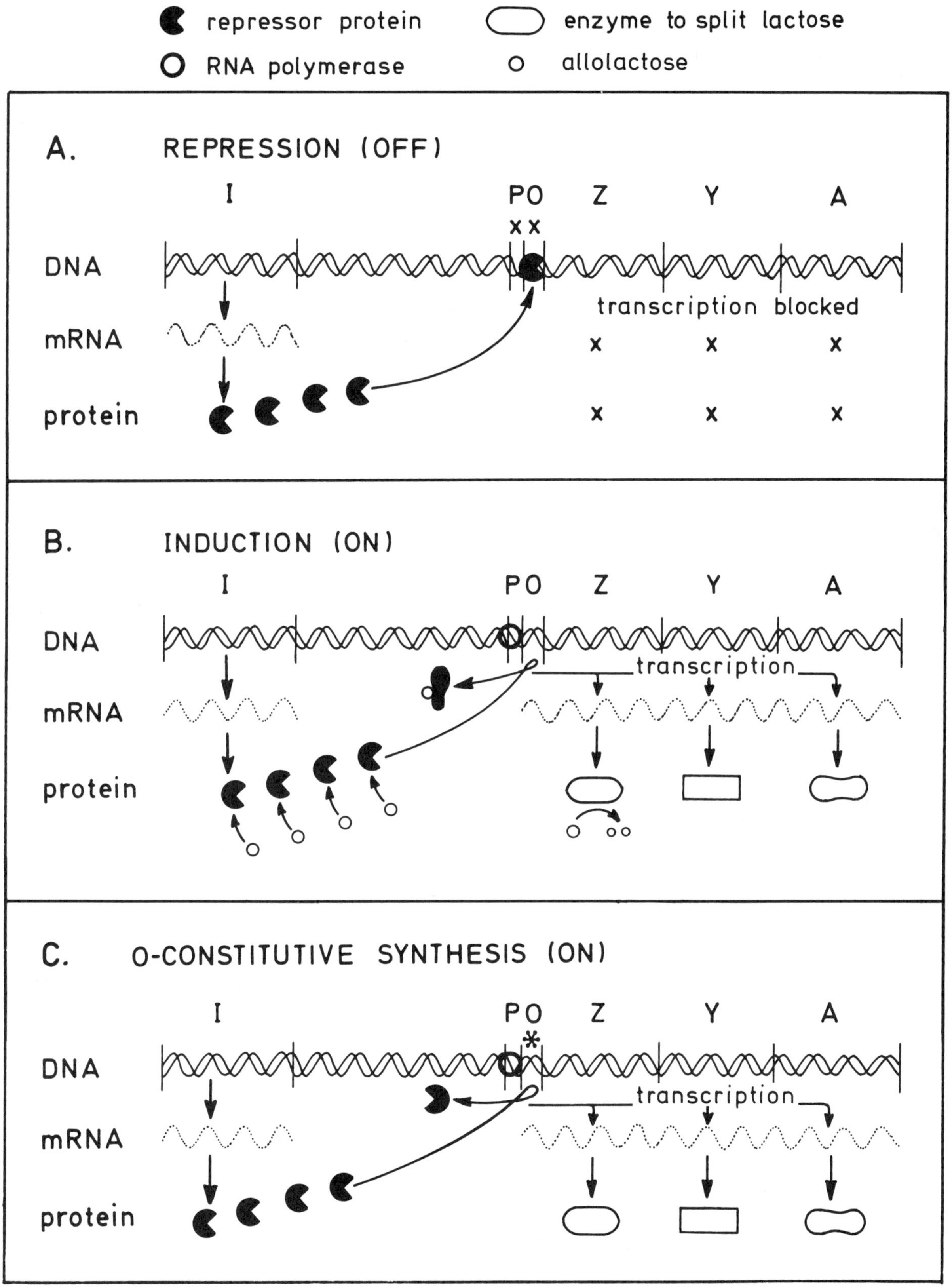

Figure 8–3. The lactose operon. Regulation of this operon involves gene I for repressor protein, gene P, the promoter (no protein), gene O, the operator (no protein), gene Z for the lactose-splitting enzyme (β-galactosidase), gene Y for lactose permease, and gene A for transacetylase.

(A) The operon is switched OFF and no enzymes are made in the absence of lactose.
(B) Transcription is switched ON by allolactose, enzymes are made and lactose is metabolized.
(C) A mutation (*) in the operator base sequence blocks repressor binding. Transcription is switched ON whether or not lactose is present.

repressor can no longer bind to DNA, and so transcription and translation of structural genes proceed. The operon is ON and lactose is metabolized.

C. **Constitutive Synthesis.** One of many interesting mutations that changes regulation is an operator constitutive (O^c) mutation. It alters the operator sequence so that repressor fails to recognize it and cannot bind. The result is the same as if the repressor had been inactivated: The operon is ON, structural genes are made, and lactose could be metabolized, if present. Induction is not possible because the genes are on all the time.

- Turn to Table 8–1 at the end of this topic and be sure you understand how to complete it for each of these situations.

DNA IN GIANT CHROMOSOMES

- Obtain two slides, two teasing needles, and physiological saline.
- Place a large *Drosophila* larva on a slide and add a drop of saline.
- Locate the anterior end of the larva and estimate the position of the salivary glands; they extend back about one third of the body length (see Figure 8–4).
- Hold the body about midway back with one teasing needle, laid flat across the body; lay the other needle flat across the head and slowly pull off the head. The salivary glands may be attached to the exposed gut or be attached to the head.
- Identify the transparent, glassy salivary glands, distinct from the grainy gut and body tissue.
- Separate the salivary glands from other tissue and transfer them to a clean slide.
- Put a drop of acetocarmine stain (for DNA) on the salivary glands and add a coverslip.
- Take two or three paper towels and fold them over several times to make a thick pad.
- Place the slide on a very flat, hard surface; cover the slide with the pad, then pound the slide once with your fist: **BANG!**
- Remove the pad and inspect your slide. If you are lucky, it will have remained intact and you will now have a nice **squash** of the salivary gland cells. If your slide broke, try it again or share with another student.
- View the slide under low, then high power in the compound microscope. Use oil immersion if your microscope has an oil immersion lens.
- Locate the cell nuclei, which look like clumps of tangled necklaces. Try to find one in which the chromosomes are extended so that you can see the banding pattern.

Each band in a giant chromosome corresponds to a single gene. The different intensities of staining reflect the physical state of the DNA, i.e. how tightly it is coiled. The coiling in turn is correlated with the activity of the genes: Genes that are active seem to be more expanded, and the most active genes are greatly expanded into large **puffs** or **rings.** The pattern of puffs changes dramatically with time as development of the larva progresses and different gene products are required.

REGULATION OF THE LACTOSE OPERON

In the following experiment, you will study these situations:

wild type (normal) *E. coli* grown on maltose with which it will be repressed

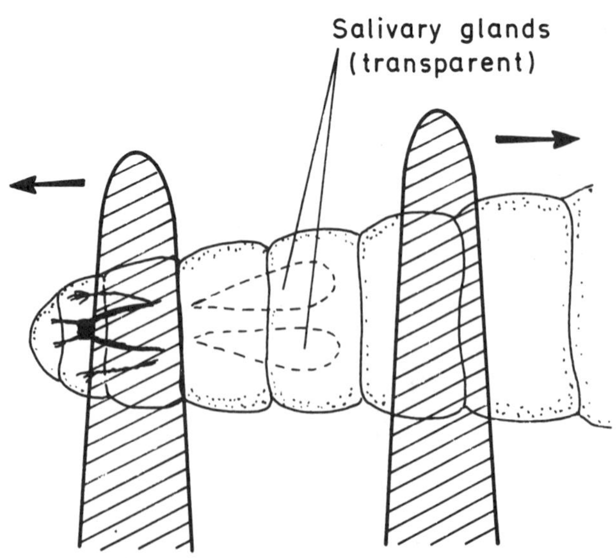

Figure 8–4. Salivary glands from a *Drosophila* larva. Press gently on the larva with two teasing needles. Then carefully separate the needles to pull the head region from the body. The salivary glands are transparent in contrast to the granular intestine.

wild type *E. coli* grown on lactose with which it will be induced and will synthesize β-galactosidase

O^c *E. coli* grown on either maltose or lactose with which it will show constitutive synthesis of β-galactosidase

The activity of the enzyme β-galactosidase can be measured in cells first made permeable by adding toluene. An artificial substrate, ONPG (*o*-nitrophenyl galactoside) is added and is split by the enzyme into galactose and *o*-nitrophenol, a bright yellow compound. The amount of yellow color will show the amount of β-galactosidase present in the cells. The color may be estimated qualitatively (++, +, or −) after incubating the toluene-treated cells for 15 min at 30°, if you do not have access to a spectrophotometer.

- Working in pairs, obtain 5-ml samples of these *E. coli* K−12 cultures:

  #1 wild type grown on lactose
  #2 wild type grown on maltose
  #3 O^c mutant grown on lactose
  #4 O^c mutant grown on maltose
  #5 unknown strain grown on lactose
  #6 unknown strain grown on maltose

- Add two drops of toluene to each tube; cover and shake vigorously.
- Allow ten minutes for the toluene to act, shaking the tubes frequently.
- Set the wavelength of the spectrophotometer to 440 nm and adjust the absorbance to ∞.
- Pour the contents of tube #1 into a spectrophotometer cuvette and use this as the control to adjust the absorbance to zero. Pour the culture back into tube #1. (Your instructor may ask you to use another prepared control tube.)
- Use a pipettor to add 0.2 mL of ONPG to tube #1, pour into a cuvette, and immediately read the absorbance (0 min). Record the value in Table 8−2 at the end of this topic.
- Read the absorbance at 1-min intervals for 6 min.
- Repeat for tubes #2−#6. Record the results.
- Plot your data in Figure 8−5 at the end of this topic, and decide whether the unknown strain was wild type or constitutive.
- Complete Table 8−2 and answer the questions.
- When you are finished with this lab, return all materials and dispose of the bacterial cultures as directed. Follow the directions of your instructor for cleaning your glassware, and store your microscope properly.

REFERENCES AND FURTHER READING

Alberts, Bruce, D. Bray, J. Lewis, M. Raff, K. Roberts, and J.D. Watson. *Molecular Biology of the Cell.* New York: Garland, 1983.

Beerman, W., ed. *Developmental Studies on Giant Chromosomes.* New York: Springer-Verlag, 1972.

Browder, L.W. *Developmental Biology,* 2nd ed. Philadelphia: Saunders College Publishing, 1984.

Watson, J.D. *The Double Helix.* New York: Atheneum, 1968.

Watson, J.D., et al. *Molecular Biology of the Gene,* 4th ed. Menlo Park, CA: Benjamin/Cummings, 1986.

Journal Articles

Beerman, Wolfgang, and Ulrich Clever. "Chromosome puffs." *Scientific American,* 50, April 1964 (#180).

Cohen, S.N. "The manipulation of genes." *Scientific American,* 24, July 1975 (#1324).

Crick, F.H.C. "The structure of the hereditary material." *Scientific American,* 54, October 1954 (#5).

Dickerson, R.E. "The DNA helix and how it is read." *Scientific American,* 94, December 1983.

Diener, T.O. "Viroids." *Scientific American,* 66, January 1981.

Drake, John W., Barry W. Glickman, and Lynn R. Ripley. "Updating the theory of mutation." *American Scientist* 71(6): 621, November-December 1983.

Federoff, Nina. "Transposable genetic elements in maize." *Scientific American,* 84, June 1984.

Felsenfeld, Gary. "DNA." *Scientific American,* 44, October 1985.

Gilbert, W., and B. Müller-Hill. "Isolation of the *lac* repressor." *Proc. Nat. Acad. Sci. 56:* 1891, 1966.

Grivell, Leslie A. "Mitochondrial DNA." *Scientific American,* 78, March 1983.

Harvey, Ronald G. "Polycyclic hydrocarbons and cancer." *American Scientist* 70(4): 386, July-August 1982.

Howard-Flanders, Paul. "Inducible repair of DNA." *Scientific American,* 72, November 1981.

Jacob, F., and J. Monod. "Genetic regulatory mechanisms in the synthesis of proteins." *J. Mol. Biol.* 3:318, 1961.

Kornberg, Roger D., and Aaron Klug. "The nucleosome." *Scientific American,* 52, February 1981.

Lewin, R. "A naturalist of the genome." [Barbara McClintock]. *Science* 222:402, 28 October, 1983.

Pederson, Thoru. "Messenger RNA biosynthesis and nuclear structure." *American Scientist* 69(1): 76, January 1981.

Ptashne, Mark, Alexander D. Johnson, and Carl O. Pabo. "A genetic switch in a bacterial virus." *Scientific American,* 128, November 1982.

Yuan, Robert, and Daniel L. Hamilton. "Restriction and modification of DNA by a complex protein." *American Scientist* 70(1): 61, January 1982.

Watson, J.D., and F.H.C. Crick. "Molecular structure of nucleic acids: A structure for deoxyribose nucleic acid." *Nature* 171:737, 1953.

The October 1985 issue of *Scientific American* is devoted to DNA and Molecular Biology.

Student Name ______________________________ **Date** ____________________

TABLE 8–1. Regulation of the Lactose Operon.

Indicate for the following situations the condition of each of these components (yes or no).

	Wild Type No Lactose	Wild Type Lactose	O^c No Lactose	O^c Lactose
Repressor Synthesized				
Repressor Binds to Lactose				
Repressor Binds to Operator				
Operator Is On				
Operator Is Off				
Promotor Binds RNA Polymerase				
Transcription of Z,Y,A Occurs				
Translation of Z,Y,A mRNA Occurs				
Lactose is Split				

TABLE 8–2. Data for Induction of the Lactose Operon

Record your absorbance data for each culture. Extra space is provided in case you need to follow your culture for longer than 6 min. Then plot your results in Figure 8–5.

	Tube #1	Tube #2	Tube #3	Tube #4	Tube #5	Tube #6
E. coli	w.t.	w.t.	O^c	O^c	unknown	unknown
Sugar	lactose	maltose	lactose	maltose	lactose	maltose
0 min						
1 min						
2 min						
3 min						
4 min						
5 min						
6 min						
min						
min						
min						
min						
ONPG split (yes/no)						
operon is ON/OFF						

Maltose is a disaccharide of glucose subunits and is structurally very similar to lactose. Explain why it does not switch on the lactose operon.

____

____

Name at least two ways in which an O^c mutant strain could mutate again so that it fails to produce β-galactosidase in the presence of lactose.

____

____

Student Name ______________________________ **Date** ____________________

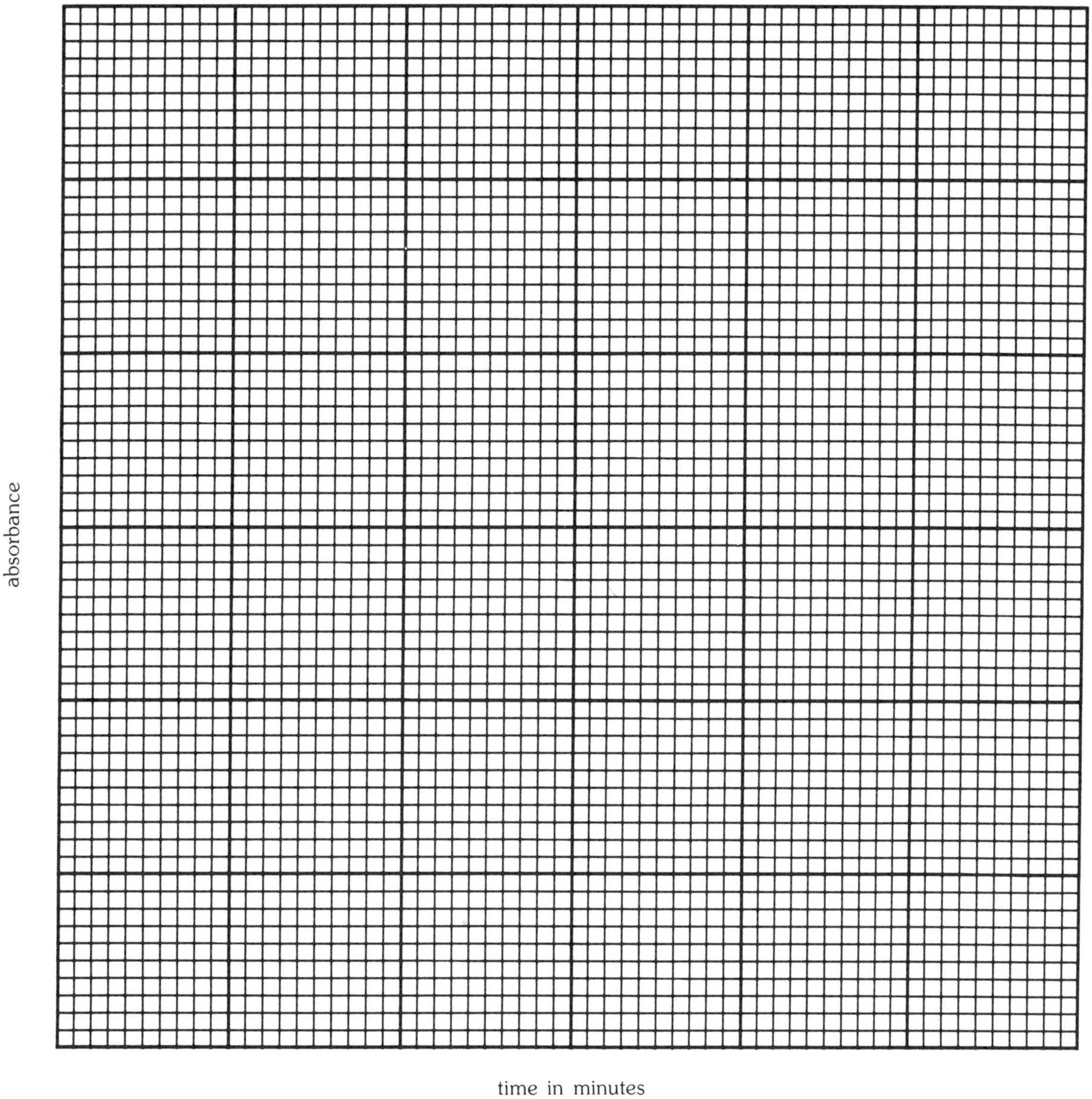

Figure 8–5. Activity of β-galactosidase. Plot the absorbance at 440 nm versus time in minutes.

TOPIC 9

Embryonic Development in Animals

OBJECTIVES

When you have completed this topic, you should be able to:

1. Describe the changes that take place in an invertebrate egg during fertilization and that distinguish an unfertilized egg from a fertilized one.
2. List the four main stages of embryonic development following fertilization.
3. Define cleavage and describe two important ways in which cleavage in a sea urchin embryo is like that in a human embryo.
4. Correctly explain and use the following words: archenteron (primitive gut), blastocoel, blastopore, blastula, cleavage, gastrula, gastrulation, morula, neurula, neurulation, zygote.
5. Draw a diagram of an embryo containing the three primary germ layers; label the germ layers, blastocoel, blastopore, and gastrocoel (archenteron), and label the embryo with its correct embryological stage.
6. List or recognize the body parts formed from each primary embryonic germ layer.
7. Define embryonic induction and give an example.
8. Define metamorphosis and explain the selective advantage to animals of going through metamorphosis at some stage of the life history.
9. Given a developing chicken egg, open it properly, and decide whether it is 24, 48, or 72 hr old by identifying the features or structures characteristic of that age.

TEXT REFERENCES Chapter 10 (J); Chapter 22 (L,M); Chapter 28.

KEY TOPICS

echinoderms	embryonic development	metamorphosis
fertilization	induction	amniote egg

INTRODUCTION

When animals reproduce sexually, each parent contributes half of the genetic material needed to make up the full genetic complement or genome of the offspring. A male parent produces male gametes or **sperm** and the female parent produces female gametes, the **eggs** or **ova** (singular: ovum). The gametes fuse during fertilization to produce a **zygote.** Since the zygote will always have twice as much genetic material as a gamete, the genetic material must be reduced by half at some stage before the next fertilization, so that the amount of nuclear material stays the same from one generation to the next. This reduction in nuclear material occurs through the process of meiosis, which we will consider in depth in Topic 10. Meiosis is carried out only by specialized germ cells in the sex organs where sperm and ova are produced. Meiosis and fertilization are thus two complementary halves of an animal's life history:

$$\text{animal} \underset{\text{fertilization}}{\overset{\text{meiosis}}{\rightleftarrows}} \text{sperm and ova}$$

Neither process makes sense without the other.

Fertilization does not, of course, give rise directly to an animal. In this part of the reproductive cycle, the unicellular zygote undergoes **development** into the many-celled adult organism, which becomes sexually mature, undergoes meiosis, and then reproduces in its turn. Today's lab will focus on embryonic development, which begins with fertilization of the egg and ends in the production of a young organism similar in form to the adult. Certain processes within the developing embryo stand out as very important: **cell division, cell differentiation,** and **cell movement.** Cell division results in the multitude of cells comprising the adult. Cell differentiation, the expression of a different set of genes in different cell types, provides the spectrum of different cell types necessary in the adult; through cell movement, the cells form a highly organized structure that permits the cells of the adult to function properly. All of these processes mutually interact.

Embryological development ends with the hatching of the embryo, or its birth when it leaves the body of its mother. Many developing organisms (including invertebrates such as the sea urchin, insects, and some amphibians) pass through an additional developmental stage called the **larva,** which is different in form from the adult. The larva usually exists independently from its parents and has its own characteristic way of life. It changes suddenly in anatomy and physiology when it undergoes **metamorphosis** into the very different adult.

STRUCTURE AND FUNCTION IN THE SEA URCHIN

If live sea urchins are available, your instructor may ask you to complete this section on the sea urchin as a representative echinoderm. Although echinoderm adults are not at all like humans, early development in these invertebrate organisms is similar to that in humans in several important ways.

- Observe the sea urchins in the aquarium or supply dish. The spines that cover the top (aboral) surface of the body are mobile, and can pivot on little bumps at the base of each spine.

If the urchin is healthy, the spines will stick straight out.

Can you tell which sea urchins are male and which are female? ______________

If the urchins are crawling on the surface of the aquarium, you will be able to see the **tube feet** sticking to the glass and the **mouth,** which is located on the flat underside. (See Color Plate 3E and Figure 9–1.)

- Use large forceps or wear rubber gloves when you are ready to take one of the animals.

Do not touch sea urchins with bare hands.

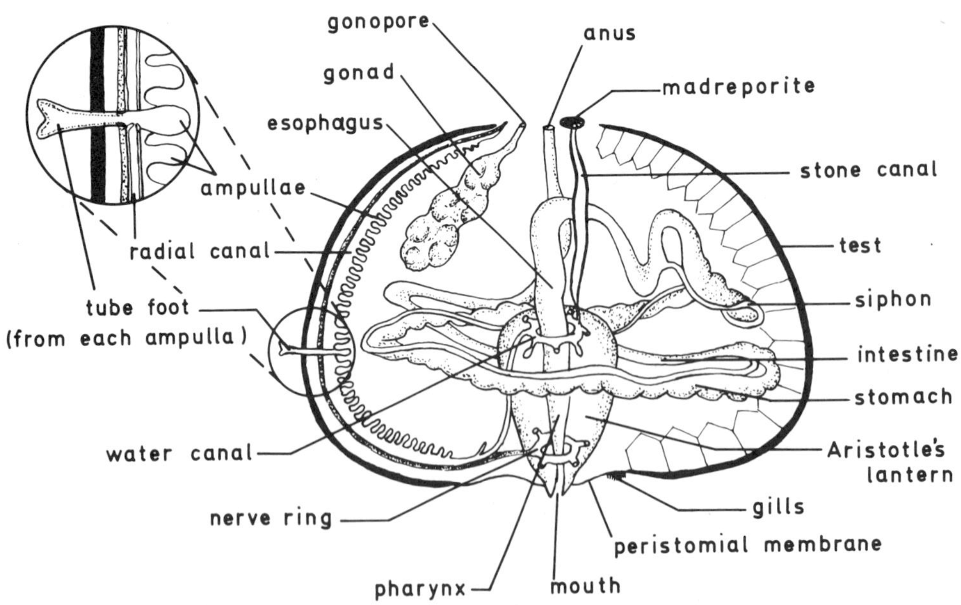

Figure 9–1. A sea urchin. Aristotle's lantern is the feeding apparatus used for ingesting kelp and seaweed. The gonads, test (skeleton), and water-vascular system show pentaradial or fivefold symmetry, although diagrammed only once. The many movable spines that cover the surface of the sea urchin are not shown.

- Place the urchin in a beaker of sea water or on a Petri dish of sea water with the flat (oral) side down.
- Observe the animal for a few minutes, and notice the tiny, clear tube feet that extend from between the spines and stick to the surface of the glass. (See Color Plate 3.)

About how many tube feet does your sea urchin seem to have? ___________

- Place a piece of lettuce or a small object on the surface of the urchin, and wait 5 or 10 min. Then gently try to remove the object.

Does the sea urchin attach its tube feet to the object?

____

Study Figure 9–1, and note how the **water-vascular system** of radial canals and ampullae connects to all of the tube feet. This is a hydraulic system that allows the feet to be extended by increasing the pressure within the ampulla at the base of each foot. When the foot contacts something, suction can be exerted so that the foot sticks to the object like a tiny suction cup. Each foot can be operated individually. Water circulates into and out of the water-vascular system through a porous plate called the **madreporite,** which is located on the aboral surface. Also located on the aboral surface is the **anus** and the five genital pores through which gametes are released from the gonads into the sea.

The shape of the sea urchin is due to a hard endoskeleton called the **test.** Both the test and the spines are covered by a delicate epidermis. Study the sea urchin test on demonstration, using the dissecting microscope when appropriate.

- Locate the five **gonopores** on the aboral surface.
- Notice the tubercles where the spines were once attached.
- Find the numerous double perforations where the tube feet extended through the test. These occur in five regions that are separated from each other by five sectors in which there were no feet and thus no perforations.

By now you should be able to describe the type of symmetry that the sea urchin shows (twofold/threefold/fourfold/fivefold/sixfold) (bilateral/radial):

____

- On the oral surface notice the large opening in the test. This is where the feeding apparatus, called **Aristotle's lantern,** originally protruded through the test, but the membranes holding it together and in position disintegrated when the test was prepared. You can see this structure in your living animal.
- Again using forceps or gloves, invert the animal so that the flat side is on top. The animal is now upside down from its normal position.
- Note the feeding structure, Aristotle's lantern, in the center of the oral surface. It contains five teeth that are used to tear the kelp on which the animal feeds, and it is held in place by a membrane that attaches it to the skeleton. The mouth is the opening in the center of the lantern.
- Take a single tooth from those on demonstration, and study its structure in the dissecting microscope.
- When you are finished observing the structure of the oral surface, leave your animal upside down for awhile, and observe its response from time to time.

How did your sea urchin react to an upside down situation?

____

What structures were used in the response?

____

When you are finished with your sea urchin, return it to the aquarium or supply pan.

SEA URCHIN DEVELOPMENT

Mature sea urchins are male or female. When the gametes are ripe, they are released through pores on the top (aboral) side of the animal, and fertilization takes place in sea water.

Is this type of fertilization internal or external?

____

In the laboratory, the urchins can be induced to release their gametes separately so that fertilization can be carried out and observed under the microscope. If the fertilized eggs are carefully washed and incubated in sea water, they will develop for many days, and will eventually form a larval stage known as the **pluteus.** The pluteus is a free swimming, feeding organism which eventually attaches to the bottom and metamorphoses into a tiny sea urchin. In this organism the main function of the larval stage is dispersion; most of the feeding and growth is done by the sessile (nonmotile) adults. Depending on the time of year, you will use an East Coast sea urchin (*Arbacia*) or a West Coast urchin (*Strongylocentrotus*) to study early development. The schedule for the development of these species is shown in Table 9–1.

Cleavage

Cleavage begins with fertilization and continues through many cell divisions without growth until the embryo is a hollow ball of many cells. Sea urchin eggs

TABLE 9–1. Development of Sea Urchin Embryos

Developmental Stage	*Arbacia punctulata* at 23°C	*Strongylocentrotus purpuratus* at 15°C
Fertilization	0	0
Fertilization membrane forms	1–2 min	1–2 min
First cleavage	50–70 min	90–110 min
Second cleavage	80–110 min	140–170 min
Blastula	5–6 hr	8–10 hr
Hatching of blastula	7–8 hr	18–20 hr
Gastrula	12–15 hr	22–26 hr
Pluteus	24 hr	72 hr
Metamorphosis	2 weeks	2–8 weeks

are very large in size because they contain everything needed for cleavage: ribosomes, mRNA information for the proteins of early development, and energy stored in the form of yolk granules. Because cleavage is so independent of outside needs, it can be induced artificially without any sperm, a phenomenon known as parthenogenesis. The genetic contribution of the sperm becomes important only later in development. You should use a depression slide or omit the coverslip while you are observing the eggs and embryos, because they are so large.

- Place a drop of stirred sea urchin egg suspension on your slide and observe under low power.

Are the eggs motile? ____________

What is the diameter of the eggs in μm?

____

What is the color and texture of the cytoplasm?

____

What does the cell membrane look like?

____

If you are studying *Arbacia* eggs, add a drop of India ink so that you can see the clear jelly coat surrounding the egg.

About how thick is the jelly coat in μm?

____

- Clean the slide and dry it before adding a fresh drop of eggs. Focus on the eggs under low power.
- Add a drop of *diluted* sperm cells provided by your instructor. Mix gently with a toothpick while observing in the microscope, and record the time and temperature in Table 9–2 at the end of this topic.

The eggs are chemically attractive to sperm and are soon surrounded by swarms of these relatively tiny cells. (See Text Figure 17–6.) When the sperm cells penetrate the jelly coat, they eventually come into contact with the membrane of the cell. Then one lucky sperm head will be drawn into the cytoplasm, leaving its tail behind, and its nucleus will combine with the egg's nucleus to form the new nucleus of the zygote. (See Text Figure 28–12.) The egg and sperm contribute an equal number of chromosomes to the zygote.

You can barely see the individual sperm cells under low power, but you can see right away that something is happening.

What change occurs within 1 or 2 min of adding sperm cells to the eggs?

____

When the egg cell is stimulated by the arrival of sperm, a chain of events begins to unfold that is very precise and almost invariably ends in the production of a normal sea urchin pluteus. First an electrical reaction begins at the site where the sperm penetrates, and spreads in the cell membrane all over the surface of the cell. Changes in the cell membrane prevent any second or third place sperm heads from entering the cytoplasm. Then Ca^{++} ions released from this region of the membrane induce changes in the egg cytoplasm, signalling that development should begin. The **vitelline membrane,** which lies between the cell membrane and the jelly coat, separates from the cell membrane and enlarges so that there is a wide space between the two membranes. The cell "membrane" now looks double and is called the **fertilization membrane.** This is the obvious change that you see whenever fertilization occurs. Any sperm still struggling through the jelly coat are lifted away from their goal and the fertilization membrane provides a protected environment for the developing embryo. Rapid protein synthesis is carried out using mRNA provided by the mother and stored in the egg. This is part of the preparation necessary for the first cell division, when the constriction known as the **cleavage furrow** will divide the zygote into two cells.

- Seal your slide of fertilized eggs by applying Vaseline with a toothpick around the coverslip, and set it aside, but observe it every 15 min for the rest of the period. You should see signs of cleavage in about 1 hr. Your instructor may ask you to sample a culture of developing eggs instead of saving the fertilized egg on your slide. Record the age, stage, and appearance of the embryos in Table 9–2 each time you observe them.
- Take an ordinary slide, and prepare a wet mount of sperm cells from the suspension indicated by your instructor. Observe under high power or oil immersion. Cut down the light by closing the diaphragm.

Are the sperm cells in motion? ____________

Can you see the flagella of the sperm cells?

____

In addition to the nucleus and the flagellum, what organelle is important for sperm cells to carry out their reproductive role? (*Hint:* This organelle provides energy needed in large amounts for sperm motility.)

____

Once the first division has taken place, cleavage will continue for many cell divisions. In sea urchins cleavage is **indeterminate:** Each of the cells formed early in development has the potential to develop into a complete organism. If the daughter cells should happen to separate after the first division, the result would be two identical organisms. When this happens in human development, the result is identical twins. Cleavage is **radial** because each cleavage plane is at right angles to or parallel to the original axis of the zygote. (See Text Figure 28–15.) In the sea urchin the yolk is in the form of cytoplasmic granules, so the cleavage furrow can pass completely through the egg without difficulty.

- Study the later stages in cleavage that are available.

How large is the embryo compared to the egg after several divisions?

____

The **morula** is a solid ball of cells that gradually develops a hollow interior called the **blastocoel.** (See Text Figure 28–17a.) The embryo is then at the **blastula** stage, and its outer cells are ciliated so that the blastula is motile, and can move about within the fertilization membrane. After several hours of cleavage, the blastula is fully developed, and "hatches" as the fertilization membrane finally breaks open and releases the embryo. Cleavage has ended and the blastula now enters the next stage of embryonic development.

Gastrulation

During gastrulation the embryo changes from a blastula with one layer of cells into a **gastrula,** which has three layers. In this process, cell movement is extremely important. Cells migrate into the interior of the blastula at a point called the **blastopore,** and form a lining so that the embryo now has two germ layers: an outer ectoderm and an inner endoderm. The ectodermal cells will form the epidermis, that is the skin, including the hair and nails in mammals. They will also form the neurectoderm, which will give rise to the whole nervous system, including the brain. The endoderm develops into the lining of the gut and of certain digestive organs and glands. The space inside both layers is called the **gastrocoel,** or **archenteron,** and will eventually form the lumen of the organism's gut. The cells in the dorsal part of the endoderm form pouches, and bud off to become the cells of the mesoderm. This type of **mesoderm formation** is a third important way (besides indeterminate cleavage and radial cleavage) in which sea urchin embryology and human embryology are alike. Mesoderm gives rise to the skeleton, muscles, sex organs, and other internal organs of the adult.

- Examine a drop of the embryos that have reached the gastrula stage.
- Record the age of the embryos and their appearance in the table.

Can you see the primitive gut? ____________

How did the embryo obtain energy before its gut formed to permit feeding?

____

Neurulation

In vertebrates such as yourself and the chick, the gastrula soon develops a dorsal groove, which pinches off and becomes part of the dorsal nerve cord and the brain. An embryo undergoing this process is at the neurula stage. (See Text Figure 28–19.)

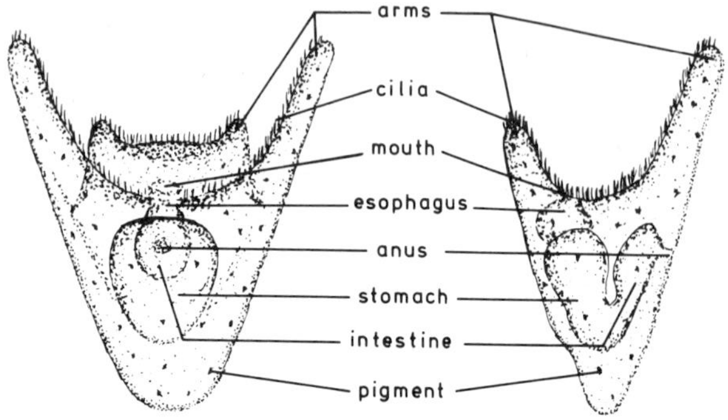

Figure 9–2. Pluteus larva. The sea urchin larva develops from the zygote within a few days. The motile larva feeds and swims about independently until it settles down and metamorphoses into a tiny adult.

Organogenesis

The process of organogenesis in which organs and organ systems are developed will be discussed in the section on the chick embryo where it can be studied more easily.

If older sea urchin embryos are available, you will be able to see a simple digestive system and skeletal system in this case as well.

Metamorphosis

The sea urchin embryo develops skeletal elements, and begins feeding as the **pluteus larva.** (See Figure 9–2.) It continues this existence for up to four months until it finally settles down on the substrate, attaches, and changes into a tiny adult sea urchin. The whole process of **metamorphosis** may take only an hour, and the result is a miniature adult only 1 mm in diameter.

- Observe the embryos that have reached the pluteus stage.
- Draw an arrow in Figure 9–2 to indicate the direction in which the pluteus swims.
- Note the skeletal spines that support the body and arms.
- Note whether colored pigment granules are present.
- Identify the complete digestive tract visible through the transparent body of the pluteus.

What is the advantage of a motile larval stage?

____

____

CHICK DEVELOPMENT

Fertilization and Early Development

The egg cell, or ovum, of a chicken consists of a yellow mass of **yolk** and a small area of yolk-free cytoplasm, the **germinal disc,** that will develop into the embryo proper. **Cleavage** and development occur only within the disc, and the yolk supplies required nutrients and energy to support the embryo.

As the egg cell passes through the reproductive tract of the female chicken, it is modified in several ways. A solution of the protein albumin is added to make up the white of the egg. This layer provides physical protection for the delicate embryo, and antimicrobial substances in it protect the embryo from bacterial infection. The albumin itself will provide the chick with a store of amino acids to draw on for protein synthesis, and the liquid is the chick's only source of water. Later on, two membranes and then a porous shell are added. The hard shell protects the embryo and its food supply but also allows gas exchange with the environment.

The egg of a bird is an **amniote** egg. Unlike a fish or amphibian egg, it is fully adapted to terrestrial reproduction. Reptiles and mammals also produce amniote eggs in which the embryo is protected in a miniature watery environment within the amnion membrane.

- Examine an unfertilized egg on demonstration. Identify the germinal disc, which is normally on the upper surface of the yolk.

Is fertilization in the chicken internal or external?

____

If the egg cell has been fertilized, it will begin cleavage and start to develop right in the hen's oviduct. The cleavage furrows will not be able to pass through the massive yolk, however, restricting cell division to the germinal disc. Early cleavage in a bird egg is shown in Text Figure 28–16. Since eggs are held in the oviduct for various lengths of time before being laid, the exact stage of development at the time of laying is different for each individual egg, but usually cleavage, blastula formation, and gastrulation have already been accomplished before the egg is laid.

How would an unfertilized egg differ from one that has been fertilized?

____

____

Is there any nutritional advantage to eating fertilized eggs?

____

____

Early in development, the cells of the **blastoderm** split into two layers: The upper layer is the **ectoderm** and the lower layer is the **endoderm.** A thickening, called the **primitive streak,** then develops in the center of the blastoderm. Here cells migrate into the space between ectoderm and endoderm, and form the embryo's **mesoderm.** Mesodermal tissue near the center of the embryo pinches off to form the **notochord,** and the pairs of **somites** that lie in rows along each side of the notochord. The notochord induces the ectoderm above it to form a **neural plate.** The plate develops folds, which eventually meet and fuse to form the **neural tube.** The neural tube in turn differentiates into the brain and the spinal cord. Finally the embryo lifts, and the ectoderm folds beneath it so that the embryo is completely covered by ectoderm and the primitive gut is enclosed by the endoderm. These changes are shown in Figure 9–3. When you have time, mark the ectoderm, mesoderm, and endoderm with different colored pencils (blue, red, and yellow, respectively) so that you can follow the developmental process more clearly.

Meanwhile, specialized cells in the area of the blastoderm surrounding the embryo form **blood islands,** which give rise to extraembryonic blood vessels and blood cells. The embryo develops a primitive heart, and blood soon begins to circulate between the embryo and the surface of the yolk.

Other cells in the blastoderm grow into protective **extraembryonic membranes.** Although these membranes are not part of the embryo proper, they are vital for its nourishment and development.

1. The **amnion** develops as a fold of tissue growing down over the embryo's head. It meets with another fold growing up over the tail, and fuses with it so that the embryo is completely enclosed in an **amniotic sac,** which is filled with amniotic fluid. The fluid acts as a protective cushion. The growing amnion is visible as a fold of tissue in the 48-hr chick.
2. The **yolk sac** grows out from the primitive gut, and eventually surrounds the whole yolk with a double membrane. Blood vessels in its walls are important for the nutrition of the growing embryo. The blood vessels you will see in the 48- and 72-hr chicks develop in the yolk sac and are important in respiration as well as nutrition at this time.
3. The **chorion** is the outer layer of the fold of tissue that forms the amnion and can be seen at 48 hr. It will eventually surround the amnion, embryo, and yolk; it protects the embryo and holds it in contact with the yolk. Later it fuses with the allantois to form the **chorioallantoic membrane,** the respiratory organ of the chick during later development. This membrane completely covers the embryo and yolk to day 12 of development.
4. Finally the **allantois** grows out as a saclike extension of the primitive gut, expands up over the embryo, and eventually surrounds the embryo and yolk with a double membrane. The outer layer of the allantois presses up against the chorion, and the inner layer against the amnion and yolk sac. Because the allantois is richly supplied with blood vessels, it allows gas exchange with the external environment. The wastes that accumulate during development, mostly in the form of uric acid, are stored in the interior of the allantois until the chick hatches. The allantois can be seen as a small sac in the 72- and 96-hr chick embryo.

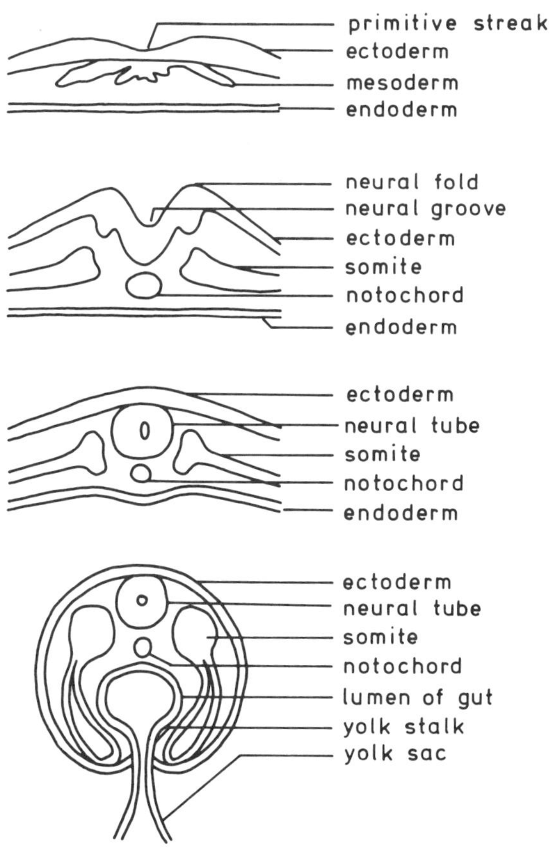

Figure 9–3. Chick embryo in cross section. The notochord of the gastrula induces part of the ectoderm to invaginate and form the neural tube. The brain and nervous system will eventually differentiate from the neural tube. You can color-code structures derived from the primary germ layers: ectoderm = blue, mesoderm = red, and endoderm = yellow.

Important cellular processes underly these visible changes during development. Cells must divide, stop dividing, and even die at the appropriate time. They must move about, recognize, and adhere to their neighbors properly. Differentiation of cells depends on their patterns of movement and position. It is often brought about by **embryonic induction** in which one cell alters the fate of another through complex interactions.

Live Chick Embryos

Each pair of students will be provided with a chick embryo that has been incubated for 33, 48, or 72 hr. You are responsible for understanding all of these stages, so study the prepared slides or another student's embryo to see all the structures.

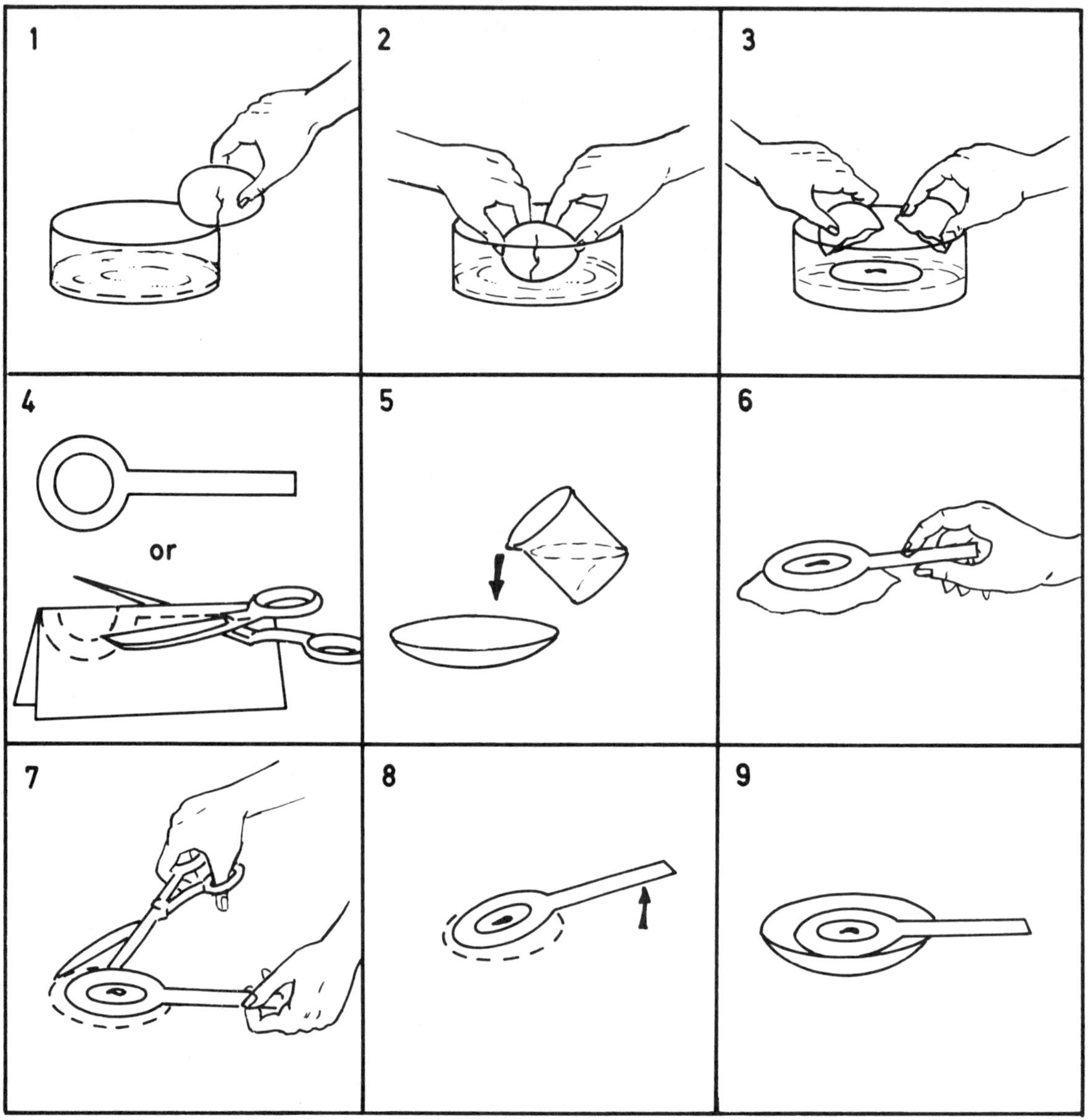

Figure 9–4. Handling live chick embryos. The delicate embryo must be kept warm and moist. The egg is opened in physiological saline and the embryo is transferred with a paper spoon.

Follow steps 1, 2, and 3 in Figure 9–4 to open your incubated egg.

- Add a little warmed saline (0.9% NaCl) to a finger bowl. (Saline may be omitted if you work quickly.)
- Take an egg from the incubator or tray, hold it in exactly the same orientation, and carry it carefully to your desk.
- Turn the egg sideways and wait 1 min for the yolk to adjust.
- Tap the egg on the edge of the dish until the shell cracks.
- Holding the egg in the saline, pry the shell apart at its bottom without disturbing the upper surface of the yolk.

If your embryo is 72 hr old, you should see the embryo and its beating heart immediately. If you do not see your embryo, ask your instructor for help. Either the egg failed to develop, or the yolk is turned upside down so that the embryo is hidden.

After you have observed as much structure as you can in the dissecting microscope, remove the embryo from the yolk, and transfer it to a watch glass so that you can see it more clearly. This process* (shown in steps 4–9 of Figure 9–4) is not difficult if you work carefully, but the top of the yolk *must not be covered by saline.*

- Fold a piece of filter paper in half, and cut out a spoon with a hole just larger than the embryo.
- Add a teaspoonful of warm saline to a watch glass.

---

* Suggested by Professor A.W. Blackler, Cornell University.

- Lay the filter paper spoon on the yolk so that the embryo shows through the hole.
- Press down gently on the paper with your probe, if necessary, so that it is wet all around and sticks to the membrane.
- Cut the membrane in a circle around the outside of the filter paper to free the embryo.
- Lift the spoon slowly so that the embryo lifts from the surface of the yolk.
- Lay the spoon in the watch glass, and observe in the dissecting microscope.
- A drop of vital stain such as 1% neutral red may be added to bring out the embryo's structure.

33-Hour Chick (Figure 9–5)

IMPORTANT FEATURES: 12 somites, heart, vitelline veins, notochord, neural tube, optic vesicles, blood islands.

The embryo is a streak in the center of the germinal disc, and you will have to transfer it to a watch glass to see much.

- The **notochord** has formed in the mesoderm and has already caused the **neural tube** to form in the neurectoderm by embryonic induction. This process begins and is more advanced in the head region. (See Text Figure 28–19.)
- The brain has begun to differentiate, and the **optic vesicles** are developing.
- There are about 12 **somites** along the sides of the notochord, which induced them to form from mesoderm. They will form muscle and part of the skeleton.
- Because the tail region develops more slowly than the head regions, it may still show the primitive streak that was the very first sign of development.
- The **vitelline veins** have joined to form the primitive **heart.** It might be feebly contracting, but there is no blood flow yet.
- If you are looking at a stained slide, notice the **blood islands** in the germinal disc outside of the embryo itself. They will give rise to the blood cells and some of the vessels.
- The **amnion** and **chorion** are beginning to develop as a fold of tissue growing down over the head region. It will eventually contact the posterior fold to enclose the embryo.

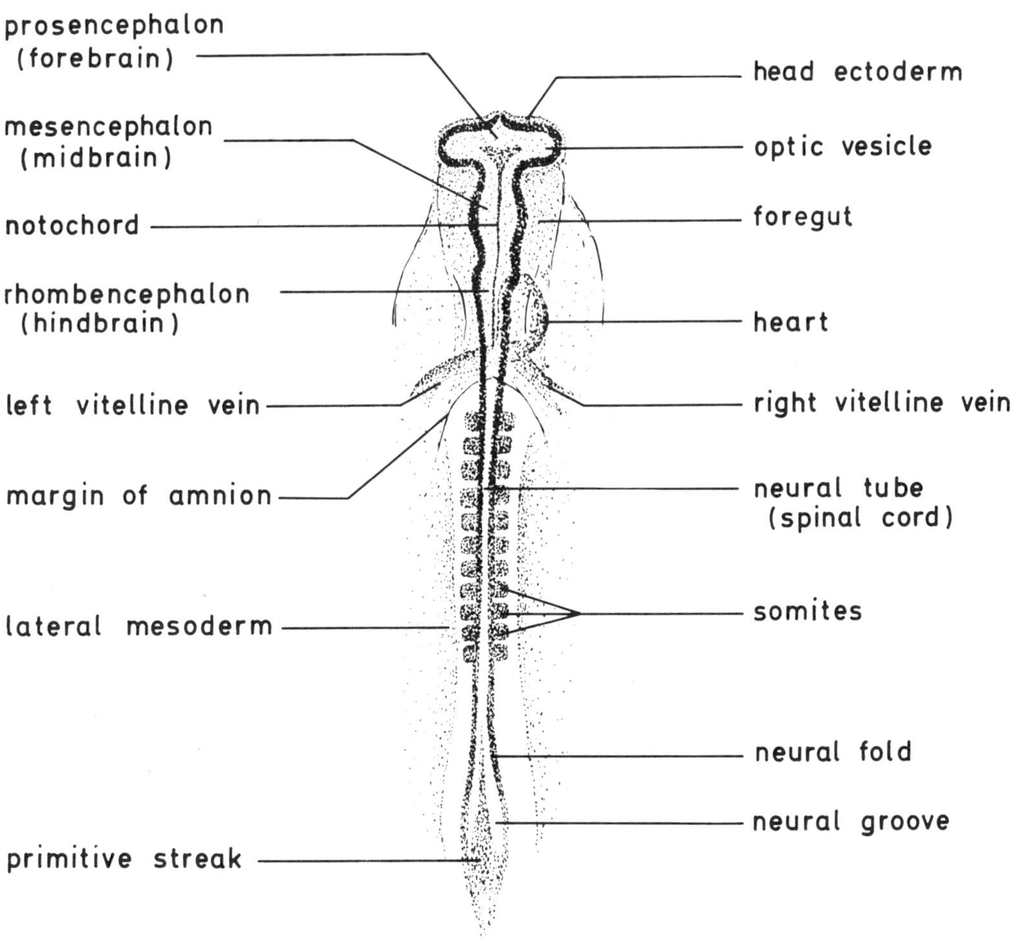

Figure 9–5. Whole mount of a 33-hour chick embryo. The anterior end of the neural tube has begun to differentiate while the posterior end has not yet formed from the neural groove.

- Finally you may be able to see the foregut, which is growing toward the head region from underneath the embryo as an outpocketing of the endoderm. It will later extend toward the tail region as well.
- Examine the demonstration of a vertebra to be sure that you understand the relationship of the notochord and the spinal cord. The notochord gives rise to intervertebral discs, and the spinal cord of the neural tube runs through the opening in the vertebra.

From which germ layer did the spinal cord develop? (ectoderm/mesoderm/endoderm) ______________

What structures will give rise to the vertebrae and back muscles? ______________

- Quiz your partner on the structures in Figure 9–5.

48-Hour Chick (Figure 9–6)

IMPORTANT FEATURES: 28 somites, eye with lens, distinct telencephalon and diencephalon, aortic arches, vitelline arteries, amnion and chorion.

This embryo has started to twist to the right so that the head region is now lying on the chick's left side. This is the process of **torsion,** which will continue until the whole embryo is lying on its left side. At the same time, the embryo is curling up in the process of **flexure** so that the neural tube takes the shape of a question mark with a bend in the region of the hindbrain. The forebrain has formed the **diencephalon** and **telencephalon.** Blood circulation usually begins at about the 16-somite stage.

- If you are viewing a living embryo, look for contractions of the heart and circulating blood.
- Trace the flow of blood from the **ventricle** of the heart through the **aorta,** and then through the **aortic arches** in the region below the head (pharynx).
- Identify the parts of the brain: The optic cup has caused embryonic induction of the **lens** of the eye by the ectoderm (see Text Figure 28–21), and the neural tube has induced the formation of the **otocyst,** or primitive ear. In each case a chemical signal called an **organizer** caused the overlying ectoderm to differentiate into an appropriate structure.

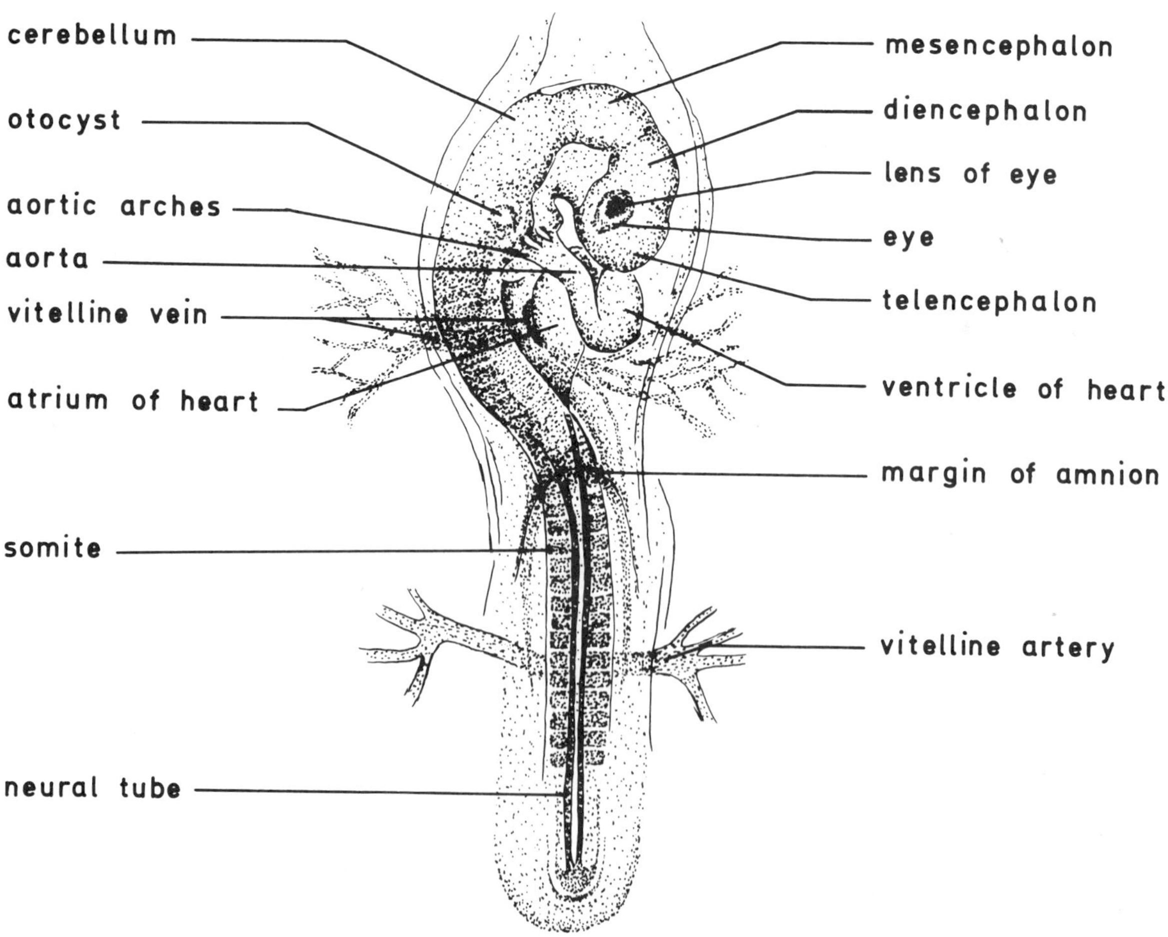

Figure 9–6. Whole mount of a 48-hour chick embryo. The embryo lies partly on its left side and is supported and nourished by an extensive extraembryonic circulation in the yolk sac. The amnion and chorion partly cover the embryo.

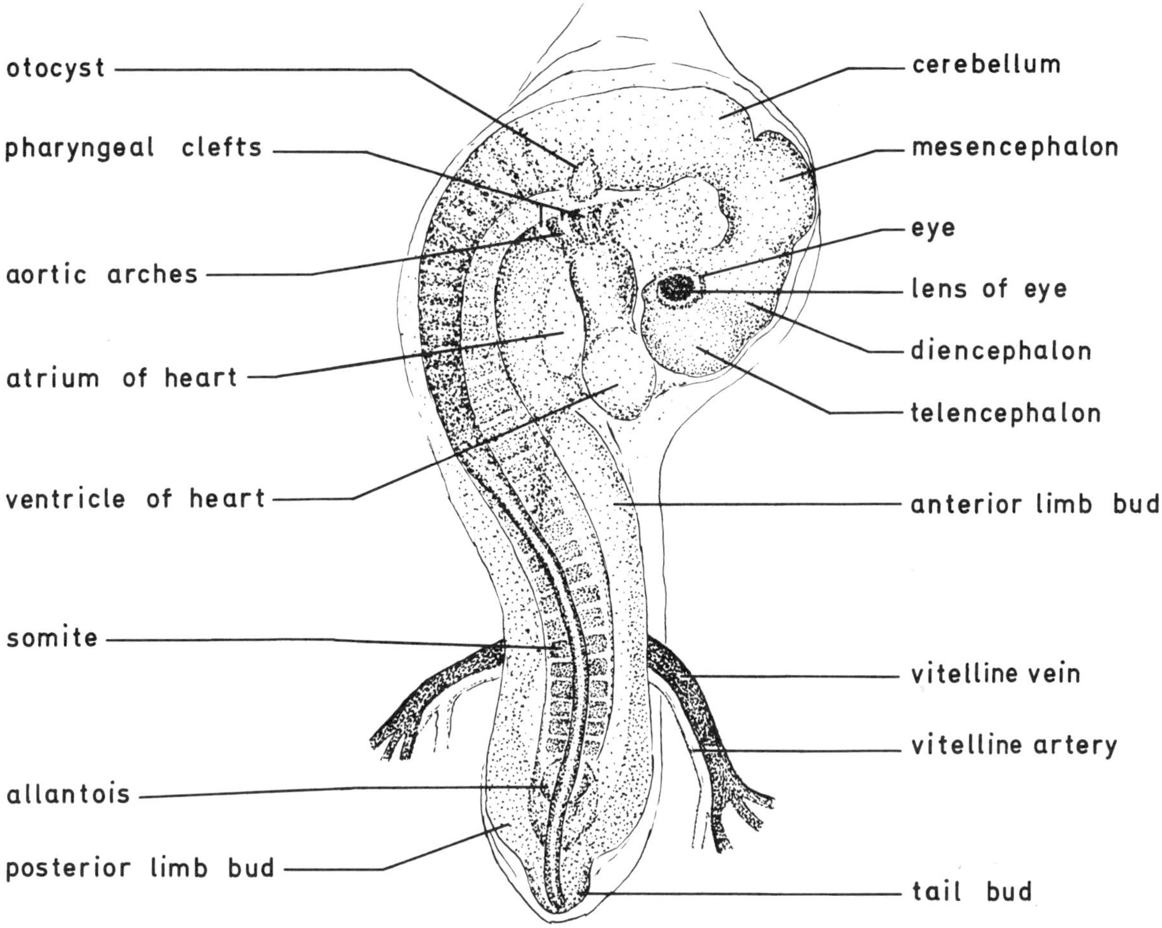

Figure 9–7. Whole mount of a 72-hour chick embryo. The embryo lies mostly on its left side and the allantois and limb buds are beginning to form.

- Note the **vitelline arteries** growing out over the yolk. They join the vitelline veins in capillary beds on the yolk surface.
- The margin of **amnion** and **chorion** can be seen above the vitelline arteries.

How many somites are still showing their dorsal sides? ______________

What structure induced the somites? ______________

- Quiz your partner on the structures in Figure 9–6.

72-Hour Chick (Figure 9–7)

IMPORTANT FEATURES: 36 somites, extraembryonic circulation in yolk sac, mesencephalon, cerebellum, otocyst, pharyngeal clefts, limb and tail buds.

The chick is now almost completely on its left side (torsion). Flexure of the brain is prominent, and it is differentiated into distinct regions. There are too many somites to count easily (about 36).

- Note the optic cup, the lens, and the **otocyst,** or primitive ear. The **atrium** and **ventricle** of the heart are greatly enlarged, and are actively pumping blood through the vitelline and embryonic arteries.
- There are several **pharyngeal clefts** and **pouches.** These structures appear at some stage in all vertebrate embryos. They are a carryover from the gills of a fishlike ancestor. In terrestrial vertebrates they will give rise to structures such as the middle ear and Eustachian tubes.

What membranes completely enclose the embryo itself?

____

- Below the heart you can see the **anterior limb buds,** which will develop into wings, and near the embryo's tail you see the **posterior limb buds,** which will form the legs.
- You may be able to see the small saclike **allantois** in the region of the posterior limb buds, but it is probably hidden by the tail, and is better seen in later embryos.
- If you can successfully transfer your embryo to a watch glass, observe circulation in the blood vessels under low power with your microscope.

Where does the circulating blood pick up oxygen and nutrients?

- Quiz your partner on the structures in Figure 9–7.

96-Hour Chick

IMPORTANT FEATURES: 41 somites, allantois, limb and tail buds.

This embryo has developed about 41 somites and the yolk is partly covered with prominent blood vessels.

- The flexure of the body posterior to the brain is now obvious. The eye is easier to see because it is pigmented, but the ear is less distinct because the tissues have become much thicker.
- Remove the membranes surrounding the embryo so that you can see the heart and the limb buds. You can use a probe or teasing needle to straighten out the embryo so that you can see the pharyngeal clefts and pouches.
- The fluid-filled sac at the posterior end is the **allantois.** It will expand and eventually fuse with the chorion to become the highly vascularized chorio-allantoic membrane. This membrane remains close to the porous shell of the egg.

What are the two functions of the allantois?

When the chick hatches, the chorioallantois will remain behind as a waste-filled membrane stuck to the shell, but its base within the embryo gives rise to the urinary bladder.

Later Development

Your instructor may demonstrate some older living embryos that are even further along in development. All of the major developmental processes happen during the first 6 days of development. The embryo then grows and matures during the rest of the incubation period until it hatches at 21 days, and begins its independent existence as a wobbly, fluffy chick.

- When you are finished with your work, discard your debris in the EGG WASTES container, clean your dissecting instruments, return the prepared slides, and store your microscopes properly.

REFERENCES AND FURTHER READING

Browder, L.W. *Developmental Biology,* 2nd ed. Philadelphia: Saunders College Publishing, 1984.

Harvey, E.B. *The American Arbacia and Other Sea Urchins.* Princeton, NJ: Princeton University Press, 1956.

Karp, G., and Berrill, N.J. *Development,* 2nd ed. New York: McGraw-Hill, 1981.

Patten, B.M. *Early Embryology of the Chick,* 5th ed. New York: McGraw-Hill, 1971.

Saunders, J.W. *Developmental Biology: Patterns, Problems and Principles.* New York: Macmillan, 1982.

Journal Articles

Caplan, A.J., and C.P. Ordahl. "Irreversible gene expression. Model for control of development." *Science* 201:120, 1978.

Davidson, E.H., B.R. Hough-Evans, and R.J. Britten. "Molecular biology of the sea urchin embryo." *Science* 217: 17, 2 July 1982.

DeRobertis, E.M., and J.B. Gurdon. "Gene transplantation and the analysis of development." *Scientific American,* 74, December 1979 (#1454).

Edelman, G.M. "Cell adhesion molecules: A molecular basis for molecular form." *Scientific American,* 118, April 1984.

Epel, David. "The program of fertilization." *Scientific American,* 128, November 1977 (#1372).

Garcia-Bellido, Antonio, Peter A. Lawrence, and Gines Morata. "Compartments in animal development." *Scientific American,* 102, July 1979 (#1432).

Goodman, C.S., and M.J. Bastiani. "How embryonic nerve cells recognize one another." *Scientific American,* 58, December 1984.

Gordon, R., and A.G. Jacobson. "The shaping of tissues in embryos." *Scientific American,* 106, June 1978 (#1391).

Levi-Montalcini, Rita, and Pietro Calissano. "The nerve-growth factor." *Scientific American,* 68, June 1979 (#1430).

Marx, J. "Electric currents may guide development." *Science* 211:1147, 13 March 1981.

Peakall, David B. "Pesticides and the reproduction of birds." *Scientific American,* 72, April 1970 (#1174).

Rahn, Hermann, Amos Ar, and Charles V. Paganelli. "How bird eggs breathe." *Scientific American,* 46, February 1979 (#1420).

Stent, Gunther S., and David A. Weisblat. "The development of a simple nervous system." *Scientific American,* 118, January 1982.

Wolpert, L. "Pattern formation in biological development." *Scientific American,* 154, October 1978 (#1409).

Student Name ______________________ **Date** ______________

TABLE 9–2. Observations on Sea Urchin Development at __________ **°C**

Genus ______________

Species ______________

Time	Time after Fertilization	Stage	Appearance
	0 min		
	min		
	min		
	min		
	min		
	min		
	min		
	min		
	min		
	min		
	hr		
	hr		
	hr		
	hr		
	hr		
	hr		

TOPIC 10

Reproduction of Eukaryotic Cells

OBJECTIVES

When you have completed this topic, you should be able to:

1. Describe the four stages of the cell cycle.
2. Draw the structure of a chromosome before replication and after replication.
3. Distinguish between sister chromatids and homologous chromosomes.
4. Describe what you look for in the microscope during interphase and in each stage of mitosis.
5. Identify in the microscope which stage of mitosis an animal or plant cell has reached.
6. Describe the main differences between mitosis in a plant cell and in an animal cell.
7. State when and in which tissues meiosis occurs.
8. Use models to show how meiosis differs from mitosis in terms of chromosome movement.
9. Explain in terms of DNA synthesis, synapsis, crossing over, and cytokinesis how meiosis differs from mitosis.
10. Identify the haploid and diploid stages of meiosis in a diploid organism such as yourself.

TEXT REFERENCES Chapter 12

KEY TOPICS

chromosomes	mitosis	spermatogenesis
haploid and diploid cells	meiosis	oogenesis
cell cycle	crossing over	fertilization

INTRODUCTION

Cells come into existence through the division of their parent cells, and most cells divide in turn to produce daughter cells. Division of the nucleus usually occurs by **mitosis,** in which the genetic material is duplicated so that exactly one copy is passed on to each daughter cell. Mitosis is generally followed by **cytokinesis,** or cytoplasmic division, in which the rest of the cell divides in half to form two new cells. Sometimes cytokinesis does not occur, and a cell with many nuclei may form. In animals such a cell is called a **syncytium,** and in plants it is termed a **coenocyte.** Mitosis is the type of cell division carried out by the cells of a developing embryo and by ordinary body or **somatic** cells.

Mitosis is part of the **cell cycle** and is the only period during which the chromosomes are visible in the light microscope. The rest of the cycle is known as **interphase.** Trace one turn of the cycle in Figure

10–1. During G_1 (first gap) of interphase, the decision to begin the process of cell division is made. DNA replication occurs in the **S** (synthesis) period. At this time each chromosome doubles so that it consists of two identical **sister chromatids** joined in the region of the **centromere.** Although the chromosomes are uncoiled and therefore invisible in the light microscope at this time, chemical techniques can be used to show that the chromosomes are replicating. The replicated chromosomes that appear at the beginning of mitosis are double-stranded; one strand will enter each of the daughter cells. The two daughter cells produced by mitosis are exactly equivalent to the parent cell and can be thought of as carbon copies. During the G_2 period preparations for division are completed and the cell enters the brief **M** period when discrete chromosomes can be observed to actually divide.

In animals and plants that reproduce sexually, two sex cells or **gametes** from the *two parents* fuse during fertilization to form the **zygote.** The zygote will always have twice as much genetic material as the gamete. Therefore, at some stage before the next fertilization, the genetic material must be reduced by half so that the amount of nuclear material stays the same from one generation to the next. This reduction in nuclear material is the process of **meiosis.** Only specialized cells, **germ cells,** carry out meiosis. These cells occur in the gonads of animals and in the sex organs of plants and fungi. The gametes or spores resulting from meiosis contain only one half of the genetic material of their parent cells. A second important consequence of meiosis is that the genetic material in the gametes or spores has been shuffled, resulting in **genetic recombination.** The offspring, like the gametes, will be genetically different from their parents. This heritable variation among offspring is the major advantage of sexual reproduction.

In this exercise you will study both mitosis and meiosis. It is essential that you thoroughly understand both processes and the difference between them before undertaking the topic of genetics.

There are two important terms associated with the genetic material that you need to understand from the outset:

Haploid: A haploid cell contains one unique set of genetic information. It includes one chromosome of each type found in a particular organism and it has the same set of chromosomes that would be found in a gamete or spore from that organism. **N** is the symbol for the haploid number and it is a characteristic number for each species (see Text Table 13–1).

Diploid: A diploid cell contains two sets of genetic information. It includes a pair (two) of chromosomes of each type found in a particular organism, and it has the same set of chromosomes that would be found in the ordinary body cells of an animal or plant. The diploid number is always **2N** (see Text Table 12–1).

Most organisms have either haploid or diploid cells, although other levels of ploidy (3N, 4N, etc.) are possible.

With these two terms, we can restate the concepts of mitosis and meiosis. In mitosis, a cell produces two daughters identical to itself in ploidy; typically a diploid cell produces two genetically identical diploid daughters. In meiosis, a diploid cell is converted to haploid cells, genetically different from itself.

Many organisms can carry out **asexual reproduction** as well as sexual reproduction. Asexual

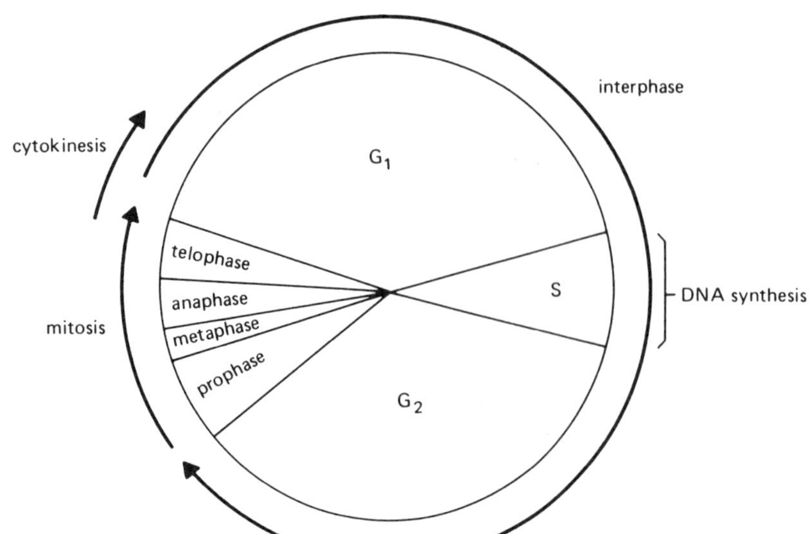

Figure 10–1. The cell cycle. The decision to divide occurs in G_1, the chromosomes are replicated in S, the preparations for cell division are made in G_2, the nucleus divides in mitosis, and the cytoplasm divides in cytokinesis.

reproduction always involves mitosis rather than meiosis and the offspring are genetically identical to the *single parent.*

MITOSIS IN AN ANIMAL CELL

An animal cell about to undergo mitosis has already replicated its DNA during the S period of interphase.

How many strands does each chromosome have?

____

What are the strands called? ______________

The **centrioles,** pairs of organelles, found only in animal cells, have also replicated just prior to S. The daughter centrioles move to the opposite ends or **poles** of the cell, and their position determines how the chromosomes will move during mitosis. The chromosomes always move toward the centrioles, and the cell always divides in a plane perpendicular to a line joining them. (See Text Figure 5–17.)

Centrioles have an interesting structure closely related to that of cilia and flagella. They are little cylinders made of nine triplets of microtubules, and each centriole is made of two such cylinders at right angles. They determine the orientation of the **mitotic spindle,** a framework of microtubules that extends from one centriole to the other. It includes **polar fibers,** which extend from the poles to the center of the cell and **kinetochore fibers** to which the chromosomes are attached. During mitosis kinetochores form on the outer surface of each chromatid in the region of the centromere and are the attachment points for the fibers that will position and then move the chromosomes to the poles of the cell. The position of the centrioles determines the orientation of the **astral rays,** microtubules extending out from the centrioles in all directions. The function of these astral rays is not known.

The centrioles, mitotic spindle, and astral rays are collectively called the **mitotic apparatus,** and take up much of the interior of the cell when it is in mitosis (see Text Figure 12–7). Mitosis requires a great deal of energy, so there are many mitochondria associated with the mitotic apparatus to provide ATP for synthesis and chromosome movement.

- View a prepared slide of animal cells undergoing mitosis, such as the cells in whitefish embryos, first under low, and then under high power in your microscope.
- Locate cells that are undergoing mitosis, some of which have large, prominent mitotic spindles and condensed chromosomes.
- Using the following descriptions and Figure 10–2 as your guide, locate cells that represent interphase, each of the stages of mitosis, and cytokinesis.

Interphase (G_1–S–G_2)

During interphase, the genetic material is in the form of greatly extended fibers of DNA that form a tangled mass called **chromatin.** The fibers are too small to be seen in the microscope so the nucleus looks homogeneous. The nucleus is bounded by the double **nuclear membrane,** and contains **nucleoli,** which are centers for the synthesis of RNA. They may not be visible in every cell.

How many nucleoli do you see in the interphase cells?

____

Prophase (M)

As the cell enters prophase, the chromatin starts to condense into discrete chromosomes, the nucleoli disappear, and the nuclear membrane breaks down. By the end of prophase, the centrioles have migrated to opposite poles, the chromosomes are moving to the center of the cell, and the mitotic spindle starts to appear.

Metaphase (M)

The spindle forms, and the chromosomes take up positions on the **metaphase plate,** an imaginary plane in the middle of the cell halfway between the two poles. They are located on the outside of the spindle, so they would be arranged in a circle if you viewed them from one of the poles. Each double-stranded chromosome is attached to kinetochore fibers in the region of its centromere, and its flexible arms are free to move about.

Anaphase (M)

Metaphase ends and anaphase begins when the centromere of each chromosome finishes dividing. Each sister chromatid now has its own centromere, and is called a chromosome, so that there are twice as many chromosomes as there were during metaphase.

If a certain cell has three pairs of chromosomes ($N = 3$), how many chromosomes does the cell have during metaphase? ______________

How many chromosomes does it have during anaphase? ______________

Kinetochore fibers interact with polar fibers to pull the chromosomes along by their kinetochores, with the

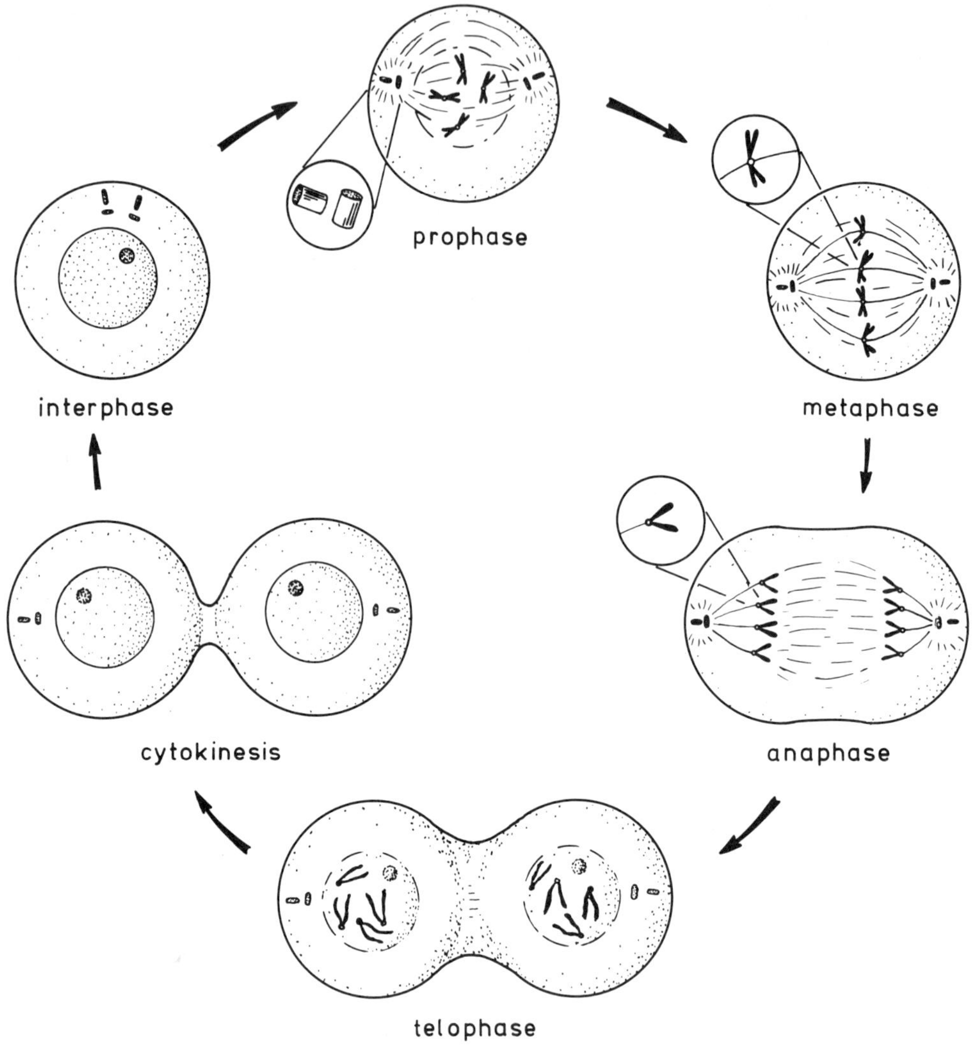

Figure 10–2. Mitosis. The stages of mitosis in an animal cell. Since mitosis is a continuous process, each stage blends into the next. Note the centrioles and the astral rays not found in plant cells.

arms of the chromosomes trailing passively behind. A chromosome without a centromere region and kinetochore will not move during anaphase, and is randomly left behind in one of the two daughter cells.

Telophase (M)

The chromosomes gather at the poles in telophase, and begin to form the daughter nuclei. Since each original chromosome in the cell contributed a chromosome identical to itself to each new nucleus, the two daughter nuclei will be identical to the nucleus of the parent cell. The nuclear membrane begins to reform, the nucleoli reappear, and the chromosomes begin to unwind and disappear from view. At the end of telophase, the nuclei look like interphase nuclei, and in fact have entered interphase of the next cell cycle.

Cytokinesis (M–G_1)

Cytokinesis, or division of the cytoplasm, usually accompanies telophase. In animal cells microfilaments attached to the membrane constrict in a ring around the middle of the cell, and eventually pinch the cell in two. This ring is called the **cleavage furrow.** The cytoplasm and all its constituents are passively divided so that each cell gets *about* half the materials and organelles in the cell.

Why couldn't the chromosomes also be passively divided when the cell pinches in two?

____

____

MITOSIS IN A PLANT CELL

- View a prepared slide of plant cells undergoing mitosis, such as the root tip cells of an onion, first under low and then under high power. In the onion root tip, look at the cells near, but not at, the very tip of the root in the region marked "X" in Figure 10–3. Avoid the cells in the very center.

- Locate cells in interphase, each stage of mitosis, and in cytokinesis. If the section does not happen to pass through the nucleus of a cell, that cell will appear to have no nucleus, and should be skipped.

Note the following differences between mitosis in the plant cells compared with mitosis in the animal cells that you have seen:

1. **Interphase.** The nuclei look similar to the animal cell nuclei, but may have a different number of nucleoli.

How many nucleoli do the onion cells have?

____

2. **Prophase.** There are no centrioles in plant cells, and you will not see any astral rays. Otherwise the events of prophase are similar.
3. **Metaphase.** Events are similar to those in the animal cell. You will see the spindle but no astral rays.
4. **Anaphase.** Events are similar to those in the animal cell.
5. **Telophase.** Events are similar to those in the animal cell. In plants, telophase cells can be easily distinguished because you will see two nuclei within a cell that is about the same size as the other cells that have only one nucleus. The two new daughter cells together are the same size as the parent cell, and look like mirror images. Growth of the cells will occur later on.
6. **Cytokinesis.** This process is completely different in plant cells. The parent cell does not pinch in or change shape at all. Instead, membrane vesicles move to the metaphase plate and coalesce to form a **cell plate** and eventually a **middle lamella** dividing the cell into two daughters. Cellulose is formed from material within the cell plate and is laid down on either side of the middle lamella to form cell walls between the two new cells. Each cell then begins to elongate in its long dimension until it is as big as the parent cell, and can enter another cycle of mitosis.

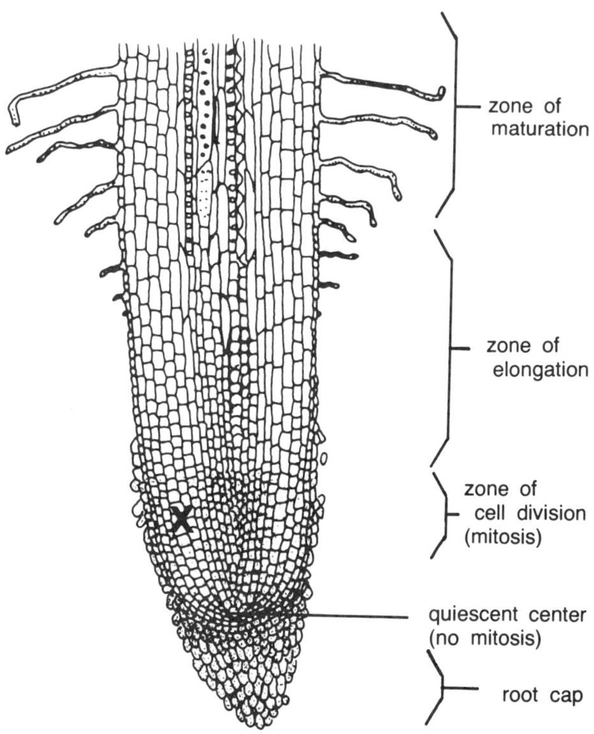

Figure 10–3. Onion root tip. Look at the position marked X to find dividing cells.

Working with your partner, determine the stage of mitosis in 100 cells, and record your results in Table 10–1 at the end of this topic.

- Move your slide until you have the field filled with an area where there are many cells undergoing mitosis (you should be viewing under high power).
- Place the pointer on a vertical row of cells, or make a mental note as to which row you are going to count.
- Indicate the stage of each cell in turn to your partner as you go down the row. Do not leave out any cells unless there is no nucleus visible at all.
- As each cell is staged, your partner should make a check next to that stage in the table.
- When you have counted about 50 cells, change places, and let your partner stage the rest.
- Continue until you have counted exactly 100 cells.

It is known that onion root tip cells take about 16 hr to complete the mitotic cycle. By determining the percentage of cells in each stage of mitosis and in interphase, you can calculate the amount of time spent in each stage. For example, if 10 cells out of 100 were found to be in prophase, the percentage of cells in prophase is $10/100 \times 100 = 10\%$. This shows that any one cell spends 10% of the time in prophase, so it spends 0.10×16 hr or 1.6 hr (1 hr and 36 min, since 0.1 hr = 6 min) in that stage.

- Calculate the percentage of cells in each stage, and enter the results in Table 10–1. (There is no calculation if you counted exactly 100 cells; just enter the number of cells.) You may be asked to use the combined totals of the class for your calculation.
- Calculate the hours and minutes spent in each stage, if the entire cell cycle takes 16 hr (0.1 hr = 6 min).

MECHANICS OF MITOSIS AND MEIOSIS

To clarify the difference in chromosome movement during mitosis and meiosis, you will use pipe cleaners of various lengths to represent the sister chromatids of the chromosomes, and beads to represent the centromeres. A given chromosome may have one or two chromatids, depending on whether it has replicated. The number of chromatids does not affect the fact that it is a single chromosome. The chromatids are not considered chromosomes in their own right until they have separated from each other and each possesses *an unshared centromere.*

The number of complete chromosomes found in a haploid cell is the **haploid number** or **N.** The number of complete chromosomes found in a diploid cell is the **diploid number** or **2N.** In Figure 10–4 you see a diploid cell with four chromosomes undergoing mitosis compared with another cell containing four chromosomes undergoing meiosis. In each case the chromosomes have first replicated in S phase so that they are double. The number of chromosomes in each cell after replication is still four. Note that in meiosis the chromosomes do not act independently as they do in mitosis but join together in two **homologous pairs** during meiosis I, the first meiotic division. In a cell with many chromosomes, all of them would join in homologous pairs. Members of a pair will look alike in size and shape and carry the same *sequence* of genes, but they are not identical at the molecular level.* During the first meiotic division, the two members of the pair separate, but the chromosomes do not split in half as they do in mitosis, and they remain double-stranded. Thus there is no need for DNA synthesis before meiosis II. Since the daughter cells receive only one set of chromosomes (i.e., two chromosomes), they are haploid (N) rather than diploid (2N). The two sister chromatids of each chromosome then separate in meiosis II so that each daughter cell receives only single-stranded chromosomes (again two chromosomes). The four daughter cells produced when a cell undergoes meiosis are usually not identical, in contrast to the carbon-copy daughters produced during mitosis. These four daughter cells are also haploid (N) rather than diploid (2N), and contain only one set of the genetic material.

* Human cells contain 22 pairs of chromosomes plus two sex chromosomes (XX or XY). The sex chromosomes X and Y do not look alike, but they act as a pair during meiosis in the male.

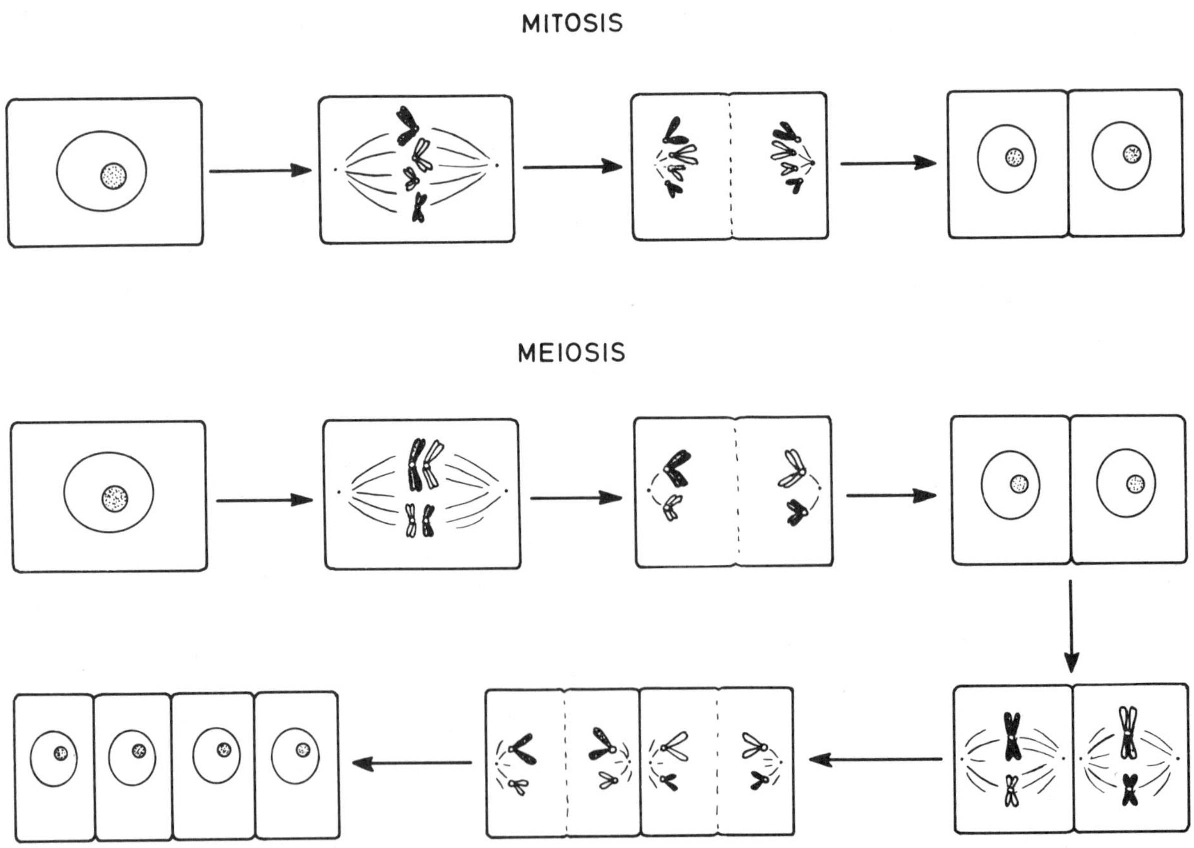

Figure 10–4. Mitosis and meiosis compared. A cell with 4 chromosomes (2N = 4) undergoes mitosis and meiosis very differently. Mitosis produces two daughter cells that are identical to the diploid parent, whereas meiosis produces four haploid cells that are genetically different from the parent cell.

- Insert two identical pipe cleaners into a bead to construct a chromosome that has already replicated.
- Make a second, **homologous chromosome** from pipe cleaners the same length as those in the first chromosome. The second chromosome does not need to be the same color as the first, but it must be the same length, and its sister chromatids must be identical to each other (they are carbon copies).
- Turn to Figure 10–8 at the end of this topic, which shows parent and daughter cells in metaphase, and place each chromosome on a *separate* spindle fiber, as it would appear in mitosis.
- Take another bead for each chromosome to represent its second centromere.
- Separate the sister chromatids and give each one a centromere so that it is a new full-fledged chromosome.
- Separate the two daughter chromosomes, placing one in each daughter cell, and repeat the process for the second homologous chromosome. Mitosis is complete.

Are the chromosomes in the daughter cells genetically identical to those that you started with except that they have not yet replicated? ____________

If not, ask your instructor for help!

During mitosis, each chromosome acts independently, and there is never any pairing of homologous chromosomes. During meiosis, there is **pairing of homologous chromosomes** and the daughter cells formed are *not identical to the parent cell.*

- Form a homologous pair of replicated chromosomes that are the same length, but different colors. The sister chromatids of each chromosome must be identical in length *and* color.
- Place the chromosomes in the center of the spindle side by side on the same spindle fiber. The two chromosomes are in **synapsis,** which is a close side-by-side position, and their four strands form a **tetrad** or four-stranded structure. Synapsis is the key event in meiosis and will result in **genetic variation** in the daughter cells. (See Text Figure 12–12.)
- Move one whole chromosome to each of the daughter cells. You do not need more beads because the chromatids **do not separate** during the first meiotic division.

How many chromosomes does each daughter cell have? ____________

- Separate the sister chromatids of one chromosome, and give each one a bead to represent its centromere. The new chromosomes can now move apart into the daughter cells formed in meiosis II. Note that there was *no chromosome replication* before meiosis II. Otherwise, meiosis II is exactly like mitosis.
- Repeat the division for the other daughter cell so that you have formed the four daughter cells that result from meiosis II.

How many chromosomes are there in each cell resulting from meiosis? ____________

Are these cells haploid or diploid? ____________

What process will restore the diploid number of chromosomes in the next generation? ____________

- Starting with four replicated chromosomes, carry out a mitotic division.
- Starting with four replicated chromosomes, in which the homologous pairs are different colors, and in which one pair is longer so that you can tell the pairs apart, carry out the two meiotic divisions. Make a note of the chromosomes found in each of the four daughter cells.
- Arrange the chromosomes differently on the spindle in meiosis I so that the products after meiosis II will now be different than before.

How many different types of daughter cells can you create using both arrangements? ____________

If you know N for cells undergoing meiosis, you can predict that they will collectively produce 2^N different kinds of gametes *through chromosome reassortment alone.* (Any single cell can produce only 2 kinds of gametes through reassortment.)

- In Figure 10–8 draw in the chromosomes of a cell undergoing meiosis. Start with four chromosomes ($2N = 4$), two long and two short, and draw one member of each homologous pair with dotted lines for chromatids, and the other with solid lines.

MEIOSIS IN AN ANIMAL CELL

In animals, meiosis occurs only in the gonads.

Where does meiosis occur in a woman? ____________

What is this process called?

____

Where does meiosis occur in a man? ____________

What is this process called?

____

In males, the sperm are formed much as you illustrated in the section on mechanics. Each cell undergoing **spermatogenesis** will produce four sperm cells. Note that each sperm will have a genetically identical twin. In **oogenesis,** however, a single, large egg cell is produced which carries the energy stores for the development of the zygote after fertilization. Nuclear division occurs in the same way, but cytoplasmic division is very unequal. A nucleus with a small amount of membrane and cytoplasm is discarded after meiosis I as the **first polar body.** A **second polar body** is discarded after meiosis II. If the first polar body undergoes division, there may be two or three polar bodies stuck onto the outside of the **ovum,** or egg cell. Thus, in the female meiosis gives rise to a single gamete, the ovum, and polar bodies that have no function and degenerate. Figure 10–5 illustrates the processes of oogenesis and spermatogenesis (see also Color Plate 4).

The roundworm *Ascaris* has large, easily removed ovaries in which the stages of meiosis can be seen. This parasitic organism is easily obtained from the intestine of horses. (Another species of *Ascaris* is a common human parasite in warm climates, and it is a serious human pest that sometimes causes intestinal blockage and lung infection.)

In *Ascaris,* as in many organisms, meiosis is not completed until after fertilization. Then meiosis I occurs, followed by meiosis II, fusion of the egg and sperm nuclei, and the first division of the zygote (see Figure 10–6). Cells in progressively later stages of

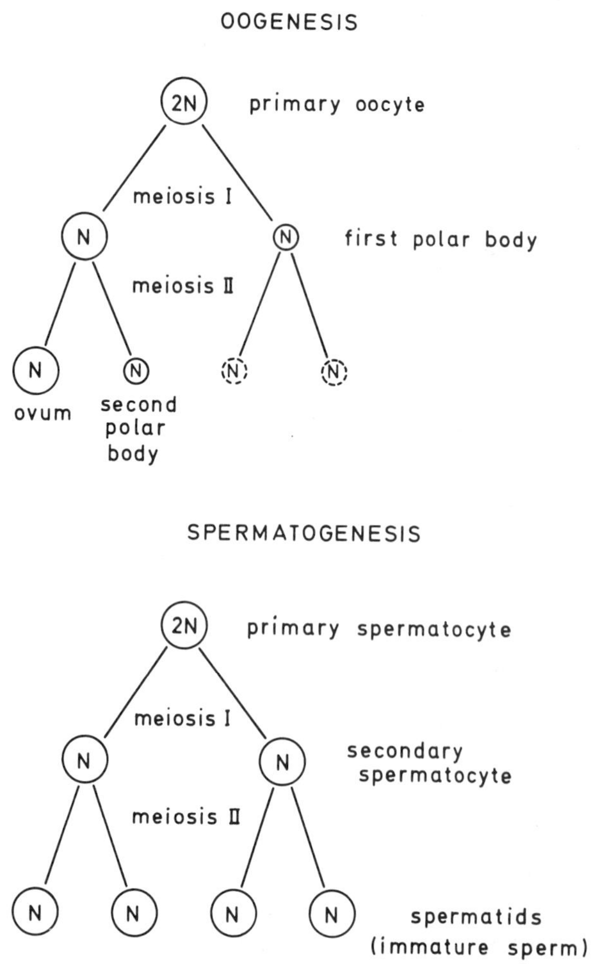

Figure 10–5. Oogenesis and spermatogenesis compared. Oogenesis produces one ovum or egg cell but spermatogenesis produces four sperm cells for each cell entering meiosis.

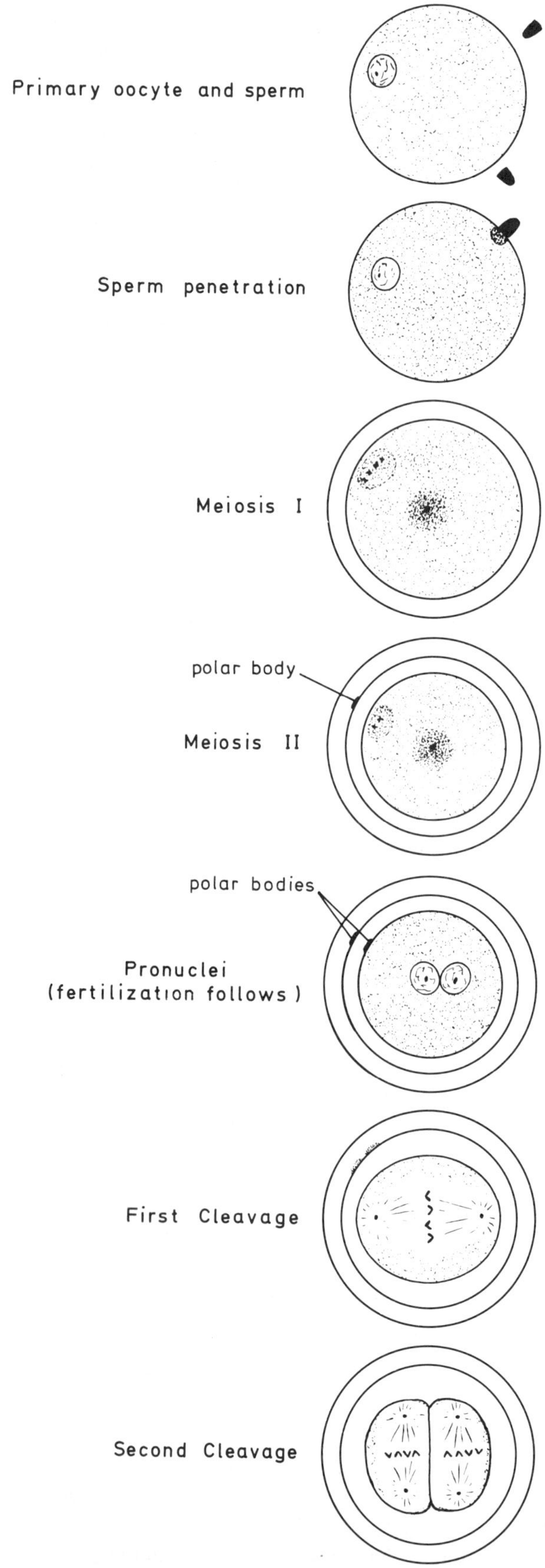

Figure 10–6. Meiosis in *Ascaris*. Fertilization precedes meiosis in this species.

meiosis and early development are found in sections of the oviduct and uterus farther and father from the ovary.

- Study meiosis in the prepared slide, noting that the earlier stages will be found in the sections at the top of the slide, which are from the part of the oviduct closer to the ovary. A set of slides rather than a single slide may be provided for your study of meiosis.

Sperm Cell

These triangular-shaped cells are small in relation to the egg cells of the female. The top of the sperm cell must enter the egg cell before meiosis can begin. *Ascaris* sperm lack flagella and reach the ova by amoeboid movement.

- Locate a free sperm cell.

Primary Oocyte (or Future Egg Cell)

This is the cell that will undergo meiosis after the sperm cell penetrates. It is a diploid cell, and has no shell around it.

- Locate a primary oocyte.
- Locate a primary oocyte into which a sperm has penetrated. The sperm head becomes modified into a sphere once it enters the oocyte.

Meiosis I

In *Ascaris* there are four chromosomes (2N = 4).

How many pairs of chromosomes will there be in meiosis I? ______________

Each pair of chromosomes forms a **tetrad** as the homologous chromosomes undergo **synapsis.** The tetrads move to one edge of the cell where the spindle forms during prophase I and metaphase I. This is the only time during the entire life cycle of a sexually reproducing organism that the members of a homologous pair of chromosomes come together. While the chromosomes are in synapsis, and the four sister chromatids are close together in the tetrad, sister chromatids of the two chromosomes frequently exchange parts of their arms in the process of **crossing over.** There is no gain or loss of genetic material, but the genes on one chromatid are exchanged one-for-one with those on the other chromatid. The shuffling of genes during crossing over results in more variety among the gametes than would otherwise occur. When crossing over is going on, there will be crosslike structures visible in the microscope, called **chiasmata.** Special techniques are required in order to see them.

Chromosomes tend to be "sticky" during crossing over and sometimes rejoin incorrectly. A chromosome may pick up a segment from a nonhomologous chromosome through **translocation** and is often defective as a result.

- Locate a cell in late prophase I in which there are prominent tetrads. The sperm nucleus may or may not be visible in the section of cell that you are viewing, but there cannot be a polar body on the cell surface.

The pairs of chromosomes separate during anaphase I and the two daughter nuclei form during telophase. Sometimes the tetrads fail to separate properly so that both members of the pair go to the same pole. This is called **nondisjunction** and has disastrous consequences for the offspring as you will see in Topic 12.

Cytokinesis I

During this process, a piece of membrane enclosing one daughter nucleus pinches off to form the first polar body. The large cell is now called the **secondary oocyte.**

- Locate a secondary oocyte that has a polar body attached to its surface.

Interkinesis

The cell prepares for meiosis II, and there is *no replication* of the chromosomes during this special interphase.

Meiosis II

The secondary oocyte enters prophase II during which the chromosomes become visible and the nuclear membrane, if it has reformed, disappears. In metaphase II the spindle forms at the cell surface, and the chromosomes line up at the metaphase plate. The sister chromatids separate and become chromosomes as the centromeres divide in anaphase II. Daughter nuclei re-form in telophase II.

Cytokinesis II

One daughter nucleus is pinched off to form the second polar body, and the large cell remaining is called the egg cell, or ovum.

- Locate an ovum with two polar bodies on its surface.

How many chromosomes does the nucleus of an *Ascaris* ovum have? ______________

Fusion of the Pronuclei

The haploid egg nucleus and sperm nucleus are called **pronuclei.** They become large and clear, and move toward each other.

- Locate an ovum in which there are two large pronuclei.

The pronuclei finally fuse to form the zygote.

- Locate a zygote with one large nucleus.

First Cleavage Division

The zygote begins the process of mitotic division, or **cleavage,** in which it begins to form the multicellular *Ascaris* embryo. After the first division there will be two daughter cells, after the second division four, and so on.

- Locate a zygote that has divided to form two cells of equal size with large nuclei.

How many chromosomes does each of these cells have? ______________

Meiosis in plants will be explained in later topics where you will see that it usually occurs in a plant sporophyte and produces spores rather than gametes.

SPERM CELLS

Your instructor will indicate the source of the living sperm that you will study. It may be from a bull, frog, or another source.

If frog sperm is used, you may be asked to remove the testes from the frog. Instructions for pithing the frog are given in Topic 25 and a diagram of the internal organs can be found in Topic 21.

- Make a wet mount of sperm suspension or a bit of tissue from the frog testis teased apart in Ringer's solution, and examine it under low, then high power in the microscope.

Are the sperm cells motile or immotile?

____

What is the shape of the sperm cells?

____

Why does a sperm cell contain many mitochondria?

____

____

____

Sperm are produced by maturation of spermatids within the seminiferous tubules of the testis. In spermatogenesis, haploid spermatids are produced from diploid spermatocytes (see Figure 10–5).

- Turn to Figure 10–7 at the end of this topic and review sexual reproduction by labeling it with these terms: chromosome, chromatid, haploid, 1N, diploid, 2N, meiosis, gamete, fertilization, and zygote.

When you are finished with your work, return all prepared slides, clean up your work area, and store your microscope.

REFERENCES AND FURTHER READING

Alberts, Bruce, D. Bray, J. Lewis, M. Raff, K. Roberts, and J.D. Watson. *Molecular Biology of the Cell.* New York: Garland, 1983.

Giese, A.C. *Cell Physiology,* 5th ed. Philadelphia: W.B. Saunders, 1979.

Journal Articles

Lazarides, Elias, and Jean Paul Revel. "The molecular basis of cell movement." *Scientific American,* 100, May 1979 (#1427).

Mazia, Daniel. "The cell cycle." *Scientific American,* 54, January 1974 (#1288).

Sloboda, Roger D. "The role of microtubules in cell structure and cell division." *American Scientist* 68:290, 1980.

Upton, A.C. "The biological effects of low-level ionizing radiation." *Scientific American,* 41, February 1982.

Student Name ______________________________ **Date** ________________

TABLE 10–1. Data on Mitotic Stages in Plant Cells

Make a check in the appropriate box for each cell staged. Then total the checks.

Number of Cells in Stage	Total	% in Stage	Hr and Min in Stage
Interphase:			
Prophase:			
Metaphase:			
Anaphase:			
Telophase:			

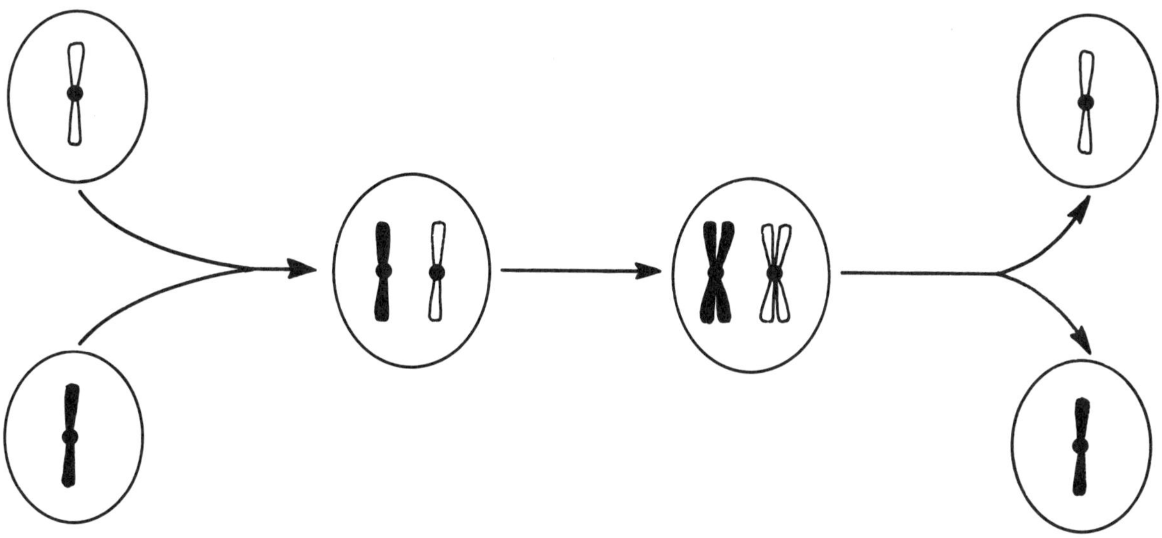

Figure 10–7. Sexual reproduction. Use the terms chromosome, chromatid, haploid, 1N, diploid, 2N, meiosis, gamete, fertilization, and zygote to label this diagram.

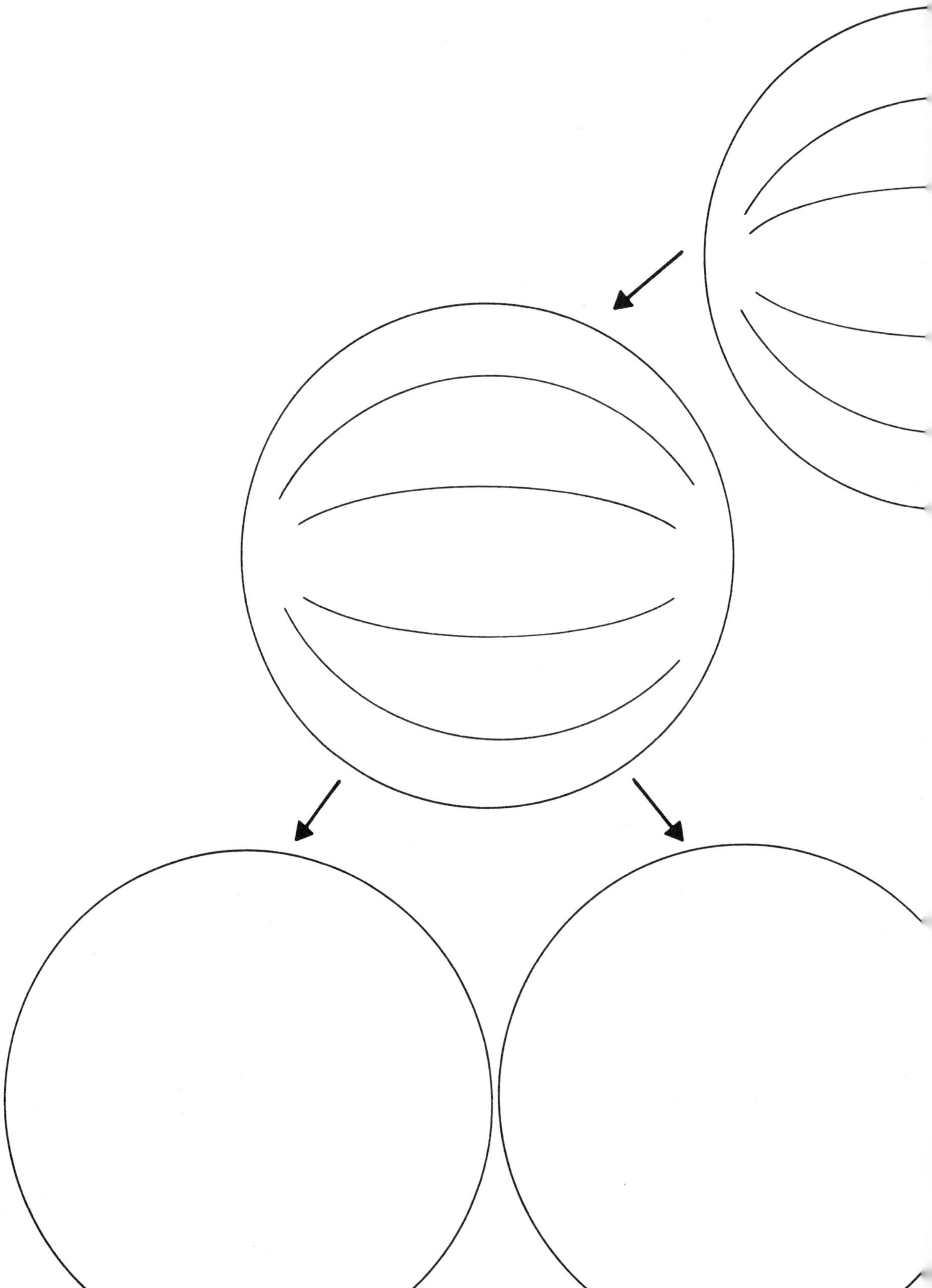

Figure 10–8. Diagram of cell reproduction.

Student Name ______________________ Date ______________

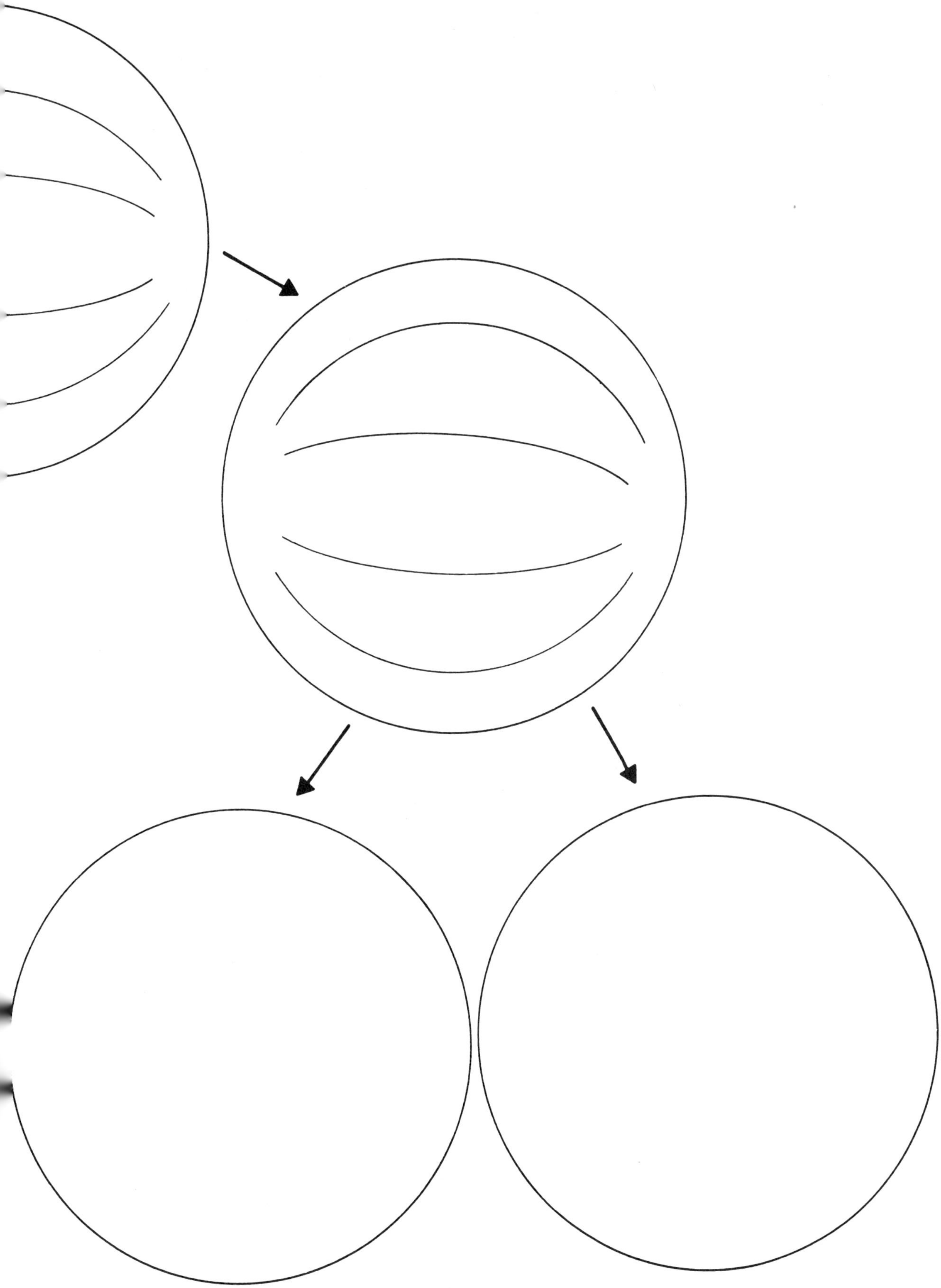

TOPIC 11

Mendelian Genetics

OBJECTIVES

When you have completed this topic, you should be able to:

1. Explain why *Drosophila* is a good experimental animal for genetic experiments.
2. Describe the morphology of or recognize each stage of the *Drosophila* life history, including the adult male and female.
3. Define the following terms: parental (P_1) generation, first filial (F_1) generation, second filial (F_2) generation, dominant, recessive, allele, homozygous, heterozygous, monohybrid cross, dihybrid cross, segregation, and independent assortment.
4. Use a Punnett square to describe a monohybrid cross and a dihybrid cross in which the two genes show independent assortment; give the genotypic and phenotypic ratios expected in each case.
5. Define linkage, and describe a dihybrid cross in which the two genes are completely linked; give the expected genotypic and phenotypic ratios for the offspring.
6. Define sex linkage; use a Punnett square to describe a cross of a homozygous female with a mutant male, and give the expected phenotype and genotype of the offspring.
7. Give the reciprocal cross and explain why the reciprocal cross is necessary when the mutation might be either dominant or recessive.
8. Given the F_1 or F_2 results from a cross, determine the mode of transmission of the genes involved.
9. Analyze the results using the chi-square test to see how closely they fit Mendel's laws.
10. Give the definition and purposes of a test cross.

TEXT REFERENCES Chapter 10 (J); Chapter 13; Chapter 14 (A, F, G); Chapter 22 (K).

KEY TOPICS

insects	dominance	sex determination
metamorphosis	law of segregation	sex linkage
corn	law of independent assortment	chi-square test
Mendelian genetics	mutation	

INTRODUCTION

Genetics is the study of the mode of transmission of genes from one generation to the next. There are two copies of each gene called **alleles** in every diploid cell, with one on each homologous chromosome. Since the alleles must follow the movement of chromosomes during meiosis, they will show **segregation** during meiosis I, with one allele going to each daughter, so that a gamete will have only one allele for each gene. The alleles of genes on different pairs of homologous chromosomes will move independently during meiosis and show **independent assortment** in the gametes and offspring.

In order to observe what is happening to the genes during reproduction, there must be a difference between the alleles. Commonly, one allele will be the usual or **wild type,** symbolized as +, and the other will be a **mutant** allele, symbolized in general as ***M*** or ***m***. The mutation is a heritable change in the DNA sequence of the allele. For experimental purposes, the mutation must cause an observable change in appearance or **phenotype.** An organism's **genotype** is its genetic makeup, and its genotype will determine its phenotype.

Genetic experiments usually begin with parents that are true-breeding or **homozygous** in that both of their alleles for the gene in question are alike: ++, *MM* or *mm*. Two homozygous parents are crossed to produce the $\mathbf{F_1}$ hybrid generation, which is **heterozygous** (***Mm***) and contains one allele from each parent. These offspring can be crossed among themselves to give the $\mathbf{F_2}$ generation. If such a cross is analyzed for one gene, it is called a **monohybrid cross (*M* × *Mm*);** if it is analyzed for two genes at the same time, it is called a **dihybrid cross (*MmNm* × *MmNn*).** Alternatively, the F_1 offspring can be crossed back to a homozygous parent in a **test cross.** Data from the test cross can be used to verify that Mendelian inheritance is involved, to detect linkage between two genes, or to determine the genotype (*MM* or *Mm*?) of an individual that shows the dominant phenotype.

The fruit fly, *Drosophila melanogaster,* and corn, *Zea mays,* are two of the organisms whose genetics has been extensively studied. In today's laboratory you will work with mutations in one or the other to investigate gene transmission. The mutations you will use will be either **dominant (*M*)** or **recessive (*m*)** compared to wild type. This can be determined by examining the results of the cross. When homozygous parents are crossed, the heterozygous cells of the F_1 offspring will receive one allele from each parent. If their appearance is like that of the mutant parent, the mutation is dominant; if the offspring look like the wild type parent, the mutation is recessive:

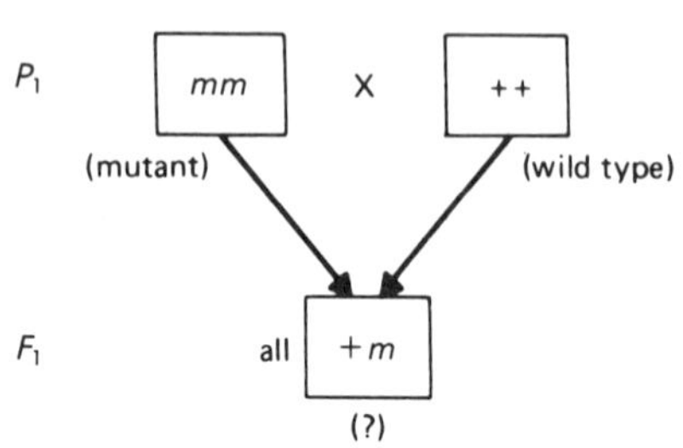

If more than one mutation is involved in the cross, you can consider each one in turn to decide whether it is dominant or recessive. You will then be able to go on to determine what kind of cross is involved and what mode of transmission applies to that mutant gene.

In *Drosophila* (2N = 8), there are three pairs of **autosomes** and a pair of **sex chromosomes** (XX or XY). Genes located on the X chromosome but not on the Y are said to be **sex-linked. Reciprocal crosses** with respect to the sex of the mutant parent will give distinguishable results for sex-linked genes, but not for autosomal genes:

Cross A: virgin wild type (++) females × mutant males

Cross B: virgin mutant females × wild type males (++)

Reciprocal crosses are exactly the same, except that the mutation is located in the parent of the opposite sex.

If the mutation is located on one of the autosomal chromosomes it will show **autosomal inheritance,** and it will not matter which parent passed the mutation to the F_1 offspring. Cross A and Cross B will give identical results:

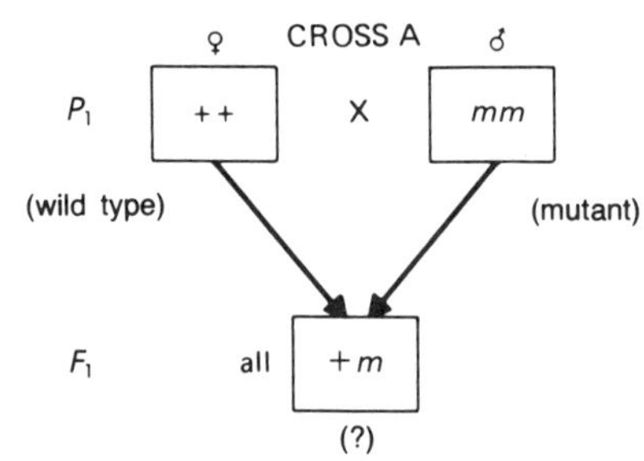

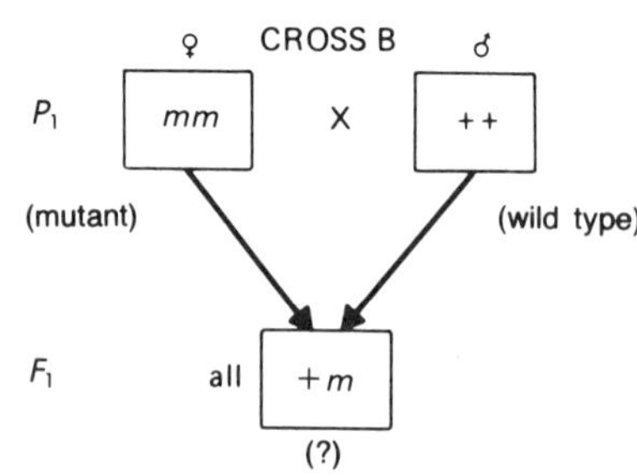

If, on the other hand, the mutation is sex-linked and located on the X chromosome, it *does* matter which parent carried the mutation, and Cross A and Cross B will give different results:

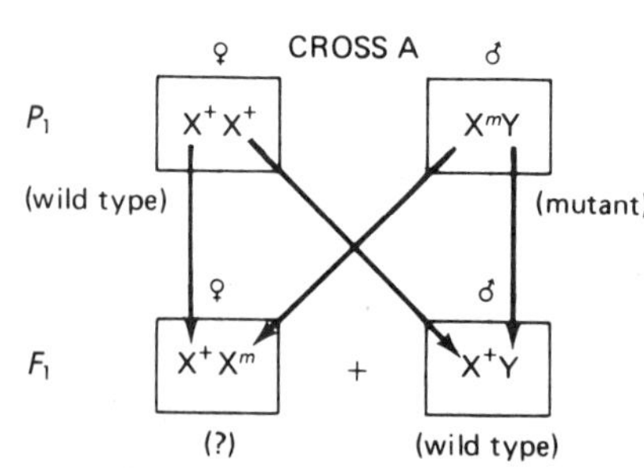

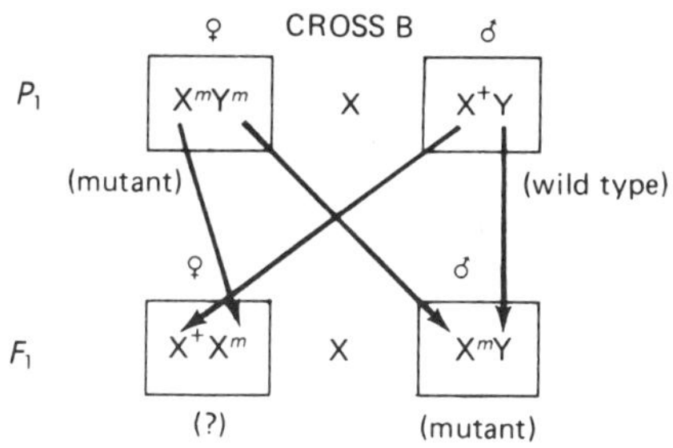

If the mutation is dominant, F_1 females, but not F_1 males, from Cross A will show the mutation, but if it is recessive, F_1 males, but not F_1 females, from Cross B will show it. In each case the reciprocal cross will not show any difference between the sexes.

It may also be possible for you to deduce the arrangement of two mutant genes on the chromosomes. Here are the possibilities:

1. Both of the mutant genes are on the same autosome, so they are inherited together, and are said to be **linked.** Any fly that inherits one mutant allele will be likely to also inherit the second.
2. Both of the mutant genes are on the X chromosome. These traits are sex-linked, and must show the pattern of sex-linked inheritance.
3. The mutant genes are on different autosomes. They are inherited independently of each other, and neither shows the sex-linked pattern of inheritance.
4. One of the mutant genes is on the X chromosome, and the second is on an autosome. The genes will be inherited independently of one another, and will show different patterns of inheritance.

If you have data from reciprocal crosses, you can identify situations 2 and 4. Cross A will identify a dominant mutation as sex-linked, while Cross B will identify a recessive mutation that is sex-linked. Since you don't know beforehand whether the mutation is dominant or recessive, it is necessary to do both reciprocal crosses. In plants such as corn, sex linkage is not an issue because there are no sex chromosomes and the same plant is both male and female. F_2 data will be needed to distinguish 1 from 2.

DROSOPHILA GENETICS

In this laboratory you will have the opportunity to work with the fruit fly, *Drosophila melanogaster.* This tiny fly is one of the most important organisms used by geneticists in studying the mechanism of inheritance. The advantages of using *Drosophila* as an experimental animal are:

1. It has a short life cycle (10 days at 25°C).
2. A single healthy female can produce several hundred offspring.
3. It is easily cultured in the laboratory in small bottles on simple culture media.
4. It is small.
5. There are many different easily recognized variations of *Drosophila.*
6. It has a small number of chromosomes (2N = 8). In addition, the cells of the salivary glands of the larva possess enlarged chromosomes which have been important in mapping the location of certain genes (See Topic 8).

In this section you will study the life history and morphology of *Drosophila melanogaster,* and analyze the results of a genetic cross involving two mutations.

LIFE HISTORY

The fruit fly is an insect of the order Diptera (two-winged flies). The Diptera all undergo complete metamorphosis, or **holometabolous** development, which means that the immature forms bear no resemblance to the adult. The flies and other insects that undergo complete metamorphosis show four different stages in their life histories (see Figure 11–1):

1. The **egg** is laid by the adult female (♀) on the food surface, and develops into the wormlike larva.
2. The white wormlike **larva,** or maggot, is characterized by its voracious appetite and rapid growth. During the course of its growth the larva will **molt,** or shed its skin, two times. The periods between molts are called **instars.** The third instar ends as the outer skin darkens and hardens, forming the pupal case.
3. During the **pupal** stage, the larval tissue breaks down, and the new winged adult is formed.

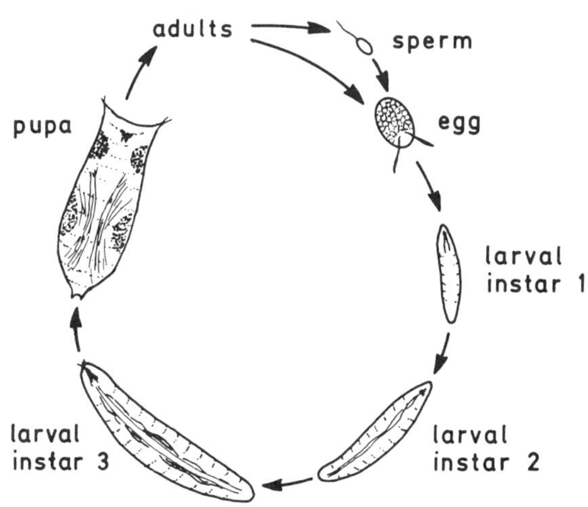

Figure 11–1. Life history of *Drosophila.* At room temperature the cycle from when the egg is laid to hatching (eclosion) of the adult from the pupa takes about 14 days.

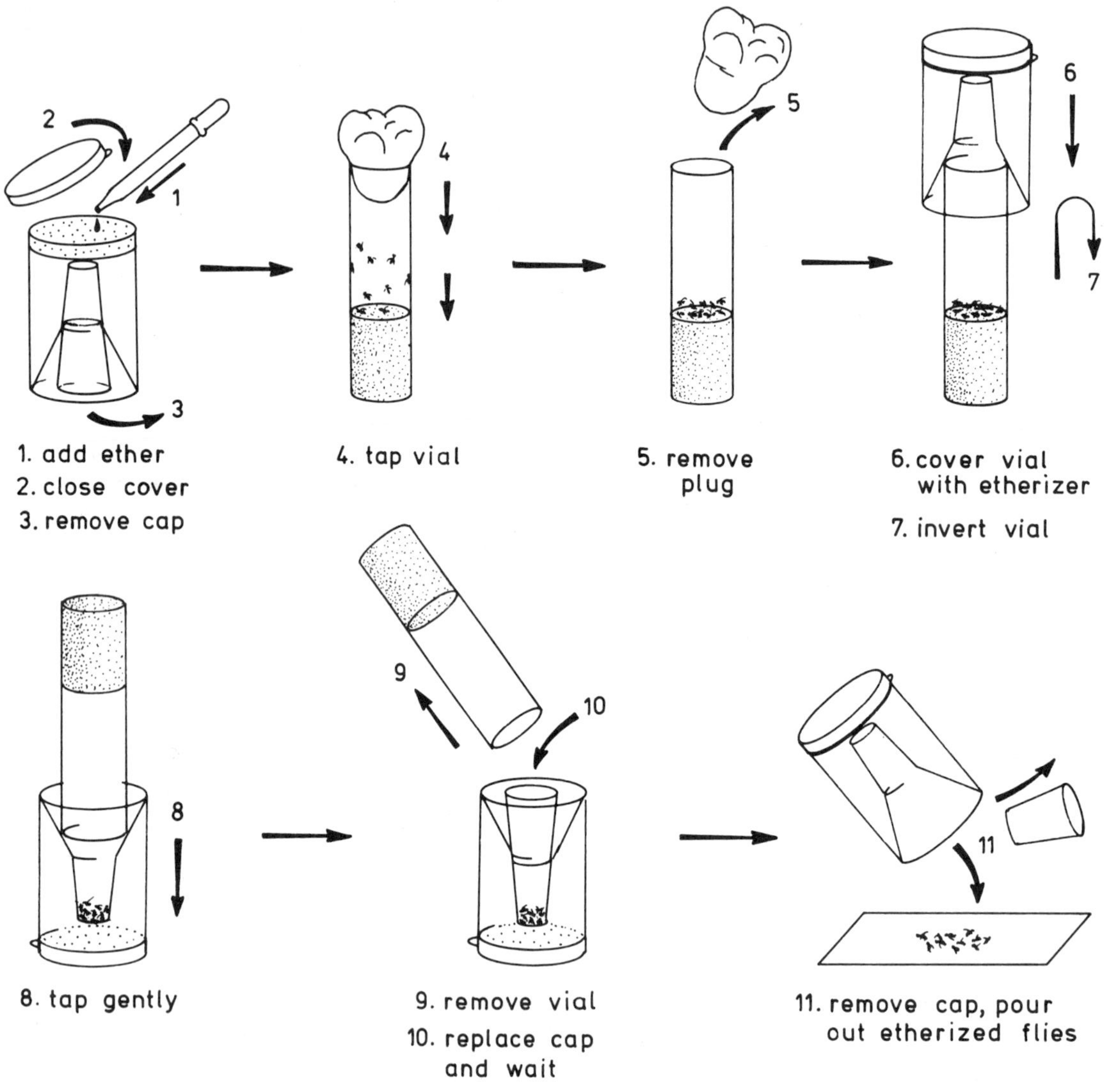

Figure 11–2. Etherizing *Drosophila*.

4. The **adult** fly that emerges from the pupal case is light-colored and elongated with unexpanded wings. The wings must expand, dry in the air, and stiffen before they can be used for flight.

The adult stage is the reproductive stage, and the adults will mate about 6–8 hr after emergence. The female (♀) receives about 4000 sperm from the male (♂), and stores these in special sacs. The sperm are released gradually as the eggs are produced. Up to a thousand eggs may be produced by a healthy female.

If a female has mated, she cannot be used in genetics experiments, because at least some of her offspring will be due to the "brother-sister" mating, not the experimental cross.

- Examine a bottle containing a *Drosophila* culture and find an example of the larva, pupa, and adult.
- View the demonstration of *Drosophila* eggs in the dissecting microscope. They are laid on the surface of the food.

Handling the Flies

Because fruit flies are small and active, it is necessary to immobilize them so they can be observed in the dissecting microscope. Ether or a similar anesthetic will be used to anesthetize them. Ether is highly explosive if ignited.

Lighting matches or lighters and smoking are forbidden in the laboratory.

- Obtain an anesthetizer, re-etherizer (glass Petri dish), ether, camel hair brush, thumper (rubber pad), dissecting needle, and a 3 × 5 in. card or filter paper.
- Take a vial of wild type (++) flies provided by the instructor. Note that (++) signifies that these flies are homozygous and carry two copies of the wild type or normal allele (+) for all traits.
- Referring to Figure 11–2, etherize the flies exactly

as follows, so that you do not overetherize and kill them.

1. Add ether to the anesthetizer bottle by putting a few drops of ether on the cotton or foam until it is moist but not wet. Recap the ether bottle immediately.
2. Close the cover on the anesthetizer bottle.
3. Remove the small cap from the cone-shaped part of the anesthetizer bottle.
4. Shake the flies down into the bottom of the glass vial by tapping it lightly on the thumper.
5. Working quickly, remove the plug from the glass vial.
6. Cover the vial with the cone-shaped part of the anesthetizer.
7. Invert the vial and anesthetizer.
8. Shake the flies into the anesthetizer by tapping it lightly on the thumper as shown in Figure 11–2. **DO NOT POUND HEAVILY** because that will loosen the food and let it drop onto the flies.
9. After the flies have all been shaken into the anesthetizer, remove the empty glass vial.
10. Working quickly, replace the small cap in the funnel and wait for the ether to take effect. Watch the flies closely because they will slow down and stop moving in about 30 sec.
11. After most of the flies have stopped moving, count to five slowly. Then remove the cap and pour the flies out onto the 3 × 5 in. card. It is important to follow this procedure exactly because overetherizing will kill the flies. You can tell that the flies are dead if their wings extend stiffly at right angles to their bodies.
12. Prepare the re-etherizer by adding a few drops of ether to the pad in the Petri dish and setting the dish on the table with the pad facing down.
13. The flies will remain immobilized for 5 to 10 min. When the flies begin to move, they must be re-etherized by covering them with the re-etherizing plate. Wait for them to stop moving, and resume your observations.

Do not allow reviving flies to escape and infest the laboratory. Dispose of all unwanted flies in the morgue provided.

- Now put the card containing the flies under the dissecting microscope. It is helpful to draw a line halfway across the card, and label the halves male (♂) and female (♀). After you learn to sex the flies, you can brush them onto the appropriate half of the card.
- Use Figure 11–3 as your guide in identifying your male and female flies.

Morphology of the Head

The members of the class Insecta have three divisions of the body: head, thorax, and abdomen.

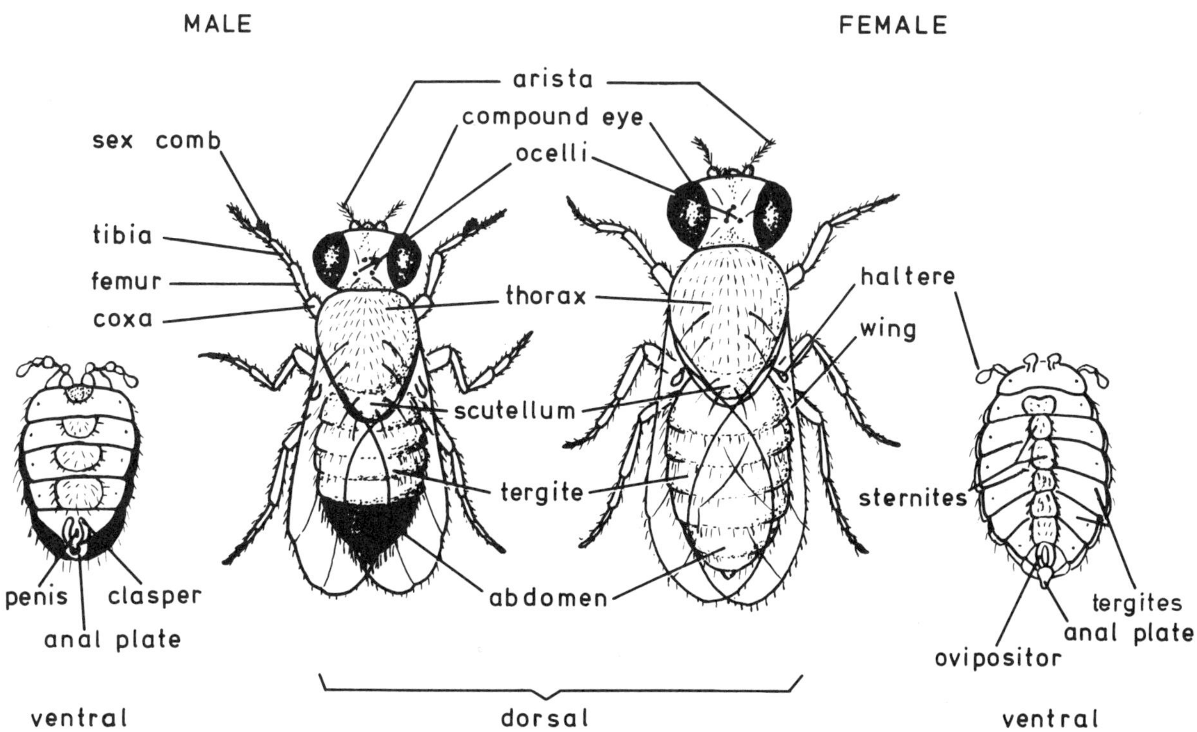

Figure 11–3. Male and female *Drosophila*. The sex comb of the male and the shape and color of the abdomen can be used to distinguish the male and female.

- Examine the head. The eyes of *Drosophila* are of two types, simple and compound. The **compound eyes** are large and made up of many different facets called ommatidia, which form a honeycomb type of arrangement. Pay particular attention to the color: This is the normal dull red or wild type (+) color (remember that the symbol for any normal characteristic is +). Two types of pigment, red and brown, occur in *Drosophila*, and together give the eye the normal color. Mutations affect the amount of each of these pigments and their distribution; many different possible eye color mutants result.
- The three simple eyes are called **ocelli,** and form a triangle on the top of the head between the two compound eyes. These apparently act to increase the brain's sensitivity to light, but they do not function in the formation of a visual image.
- Notice the **antennae.** Each antenna has a fleshy base and a featherlike **arista.**
- Examine the head from a side view, and look at the mouth parts. In the flies they have been modified to form a tubular **proboscis,** through which food is sucked up. If you press on the side of the head gently, the fly will usually protrude its proboscis.

Morphology of the Thorax

The thorax consists of three fused segments, each bearing a pair of legs.

- Examine the legs. From the body outward each leg consists of the **coxa,** the very small **trochanter,** the **femur** (the largest part), the **tibia,** and the five-segmented **tarsi** (singular, **tarsus**). Using the highest magnification on your dissecting microscope, look closely at the first tarsal segment of the foreleg. In the male there is a clump of enlarged black bristles. This so-called **sex comb** is found only in the male, and is an important characteristic used in differentiating the sex of the flies. Keep looking, while focusing up and down, until you locate the sex comb. (*Hint:* males are smaller than females.)
- Examine the **wings** of the flies, and notice the veins, which give support to the filamentous wings. Many insects have two pairs of wings that function in flight. The flies, however, have only one pair of functional wings; the second pair is represented by the **halteres,** a pair of small bulbous structures located in back of the base of the large wings.
- If you press lightly on the side of the thorax with a dissecting needle and remove the needle, the wings will become erect, and you can easily locate these rudimentary wings. The halteres probably function as balancing organs to stabilize flight.
- Look down on the dorsal (or top) surface of the fly's thorax, and identify the shield-shaped posterior part, or **scutellum,** with its four large scutellar bristles.

There are many genetically defined variations of the thorax such as differences in color of the thorax, shortness or absence of scutellar bristles, and altered wing structure.

Morphology of the Abdomen

Abdomens of male and female flies show several differences that are important in distinguishing between the sexes. Sexing the flies may seem difficult at first, but with practice you will find it quite easy. Some *Drosophila* research workers can supposedly sex flies on the wing! While you are not expected to gain such proficiency, you will have to learn to sex the flies accurately. Here are the differences to look for (in addition to the male sex comb, which you have already seen):

1. **Claspers.** The male has dark-colored claspers on the underside of the abdomen which are not present in the female.
2. **Size.** The male is smaller than the female.
3. **Shape.** The male has a blunt and rounded posterior, whereas that of the female is longer and pointed.
4. **Color.** The entire rear portion of the male abdomen is black, whereas in the female the alternating light and dark bands continue to the tip.

- First examine the abdomen of a female more closely. On the dorsal (or top) surface the segmented plates, or **tergites,** will be seen as seven bands. The last segment of the abdomen is modified by the presence of the **anus** and the **ovipositor,** which forms the pointed posterior mentioned in (3) above. It is not unusual for a female to lay an egg while anesthetized.
- On the ventral surface find the small **sternites,** the midventral plates (one per segment). The tergites and sternites form a protective armor for the abdomen. Note that the tip of the female's abdomen is pointed and very light in color underneath.
- Now examine the abdomen of a male fly. Notice that the male has six tergites, compared with the female's seven. The last segments are modified to form the genitalia.
- Brush the male fly onto his back to observe the genitalia. Just in front of the anus, find the **claspers,** hooklike appendages on the anal plate that aid in copulation.
- If you press lightly on the abdomen with a needle, you may be able to see the **penis** protruding between the claspers.

The dark, blunt abdomen, with its dark-colored claspers underneath, in the male, is the best trait to use in distinguishing male from female flies.

- After you have finished examining the wild type (normal ++) flies, obtain a vial of mutant flies from your instructor. It is helpful to examine the mutant and wild type side by side so the differences can be more clearly seen.

When you are ready to return the flies to the vial (if they are to be saved), do the following:

- Bend the 3 × 5 in. card slightly, and brush the flies gently onto the glass of the *horizontal* vial as shown in Figure 11–4. Use a creased paper funnel if possible (Figure 11–4).

Leave the vial on its side until the flies recover before moving the vial.

This precaution is important because the surface of the food source is moist, and the flies may stick to the food and suffocate.

If the flies are dead, or if they are not to be saved, dispose of them in the morgue.

The *Drosophila* Crosses

The purpose of this experiment is to determine whether the mutant alleles are dominant or recessive, and how they are inherited.

You will be asked to study either cross A or cross B, its reciprocal cross:

Cross A: virgin wild type (++) females × mutant males
Cross B: virgin mutant females × wild type males (++)

Why must the female flies be virgin?

____

____

Remember that all parent flies used in this experiment are true-breeding, i.e., homozygous for all of their genes. The wild type parents are homozygous for normal alleles, and the mutant stock is homozygous mutant for two genes.

The procedure for setting crosses is as follows:

1. Remove all flies from stock bottles of wild type and mutant flies. Discard these flies because they have already mated.
2. Within the next 8 hr, collect and etherize the flies that have emerged. The females collected within 8 hr of emerging will be physiologically incapable of mating, and will be virgin.
3. Sex the flies, and separate the males from the females. Store them in separate vials if they are not to be used immediately. Discard any flies of doubtful sex.
4. Place three virgin females and three males of the type called for in the cross into a horizontal food vial, and label the vial with the cross, the date, and your initials.
5. Wait until all of the flies have recovered from the ether, and then incubate the vial at 24–26°C for two weeks.

If you are going to set the cross yourself, your instructor will provide you with stock bottles of the parental flies and instructions as to when you will collect the virgins and set the cross. Cooperate with students doing the other cross so that you do not waste the flies that you do not need.

If the cross has been done for you, your instructor will give you a bottle or vial containing the adult flies resulting from a cross made two weeks ago. These flies represent the F_1 generation from the homozygous wild type and mutant parents.

Analyzing the *Drosophila* Crosses

- Anesthetize the flies, and examine them under the dissecting microscope.
- Brush the flies onto the center line on the card, and

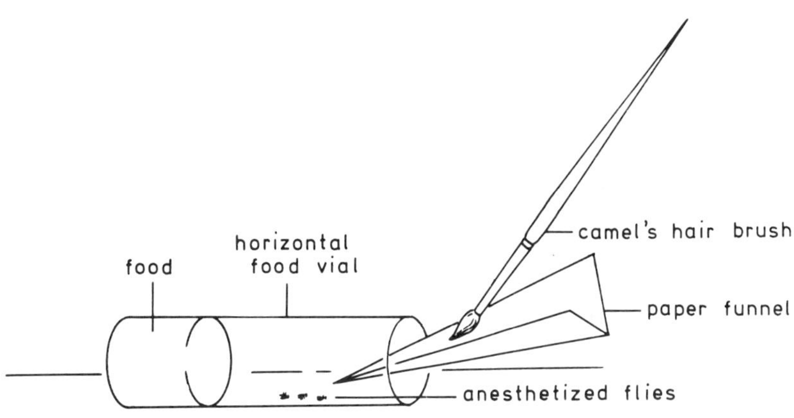

Figure 11–4. Returning flies to the vial. Sleeping flies must never come in contact with the sticky food, so keep the vial horizontal.

sex them one by one, pushing them to the proper side of the card as you determine the sex.

Are all the males similar? Are all the females alike? Do the males resemble the females? Do the offspring resemble the mutant or wild type parents for each mutant trait?

- Record your observations of the F_1 sample under cross A or cross B (whichever applies) in the analysis chart at the end of this topic. Put a check in the appropriate column for each fly, and total the classes.
- **Check the results with your instructor before you discard the flies.**
- When you are satisfied with your results, follow instructions for disposing of the flies, and put your work place and equipment back in order. Before starting your analysis, consult with the students who did the reciprocal cross, and record these data also.
- Analyze your data by answering the questions in the *DROSOPHILA* ANALYSIS CHART. Diagram your cross for both traits.

Report and Statistical Analysis

Your instructor may ask you to prepare a formal report on your results. Follow the instructions in Appendix A for preparing this report.

The data that you obtained may be tabulated for the class as a whole, and statistically analyzed to see how well they fit the results expected from the most likely pattern of transmission. Instructions for using the chi-square test are given in Appendix B. In your report, show that you understand this test by applying it to the data in the cross that you studied. Use the STATISTICAL ANALYSIS CHART at the end of this topic to analyze your data.

CORN GENETICS

Plant genetics, particularly in corn and other crops, has been extensively studied, both to improve our agricultural strains, and to understand genetic mechanisms. Corn plants, like *Drosophila,* are diploid (2N = 20) and produce both male and female gametes. Each plant is both male and female, however, so there is no sex linkage to complicate the results. Large numbers of offspring can be obtained. Some traits appear as altered phenotypes in the corn kernels, and for these it is not necessary to wait a whole generation to find out the results of a cross.

Many traits are available for genetic studies. Some of those commonly used are:

Kernel Color: **red** (red pigment in outer cells)
purple (purple pigment in outer cells)
yellow (carotenoids in inner kernel)
white (no pigments present)

Kernel Shape: **smooth** (starch present in kernel)
wrinkled (sugar present in kernel)

Plant Color: **green** (normal photosynthetic pigments)
white albino (no photosynthetic pigments present—lethal in sprouts)

The traits you will study are either dominant or recessive (*A* or *a* in general notation). Your instructor will provide information about the specific traits you will study.

The Corn Crosses

You will be asked to work with a partner to study one or more of the following crosses in which homozygous P_1 parents differing in one or more traits were crossed and the resulting F_1 progeny were again crossed to give the F_2. In each case give the genotypes of the P_1 parents and the phenotype ratios expected in the F_2 offspring (symbol for dominant phenotype = A−, recessive phenotype = aa).

1. **Monohybrid cross** of F_1 progeny:

$$Aa \times Aa \rightarrow 1\ AA : 2\ Aa : 1\ aa$$

P_1 genotypes: ____________ × ____________

F_2 phenotypic ratio expected: ____________

2. **Monohybrid test cross** of F_1 progeny with recessive parent:

$$Aa \times aa \rightarrow 1\ Aa : 1\ aa$$

P_1 genotypes: ____________ × ____________

F_2 phenotypic ratio expected: ____________

3. **Dihybrid cross** of F_1 progeny: $AaBb \times AaBb \rightarrow$

$$1AABB : 2AABb : 1AAbb:$$
$$2AaBB : 4AaBb : 2Aabb:$$
$$1aaBB : 2aaBb : 1aabb$$

P_1 genotypes: ____________ × ____________

F_2 phenotypic ratio expected: ____________

____

4. **Dihybrid test cross** of F_1 progeny with recessive parent:

$$AaBa \times aabb \rightarrow 1\ AaBb : 1\ Aabb : 1\ aaBb : 1\ aabb$$

P_1 genotypes: ____________ × ____________

F_2 phenotypic ratio expected: ____________

These crosses and the progeny predicted by assuming segregation and independent assortment of alleles during meiosis are the possible hypotheses available for testing. The F_2 progeny will be either ears of corn or young seedlings grown from F_2 kernels. The progeny from the crosses will be labeled CROSS C, CROSS D, CROSS E, or CROSS F, but the nature of the cross will not be given. It is up to you to determine for each cross which traits were involved and the type of cross that was done.

Analyzing the Corn Crosses

Here is an overview of how you should proceed:

FIRST: Determine how many phenotypic classes there are in the progeny.

SECOND: Qualitatively inspect the progeny and decide which of the possible phenotypic ratios for that number of classes is more likely to apply.

THIRD: Select the corresponding hypothesis to be tested.

FOURTH: Count the number of offspring in each phenotypic class.

FIFTH: Use the chi-square test to analyze your data and determine whether it supports your hypothesis.

- Obtain an ear of F_2 corn progeny from one of the crosses and 2 pins, or a flat of F_2 seedlings and a paper towel.

Please handle easily damaged ears of corn with care. Do not drop.

- Inspect the progeny to determine how many phenotypes are present and their approximate ratio.
- Record these phenotypes and the appropriate hypothesis to be tested in the CORN ANALYSIS CHART at the end of this topic.
- For ears, mark the starting row with a pin and count off a minimum of 160 consecutive kernels. (160 is a convenient number, since then $\frac{1}{16} = 10$, $\frac{3}{16} = 30$, etc.) Count more than the minimum, if time permits.
- For flats, unless your instructor gives other directions, count all of the seedlings in the flat.

Do not remove or damage the seedlings.

- Count and record the number of progeny in each phenotypic class.
- When you have obtained your data, promptly return the ear or flat as directed. The ear should be wrapped in a paper towel and carefully placed in its proper storage container.
- Assuming you are still satisfied with your hypothesis, diagram the cross in the appropriate Punnett square and complete the analysis chart.
- Repeat for the other crosses.

Report and Statistical Analysis

Your instructor may ask you to present your results in a formal report. Follow the instructions in Appendix A for preparing this report.

The data you obtained may be statistically analyzed to see how well they fit the results expected for the type of cross involved. Instructions for using the chi-square test are given in Appendix B. Use the STATISTICAL ANALYSIS CHART at the end of this topic to analyze your data and calculate χ^2. In your report, show that you understand the test by applying it to the data for at least one cross.

REFERENCES AND FURTHER READING

Buchsbaum, R. *Animals Without Backbones*, 2nd rev. ed. Chicago: University of Chicago Press, 1975.

Goodenough, Ursula. *Genetics*, 3rd ed. Philadelphia: Saunders College Publishing, 1984.

Keller, Evelyn Fox. *A Feeling for the Organism. The Life and Work of Barbara McClintock.* San Francisco: W.H. Freeman, 1983.

Manglesdorf, Paul C. *Corn. Its Origin, Evolution and Improvement.* Cambridge, MA: Belknap Press of Harvard University Press, 1974. Pages 114–118 describe the color of corn kernels.

Strickberger, M.W. *Genetics*, 3rd ed. New York: Macmillan, 1984.

Srb, A.M., R.D. Owen, and R.S. Edgar. *General Genetics*, 2nd ed. San Francisco: W.H. Freeman, 1965.

Journal Articles

Beadle, George W. "The ancestry of corn." *Scientific American*, 96, January 1980 (#1458).

Bennet-Clark, H.C., and A.W. Ewing. "The love song of the fruit fly." *Scientific American*, 84, July 1970 (#1183).

Benzer, S. "Genetic dissection of behavior." *Scientific American*, 24, December 1973 (#1285).

Crow, James F. "Genes that violate Mendel's rules." *Scientific American*, 134, February 1979 (#1418).

Eigen, M., W. Gardiner, P. Schuster, and R. Winkler-Oswatitsch. "The origin of genetic information." *Scientific American*, 88, April 1981.

Federoff, Nina V. "Transposable genetic elements in maize." *Scientific American*, 84, June 1984.

Hadorn. "Fractionating the fruit fly." *Scientific American*, 100, April 1962 (#1166).

Prokopy, Ronald J., and Bernard D. Roitberg. "Foraging behavior of true fruit flies." *American Scientist* 72(1): 44, 1984.

Student Name ______________________________ **Date** ____________________

DROSOPHILA ANALYSIS CHART

Record the results from your cross and from the reciprocal cross in the following tables; then analyze the results by completing the rest of the form.

Cross A

Data from a cross of wild type females × ____________________ males.

Phenotype	Females	Males	Totals
Totals			

Cross B

Data from the reciprocal cross which was ____________________ females × wild type males.

Phenotype	Females	Males	Totals
Totals			

Now, keeping in mind the information about the parental generation and your F_1 data from both crosses, draw your conclusions concerning the dominance and mode of transmission of the mutant genes you are studying. To guide you in your analysis, ask yourself the following questions.

Trait 1

Trait 1 = ____________________

Is the mutant dominant or recessive to wild type? ____________________

Did the mutant show a sex difference in either cross? ____________________

If so, in which cross? ____________________

This trait is inherited as

(a) autosomal dominant ______

(b) autosomal recessive ______

(c) sex-linked dominant ______

(d) sex-linked recessive ______

Trait 2

Trait 2 = ____________________

Is the mutant dominant or recessive to wild type? ____________________

Did the mutant show a sex difference in either cross? ____________________

If so, in which cross? ____________________

This trait is inherited as

(a) autosomal dominant ______

(b) autosomal recessive ______

(c) sex-linked dominant ______

(d) sex-linked recessive ______

Arrangement on Chromosomes

Your data may allow you to tell whether these two mutations are linked, that is, carried on the same chromosome.

Do both mutations show sex-linked inheritance? ____________________

If so, they must both be on the X chromosome, and are therefore linked to each other.

Does one but not the other show sex-linked inheritance? ____________________

If so, they must be unliked because one is on the X chromosome, and the other is on an autosome.

To decide whether the two mutant traits are always inherited together in an autosomal pattern, you must do another cross.

What kind of cross should this be?

____

If so, they are located on the same autosomal chromosome.

Which conclusion(s) can you draw from your data?

(a) the mutations are linked and are on the X chromosome ______

(b) the mutations are unlinked because one is on the X chromosome ______

(c) the mutations are linked and are on an autosomal chromosome ______

(d) the mutations are unlinked and neither one is sex-linked ______

(Conclusions (c) and (d) cannot be distinguished on the basis of your F_1 results alone.)

Student Name ______________________ **Date** ______________

Cross Diagrams

When you have decided on the mode of inheritance for each of the two mutant genes, assign symbols for each gene, and diagram your cross (A or B) below. Show your work in the Punnett square.

What types of progeny would be expected?

____

____

Why do you have to do the reciprocal cross?

____

____

If you knew ahead of time that the mutation was recessive, which cross (A or B) would you do to decide whether the mutation was sex-linked or not? ______________________

Diagram of Cross (A or B) ________

Trait #1: ______________________

Type of inheritance for Trait #1: ______________________

Symbol for mutant allele: ____________

Symbol for normal allele: ____________

P_1 ♀ × ♂

Genotype

Phenotype

F_1 ♀ + ♂

Genotype

Phenotype

Trait #2: ______________________

Type of inheritance for Trait #2: ______________________

Symbol for mutant allele: ____________

Symbol for normal allele: ____________

P_1 ♀ × ♂

Genotype

Phenotype

F_1 ♀ + ♂

Genotype

Phenotype

Student Name ______________________________ **Date** ________________

CORN ANALYSIS CHART

Record the results for each cross in the following tables; then analyze the results by completing the rest of the chart. Note that not all crosses will have four phenotypic classes.

Cross: ________ Hypothesis Tested: ______________________

Phenotype	No. of Individuals	Total in Class
	Total	

Cross: ________ Hypothesis Tested: ______________________

Phenotype	No. of Individuals	Total in Class
	Total	

Cross: ________ Hypothesis Tested: ______________________

Phenotype	No. of Individuals	Total in Class
	Total	

Cross: ________ Hypothesis Tested: ______________________

Phenotype	No. of Individuals	Total in Class
	Total	

Student Name ______________________________ **Date** ________________

Monohybrid Cross:

Which cross was the monohybrid cross? ________

Dominant trait phenotype: ____________ Symbol: ____________

Recessive trait phenotype: ____________ Symbol: ____________

Diagram the cross:

Give the expected F_2 phenotypes and their ratio: ______________________________

What percentage of the F_2 offspring are homozygous? ____________

Monohybrid Test Cross

Which cross was the monohybrid test cross? ____________

In order to tell which trait was dominant, you need to know the F_1 phenotype.

What was the F_1 phenotype (given by instructor): ____________________

Dominant trait phenotype: ____________ Symbol: ____________

Recessive trait phenotype: ____________ Symbol: ____________

Diagram this cross:

Give the expected F_2 phenotypes and their ratio: ______________________________

Why would a breeder perform a test cross? ______________________________

____

____

Why is a test cross never performed on an individual with the recessive phenotype?

____

Dihybrid Cross

Which cross was the dihybrid cross? ____________

Dominant trait 1 phenotype: ____________ Symbol: ____________

Recessive trait 1 phenotype: ____________ Symbol: ____________

Dominant trait 2 phenotype: ____________ Symbol: ____________

Recessive trait 2 phenotype: ____________ Symbol: ____________

Diagram the cross:

Give the expected F_2 phenotypes and their ratio:

What percentage of the F_2 offspring are homozygous or truebreeding?

What percentage have the phenotypes of their P_1 grandparents?

Are traits 1 and 2 located on the same chromosome?

How do you know?

Dihybrid Test Cross

Which cross was the dihybrid test cross?

Dominant trait 1 phenotype: Symbol:

Recessive trait 1 phenotype: Symbol:

Dominant trait 2 phenotype: Symbol:

Recessive trait 2 phenotype: Symbol:

Diagram the cross:

Give the expected F_2 phenotypes and their ratio:

What is the usual purpose of doing a dihybrid test cross?

Suppose you did a dihybrid test cross for two other traits and obtained only two phenotypic classes in a ratio of 1 : 1. How would you explain this result?

If you are analyzing your data statistically, transfer your observed values to the STATISTICAL ANALYSIS CHART that follows.

Student Name ______________________ **Date** ______________

STATISTICAL ANALYSIS CHART (SEE APPENDIX B)

After each cross has been tentatively diagrammed (as autosomal recessive, sex-linked dominant, monohybrid test cross, etc.), record your observed values (o) for each phenotypic class.

- Record the values expected (e) for each class based on your hypothetical diagram.
- Calculate the deviation (d = o − e) from expected for each value, d^2, and d^2/e.
- Calculate χ^2 for the cross by adding up all the values of d^2/e.
- Note the number of classes of data (n) and the degrees of freedom (n − 1).
- Consult the appropriate line of Table B–3 and match your value of χ^2 with the values in the table to determine P. Your value will probably lie between two values in the table, so estimate the corresponding value of P or use the lower value. P shows how much your data differed from the ideal data for your hypothesis. Your data support the hypothesis unless your χ^2 value falls in the 0.05 or 0.01 probability range.

Cross: ____________ **Hypothesis for Transmission:** ______________________

Class	Observed o	Difference d = o − e	d^2 $(o - e)^2$	d^2/e $(o - e)^2/e$

NO. OF CLASSES: ____________ DEGREES OF FREEDOM: ____________

VALUE OF χ^2: ____________ PROBABILITY VALUE P: ____________

HYPOTHESIS SUPPORTED (YES OR NO): ____________

Cross: ____________ **Hypothesis for Transmission:** ______________________

Class	Observed o	Difference d = o − e	d^2 $(o - e)^2$	d^2/e $(o - e)^2/e$

NO. OF CLASSES: ____________ DEGREES OF FREEDOM: ____________

VALUE OF χ^2: ____________ PROBABILITY VALUE P: ____________

HYPOTHESIS SUPPORTED (YES OR NO): ____________

Cross: ____________ **Hypothesis for Transmission:** ________________________

Class	Observed o	Difference d = o − e	d^2 $(o - e)^2$	d^2/e $(o - e)^2/e$

NO. OF CLASSES: ____________ DEGREES OF FREEDOM: ____________

VALUE OF χ^2: ____________ PROBABILITY VALUE P: ____________

HYPOTHESIS SUPPORTED (YES OR NO): ____________

Cross: ____________ **Hypothesis for Transmission:** ________________________

Class	Observed o	Difference d = o − e	d^2 $(o - e)^2$	d^2/e $(o - e)^2/e$

NO. OF CLASSES: ____________ DEGREES OF FREEDOM: ____________

VALUE OF χ^2: ____________ PROBABILITY VALUE P: ____________

HYPOTHESIS SUPPORTED (YES OR NO): ____________

TOPIC 12

Human Genetics

OBJECTIVES

When you have completed this topic, you should be able to:

1. Determine the most likely mode of inheritance of a genetic trait from the information in a human pedigree.
2. Determine your own possible genotypes for several physical traits.
3. Define antigen, antibody, and antiserum.
4. Give the possible genotypes of persons with the following blood types: A, B, AB, O, Rh^+, Rh^-.
5. Give the blood types that would be compatible in a transfusion with each of the blood types in Objective 4.
6. Describe a Barr body and explain why it is found in normal female, but not in normal male, somatic cells.
7. Explain why sex-linked recessive traits in mammals occur much more frequently in males than in females.
8. Define karyotypes, centromere, chromosomal long arm, and chromosomal short arm.
9. Use diagrams to explain how nondisjunction in the male or female can lead to the following karyotypes: XO, XXY, XXX, XYY.
10. Analyze an unknown karyotype and determine whether it represents a normal male, a normal female, or an abnormal genotype.

TEXT REFERENCES Chapter 14.

KEY TOPICS

human pedigrees
sickle cell anemia
metabolic defects
multiple alleles
human blood groups
human sex chromosomes
Barr bodies
nondisjunction
normal and abnormal karyotypes

INTRODUCTION

Humans may be of much more interest to the average person than fruit flies, but it is not easy to study variation in human genes by conventional methods. There are problems in finding traits that are controlled by a single gene because it is unethical to deliberately mutagenize human populations. There are also problems in finding the appropriate matings, reciprocal crosses, and so forth. Few are eager to donate blood, tissues, or information about their families and themselves in order to push back the frontiers of human genetics. Much of the early work in this field depended, therefore, on the painstaking analysis of **pedigrees** in which the pattern of appearance of a certain trait in several generations of a family showed how the allele for that trait was inherited. Family photographs, diaries, and the recollections of older family members were used to identify the phenotypes of as many family members as possible. If enough information existed, it was possible to determine the pattern of inheritance of the allele by eliminating those patterns not consistent with the pedigree.

A human pedigree is illustrated in Figure 12–1. Males are indicated by the squares and females by

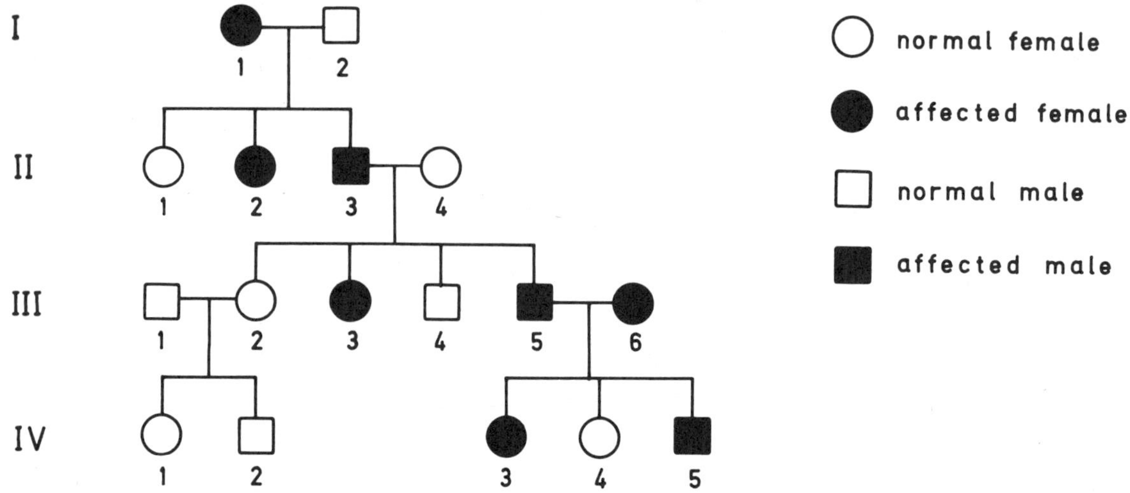

Figure 12–1. A human pedigree. Open symbols represent individuals with a normal phenotype and solid symbols represent affected individuals.

the circles. Open symbols are used for individuals with a normal phenotype and filled symbols for individuals who show the trait phenotypically. A T connecting the symbols for a man and woman represents their marriage and offspring.

- Use what you have learned about Mendelian genetics to determine whether this trait is inherited as a dominant or recessive and whether or not it is sex-linked. One approach is to check each possibility in turn, one at a time, until you either reach an inconsistent result or show that your choice is consistent with the entire pedigree.
- Repeat for each of the other possibilities. Record your conclusions in Table 12–3 at the end of this topic.
- If you have not already done so, be sure to do problem 4 in the Self-Quiz at the end of Chapter 14 in your text.

It is now known that in addition to many physical characteristics, single genes control many biochemical traits such as blood type and the presence of certain enzymes. More than 150 human diseases are known to be caused by mutant alleles for important proteins and enzymes. Some examples are hemophilia, sickle-cell anemia, Tay-Sachs disease, cystic fibrosis, phenylketonuria (PKU), and albinism. Text Figure 15–6 shows how defective enzymes coded for by the mutant alleles for the latter two traits result in the abnormal phenotypes. Some genetic diseases can be cured or controlled by medical treatment, but many cannot.

Now that the genetic basis of these diseases is understood, it is sometimes possible to design tests to determine whether prospective parents carry mutant alleles that can cause a genetic disease in their child. If there is a risk of genetic disease, the parents may ask for genetic counseling to maximize their chances of having a normal child. In some cases **amniocentesis** may be done at about the 16th week of pregnancy. In this procedure a sample of amniotic fluid is withdrawn from the amniotic sac surrounding the fetus to obtain a sample of fetal cells (See Text Essay, p. 242). Embryonic cells may also be obtained earlier in pregnancy (8th to 10th week) by **chorionic villus biopsy** done through the mother's cervix. The chromosomes from these cells can be studied by **karyotype analysis** to see whether there is a normal complement of chromosomes and whether the fetus is male or female. If the disease is known to be caused by a sex-linked recessive allele, males will have a much greater risk of having the disease than females: A single mutant allele on the male's X chromosome will cause him to be affected, whereas a single mutant allele in the female will be masked by the normal allele on her second X chromosome and her phenotype will be normal. A female must have the recessive mutation on both X chromosomes in order to have the disease, and that is not likely to happen, since she inherits a normal X from her father. Thus a male phenotype predicts a serious risk of the disease and a female phenotype is assurance that the fetus is almost certainly normal. The fetal cells can also be tested for **biochemical genetic defects,** such as the one causing Tay-Sachs disease. If karyotype or biochemical analysis shows a serious untreatable or fatal genetic disease, the parents may decide to terminate that pregnancy and hope that the next one will be free of the disease.

PHYSICAL GENETIC TRAITS

Many human physical traits are genetically controlled. The genes on the Y chromosome of males seem to determine that the child will become male, and other genes determine eye color, hair color, height, skin color, and many other traits that distinguish individuals from one another. Several characteristics are known to be controlled single gene differences. Some of these are listed below and illustrated in Figure 12–2.

1. **Pigmented Iris:** The *P* allele for a pigmented iris (green, hazel, brown, or black eyes) is dominant over the *p* allele for lack of pigment (gray or blue eyes).
2. **Tongue Rolling:** The *R* allele for the ability to roll the tongue into a U shape is dominant over the *r* allele for lack of this ability.
3. **Bent Little Finger:** The *B* allele for a bent little finger is dominant over the *b* allele for a straight little finger.
4. **Widow's Peak:** The *W* allele for a widow's peak is dominant over the *w* allele for a straight hairline.
5. **Thumb Crossing:** The *C* allele for crossing the left thumb over the right thumb when you interlace your fingers is dominant over the *c* allele for crossing the right thumb over the left.
6. **Attached Ear Lobes:** The *a* allele for attached ear lobes is recessive to the *A* allele for unattached ear lobes.
7. **Hitchhiker's Thumb:** The *h* allele for the ability to bend the last joint of the thumb back at an angle of 60° or more is recessive to the *H* allele for the lack of ability to bend the thumb back this far.

- Determine your own phenotype for each of the above characteristics and record your phenotypes in Table 12–4 at the end of this topic.
- Fill in your possible genotypes for each of the traits.

If you have the trait determined by the recessive allele, there is only one possible genotype. If your trait is determined by the dominant allele, you should list both possible genotypes, unless one of your parents is homozygous recessive. In that case you are heterozygous.

HUMAN BLOOD GROUPS

The first chemical genetic trait in humans to be understood was the determination of the ABO antigens on red blood cells. Any substance that causes the immune system to produce **antibodies** is an **antigen** or **anti**body-**gen**erating substance. Many different types of substances can be antigens. When a foreign antigen enters the body, antibodies corresponding specifically to the antigen will rapidly increase. Antibodies are proteins that are specific for one particular antigen and bind molecules of the antigen together into clumps. If the antigens are attached to red blood cells, as are the ABO antigens, the red blood cells will clump in the presence of the antibody:

A antigen will clump with anti-A antibodies.
B antigen will clump with anti-B antibodies.
AB antigen will clump with both anti-A and anti-B antibodies.
O antigen clumps with neither anti-A nor anti-B antibodies.

Another antigen, which is also present on red blood cells, is called the Rh antigen. Blood cells with this antigen are Rh^+, and those without it are Rh^-.

Rh^+ cells will clump with anti-Rh antibodies.
Rh^- cells will not clump with anti-Rh antibodies.

There are many other antigens on human blood cells that are also genetically determined.

Because it is possible to obtain a small sample of blood from a large number of people, the genetic basis of blood groups was soon well understood. The type of antigens of the ABO group that are present are determined by **multiple alleles,** any one of which may occur at the **ABO locus,** which is the gene's position on a human chromosome. Instead of just two possible alleles at this locus, there are three—namely I^A, I^B, and i. The I^A allele determines the A antigen, the I^B allele determines the B antigen, and the i allele determines neither antigen. In biochemical terms, I^A codes for an enzyme that attaches acetyl galactosamine to red blood cells, I^B codes for an enzyme that

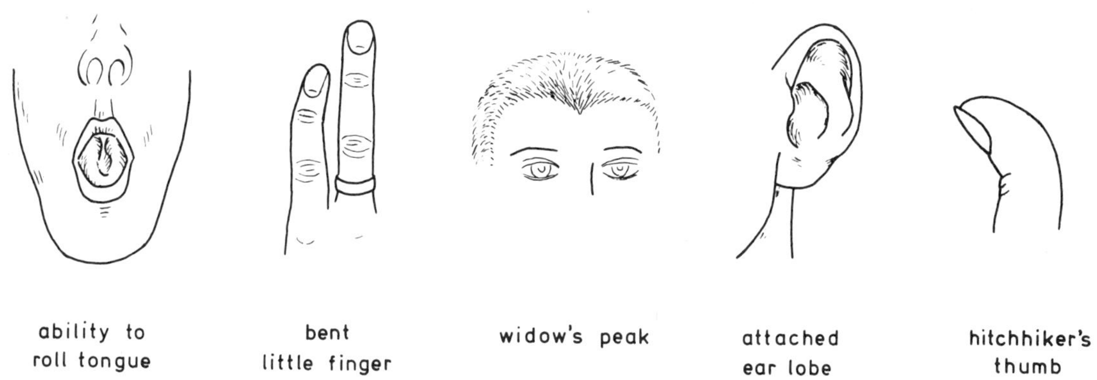

Figure 12–2. Human genetic traits. These traits are each controlled by a single gene.

attaches galactose instead, and *i* doesn't code for any enzyme. An individual person has two chromosomes and thus two alleles for ABO blood groups, one at the ABO locus on each chromosome. The following genotypes are possible.

Genotype	Blood Type (Phenotype)	Antigen Present	Antibodies in the Blood
I^AI^A	A	A	Anti-B
I^Ai	A	A	Anti-B
I^BI^B	B	B	Anti-A
I^Bi	B	B	Anti-A
I^AI^B	AB	A and B	None
ii	O	None	Anti-A, Anti-B

Because a normal person never makes antibodies against his own antigens, the antibodies present in the blood never react with the antigens present on the red blood cells. A person with A antigen on his red blood cells will have anti-B antibodies but never anti-A antibodies. Otherwise the red blood cells would clump together.

Blood groups are medically important because only compatible blood can be used during a blood transfusion. The transfusion will be incompatible if the red blood cells of the donor can clump by reacting with the antibodies of the recipient. The safest type of transfusion is between individuals of the same blood type. In this case, the red blood cell antigens and the antibodies are identical, and there is no possibility of a clumping reaction. If the donor's red blood cells contain any antigen, however, the blood can be given only to individuals who lack the corresponding antibodies. Type A blood, for example, can be given only to persons with type A blood or type AB blood. If the recipient's blood were of type B or type O, there would be anti-A antibodies in it that would clump the donor red blood cells. The antibodies of the donor are greatly diluted during a transfusion and can be ignored.

Use the following procedure to determine your ABO blood type. The anti-A serum that you will use contains anti-A antibodies and will clump or agglutinate red blood cells that have A antigen. The anti-B antiserum contains anti-B antibodies and will clump cells that have B antigen.

Do not prick your finger if you have a blood clotting disorder or have been exposed to hepatitis. To avoid the risk of serum hepatitis, do not touch any materials that might be contaminated with another person's blood. Never use a blood lancet already used by another person.

If you are unable or unwilling to volunteer to have your finger pricked, work with a partner who doesn't mind doing it. Work quickly so your blood doesn't clot during the test.

- Obtain a clean slide. Mark it with two circles labeled A and B.
- Squeeze one of your fingers to increase the circulation, and wipe it with sterile cotton moistened with 70% ethanol.
- Without touching the "sterile" finger, open a sterile lancet packet from the correct end, and prick your finger or have someone prick it for you.
- Squeeze out a drop of blood onto each of the circles on your slide.
- Obtain a toothpick and the anti-A antiserum. Add one drop of antiserum to the blood drop in the circle marked A without touching the dropper to the blood, and stir the drop with the toothpick.
- Obtain another toothpick and the anti-B antiserum. Add one drop of antiserum to the circle marked B, and stir the second drop with the toothpick.

If there is a positive reaction, the blood cells will form large clumps within 1–2 min, and the drop will appear speckled. If the reaction is negative, the drop will remain cloudy, without clumps.

Did your red blood cells react with the anti-A antiserum? ____________

Did they react with anti-B antiserum?

____

What is your ABO blood type? ____________

What is your ABO genotype(s)? ____________

The frequency of the alleles for the ABO blood groups varies among different ethnic groups. The I^B allele, for example, has a high frequency in oriental populations and lower frequencies in European populations (see Text Figure 16–11).

Another antigen that may be present on red blood cells is the Rh antigen. There are several dominant alleles (*Rh*) that determine the presence of this antigen and a recessive allele (*rh*) that determines the absence of the antigen. Persons normally do not have anti-Rh antibodies whether they have the antigen or not. These are the possible genotypes and blood groups:

Genotype	Blood Type (Phenotype)	Antigen Present	Antibodies in the Blood
Rh/Rh	Rh^+	Rh	None
Rh/rh	Rh^+	Rh	None
rh/rh	Rh^-	None	None

If an Rh^- person comes in contact with Rh^+ blood cells, the immune system will begin producing anti-Rh antibodies. This might happen after a blood transfusion or after a pregnancy in which an Rh^- mother carries an Rh^+ child. If the person then comes in contact with Rh^+ blood cells a second time, the antibodies will cause the cells to clump. This might happen during a second pregnancy of an Rh^- woman carrying a second Rh^+ child. The antibodies of the mother might enter the circulation of the baby and cause a disease called **erythroblastosis fetalis.** The baby must have an immediate transfusion of antibody-free blood when it is born to stop the clumping reaction. An Rh^- woman should receive an injection of anti-Rh antibodies after an interrupted pregnancy or delivery of an Rh^+ baby to clear her system of all Rh^+ antigens and minimize the chance of difficulties with later pregnancies.

Determine your Rh blood type using anti-Rh antiserum, which clumps red blood cells containing Rh antigen.

- Obtain a clean slide.
- Prick your finger as you did before, or use the same finger, and squeeze a drop of blood onto the slide.
- Obtain the anti-Rh antiserum and a toothpick.
- Add 1 drop of the serum to the blood without touching it, and stir the drop with a toothpick.
- Note whether there is a clumping reaction.

Did your blood clump with anti-Rh antiserum?

____

What is your Rh blood type? ______________

What is your Rh genotype(s)? ______________

When you are finished with these tests, dispose of all lancets, slides, and debris in the container marked **BLOOD TEST WASTES.**

SICKLE CELL ANEMIA

The presence of normal beta chains in hemoglobin A is determined by the *A* allele. The mutant allele at this locus that causes sickle cell anemia is called *S*. The mutant hemoglobin S molecules have a greater tendency to aggregate when the concentration of oxygen is low. The aggregates that form distort the shape of the red blood cells and cause them to block capillaries in the joints and internal organs (see Figure 12–3). The lack of circulation that results causes a painful **sickle cell crisis,** which can be dangerous and even fatal. The disease of sickle cell anemia, which occurs mainly in people of African descent, can be controlled to a certain extent by medical treatment, but there is no cure at the present time.

The allele that causes this disease shows incomplete dominance with the normal allele so that a person who is heterozygous has both hemoglobin A and hemoglobin S in his red blood cells. The heterozygous person is a **carrier** for the disease, but usually does not have any symptoms. The normal hemoglobin can apparently compensate for the abnormal hemoglobin S except under conditions of extreme stress. A person who is homozygous for the *S* allele is **affected** and has **sickle cell anemia.** Until recent advances in medical treatment became available, most homozygous children would die early in childhood.

If two persons heterozygous for the *S* allele marry, what is the chance of their child having sickle cell anemia? ______________

- Examine the prepared slide of normal blood cells and cells from a person with sickle cell anemia.
- Draw a few of the abnormal cells in the space provided in Figure 12–5 at the end of this topic.

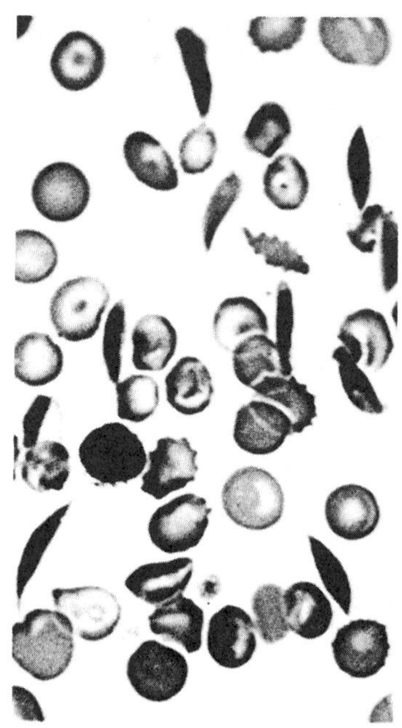

Figure 12–3. Sickled red blood cells. Cells containing hemoglobin S will assume bizarre shapes when they are deprived of oxygen. As a result, capillaries will become blocked and cause painful symptoms in the patient. (Photo by Carolina Biological Supply Company).

HUMAN SEX CHROMOSOMES

In humans, sex and other traits are determined by the sex chromosomes (see Text Figure 14–10). Males have one X chromosome and a tiny Y chromosome that contains very few genes, of which only 3 have been identified. Embryos with one or more Y chromosomes will develop into males, as long as there is also at least one X chromosome (YO is a lethal combination). Females normally have two X chromosomes and develop the female phenotype as long as the Y chromosome is absent. However, early in the development of a female embryo, one of the X chromosomes becomes inactive in each cell, so that only one of the two X chromosomes is actually active. If the female is heterozygous for a sex-linked trait, one allele will be active in about half of the cells and the other allele in the remaining cells. The inactive chromosome becomes condensed, but is still replicated and passed on to all the daughter cells. The condensed X chromosome can be seen in certain cells as a **Barr body.** In epithelium, for example, many of the cells will show the Barr body as a darkly staining mass just inside the nuclear membrane. In other cells, such as certain white blood cells, the condensed chromosome forms a **drumstick,** which protrudes from the nucleus.

Preparations of epithelial cells from the oral cavity are easy to obtain and can be used to screen for chromosomal abnormalities in the sex chromosomes. If a person is a normal female, there will be one Barr body in many of the cells. There should be no Barr bodies in a normal male. If there is an extra X chromosome, it will be condensed *because there is always only a single active X chromosome in human cells.* Apparently both males and females must have one and only one functioning X chromosome in order to survive. Thus the human male and female genotypes are much more similar than we originally supposed: X plus a few Y genes in the male, and X plus possibly a few X genes in the Barr body for the female. An abnormal male with an extra X chromosome (XXY) will have one Barr body, and an abnormal female with an extra X chromosome (XXX) will have two Barr bodies.

How many Barr bodies would you expect to find in a person with four X chromosomes (XXXX)?

____

Determine whether your epithelial cells contain Barr bodies using this procedure. About 35 % of the cells should have visible Barr bodies if you are female.

- Obtain a clean slide and a tongue depressor or clean toothpick.
- Add a drop of methylene blue dye to the slide.
- Scrape the inside of your cheek with the tongue depressor or toothpick, and stir the cells into the dye.
- Add a coverslip and examine the cells under oil immersion or high power in the microscope.

Do your cells have Barr bodies? ____________

- Exchange slides with someone of the opposite sex, and note whether Barr bodies are present.

Draw a cell with a visible Barr body and one with no Barr body in the space provided in Figure 12–5 at the end of this topic.

This type of test is used to determine whether athletes competing in certain important sporting events such as the Olympic games are genetically of the appropriate sex for the events they are entering. Persons are usually not allowed to compete in women's events unless they are genetically female.

HUMAN KARYOTYPES

A **karyotype** is the display of a person's chromosomes. Because each cell is supposed to have the same set of chromosomes, in theory, a karyotype could be made from any cell in metaphase. Karyotypes from adults can be made easily from cheek cells or white blood cells. The karyotype can also be determined during fetal development by studying cells removed from the amniotic fluid during amniocentesis. In that case, care must be taken not to confuse cells of the mother with cells from the fetus.

After the cells have been stained and squashed, the individual chromosomes are apparent and can be photographed. The chromosomes are cut out, matched in pairs, and arranged in a chart according to size and shape. The karyotype of a normal male is shown in Figure 12–4. A special staining technique is used to produce the banding patterns that you see in which each band corresponds to many genes. Such patterns help to identify the individual chromosomes and to locate large chromosomal defects.

The pairs of chromosomes are first sorted into groups of about the same shape and size. Note that each chromosome consists of two sister chromatids joined in the region of the **centromere.** The position of the centromere divides the chromosome into two **long arms** and two **short arms.** Sometimes the arms are almost equal in length, and sometimes the long arms are much longer than the short arms. The pairs are then arranged within the groups in order of decreasing size. The X chromosome is a large chromosome, about as large as pairs 6–7, with arms of about equal length. It is placed over the X in the chart. The Y chromosome is the smallest one, even smaller than those in group G, and it is placed over the Y. A normal female would have two X chromosomes and no Y chromosome. Note that the pairs within a group are very similar and cannot be told apart easily. The exact order of the pairs within the group doesn't matter

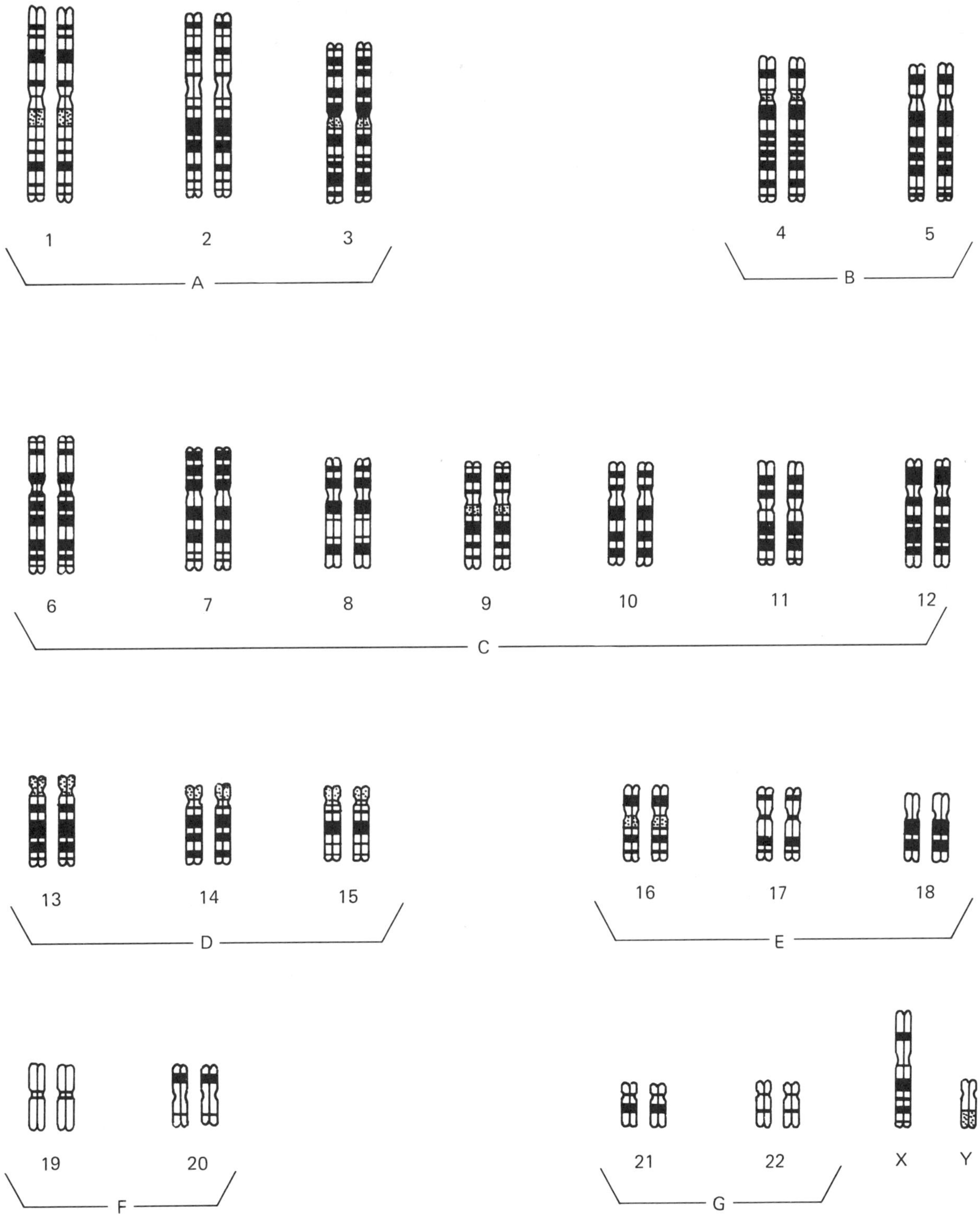

Figure 12–4. Human male karyotype. These are the chromosomes found in each cell of a normal male diagrammed to show the banding pattern after staining. Length and centromere position are important in sorting the chromosomes. (Redrawn from Jorge J. Yunis. *Science* 191:1269, 1976. © American Association for the Advancement of Science.)

much in the karyotype analysis that you will carry out, but the chromosomes must be placed in the correct group according to their size and the relative lengths of their arms.

Karyotype analysis is often used to determine whether a particular disease or syndrome is due to an abnormal complement of chromosomes. In the course of studies carried out for this purpose, or **medical**

TABLE 12–1. Characteristics Associated with Defective Karyotypes

Syndrome	Karyotype	Characteristics
Klinefelter's male	XXY	Unusual body proportions and sterility; subnormal mental ability
"XYY" male	XYY	Tall, prone to acne, impaired fertility, mentally normal
Turner's female	XO	Short stature; webbing of the neck; may have low mental ability and sterility
Super female	XXX	May have low mental ability; fertile
"Cri du chat"	# 5 defective in short arms	Catlike cry; severe physical and mental abnormalities; nonlethal
Patau's syndrome	extra #13	Physical abnormalities; lethal soon after birth
Edward's syndrome	extra #18	Unusual features of the head and fingers; often dies in infancy
Down's syndrome	extra #21	Characteristic facial features; low mental ability; stocky build; sometimes heart defects

genetics, it has become clear that there is a lot of variation among apparently normal people in the exact arrangement of their chromosomes. This discovery has opened up the field of **somatic cell genetics,** in which the complexity of the interactions of normal human genes is being investigated.

In 1959 Lejeune discovered that children with Down's syndrome (Mongolism) have three copies of chromosome #21. Since then many human genetic disorders have been traced to an abnormal chromosome complement. It has also been found that up to 40% of spontaneously aborted fetuses have abnormal chromosomes, with the XO karyotype (Turner's syndrome) being particularly common (see Stern, p. 131, for example). The physical and mental defects associated with these karyotypes are shown in Table 12–1.

An abnormal number of chromosomes in an individual is often the result of an error in meiosis during which an egg or sperm cell with an incorrect number of chromosomes is produced. When the homologous chromosomes pair during meiosis I, it is possible that the members of the pair fail to separate properly during metaphase I, and so both members of the pair eventually end up in the same gamete. The failure of the chromosomes to separate is called **nondisjunction,** and it will result in gametes with extra or missing chromosomes. When the gametes are fertilized, the offspring will have extra or missing chromosomes also. Suppose, for example, that the X chromosomes stayed together during meiosis I in a female. The resulting egg might have two X chromosomes. After fertilization of such an egg, the zygote would have either three X chromosomes or two X chromosomes and one Y chromosome, depending on the type of sperm that fertilized the egg. Alternatively, the egg might have no X chromosomes after nondisjunction.

What types of zygotes could such an egg form?

- Obtain from your instructor a copy of a chromosome squash of a human cell.
- Record the number of this squash in Figure 12–6 at the end of this topic.

The karyotype of the cell corresponds to one of the following patterns:

TABLE 12–2. Human Karyotypes

Phenotype	Karyotype	Number	✓
Normal male	XY	46	
Normal female	XX	46	
Klinefelter's male	XXY		
"XYY" male	XYY		
Turner's female	XO		
Super female	XXX		
"Cri du chat" male	Partial deletion of short arm of #5		
Patau's female	3 copies of #13		
Edward's female	3 copies of #18		
Down's male	3 copies of #21		

Normal males and females have 46 chromosomes in each cell. There are 22 pairs of autosomes and either a pair of X chromosomes or one X and one Y.

- Determine the number of chromosomes that you would find in cells of persons with the chromosomal defects listed, and enter the numbers in Table 12–2.
- Count the number of chromosomes in your chromosome squash and record the number in Figure 12–6.
- Check off the karyotypes that the squash might represent in Table 12–2.
- Cut out the chromosomes and arrange them in order of decreasing size. Use a ruler if possible, to accurately determine the relative lengths of the chromosomes.
- Match up pairs of similarly shaped chromosomes and place them in the correct groups of Figure 12–6.
- Rearrange the pairs of autosomes and the sex chromosomes until you have a reasonably good version of one of the karyotypes that you have checked off in Table 12–2. The order of chromosomes within each group is not important, but the chromosomes within a group must have the same general shape with respect to the position of the centromere and must be of approximately the same size. A chromosome in group B, for example, must be larger than any in group C, smaller than any in group A, and its centromere located toward the top rather than in the middle.

Note that the sex of each karyotype listed is specified.

- Determine the sex of your karyotype and check that you have arrived at one of the possible phenotypes.
- Tape the chromosomes in place on the chart.
- Ask your instructor to check your work, or hand in your chart before you leave.

REFERENCES AND FURTHER READING

Bodmer, W.F., and L.L. Cavalli-Sforza. *Genetics, Evolution and Man.* San Francisco: W.H. Freeman, 1976.

deGrouchy, J., and C. Turleau. *Clinical Atlas of Human Chromosomes,* 2nd ed. New York: John Wiley and Sons, 1984.

McKusick, Victor A. *Mendelian Inheritance in Man: Catalogs of Autosomal Dominant, Autosomal Recessive and X-linked Phenotypes,* 6th ed. Baltimore: Johns Hopkins, 1983.

Saxen, L., and J. Rapola. *Congenital Defects.* New York: Holt, Rinehart and Winston, 1975.

Stern, C. *Principles of Human Genetics,* 3rd ed. San Francisco: W.H. Freeman, 1973.

Sutton, H.E. *An Introduction to Human Genetics,* 3rd ed. New York: Holt, Rinehart and Winston, 1980.

Winchester, A.M. *Heredity, Evolution and Humankind.* St. Paul, MN: West, 1976.

Journal Articles

Anderson, W. French, and Elaine G. Diakumakos. "Genetic engineering in mammalian cells." *Scientific American,* 106, July 1981.

Bishop, J.M. "Oncogenes." *Scientific American,* 80, March 1982.

Brady, R.O. "Hereditary fat metabolism diseases." *Scientific American,* 88, August 1973.

Cerami, Anthony, and Charles M. Peterson. "Cyanate and sickle-cell disease." *Scientific American,* 44, April 1975 (# 1319).

Clarke, C.A. "The prevention of 'rhesus' babies." *Scientific American,* 46, November 1968 (#1126).

Croce, Carlo M., and George Klein. "Chromosome translocations and human cancer." *Scientific American,* 54, March 1985.

D'Eustachio, Peter. "Gene mapping and oncogenes." *American Scientist* 72(1): 32, January–February 1984.

Friedman, Milton J., and William Trager. "The biochemistry of resistance to malaria." *Scientific American,* 154, March 1981.

Fuchs, Fritz. "Genetic amniocentesis." *Scientific American,* 47, June 1980 (#1471).

Glass, H. Bentley. "The genetics of the Dunkers." *Scientific American,* 76, August 1953 (#1062).

Kolata, G. "Fetal hemoglobin genes turned on in adults." *Science* 218: 1295, 24 December 1982. Treatment for sickle cell anemia.

Lejeune, J., R. Turpin, and M. Gautier. "Le Mongolisme. Premier Exemple d' Aberration Autosomique Humaine." *Ann. de Genet.* 1: 41, 1959.

Macalpine, Ida, and Richard Hunter. "Porphyria and King George III." *Scientific American,* 38, July 1969 (#1149).

Marx, J. "New sickle cell test." *Science* 230: 1365, 20 December 1985.

Patau, K. "The identification of individual chromosomes, especially in man." *Am. J. Human Genetics* 12: 250, 1960.

Pines, M. "In the shadow of Huntington's." *Science 84:* 32, May 1984.

Roth, E.F., Jr. *et al.* "Sickling rates of human AS red cells infected in vitro with *Plasmodium falciparum* malaria." *Science* 202: 650, 10 November, 1978.

Ruddle, Frank H., and Raju S. Kucherlapati. "Hybrid cells and human genes." *Scientific American,* 36, July 1974 (#1300).

Trost, C. "The blue people of Troublesome Creek." *Science 82:* 35, November 1982.

Weinberg, R.A. "A molecular basis of cancer." *Scientific American,* 102, November 1983.

Student Name ______________________ **Date** ______________

TABLE 12–3. Human Pedigree Analysis

Referring to individuals in the pedigree by generation and number, explain why the pedigree in Figure 12–1 is consistent or inconsistent with each of the following possibilities:

Autosomal Dominant: ______________________

Autosomal Recessive: ______________________

Sex-linked Dominant: ______________________

Sex-linked Recessive: ______________________

Y-linked: ______________________

TABLE 12–4. Human Physical Genetic Traits

Determine your phenotypes and genotypes for the traits listed in the following table.

Trait	Dominant Allele	Your Phenotype	Your Genotype(s)
Pigmented iris	P = pigment present		
Tongue roller	R = ability to roll		
Bent little finger	B = bent little finger		
Widow's peak	W = widow's peak present		
Thumb crossing	C = left over right		
Attached ear lobes	A = unattached ear lobes		
Hitchhiker's thumb	$H = < 60^\circ$ angle of the thumb		

red blood cells in sickle cell anemia

cell with Barr body

cell without Barr body

Figure 12–5. Microscopic observations.

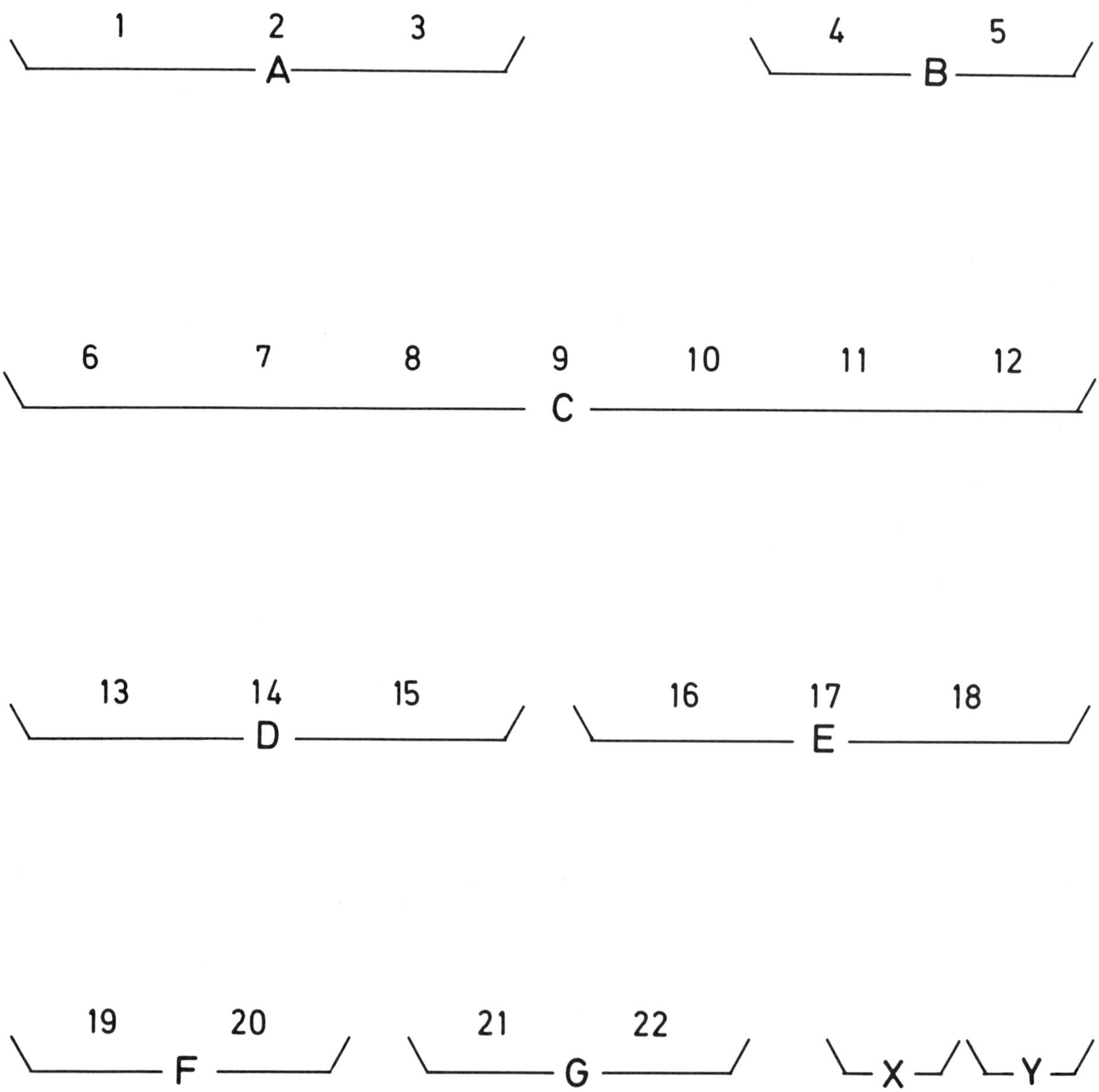

Figure 12–6. Karyotype analysis chart.

Chromosome squash # ______________

Karyotype of the squash (See Table 12–2): ______________________________

Corresponding phenotype (See Table 12–2): ______________________________

TOPIC 13

Evolution and Population Genetics

OBJECTIVES

When you have completed this topic, you should be able to:

1. Define evolution in genetic terms.
2. State the Hardy-Weinberg Law and the conditions under which it applies.
3. Determine the homozygous frequency of a recessive allele in a sample assumed to be in Hardy-Weinberg equilibrium and calculate the approximate numbers of heterozygous and homozygous dominant members expected in the sample.
4. Define industrial melanism in genetic terms and explain why it increased and then decreased in England.
5. Describe a model experiment with two alleles in which the selection against the alleles is different, and predict the change in frequency of the less advantageous allele.
6. Define genetic drift.
7. Describe a model experiment to test whether a certain population is small enough to show genetic drift.

TEXT REFERENCES Chapter 15 (Introduction, C); Chapter 16 (A, B, C).

KEY TOPICS

evolution	natural selection
gene pool	genetic drift
Hardy-Weinberg Law	

INTRODUCTION

Evolution, the idea that organisms arise by modification from previously existing forms of life, is the theme underlying all of the topics in this book. We have already touched on some of the evidence supporting this concept from embryology and comparative anatomy. The mechanism of evolution is closely connected with genetics, since only traits that are inherited can evolve. In this topic you will study the genetic mechanisms underlying the evolutionary process.

All of the alleles present in the members of a population form its **gene pool.** Evolution can be defined in genetic terms as a change in the frequency of one or more alleles in the gene pool from one generation to the next. Investigators study how the gene pool changes over time in order to understand how evolution works. Since the frequency of an allele refers to its frequency in the population, evolution will occur in the population as a whole, not in individuals. There must be **genetic variation,** that is, some heterozygosity among the members of the population, for its allele frequencies to change and for it to evolve.

Hardy and Weinberg discovered that sexual reproduction alone will do nothing to change the frequencies of alleles in a population provided that certain conditions are met:

1. **No Mating Preference.** Mating between the sexually reproducing members of the population must be random (panmictic).
2. **No Mutation.** There must be no net mutation of one allele to the other.
3. **Isolation.** There must be no gene flow due to individuals entering or leaving the population through emigration or immigration.
4. **Large Size.** The population must be large enough in size so that each generation represents a random sample.
5. **No Selective Pressure.** There must be no natural selection against either allele, or no difference in selection.

If these conditions are perfectly met, the frequency of each allele will remain exactly the same from generation to generation, so there will be no evolution. Furthermore, the frequencies of the dominant and recessive phenotypes can be determined if the frequency of either phenotype or of either allele is known.

Suppose that the frequency of the *A* allele is X, and there are two alleles for this gene. The frequency of the second allele, *a*, has to be equal to 1 − X because the sum of the two alleles must be 100%, or a frequency of 1. The chance of getting the genotype *AA* is the chance of getting one *A* allele times the chance of getting a second or X^2. To avoid confusion, the frequency of the dominant allele is called p, so the frequency of the *AA* genotype is p^2.

The chance of getting the *aa* genotype is the chance of getting one *a* allele times the chance of getting the second. The frequency of the recessive allele is called q, so the chance of the *aa* genotype is q^2. A heterozygote can be formed by getting a recessive, then a dominant allele ($q \times p$); or by getting a dominant, then a recessive allele ($p \times q$). Therefore the chance of ending up with the *Aa* genotype is $qp + pq$ or 2pq. In summary, the frequencies of genotypes in each and every generation of a population meeting Hardy-Weinberg conditions can be predicted, once p and q are known, as follows:

$$\begin{array}{rccc} \text{Frequency:} & p^2 & + 2pq + & q^2 \\ \text{Genotype:} & AA & Aa & aa \end{array}$$

Note that the frequency terms represent the terms of the binomial expansion $(p + q)^2$. If more than two alleles are involved, more unknowns can be added, and each term will represent the frequency of a corresponding genotype.

The frequency of phenotypes can then be stated once the frequencies of the genotypes are known:

Dominant phenotype ($A-$) $= AA + Aa = p^2 + 2pq$
Recessive phenotype (aa) $= aa = q^2$

Because the frequencies of the dominant and recessive phenotypes must add up to 1, the second frequency is known as soon as one or the other is calculated.

TASTE TEST

In actual populations it is usually not possible to tell the genotypes of all the individuals. If the *A* allele for a certain trait is dominant over the *a* allele, the *AA* and *Aa* genotypes will have the same phenotype. The **phenotype** is what can be easily observed or measured. If the frequency of the recessive phenotype (*aa*) is measured ($aa = q^2$), q^2 is known. If the frequency of the dominant phenotype is known ($AA + Aa = p^2 + 2pq$), q^2 can be calculated because the dominant and recessive phenotypes must add up to 1:

$$q^2 = 1 - (p^2 + 2pq)$$
$$aa = 1 - (AA + Aa)$$

This means that the recessive phenotype frequency is equal to 1 minus the dominant phenotype frequency.

In this experiment you will determine your phenotype for a certain trait and use the results from the entire class to calculate the approximate number of individuals in the class who are homozygous dominant or heterozygous for the trait, assuming the class is a population meeting the Hardy-Weinberg conditions.

Phenylthiocarbamide (PTC) is a chemical that tastes bitter to some people (tasters) but not to others (nontasters).

- Place a piece of control paper on your tongue and chew it a bit. This is the taste of paper alone.
- Now place a piece of PTC-impregnated paper on your tongue and moisten it.

Do you detect a bitter taste other than the taste of the paper itself? ______________

- Chew the paper a bit and roll it around on your tongue.
- Classify yourself as a taster or nontaster, and record the results of this test in Table 13–2 at the end of this topic.
- Also record the class results and the total number of students in the class.

The ability to taste PTC is genetically determined by the dominant alleles T^e (early tasters) and T^l (late tasters). (Late tasters can detect PTC only after chewing the paper slightly.) These dominant alleles will be lumped together as *T*. Those people homozygous for the recessive allele *t* are nontasters. Assume that your class is a reasonably accurate sample of an ideal randomly mating population, and use the Hardy-Weinberg Law to calculate the frequencies of *t* and *T* in the class. Once you know *t* and *T*, calculate the percentage of *TT* and *Tt* genotypes in the class.

How many individuals are expected to be homozygous dominant (*TT*)? ____________

How many are heterozygous (*Tt*)? ____________

- Record your results in Table 13–2.

Sodium benzoate (SB) is another chemical that only some people can taste. Its taste may seem sour, sweet, or bitter.

- Repeat the test with a piece of paper impregnated with sodium benzoate.

Are you a taster for sodium benzoate?

____

Which type of taste did you detect, if you are a taster?

____

- Record your observation and the class results in Table 13–2.

What is the frequency of the *t* allele, assuming *t* is a recessive allele for nontaster? ____________

Is this frequency different from that of *t* for PTC tasting? ____________

Variation in the ability to taste PTC and sodium benzoate has no obvious survival value today, and yet there is ethnic variation in allele frequencies between different groups. For instance, 63% of Arabs are tasters for PTC, whereas 98% of Native Americans can taste it. At least one condition for the Hardy-Weinberg Law was not met with respect to this gene in the world population.

How do you think this ethnic variation came about?

____

____

____

____

How could this variation in allele frequency be associated with survival potential for the ability to detect a bitter taste?

____

____

____

____

____

Assuming there was no selective advantage for this trait, what other factor might account for the ethnic differences?

____

NATURAL SELECTION

As you have seen, if a population meets the Hardy-Weinberg conditions, it is in genetic equilibrium, so there will be no change in gene frequency and therefore no evolution. The usefulness of the concept of genetic equilibrium is that we can measure changes in gene frequency that do occur and try to discover which of the stipulated conditions has been altered in order to give that change. Is the observed change in gene frequency over a period of time or in different populations due to mutation, small population size, selective mating, migrating individuals, or natural selection? For most genes, natural selection is almost always by far the most important factor in bringing about changes in gene frequency.

Natural selection is the differential reproduction of individuals with different phenotypes and genotypes. It can act only on phenotypes and will result in a change in allele frequencies if the selection against one of the phenotypes is stronger, so that another phenotype reproduces more successfully. In the most extreme case, where one allele is **lethal,** selection against the corresponding phenotype will be 100%, since the affected individuals never reproduce. The frequency of a recessive lethal allele will drop rapidly, but it will persist in the population because there is no selection against the heterozygote carriers (see Text Figure 16–3). Often selection may be only slightly stronger against one phenotype, but the allele frequencies will nevertheless change over many generations. In the model experiment below, you will test how rapidly strong natural selection can work.

In the English peppered moth, *Biston betularia,* the normal moth is a mottled color, but there is a dominant mutation that gives the moth a black color. Dark individuals with the mutation are called the **melanic** form, or *Biston betularia carbonaria.* The dark moths were much more conspicuous when they rested during the day on the light-colored background of woodland trees, and they were more likely to be eaten by the birds (see Figure 13–1 (a)). In other words, natural selection against the dark form was greater than that against the light form. As a result of the widespread pollution that followed in the aftermath of the Industrial Revolution, however, the trees became darkened due to the loss of light-colored lichens. Lichens are extremely sensitive to air pollution, and cannot survive under polluted conditions. Eventually the dark forms were no longer at a selective disadvantage. In fact, their dark coloration was now better

Figure 13–1. The light and dark forms of the peppered moth. (A) In unpolluted areas where the trees are covered with light-colored lichens, selection against the dark form of the moth (genotype *BB* or *Bb*) is strong, whereas the light form to its lower right is well camouflaged. (B) In polluted areas where the bare and sooty tree trunks are dark, the dark form of the moth is well camouflaged and has a selective advantage. (Bernard Kettlewell)

protection against predation than the normal coloration (see Figure 13–1 (b)). As a result, the frequency of the dark allele and the percentage of dark moths increased, a phenomenon known as **industrial melanism.**

The elegant experiments of H. B. D. Kettlewell in the 1950s showed that dark moths really do have an advantage over normal moths in actual field studies. When normal moths were released in a polluted environment in one experiment, 25% of them were recovered, but when dark moths were released under the same conditions, 53% of them were recovered. He went on to show that more white moths than black moths were actually eaten by birds. Air pollution abatement measures over the next two decades reversed the darkening of woodland trees and, predictably, increasing selection against the melanic form of the moth decreased its frequency.

First Round

In this experiment we will exaggerate the difference in selection against the two phenotypes:

Selection against the normal phenotype (*bb*) = 90%

Selection against the dark phenotype (*BB* + *Bb*) = 50%

- Obtain a small plastic bag with 50 white beans and 50 dark beans. You may also need a coin to flip. These 100 beans represent your **initial gene pool.** This is the gene pool that would be present if you mixed 25 homozygous dark (*BB*) and 25 homozygous light (*bb*) moths together and allowed them to mate randomly.
- Shake up the beans to simulate random mixing of the gametes during the first season of mating.
- Reach into the bag, and (without looking!) take out two beans: This is the first individual.
- Set these beans aside, and repeat 49 times until you have used up all the beans, arranging the *BB*, *Bb*, and *bb* genotypes in groups as you draw.
- Select against the phenotypes by removing 50% of the *BB* genotype and 50% of the *Bb* genotype (birds can't tell the difference). Then remove 90% of the *bb* genotype. Round off the number of individuals to be removed to the nearest integer. For example, if you happen to calculate that 4.5 individuals should be removed, flip a coin to decide whether to remove 4 or 5 individuals.
- Count up the number of light and dark beans left, and calculate the percentages of each.
- Enter the results in Table 13–3 under Generation 1.

If you are not confident about how to proceed, follow the example shown in Table 13–1.

The numbers corresponding to the boxed numbers in the example are the ones you should enter in Table 13–3 under Generation 1 (at the end of this topic).

Second Round

Begin the second round by replenishing your beans so that you have 100 beans with the same percentages of light and dark alleles that you had among the survivors after the first round.

- Shake up the beans.
- Select 50 individuals.
- Remove 50% of the *BB* and *Bb* genotypes and 90% of the *bb* genotype.

TABLE 13–1. Change in Gene Pool for First Round

		B		*b*		Total
Initial gene pool:		50 alleles 50% .50		50 alleles 50% .50		100 alleles 100% 1.00
First draw:	9 *BB*		32 *Bb*		9 *bb*	100 alleles
Selection against "moths": (50% *BB*, *Bb*; 90% *bb*):	−5 *BB*		−16 *Bb*		−8 *bb*	−58 alleles
Survivors:	4 *BB*		16 *Bb*		1 *bb*	42 alleles
Selected gene pool:		24 alleles 57% .57		18 alleles 43% .43		42 alleles 100% 1.00
Add in:		+33 alleles		+25 alleles		+62 alleles
Replenished gene pool with new frequencies		57 alleles		43 alleles		100 alleles
Second draw:	YOU ARE NOW READY FOR THE SECOND DRAW					

■ Calculate the percentages of light and dark beans left after selection, and enter the results in the table under Generation 2.

Third Round, etc.

■ Begin the third round by replenishing your beans so that you have the new percentages, and repeat the procedure as described in the second round until you have drawn and selected against five generations.

When you have finished five rounds, you will be very tired of drawing beans, but will have enough data to make a meaningful graph of the decline in the frequency of the *b* allele over five generations.

■ Use the grid in Table 13–3 to plot the *frequency of b* against the *number of generations.*

GENETIC DRIFT

Even if there is no natural selection, allele frequencies can change. In very small populations the selection of alleles making up the individual genotypes in the next generation will not be completely random, and the frequencies of the alleles will fluctuate even without mutation, migration, or selection. This type of evolution is due to **genetic drift** and it becomes more important as population size decreases (see Text Figure 16–8).

It is possible that the frequency of one allele may drop so low that it disappears from the next generation altogether. The situation where only one of the two alleles is left is termed **gene fixation.**

Measure genetic drift in the following experiment without natural selection.

First Round

■ Put 50 light beans and 50 dark beans in the bag, and shake it up to simulate mating.

■ Draw 10 pairs of beans from the bag to represent 10 individuals.

■ Count the light and dark beans and calculate the percentages. Enter the results in Table 13–4 under Generation 1 (at the end of this topic).

Second Round

■ Begin by replenishing your beans so that you have 100 beans in the percentages that you had among the 20 beans at the end of the first round.

■ Mix the beans.

■ Draw 10 pairs of beans from the bag to represent 10 individuals.

■ Count the light and dark beans, and calculate the percentages. Enter the results in Table 13–4 under Generation 2.

Third Round, etc.

■ Replenish your beans, and repeat as in the second round until you have completed five rounds. Enter the results in the table.

■ Plot the *frequency of b* as a function of *number of generations* on the grid below Table 13–4. This is the kind of genetic drift that you might expect to see in a real population that is too small to avoid sampling error due to small population size.

Did either your light or dark alleles disappear?

____

What factors could reverse gene fixation in a real population?

___

___

___

Return your bean bag, and dispose of any taste paper debris at the end of this lab.

REFERENCES AND FURTHER READING

Ayala, F.J., and J.A. Kiger. *Modern Genetics,* 2nd ed. Menlo Park, CA: Benjamin/Cummings, 1984.

Dobzhansky, T. *et al. Evolution.* San Francisco: W.H. Freeman, 1977.

Johnson, C. *Introduction to Natural Selection.* Baltimore: University Park Press, 1976.

Stern, C. *Principles of Human Genetics.* San Francisco: W.H. Freeman, 1973.

Stone, Irving. *Origin.* New York: Plume/New American Library, 1981. A fascinating historical novel based on the life of Charles Darwin.

Wilson, E.O., and W.H. Bossert. *Primer of Population Biology.* Sunderland, MA: Sinauer, 1971.

Journal Articles

The September 1978 issue of *Scientific American* is devoted to Evolution.

Alcock, John. "Natural selection and the mating system of solitary bees." *American Scientist:* 68 146, March 1980.

Allison, A.C. "Sickle cells and evolution." *Scientific American,* 87, August 1956 (#1065).

Ayala, F.J. "The mechanisms of evolution." *Scientific American,* 56, September 1978 (#1407).

Bishop, J.A., and L.M. Cook. "Moths, melanism and clean air." *Scientific American,* 90, January 1975 (#1314).

Brower, L.P. "Ecological chemistry." *Scientific American,* 22, February 1969 (#1133).

Cook, L.M., G.S. Mani, and M.E. Varley. "Postindustrial melanism in the peppered moth." *Science* 231: 611, 7 February, 1986

Eckhardt, Robert B. "Population genetics and human origins." *Scientific American,* 94, January 1972 (#676).

Kimura, Motoo. "The neutral theory of molecular evolution." *Scientific American,* 98, November 1979 (#1451).

Lewontin, Richard C. "Adaptation." *Scientific American,* 212, September 1978 (#1408).

Mayr, Ernst. "Darwin and natural selection." *American Scientist* 65:321, May 1977.

Mayr, Ernst. "Evolution." *Scientific American,* 46, September 1978 (#1400).

Stebbins, G.L., and F.J. Ayala. "The evolution of Darwinism." *Scientific American,* 54, July 1985.

Wills, C. "Genetic load." *Scientific American,* 98, March 1971 (#1172).

Student Name ______________________________ **Date** ______________

TABLE 13–2. Data on PTC and SB Tasting

Data from the PTC Test

Individual results: Taster ____________ Nontaster ____________

Class results:

Number of tasters ______ = ______ % tasters (*TT* + *Tt*)
Number of nontasters ______ = ______ % nontasters (*tt*)
Total ______

q^2 = ______ ; q = ______ ; frequency of *t* allele = ______
1 − q = p = frequency of *T* allele = ______
p^2 = genotype of *TT* = ______ % of the class
2pq = genotype of *Tt* = ______% of the class

If a class member is known to be a taster, what is the chance that he or she is homozygous for the *T* allele?
% *TT* × 100/(% *TT* = % *Tt*) = ____________ %

Data from the SB Test

Individual results: Taster ____________ Nontaster ____________

Class results:

Number of tasters ______ = ______ % tasters (*TT* + *Tt*)
Number of nontasters ______ = ______ % nontasters (*tt*)
Total ______

q^2 = ______ ; q = ______ ; frequency of *t* allele = ______
1 − q = p = frequency of *T* allele = ______
p^2 = genotype of *TT* = ______ % of the class
2pq = genotype of *Tt* = ______% of the class

If a class member is known to be a taster, what is the chance that he or she is a carrier for the *t* allele?
% *Tt* × 100/(% *TT* + % *Tt*) = ____________ %

TABLE 13–3. Fluctuation in Frequency of *b* Allele Due to Increased Selection Against It

	B			*b*		
Rounds	**#**	**%**	**Frequency**	**#**	**%**	**Frequency**
Initial	50	50	.5	50	50	.5
Generation 1						
Generation 2						
Generation 3						
Generation 4						
Generation 5						

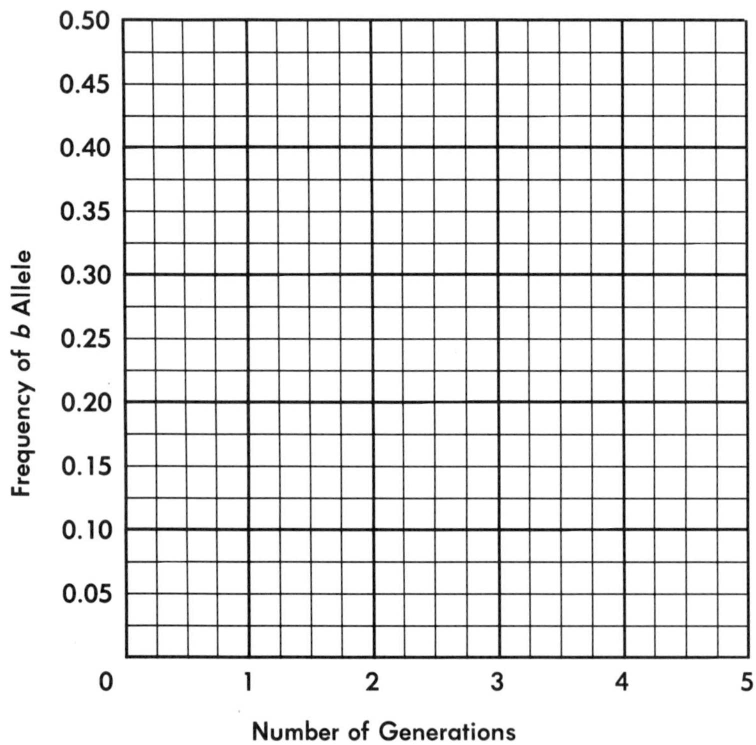

Student Name ______________________ **Date** ______________

TABLE 13–4. Fluctuation in Frequency of *b* Allele Due to Genetic Drift

	B			*b*		
Rounds	#	%	Frequency	#	%	Frequency
Initial	50	50	.5	50	50	.5
Generation 1						
Generation 2						
Generation 3						
Generation 4						
Generation 5						

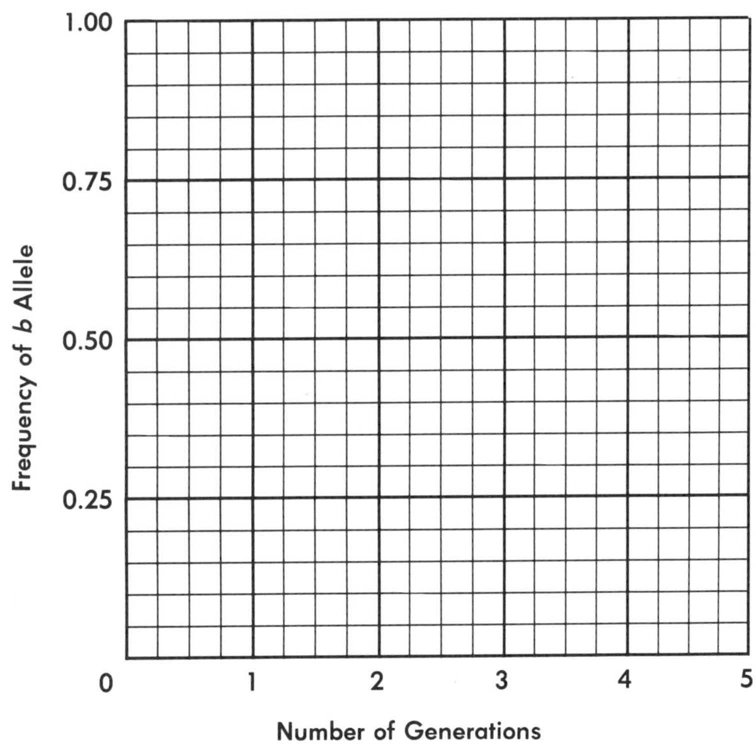

Introduction to the Diversity of Life

TEXT REFERENCES Chapter 19 (D).

KEY TOPICS Five Kingdoms: Monera, Protista, Fungi, Plantae, Animalia
prokaryotes
eukaryotes
fossil record

As more information about the evolutionary history and biochemistry of organisms has become available, it has become apparent that the old division of organisms into plants, animals, and protists is no longer sufficient. Although the boundaries between the kingdoms listed below are somewhat unclear, the following scheme first proposed by R.H. Whittaker is used in Topics 14–21:

Kingdom Monera: All prokaryotic organisms (bacteria, cyanobacteria).

Kingdom Protista: Eukaryotic unicellular heterotrophs and most unicellular autotrophs (protozoa, diatoms, dinoflagellates, euglenoids).

Kingdom Fungi: Eukaryotic heterotrophs that absorb food, mostly multicellular (yeasts, molds, mushrooms).

Kingdom Animalia: Eukaryotic multicellular heterotrophs that ingest food (sponges to vertebrates).

Kingdom Plantae: Eukaryotic autotrophs: usually multicellular, with cell walls of cellulose (algae, mosses, ferns and fern allies, seed plants).

The characteristics of the kingdoms given here are to provide a general idea of the type of organism that you might find in each one, and are by no means definitive. (A yeast, for example, is a unicellular heterotroph, but it is placed in Kingdom Fungi because it is much more closely related to fungi than to the protozoans or unicellular algae.)

The differences between prokaryotes and eukaryotes are so significant, that the prokaryotes are given a kingdom of their own, Kingdom Monera. The other four kingdoms are grouped together as the eukaryotes. The multicellular forms are divided into kingdoms primarily on the basis of their means of nutrition: absorption, ingestion, or photosynthesis. Most of the unicellular organisms are lumped together in Kingdom Protista. The relationship between the five kingdoms is shown in the diagram on p. 154.

In the discussion of the five kingdoms in Topics 15–21 you will become familiar with the diversity of living organisms. The structures and life histories are given in considerably more detail than you may be required to learn. This will allow you some flexibility to concentrate more on those organisms that you find most interesting and that the instructor is able to provide. The point of these laboratories is not to have you learn a stupendous number of biological "facts," but rather to introduce you to the different ways that organisms cope with their common biological needs: protection, nutrition, excretion, reproduction, and dispersal.

REFERENCES

Margulis, L., and K.V. Schwartz. *Five Kingdoms: An Illustrated Guide to the Phyla of Life on Earth.* San Francisco: W.H. Freeman, 1982.

Whittaker, R.H. "New concepts of kingdoms of organisms." *Science* 163:150, 10 January, 1969.

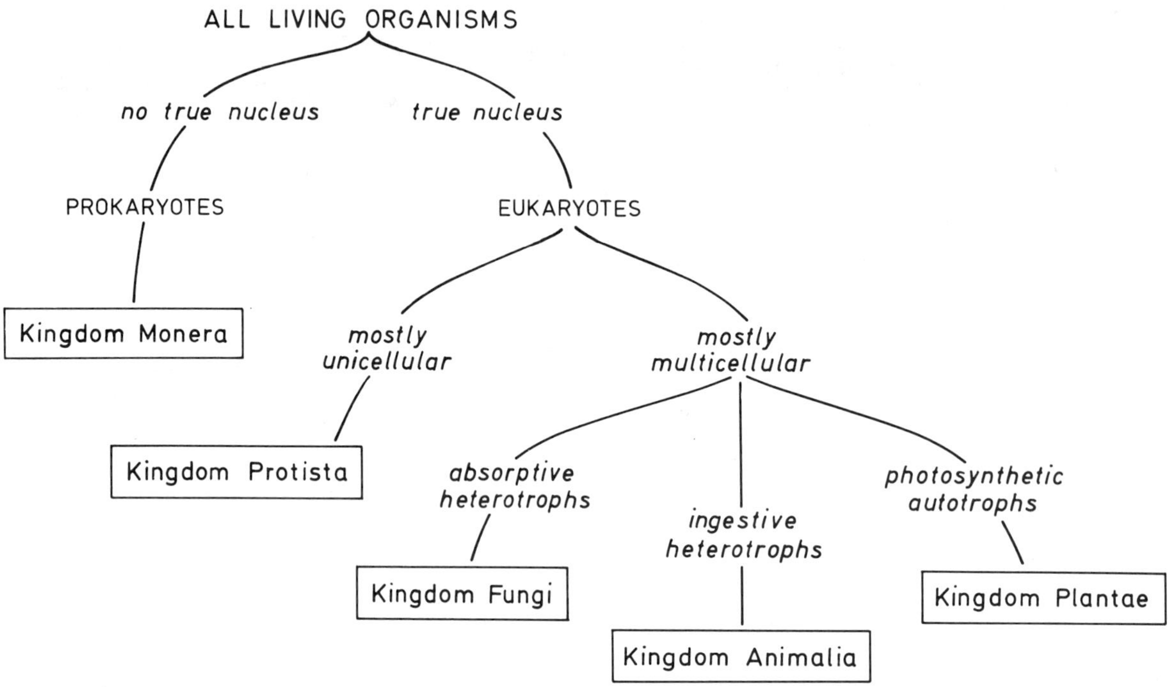

THE DIVERSITY OF LIFE

TOPIC 14

Classification of Organisms

OBJECTIVES

When you have completed this topic, you should be able to:

1. Define taxonomy and taxon.
2. List the major taxa in the correct hierarchy from the most general to the least general.
3. Explain the difference between phenetic taxonomy and phylogenetic taxonomy.
4. Correctly classify an unknown organism using a simple dichotomous key.
5. Define numerical taxonomy.
6. Arrange a set of imaginary organisms into a taxonomic array based on phenotype to show their possible evolutionary relationships.

TEXT REFERENCES Chapter 16 (E); Chapter 19

KEY TOPICS

species
binomial nomenclature
taxonomy
systematics
Five Kingdoms

INTRODUCTION

Taxonomy is simply the naming and ordering of a set of organisms according to some system of classification. The basic unit for classification is the **species,** similar species are grouped together in a **genus,** and so forth. Any unit of classification, such as genus or species, is called a **taxon.** Animals, for example, are classified according to the following scheme of progressively smaller and more specific taxons:*

Kingdom
Phylum
Class
Order
Family
Genus
Species

An organism is always designated by a **binomial,** which is its genus name and species name. Can you classify yourself (*Homo sapiens*) according to this scheme? (See Text Table 19–1.)

Kingdom ______________
Phylum ______________
Class ______________
Order ______________
Family ______________
Genus ______________
Species ______________

Members of the other kingdoms are classified in the same way, except that Division replaces Phlyum in Kingdoms Monera, Fungi, and Plantae. The accepted conventions for using binomials and the other taxa are given on p. 321 in your text.

The members of each taxon all share certain characteristics. Within the taxon, the members may be divided into smaller groups whose members share certain characteristics that are not found in the members of the other groups. The presence or

*An easy way to remember this scheme is by a mnemonic device such as "**K**indly **P**rofessors **C**annot **O**ften **F**ail **G**ood **S**tudents.

absence of certain traits will allow you to determine whether an organism belongs to a certain taxon, which more specific taxon it belongs to within the first taxon, which taxon within the second taxon, and so on.

The central question in all taxonomic systems is *What characteristics are important for each taxon?* The only taxon that has a firm biological basis is the species: A **species** is a group of organisms sharing a common gene pool. Members of a species have at least a potential for interbreeding and do not interbreed with members of other species. These criteria, however, are extremely difficult to apply in practice and most classification is based on anatomical features. Linnaeus worked out the first system of classification solely on the basis of the morphological traits he could observe. The current system of **phenetics** is similar in that it uses the phenotypic similarity of organisms as the criteria for classification. In most taxonomic systems, however, the traits used for classification are not always the ones that are obvious in an intact animal or plant. Assignment to a certain taxon may depend on the careful study of an organism's biochemistry, embryology, or parts of its life cycle that are difficult to observe. This is because the basic principle underlying most taxonomic systems is that organisms should be classified according to their **phylogeny** or the actual history of their evolution. The study of these evolutionary relationships and the construction of phylogenies is the discipline of **systematics.** The results of studies in systematics are used by taxonomists to construct a **phylogenetic system** of classification.

It is clearly impossible for an actual taxonomic system to perfectly reflect the evolutionary history of a group of organisms, because evolution is a continuum with a spectrum of organisms, whereas the taxa are discrete, separate units. One approach is to make each taxon **monophyletic,** so that all of its members would be the descendants of a common ancestor. Such a taxon is called a **clade,** and **cladistics** is a system based solely on such evolutionarily related groups. In this type of system the actual characteristics of the organisms are put aside in favor of their known evolutionary history. In practice, most taxonomists consider both evolutionary relationships and shared characteristics and advocate a taxonomy based on **evolutionary systematics.** In this case, the taxa may be **polyphyletic** because they include groups with different ancestors. Such a taxon represents a **grade** of organization attained more than once by related, but different evolutionary lines, rather than a monophyletic clade (see Text Figure 19–5).

An interesting approach to the problem of determining evolutionary relationships is **numerical taxonomy,** described later in this topic. In this system great importance is given to the number of shared phenotypic characteristics, as in phenetics, but the intent is to reveal evolutionary history, rather than to produce a convenient taxonomic scheme.

In order to identify the taxon to which a certain organism belongs *in the field* or *without extensive study,* we use a shorthand guide to readily discernible features called a **dichotomous key** (dichotomous means "two forks"). Such a key divides a certain group of organisms into two groups on the basis of a certain trait, then divides each of those groups into two groups, and so on, until the desired level of identification is reached. These groups are for convenience in identifying organisms and are not necessarily the same as the taxa described above. Of course the ultimate result, namely the identification of the organism, is deciding to which taxon it belongs.

Because the purpose of a key is to identify something quickly, each key will be limited in scope. If it starts out with a broad group of organisms, say, invertebrates or algae, it will not be able to identify the species of each organism but perhaps only the family or order. If the key is more limited in content, limited, for example, to spring wildflowers of a certain geographical area, it will be able to include genus- and species-level identification. The person using a key can use his prior knowledge or the constraints of geographical region or season to choose a key that will give as precise an identification as possible. Your text shows an example of a dichotomous key for members of Phlyum Arthropoda in Figure 16–13.

Dichotomous keys can be used in series to identify organisms of special interest when the first key is very general. A person studying seashore life might use a key to marine invertebrates in conjunction with a key to bivalves to identify a certain clamlike organism.

Consider the dichotomous scheme shown in Figure 14–1 for grouping all organisms. Each step in keying out an organism is simply a yes or no answer to successive questions in the key, part of which would look like this:

1a. Is the organism motile? (2a)
1b. Is the organism stationary? (3a)
2a. Is the organism aquatic? (4a)
2b. Is the organism terrestrial? (5a)
3a. Is the organism aquatic? (6a)
3b. Is the organism terrestrial? (7a)
4a. Does the organism have ray fins? (class Osteichthyes, or bony fish)
4b. Does the organism lack ray fins? (9a)

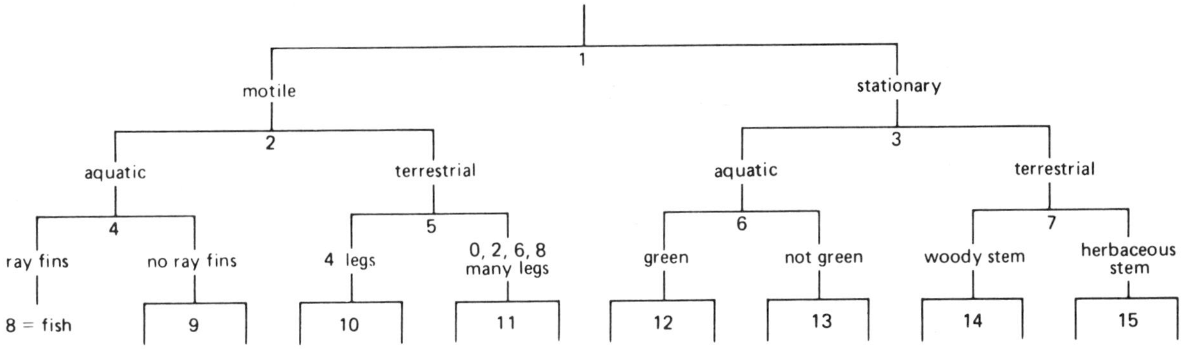

Figure 14–1. A simple dichotomous key.

5a. Does the organism have four legs? (10a)
5b. Does the organism have zero, two, six, eight, or many legs? (11a)
6a. Is the organism green? (12a)
6b. Is the organism not green? (13a)
7a. Does the organism have a woody stem? (14a)
7b. Does the organism have a herbaceous stem? (15a)

Note that many of the characteristics used for the key, such as motile versus stationary or green versus not green, are not important in defining taxa.

To use the key, you would begin with question 1a. If the answer to this question were yes, you would proceed to question 2a; if the answer were no, you would go to question 1b. You would then continue from 1b to 3a to 6a, and so on. Three consecutive yes answers would identify an organism as a fish in the class Osteichthyes, but the identification of most organisms, including other fish, would require several more questions. A more detailed key would be necessary to identify the fish more specifically.

In today's laboratory exercise you will use simple dichotomous keys to identify members of a class that are similar in their general features but easily distinguished when you know what to look for.

CONIFER KEY (DIVISION TRACHEOPHYTA, CLASS CONIFEROPSIDA)

Conifers are seed plants that usually have evergreen leaves and produce gametes in male and female cones.

- Select one of the specimens of a conifer on demonstration in the lab, and key it out using the key in Table 14–1. This key includes some of the conifers that will be found in your region of the country. Your instructor may give you additional information about species that may be especially common in your area. To use the key, begin with question 1a. If this item applies, go to question 2a; if not, go to 1b. Continue from 1b to 14a, from 14a to 15a, and so forth.

Specimen number: ______________

Identification: ______________

- If living conifers are available in the field for class study, select a specimen and key it out in your spare time.

Characteristics of field specimen:

____

____

____

____

____

____

____

____

Identification: ______________________________

Your instructor may ask you to key out other preserved or live specimens.

NUMERICAL TAXONOMY

It is very difficult to decide which characteristics should be considered in setting up a given category, or taxonomic group. All the organisms within the category will have some characteristics in common, but none of the organisms would be expected to have every one of the characteristics associated with the category.

The introduction of computers has made it possible to consider large numbers of characteristics in determining the degree of relatedness of different organisms and to sidestep the problem of deciding weights for various characteristics. Classification can then be based not on the relative *importance* of certain characteristics, but rather on the *number* of characteristics held in common. This type of classification is called

TABLE 14–1. A Key to the Conifers

After you have identified a given conifer, refer to a field guide and check that the description matches the characteristics of your specimen.

1a. Leaves are in clusters	2a
1b. Leaves are borne singly along the stem.	14a
2a. More than 5 leaves in a cluster.	3a
2b. Leaves occur in clusters of 2–5; persistent for several seasons.	4a
3a. Leaves are deciduous (fall off in winter), soft, flattened.	LARCH or TAMARACK
3b. Leaves are persistent, stiff, 4-sided (ornamental in warmer areas).	TRUE CEDAR
4a. Leaves mostly 5 in a cluster.	WHITE PINE
4b. Leaves 2–3 in a cluster.	5a
5a. Leaves mostly 3 in a cluster.	6a
5b. Leaves mostly 2 in a cluster.	8a
6a. Twisted leaves, less than 5 in. long	PITCH PINE
6b. Straight leaves, more than 5 in. long	7a
7a. Leaves 5–10 in., cones very thorny (South)	LOBLOLLY PINE
7b. Leaves mostly over 11 in. long, cones not thorny (South)	LONGLEAF PINE
8a. Leaves longer than 3 in.	9a
8b. Leaves shorter than 3 in.	11a
9a. Leaves rigid, bark grayish.	AUSTRIAN PINE
9b. Leaves narrower than $\frac{1}{16}$ in., bark reddish brown or brown.	10a
10a. Twigs brown, cones thornless (North).	RED PINE
10b. Twigs whitish, cones strongly thorny (South).	SHORTLEAF PINE
11a. Leaves wider than $\frac{1}{16}$ in.	12a
11b. Leaves narrower than $\frac{1}{16}$ in.	13a
12a. Leaves longer than $1\frac{1}{2}$ in.	MOUNTAIN PINE
12b. Leaves shorter than $1\frac{1}{2}$ in.	JACK PINE
13a. Twigs whitened.	SCRUB PINE
13b. Twigs not whitened.	SCOTCH PINE
14a. Leaves are scalelike or triangular and opposite or whorled.	15a
14b. Leaves needlelike or flattened.	18a
15a. Branchlets flattened; all leaves scalelike, opposite	WHITE CEDAR or ARBOR VITAE
15b. Branchlets not flattened	16a
16a. Leaves needlelike to triangular in tight spirals (frost-free areas).	MONKEY PUZZLE TREE
16b. Leaves needlelike in whorls or pairs.	17a
17a. Leaves 3-sided in whorls of 3 on 3-sided twigs.	DWARF JUNIPER
17b. Leaves 3-sided in pairs in 4 rows on 4-sided twigs	RED CEDAR
18a. Leaves soft, on green branchlets (like compound leaves): (South or West).	19a
18b. Leaves rigid on brownish branchlets (solitary).	21a
19a. Leaves dark green, persistent.	REDWOOD
19b. Leaves light yellow-green, deciduous.	20a

Plate 1. Close-ups of the Five Kingdoms

At the cellular level, unsuspected similarities appear between the kingdoms, but important differences exist as well.

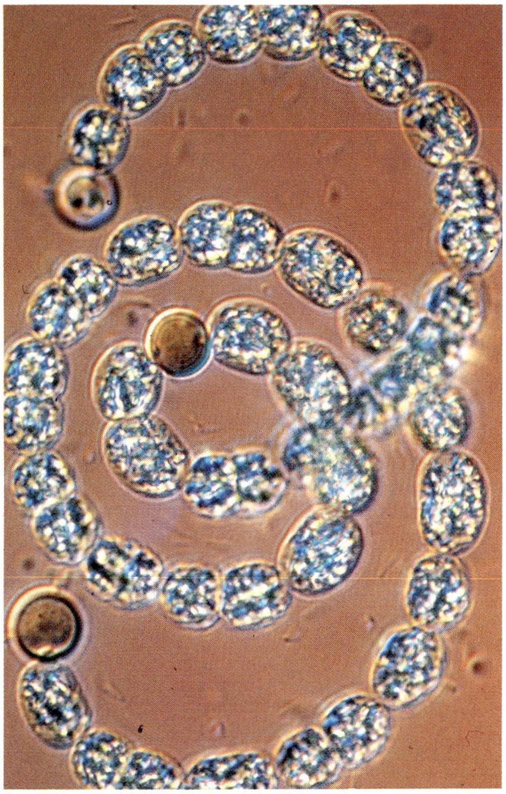

Cyanobacterium. Prokaryotes lack membrane-bound organelles, but some are nevertheless photosynthetic. *Anabaena* has a system of internal photosynthetic membranes instead of chloroplasts, and can carry out both photosynthesis and nitrogen fixation. (Biophoto Associates)

Plant Cells. The chloroplasts in these moss cells may be derived from ancient endosymbionts that could carry out photosynthesis and thus became associated with ancient eukaryotic heterotrophs. (Biophoto Associates)

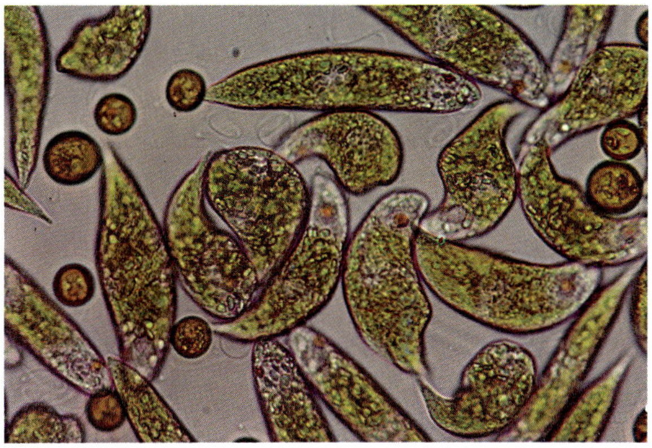

Euglena. This complex protistan is both motile and photosynthetic. It has very close nonphotosynthetic relatives that are similar to the flagellates that gave rise to animals. (Tom Adams)

Human Blood Cells. The red blood cells of this liquid tissue are highly specialized for oxygen transport, and the white blood cells for defense. The two neutrophils shown here phagocytize bacteria and dead cells. Unlike protists, most animal cells cannot live independently and must depend on the rest of the body for many of their needs. (Biophoto Associates)

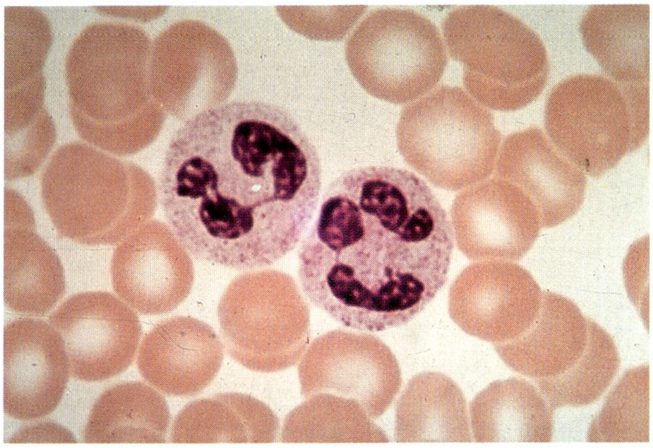

Physarum. This fungus is an enormous cell, called a plasmodium, with many nuclei. Like a giant amoeba, it engulfs bacteria and organic matter and shows cytoplasmic streaming. (From Arms and Camp, *Biology*)

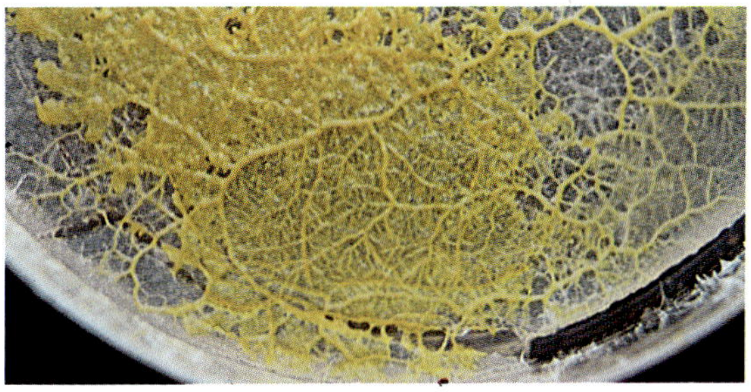

Plate 2. Is Sex Necessary?

Although sexual reproduction occurs in all five kingdoms, asexual or vegetative reproduction is common, especially among plants. Since all offspring are genetically identical to the parents, this type of reproduction is highly effective during "good times," when environmental conditions are stable and favorable to the organism.

Dandelion. This familiar flower is actually concerned with asexual rather than sexual reproduction. There is no meiosis or fertilization during the formation of the seeds. Thus, all the seeds within the fluffy fruits are genetically identical and will flourish under the disturbed, pioneer conditions for which this plant is adapted. (Biophoto Associates)

Strawberry. An individual plant sends runners over the surface of the ground; these take root and send up the shoot of a new plant. The offspring is nourished through the runner during its early development and has an excellent chance of surviving compared to a seedling. (From Arms and Camp, *Biology*)

Hydra. A new offspring can be produced by budding. It is genetically identical to the parent, and remains attached until it can feed and survive on its own. (Biophoto Associates)

Liverwort. These plants are small like their relatives, the mosses, because they lack vascular tissue. They can reproduce asexually by forming gemmae cups, which contain balls of special reproductive cells. When raindrops hit the cups, these cells are splashed out, land nearby, and grow into new liverwort plants. (From Arms and Camp, *Biology*)

Paramecium. Cell division in this protist is not a simple matter. Both the macronucleus and the micronucleus must divide by mitosis before the rest of the cell splits in two during cytokinesis. Two new individual organisms result. (Biophoto Associates)

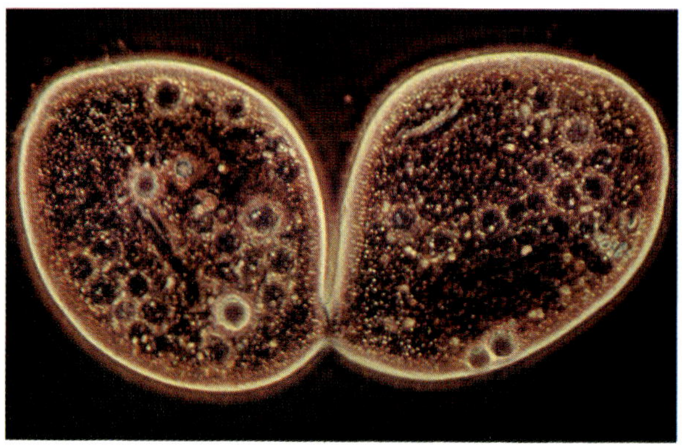

Plate 3. Form and Function in Invertebrates

About 95% of animal species are invertebrates. They have evolved adaptations for surviving in all environments in the biosphere.

Squid. Squid are active predators. They use "jet action" to swim by squirting water out of a siphon with powerful muscles. Their beak is strong enough to attack and eat their prey, and they have excellent vision. (Charles Seaborn)

Hydrozoans Feeding at Night. These delicate cnidarians avoid being seen and eaten by predators because they hide by day and come out to feed only at night. (Steven Webster)

Larvacean. The body of this tiny planktonic filter feeder resembles the larva of a sessile tunicate and secretes a filmy "house" used to trap food particles. Larvaceans may represent a branch of urochordates in which the motile larva became sexually mature and the sessile "adult" stage disappeared. Thus they may be an important link between the chordates and the earliest vertebrates. (Alice Alldredge)

Sea Urchin. This echinoderm is protected by a calcareous skeleton and its "spiny skin." In between the spines are the delicate and versatile tube feet by which it holds tightly to the substrate, or creeps about or grasps food items and moves them to the mouth on its underside. (Charles Seaborn)

Bristleworm. A marine annelid worm that uses its paddle-like parapodia for locomotion. Most annelids have a true coelom and a closed circulatory system and can be much larger than flatworms. (Charles Seaborn)

Sponge. This vase sponge is a filter feeder that extracts particles of food from the current of water that constantly flows through tiny pores in the body wall into the central cavity. Sponges are an evolutionary offshoot and never reached the tissue level of organization. (Charles Seaborn)

Marine Flatworm. This freeliving flatworm is a turbellarian. Gases are exchanged across the thin skin, and nutrients pass directly from the gastrovascular cavity to the body cells. Every cell is close enough to the inner and outer surfaces to meet its physiological needs without benefit of a circulatory system. (Biophoto Associates)

Plate 4. Cells That Make Your Body Work

A simple handshake, the beating of your heart, or the recognition of a friend's face depends on the fine structure and activity of cells we cannot see—except under the microscope.

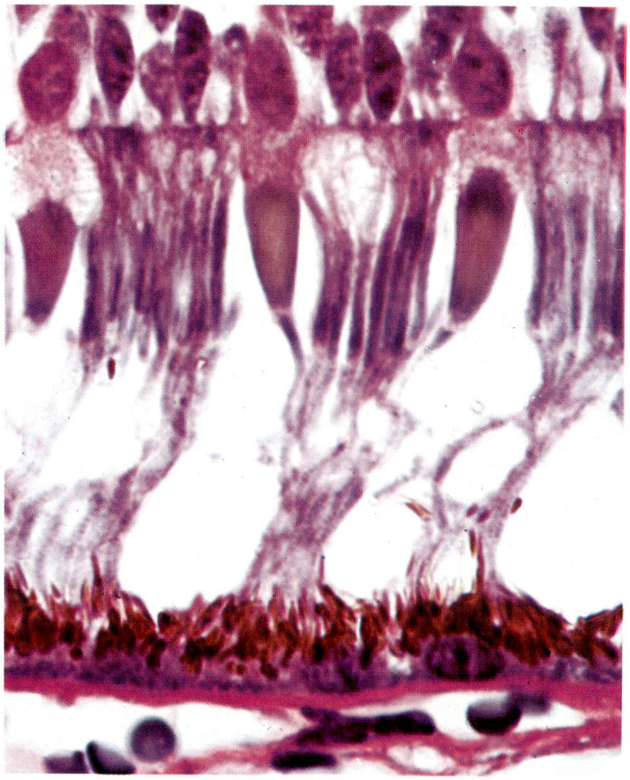

Vision. The retina of your eye contains rod cells, cone cells, and neurons. Visual perception begins with a photochemical reaction of the visual protein rhodopsin. Nerve impulses triggered as rhodopsin reacts are processed in the retina and on their way to the brain where the image is registered. (Biophoto Associates)

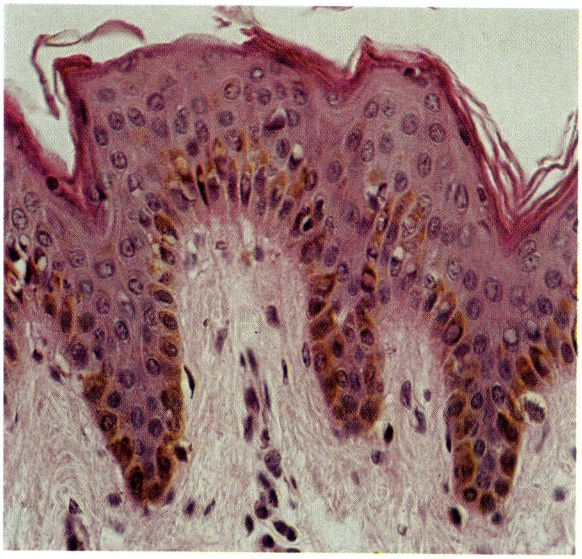

Protection. Stratified squamous epithelium tissue covers and protects the body surface and lines the mouth and vagina. In this cross section of skin, dividing cells in the basement membrane become filled with the protein keratin, flatten, and die as they are pushed upward toward the surface. (Biophoto Associates)

Female Gametes. Ovaries contain about 300,000 potential egg cells. Each month one matures within its follicle and is released at ovulation midway through the cycle. Here you can see a mature follicle in a rabbit ovary. It swells with fluid on the ovary surface and bursts to expel the egg at ovulation. (Biophoto Associates)

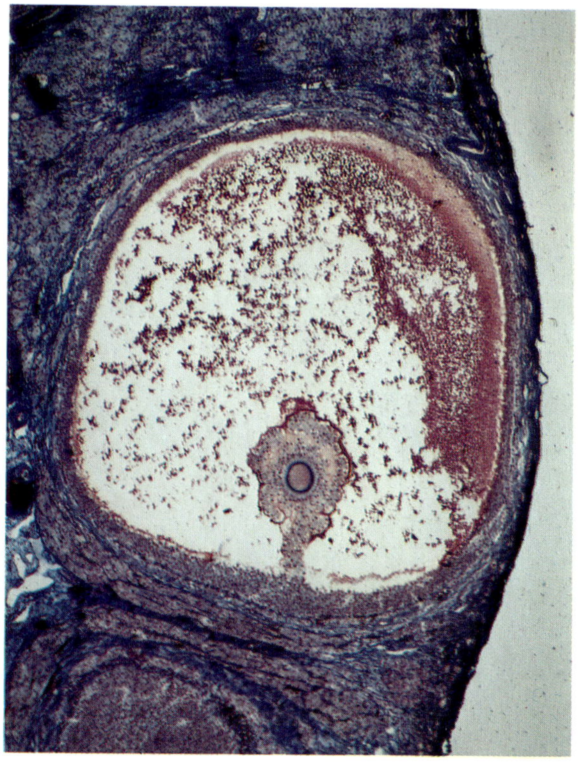

Transport. Blood travels from the heart to the capillaries under high pressure in the thick-walled arteries and returns under low pressure in the thin-walled veins. Aging or damage to the delicate arterial linings can lead to the hard, fatty deposits inside called atherosclerosis. Excessive pressure on the veins may stretch them into varicose veins, which may cause their valves to leak. (Biophoto Associates)

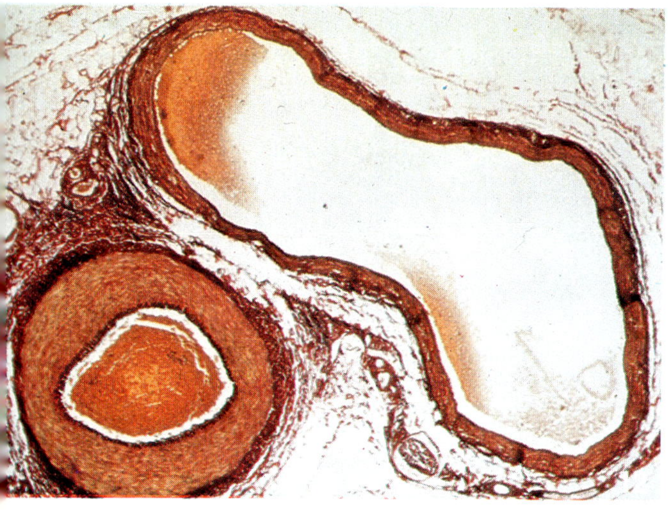

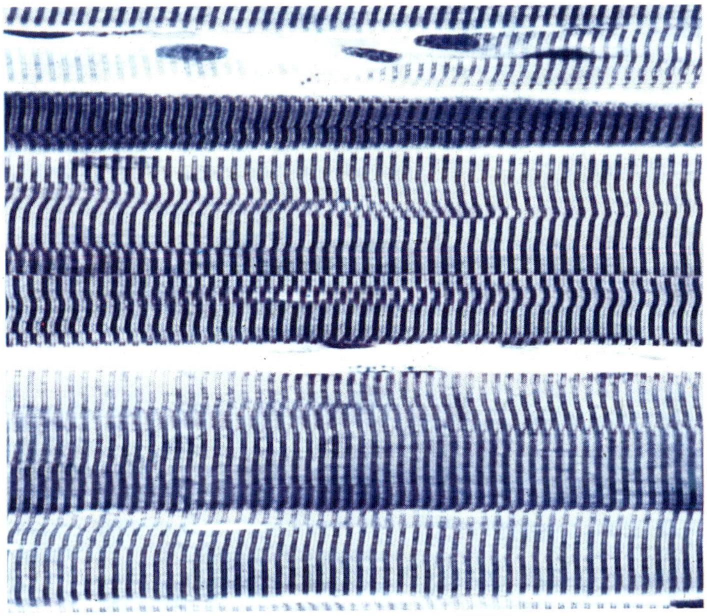

Movement. Our skeletal muscles contain giant cells packed with dense arrays of myofibrils made of protein. To make a muscle cell contract, the noncontractile myofibrils slide past each other in a complex process driven by ATP energy. (Biophoto Associates)

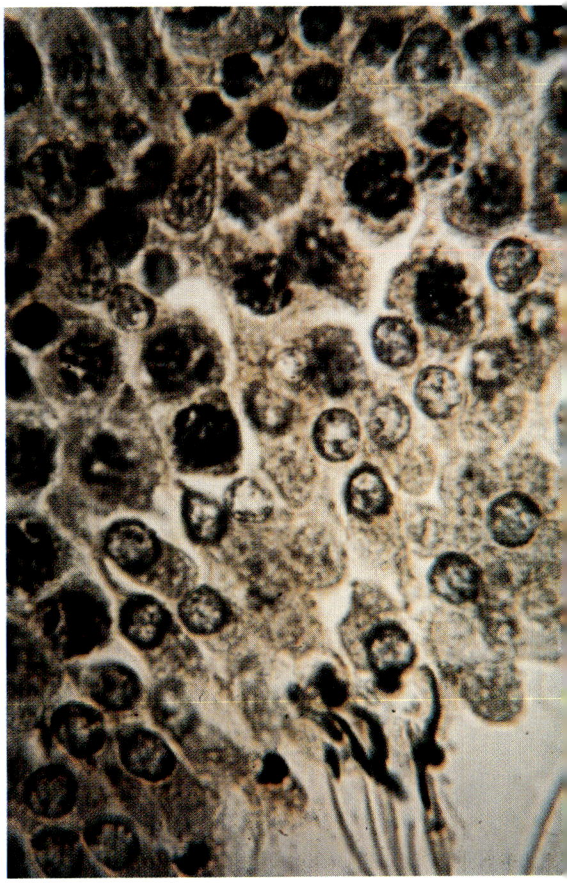

Male Gametes. Sperm cells are continuously formed within the seminiferous tubules of the mature testis. In this cross section of one rat tubule, the youngest immature spermatocytes are next to the basement membrane (outermost cells at top) and move toward the lumen (bottom) as they mature. Sperm within the tubules are not yet motile and are pushed along by tubule contractions. (Biophoto Associates)

Waste. In the kidney, blood plasma (without cells) is forced into the kidney tubules where water and useful nutrients are resorbed and noxious or excess substances are excreted to eventually form the urine. Active transport in the tubule cells accounts for the extremely high metabolic rate of kidney tissue. (Biophoto Associates)

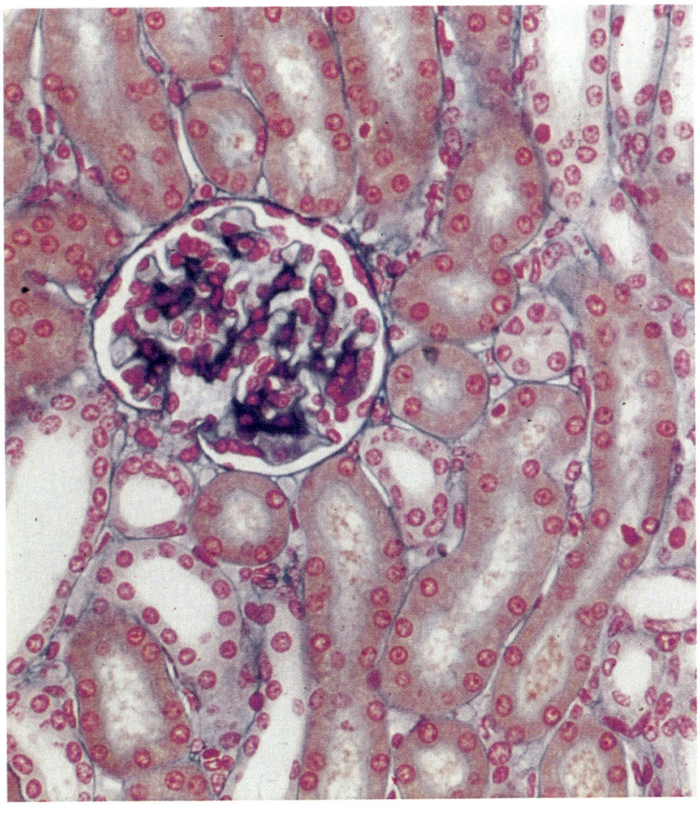

Gas Exchange. Delicate simple epithelium only one cell thick forms the alveoli of the lungs where oxygen and carbon dioxide are exchanged with blood in the capillary beds. (Biophoto Associates)

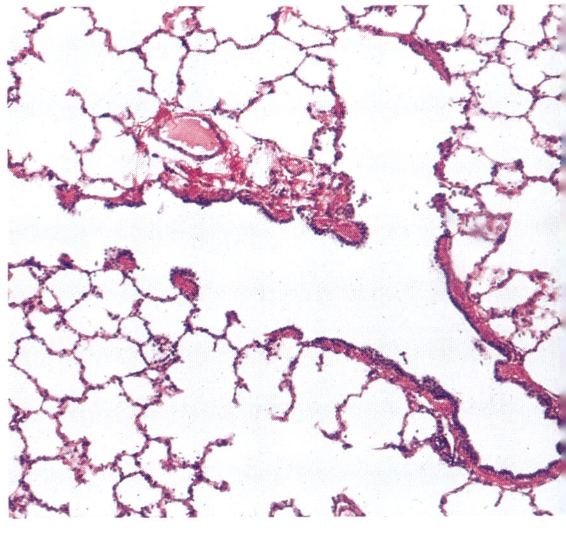

Plate 5. How Plants Cope With Physiological Problems

Plants face most of the basic physiological problems that animals do, but their solutions are sometimes very different.

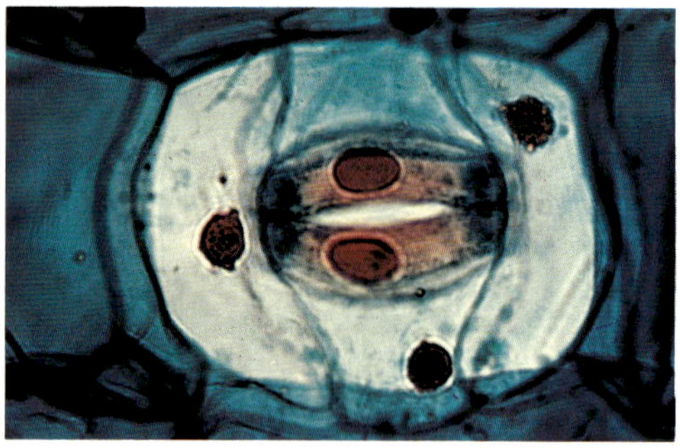

Gas Exchange in a Leaf. Gases pass to and from the spongy leaf interior through pores called stomata. Two guard cells control the size of the stoma and thereby regulate the rate at which gases are exchanged and water is lost from the leaf. (Carolina Biological Supply Company)

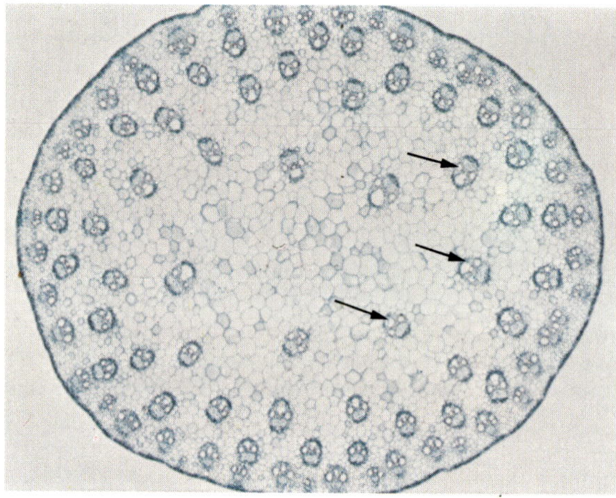

Transport. The vascular bundles (arrows) in this corn stem contain xylem for transporting water and minerals, and phloem for transporting organic nutrients. (Carolina Biological Supply Company)

Water Supply. The tap root system (left) grows deep to obtain water, while the fibrous root system (center) maximizes the surface area for absorption Why does the root system on the right have this bizarre form? (Biophoto Associates)

Reproduction. Milkweed seeds have a tuft of silky fibers to float them on the wind away from competition with the parent plant. Since plants can't move, many of them depend on wind or animals to disperse their seeds. (From Arms and Camp, *Biology*)

Gas Exchange in a Stem. Even the thin bark of a twig is too thick for gases to penetrate, so spongy areas called lenticels allow oxygen to enter and CO_2 to exit. (Carolina Biolgical Supply Company)

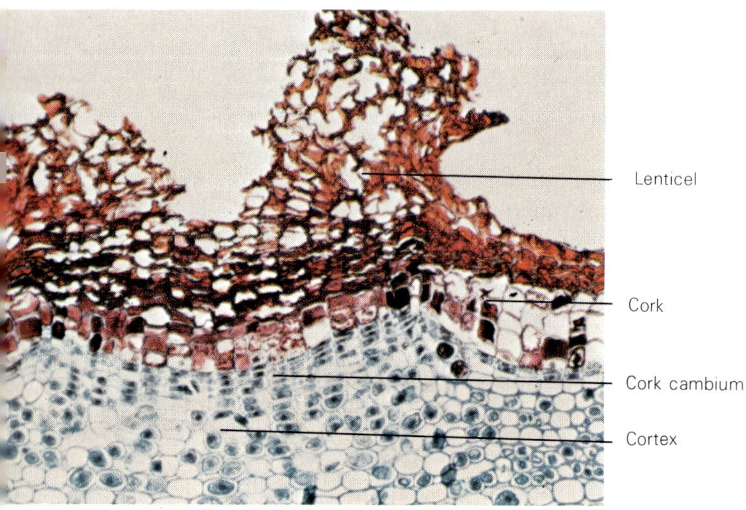

Plate 6. How Plants Cope With Their Environment

Plants are usually restricted to the same location throughout their lives, *but,* as autotrophs, they do not have to spend much energy to locate or digest their food. Their big problems are getting enough sunlight and water and avoiding ingestion by predators.

Venus Flytrap. This carnivorous plant grows in acid bogs where minerals, especially nitrogen, are in such short supply that other plants cannot survive. The trap is a highly modified leaf that bears three trigger hairs and can fold shut when an insect lands and touches the triggers. The trap then secretes fluid to digest the insect and absorbs the nutrients through the leaf surface. (Biophoto Associates)

Euphorb. This is the Old World equivalent of a cactus, adapted to survive under dry conditions. The green stem can carry out photosynthesis and is also adapted to store water after a rain. The leaves are reduced to spines to cut down on water loss and to protect the succulent stem from hungry, thirsty predators. (From Arms and Camp, *Biology*)

Chemical Defense. Many plants contain poisonous or noxious chemicals to discourage insect pests. *Ageratum* protects itself in a sophisticated fashion by producing the chemical precocene. Insects that eat this compound fail to make juvenile hormone and so metamorphose prematurely. (From Arms and Camp, *Biology*)

Predation. Plants are vulnerable to a host of herbivorous predators. This leafhopper sucks fluids from the plant with its beaklike mouthparts. (From Arms and Camp, *Biology*)

Epiphyte. Epiphytes living far off the ground cannot get water from the soil: Other adaptations are necessary. In this case the leaves have pockets at their bases to trap rainwater and dust for the plant's nutrition. (Biophoto Associates)

Plate 7. Coevolution of Flowers and Pollinators

A close, mutually beneficial relationship exists between many flowers and animal pollinators. The flowers are assured of timely, reliable, and sometimes species-specific pollination, and the animal is rewarded with nectar or pollen.

Bird. A bird flower is typically red and has no landing platform. A visiting hummingbird will hover while drinking the nectar and picking up pollen on the top of its head. (From Arms and Camp, *Biology*)

Bee. Bee flowers are often yellow to violet in color with abundant pollen. This wild bee will transfer a generous amount of pollen to the next flower she visits. (Biophoto Associates)

Butterfly. A butterfly sucks up its nectar reward through its long, tubular proboscis. Delicate, tubular flowers without landing platforms, such as thistles, are adapted to these pollinators. (Biophoto Associates)

Orchid. These highly evolved flowers have several strategies for encouraging insects to pollinate a specific species. This particular orchid resembles the rear end of a female bee. The male picks up pollen when he attempts to copulate with the flower, and carries it to other flowers of the same species. (Biophoto Associates)

Beetle or Fly. Flowers depending on these pollinators must have a flat surface on which insects can land or crawl, and they often have a disgusting fragrance that only a beetle or fly would find attractive. These marsh marigolds bloom early in spring and are adapted to pollinators that are active in cooler weather. Their nectar guides are prominent in the ultraviolet, but are not visible to us. (From Arms and Camp, *Biology*)

TABLE 14–1 (continued)

20a. Leaves to 1 in., subopposite.	DAWN REDWOOD
20b. Leaves less than 1 in., alternate (South).	BALD CYPRESS
21a. Leaves 4-sided with persistent woody petiole.	22a
21b. Leaves flattened in cross section.	26a
22a. Leaves dark green or yellow green.	23a
22b. Leaves blue-green.	24a
23a. Branchlets droop.	NORWAY SPRUCE
23b. Branchlets do not droop.	RED SPRUCE
24a. Leaves are at right angles to the stem.	BLUE SPRUCE
24b. Leaves point forward.	25a
25a. Leaves have whitish powder; twigs hairy.	BLACK SPRUCE
25b. Leaves lack whitish powder; twigs not hairy.	WHITE SPRUCE
26a. Leaves lack stomata underneath; seed in red berry; shrub.	27a
26b. Leaves have 2 lines of stomata underneath; tree.	28a
27a. Leaves $\frac{3}{4}$ in. long or less with inconspicuous midrib.	AMERICAN YEW
27b. Leaves 1 in. or longer with prominent midrib.	JAPANESE YEW
28a. Leaves round-tipped, whitened beneath.	29a
28b. Leaves pointed, green beneath	30a
29a. Leaves shorter than $\frac{3}{4}$ in., attached by slender stalks; twigs rough.	EASTERN HEMLOCK
29b. Leaves longer than $\frac{3}{4}$ in., unstalked, alternate; twigs smooth.	BALSAM FIR
30a. Leaves are dark green and straight; bracts with prongs on cones.	DOUGLAS FIR
30b. Leaves are light bluish-green and strongly curved.	FIR

numerical taxonomy and is an example of phenetic taxonomy. It has been widely used in the classification of organisms, diseases, plant communities, politicians, psychological types, socioeconomic neighborhoods, and so forth.

An experimental system to test the procedure of numerical taxonomy was devised by Joseph H. Camin. It involves a set of imaginary animals, christened "Caminalcules," which was generated "according to rules known so far only to him, but believed to be consistent with what is generally known of evolutionary principles."*

Study the set of 29 Caminalcules illustrated in Figure 14–2, and try to arrange the set into some meaningful pattern of groups with anatomical features in common. For example, individuals with eyes might be grouped separately from those without eyes. Then animals with two eyes might have given rise to animals with a single eye or vice versa.

Your instructor will either provide you with a set of cutouts corresponding to Figure 14–2 or will ask you to cut up Figure 14–2 yourself so that you can try out different arrangements.

Arrange all of the individuals into a branching tree called a **phenogram.** Make adjustments to the tree to show how its individual members might have arisen through evolution from a common ancestor. The goal is to construct the tree so that it requires the *fewest number of postulated evolutionary steps.* A change from organism #7 to #15 (or vice versa) requires a single step, whereas a change from #7 to #8 would require two steps. When you are satisfied with the arrangement that you have worked out, record your phenogram in Table 14–2 at the end of this topic. Use numbers to identify the individual Caminalcules, and work with a pencil so that you can change your mind if necessary. Consult your instructor to see how your tree compares with others produced by the class. Your instructor also has a computer-generated phenogram which you can check when you have finished your own scheme.

Return all materials when you are finished with your work for this laboratory.

* From "Numerical taxonomy" by Robert R. Sokal.

REFERENCES AND FURTHER READING

Blackwelder, R.E. *Taxonomy: A Text and Reference Book.* New York: John Wiley and Sons, 1967.

Margulis, L., and K.V. Schwartz. *Five Kingdoms: An Illustrated Guide to the Phyla of Life on Earth.* San Francisco: W.H. Freeman and Company, 1982.

Miller, H.A., *How to Know the Trees,* 3rd ed. Dubuque, IA: W.C. Brown, 1978.

Petrides, George A. *Field Guide to Trees and Shrubs.* New York: Houghton-Mifflin, 1973.

Preston, R.J. *North American Trees.* Ames, IA: Iowa State University Press, 1976.

Savory, T. *Animal Taxonomy.* London: Herremann Educational, 1970.

Journal Articles

Cracraft, Joel. "Cladistic analysis and vicariance biogeography." *American Scientist* 71(3): 273, May 1983.

Leedale, G.F. "How many are the kingdoms of organisms?" *Taxon* 23:261, 1974.

Lowenstein, Jerold. "Molecular approaches to the identification of species. *American Scientist* 73(6): 541, November–December 1985.

Sokal, Robert R. "Numerical taxonomy" *Scientific American,* 106, December 1966 (#1059).

Whittaker, R.H. "New concepts of kingdoms of organisms." *Science* 163: 150, 1969.

Whittaker, R.H., and L. Margulis. "Protist classification and the kingdoms of organisms." *Biosystems* 10: 3, 1978.

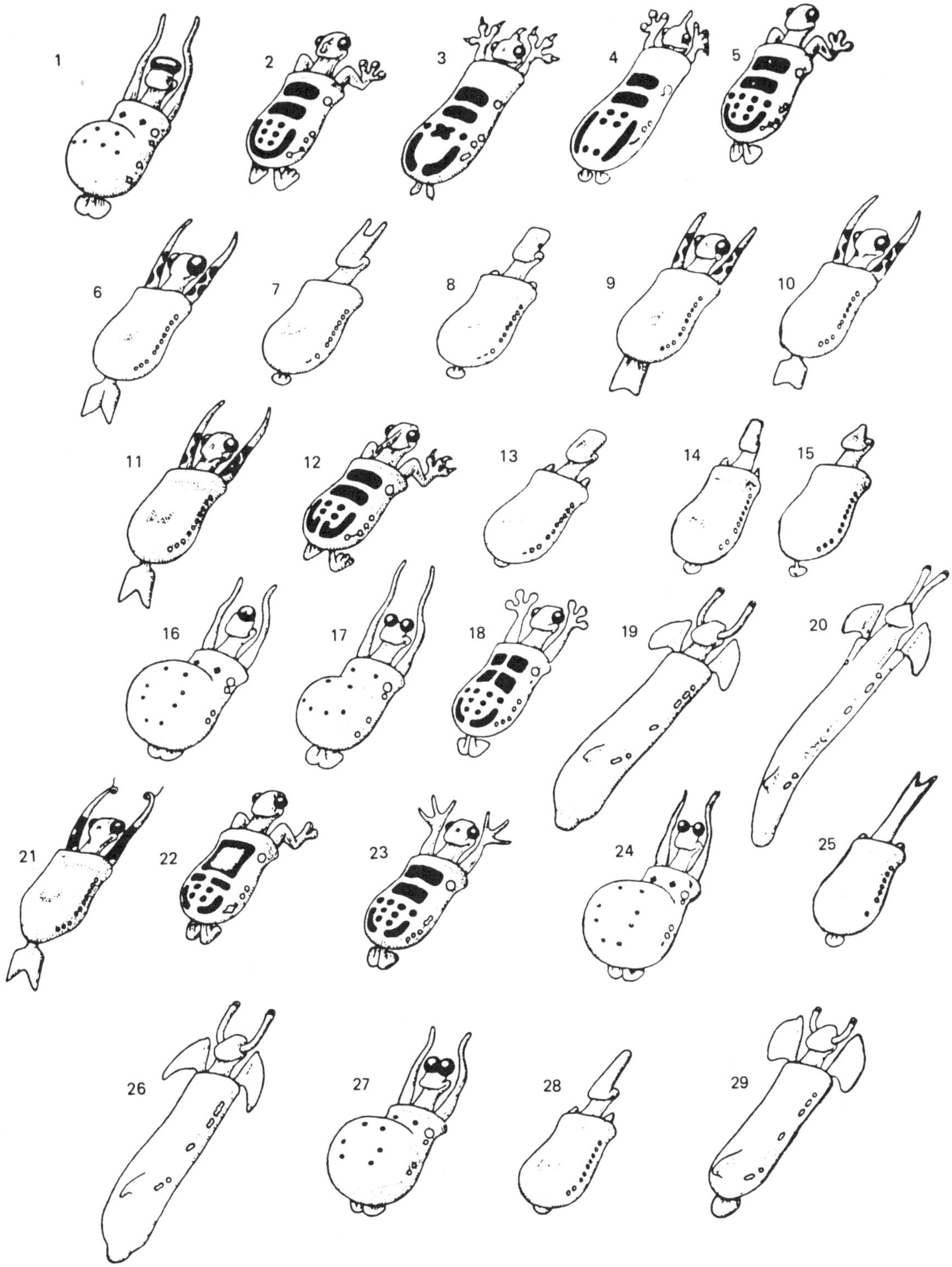

Figure 14–2. Caminalcules. Imaginary organisms ready to be classified. (From "Numerical Taxonomy" by Robert R. Sokal.)

Student Name ______________________________ **Date** ____________________

TABLE 14–2. Phenogram of Caminalcules

Use the identifying numbers to record your classification scheme in the form of a branching diagram.

TOPIC 15

Bacteria and Protista

OBJECTIVES

When you have completed this topic, you should be able to:

1. Describe the characteristics of prokaryotic organisms.
2. List the types of organisms that are prokaryotic.
3. Demonstrate the occurrence of bacteria in the environment by culturing samples from locations where they might be found.
4. Using bioluminescent bacteria as an oxygen indicator, explain the effect of bacteria on the level of oxygen in a closed system.
5. Describe the features of cyanobacteria that distinguish them from other bacteria, and discuss their role in nature.
6. List the features of eukaryotic cells that distinguish them from prokaryotic cells.
7. Distinguish dinoflagellates from diatoms, and state the importance of these organisms in nature.
8. Describe the animal-like and plant-like characteristics of *Euglena,* a photosynthetic protist that also has features of a flagellate.
9. Describe the most common eukaryotes inhabiting the termite gut in terms of their appearance and locomotion and the phylum to which each belongs.

TEXT REFERENCES Chapter 20.

KEY TOPICS

prokaryotes	eukaryotes	protozoa
eubacteria	algae	flagellates (Zoomastigina)
cyanobacteria	dinoflagellates (Pyrrophyta)	sarcodines (Sarcodina)
pathogenic bacteria	*Euglena* (Euglenophyta)	ciliates (Ciliophora)
symbiotic bacteria	diatoms (Chrysophyta)	

INTRODUCTION

In this exercise you will study the variety of single-celled organisms and learn something about their importance to larger organisms like ourselves. Some single-celled organisms are prokaryotes and belong to Kingdom Monera. These are the **archaeobacteria** and all of the **eubacteria,** including the **cyanobacteria** (blue-green photosynthetic bacteria). As a taxonomic group, the bacteria are set aside from eukaryotes by virtue of the unique structure of their **cell walls,** by a lack of membrane-bound **organelles,** and by DNA located in a **nuclear area** (lack of a true nucleus). Other important differences are their **small size** (1–10 μm), rapid rate of **reproduction,** and their metabolic and physiological **diversity.** Bacteria are so different from eukaryotes that as the Russian biologist Oparin has pointed out, a bacterium is as different from an amoeba as that amoeba is from a human. (See Text Figure 5–2.)

All other organisms belong to the eukaryotes and are placed in one of the other four kingdoms. Eukaryotic cells have a membrane-bound **true nucleus** in which DNA is complexed with protein in linear chromosomes, membrane-bound **organelles** such as mitochondria and chloroplasts, and complexity possible because of their **large size, microtubules,** and **microfilaments.** They divide by **mitosis** (or meiosis). Most of those that are single cells are grouped together in Kingdom Protista, although a few are classified with closely related organisms in one of the other kingdoms. Those protists that are generally motile and ingest their food are called **protozoa.** A few protozoans do have chloroplasts and can carry out photosynthesis. Most of the photosynthetic autotrophs, however, are considered **algae** and are grouped separately in their own phyla, depending on their type of cell wall and chlorophyll. These relationships are shown in Figure 15–1. The division of protists into protozoa and algae is unfortunately not at all clear-cut, and certain heterotrophs are obviously very closely related to certain autotrophs. Sometimes an autotroph can ingest food if kept in the dark and sometimes it may even lose its chloroplasts altogether and continue to exist perfectly well as a colorless carbon copy of its former self.

Eukaryotes arose from prokaryotes about 1.45 billion years ago and gave rise to the immense variety of higher organisms that we have today. (See Text Figure 18–7.) The prokaryotes, however, existed for some 3 billion years before that time and also underwent extensive evolution, so that they are an equally diverse group at present. They continue to compete successfully with eukaryotes and are vital in maintaining the stability of our biosphere. Eukaryotes and prokaryotes are compared in Text Table 5–2, p. 77.

KINGDOM MONERA

Bacteria are so small (often only about 1 μm in size) that their internal structure is not visible in the light microscope. For this reason, their identification depends instead on their shape and whether they stain positively or negatively in the Gram test for type of cell wall. Their arrangement in groups, the appearance of their colonies on culture plates, and the metabolic reactions that they can carry out are also aids in identification.

Bacteria in the Environment

SHAPE

The form of a bacterium may be a sphere (**coccus**), a rod (**bacillus**), or a spiral (**spirillum**). Any of these forms may have one or more **flagella** and thus be motile.

- Review the basic bacterial forms by examining a prepared, stained slide under oil immersion or high power.

COLONIES: AIRBORNE BACTERIA

Any organism of sufficient size is covered with a collection of bacteria called its "natural flora." Most of these symbiotic bacteria are harmless to the host (**commensal**), or even helpful (**mutualistic**), but some of them are **pathogens** which have the potential of causing disease. Even otherwise harmless bacteria can cause infection and disease if they gain access to parts of our body that are normally protected. The harmless bacteria of our natural flora help protect our bodies by maintaining conditions that are unfavorable for the growth of pathogens.

Many bacteria on our bodies, in the air, and in our immediate environment can be grown in the laboratory or **cultured** by providing them with a suitable medium that simulates the conditions to which they are accustomed. The medium can be either a liquid broth or a solid gel that is made by adding agar to a liquid medium and pouring the sterilized liquid into a flat Petri plate where it hardens and can be kept sterile. When sterile medium is **inoculated** by adding a sample of bacteria, the bacteria will multiply rapidly due to the lack of competition and will soon exhaust the nutrients. A liquid broth will become cloudy due to the presence of billions of cells, and an agar Petri plate will become covered with dots of **colonies,** or clones, each consisting of 10^9 or so cells all descended from a single cell. The size, shape, color, and appearance of the colony help identify the type of bacterium producing it. **Indicators** may be added to the agar to show whether a certain reaction has been carried out. Any bacterium can be identified if you can find conditions under which the bacterium will grow in the laboratory and have the patience to carry out a sufficient number of tests.

In the following experiment, follow the directions given by your instructor for sealing the culture plates and for disposing of them properly. Never open plates that have been incubated because they may contain pathogenic bacteria.

Using nutrient agar plates, test for the presence of airborne bacteria.

- Obtain 3 agar plates and a marker.
- Label one plate AIR, and add your initials and the date.
- Place the plate near a source of moving air such as an open window, a hallway with traffic, or an air conditioner, and open it fully for 5 min.
- Close the plate and place it upside down in the incubator at 37°C for 24–48 hr.
- Label a second plate COUGH, and add your initials and the date.

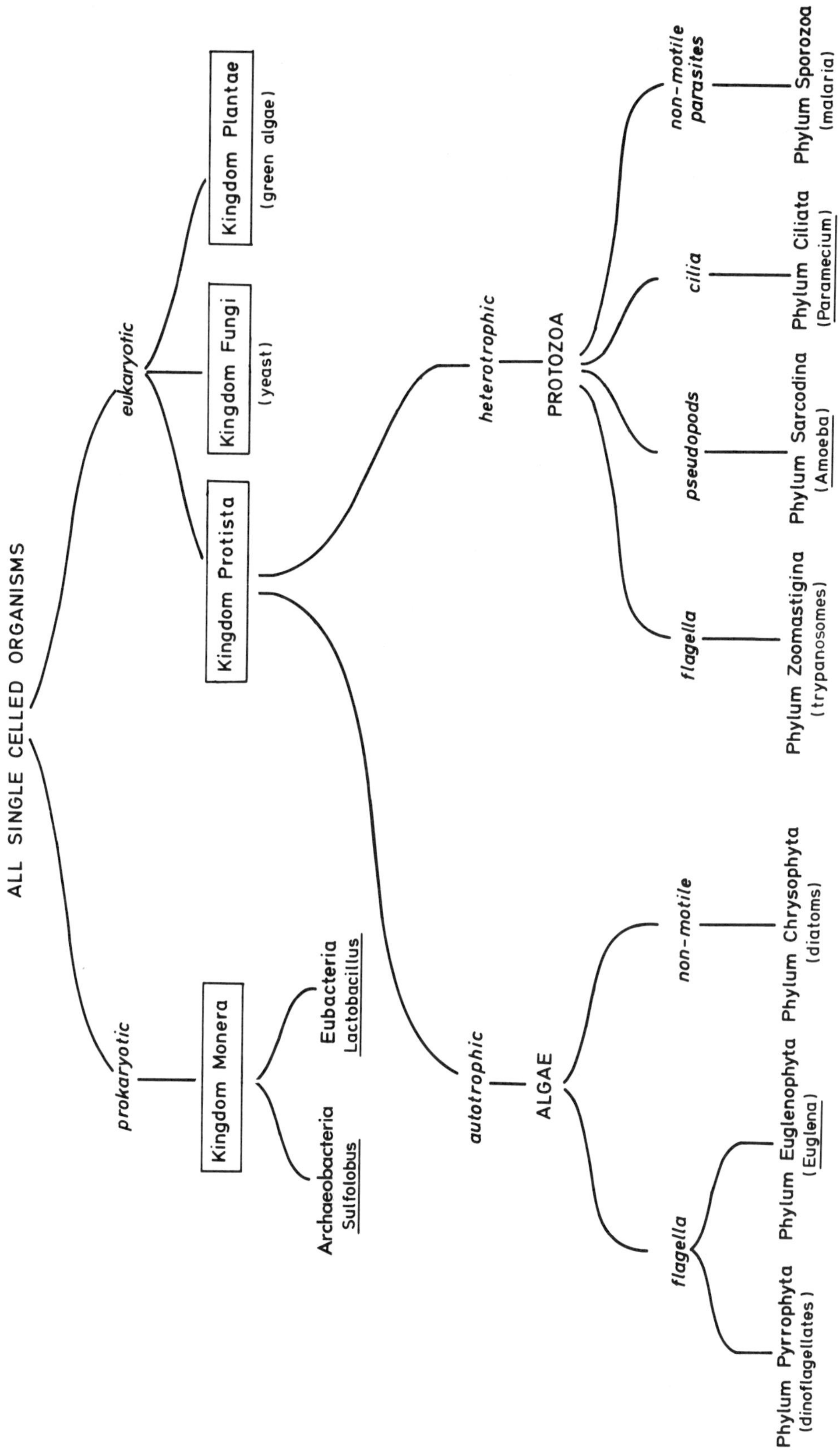

Figure 15–1. Classification of single-celled organisms.

- Open the plate and cough over it with your mouth open three times.
- Close the plate and place it upside down in the incubator at 37°C for 24–48 hr.
- Label a third plate SUPPRESSED COUGH, and add your initials and the date.
- Cough over the plate three times but either keep your mouth closed or use a handkerchief. Close the plate.
- At the end of the experiment record the number of colonies on each plate, and enter the results in Table 15–1 at the end of this topic. If there are too many colonies on a plate, count only a sector such as one quarter or one tenth; multiply by the appropriate factor to determine the number of colonies estimated for the total plate.
- Dispose of the plates properly.

Are you or the air in general a better (or worse!) source of bacteria? ________________

____

Did suppressing your cough cut down on the number of bacteria that you spread? ________________

____

METABOLISM: ORAL BACTERIA

The natural flora of your mouth is important in determining the susceptibility of your teeth to decay (see Text Figure 20–4). *Lactobacillus* is a type of Gram-positive bacterium that contributes to conditions favoring decay by producing acid that lowers the pH at the tooth surface. Even though this bacterium does not actually cause the decay (the culprit may be *Streptococcus mutans*) there is a strong statistical correlation between the activity of *Lactobacillus* and susceptibility to decay. It is possible to test for the presence of this bacterium by using Snyder's medium, which contains an indicator (bromcresol green) that turns yellow when acid is produced.

- If you have not looked at some of your own bacteria, turn to Topic 5 and do so now, following the directions for obtaining and staining a sample.
- Obtain 3 tubes of Snyder's medium and label them 1, 2, and 3. Also write your initials and the date on each tube.
- Chew some sugarless gum without swallowing and drool in Tube 1 (BEFORE) until the surface of the agar is completely covered with liquid saliva.
- If you have brought your toothbrush, brush your teeth thoroughly, and repeat the saliva collection in Tube 2 (AFTER).
- Incubates Tubes 1 (BEFORE), 2 (AFTER), and 3 (uninoculated CONTROL) at 37°.
- As your schedule permits, observe the tubes each day for 4 days and record the progress of the yellow sector in Table 15–2 at the end of this topic. Score the tubes as follows:

Observation	Score
No color change	–
Color changing to yellow at top	+
Top half of tube yellow	++
Top $\frac{3}{4}$ of tube yellow	+++
Entire tube yellow	++++

- Use the interpretation chart in Table 15–2 to determine your susceptibility to tooth decay with and without brushing.

Bioluminescent Bacteria

Some marine bacteria can produce light under favorable conditions in the presence of oxygen. These Gram-negative bacteria are quite plentiful in sea water and are also symbiotic in the light organs of certain fish. The fish provide the bacteria with nutrients and oxygenation and then utilize the light produced by the bacteria for their own purposes—for example, communicating, attracting prey, and defending against predators.

The reaction requires free **oxygen,** the enzyme **luciferase,** and reduced **flavin mononucleotide** (FMN_{red}):

$$\text{aldehyde} + FMN_{red} + O_2 \xrightarrow{\text{luciferase}} \text{acid} + FMN_{ox} + \text{light}$$

The bacteria must be actively growing to produce enough enzyme and reduced FMN to carry out the reaction in which aldehyde and FMN are oxidized by oxygen.

- Fill a test tube two-thirds full with an active culture of luminescent bacteria such as *Vibrio fischeri,* also known as *Photobacterium fischeri.*
- Cork the tube, and take it into a dark closet or darkroom.
- Shake the tube gently to mix it well.

What color of light is produced? ________________

- Hold the tube very still and watch the light fade away.

What part of the tube becomes dark most quickly?

____

____

Which component of the reaction is being used up so that the cells become dark? ________________

____

■ Test your answer by gently swirling the tube again.

Why does mixing the tube allow the cells to light up?

____

____

____

____

Do the cells at the air interface ever become dark?

____

These bacteria are indicators of the amount of oxygen in the environment. There is very little oxygen in the part of the tube that is dark, and this is not enough to support bioluminescence.

How is oxygenation of bodies of water such as ponds and streams maintained so that fish and other aerobic organisms can survive?

____

____

____

____

Cyanobacteria (Blue-green Bacteria)

Cyanobacteria are widespread in the environment and can exist under extremely harsh conditions in which organic nutrients are entirely lacking. They are a problem in the laboratory because they are able to contaminate distilled water supplies that are exposed to light by using the light for energy and obtaining minerals from the traces that remain in the distilled water. Most cyanobacteria carry out **photosynthesis** with the production of **oxygen** and thus help recycle CO_2 into organic compounds. Many also carry out **nitrogen fixation** by converting atmospheric N_2 into ammonia (NH_3), which can be used to synthesize amino acids (see Color Plate 1 and Text Figure 20–8). The ability of these forms to obtain energy, carbon, and nitrogen from the air makes them the most nutritionally independent organisms on earth. In addition, their prokaryotic features make them very hardy and allow them to compete successfully with eukaryotes. They lack cilia and flagella, but some forms are capable of a gliding movement.

When conditions favor their rapid multiplication, cyanobacteria tend to accumulate and form a mat on the surface of a body of water, because their gelatinous capsules and toxins make many of them poor food for predators. This depletes the oxygen in the water and causes the suffocation of the aerobic organisms beneath. Finally, anaerobic bacteria flourish and produce foul-smelling gases, such as hydrogen sulfide, that make the water unfit for drinking or recreation.

■ Examine a drop of *Anabaena* culture under high power in your microscope.

Is this organism in the form of single cells, colonies, or filaments? ______________

Are all the cells alike? ______________

The larger cells are called **heterocysts** and are specialized for nitrogen fixation. (They may not be present in your sample.)

What color are the majority of the cells?

____

What color are the heterocysts? ______________

Do you see chloroplasts? ______________

Cyanobacteria are not always blue-green in color, but they always contain chlorophyll *a* as do green plants. The accessory pigments **phycoerythrin** and **phycobilin** may give them a red, purple, or brown color. The pigments are located on membranes, and there are no membrane-bound chloroplasts.

KINGDOM PROTISTA

Phylum Pyrrophyta: Dinoflagellates

This phylum contains the **dinoflagellates,** which are simple autotrophs and most of which contain chlorophylls *a* and *c* for photosynthesis. Their accessory pigments often give them a red color. Dinoflagellates are an important component of the marine phytoplankton and make a substantial contribution to replenishing atmospheric oxygen. Their chromosomes contain DNA but not protein, and the division of the nucleus during reproduction of the cell is a bizarre intermediate between prokaryotic and eukaryotic nuclear division. Most have two flagella, and some have distinctively formed cellulose armor plates just under the cell membrane. Figure 15–2 illustrates some dinoflagellates.

When conditions favor their reproduction, dinoflagellates "bloom" in large numbers causing a **red tide.** One species of *Gonyaulax* produces a toxin that kills fish and mussels that ingest it in large numbers. Other species are bioluminescent and cause "sparks" in sea water at night during the hot summer months. *Noctiluca,* as its name (= night light) implies, also causes flashes of bioluminescence at night. Each cell produces repeated bursts of light in response to the agitation of wave action or the movements of a nighttime swimmer.

■ Examine a prepared slide of some dinoflagellates in your microscope.

■ Note the presence of flagella.

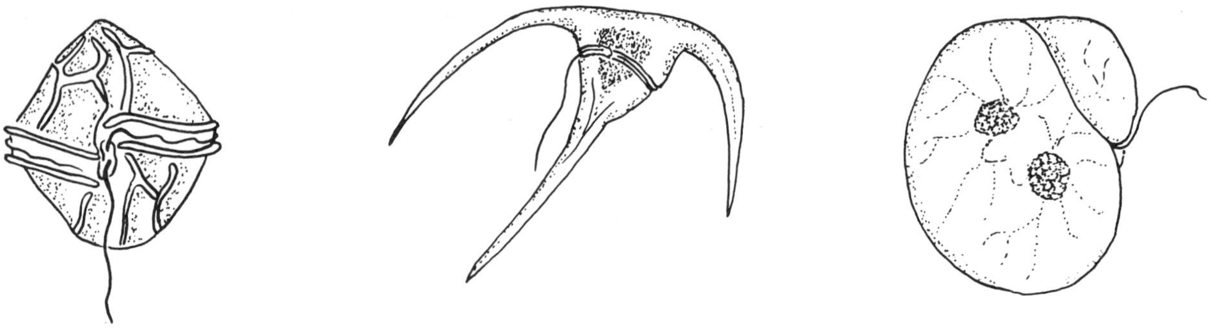

Figure 15–2. Dinoflagellates. These motile autotrophs belong to Phylum Pyrrophyta and may be toxic to fish and shellfish.

How big are the organisms? ______________ μm

What advantage would a dinoflagellate have if it stored its reserve food as oil rather than starch?

____

If a mutant dinoflagellate lacking flagella arose, how would its probability of biological success be affected?

____

____

Phylum Euglenophyta: *Euglena*

In Topic 5 you examined *Eremosphaera,* which has so much in common with multicellular autotrophs that it is classified in Kingdom Plantae. Photosynthetic protists, such as *Euglena,* are not as closely related to green plants; in fact *Euglena* may be related to the simple ancestral flagellates that gave rise to both modern plants and animals (see Color Plate 1). A diagram of *Euglena* is shown in Figure 15–5 at the end of this topic.

Euglena has specialized organelles in which carbohydrate is stored, a characteristic of green plants. It has a photosensitive **eyespot** which allows it to perceive and respond to light. Also, like *Eremosphaera, Euglena* has chlorophylls *a* and *b* contained in chloroplasts, but *Euglena* has two flagella and lacks a thick cell wall composed of cellulose. Instead it has a tough elastic **pellicle** that incorporates the cell membrane. This allows *Euglena* to move by changing its shape and wriggling along the substrate in a unique form of locomotion called **euglenoid movement.** Furthermore, this organism is capable of absorptive nutrition just like a heterotroph. If it is deprived of light, but given a rich medium, it will survive. Its photosynthetic pigments will be lost, and it will not be able to resume autotrophy if the nutrients are removed. This conversion of an autotroph to its sister heterotroph can also be caused artificially by treatment with certain drugs.

Would you expect *Euglena* to show a positive or negative response to light? ______________

- Cover a culture dish of *Euglena* with black paper or plastic, and make a small hole on one side.
- Place the dish so that the hole is near a source of strong light such as a lamp or sunlit window.
- Examine the dish at the end of the period, and observe whether the *Euglena* have migrated toward or away from the source of light.

What kind of response does *Euglena* show?

____

- Make a wet mount of a *Euglena* culture, add a drop of Protoslo, and observe under high power.
- Try to locate the **eyespot** (stigma) near the attachment of the flagellum.

What color is the eyespot? ______________

- Locate the **reservoir** (gullet) in which the base of the flagella are attached.
- Locate the **contractile vacuole** near the reservoir.
- Identify the green **chloroplasts** that take up much of the space within the cell.
- Locate the large **nucleus.**
- Locate the **pellicle,** which gives the cell support at the outer boundary of the cell. Note the location of the **cell membrane** within it.
- Scan the slide for cells that may have become attached to the substrate. Observe their method of locomotion.

Does *Euglena* have a rigid cell wall? ______________

- Stain the slide with iodine solution, and allow a few minutes for the stain to penetrate.
- Locate the **paramylum bodies,** which stain dark. Paramylum bodies contain paramylum, a storage form of glucose.
- Study the flagella, which should now be clearly visible.

■ Label these structures in Figure 15–5, at the end of this topic.

Phylum Chrysophyta: Diatoms

Most members of this phylum are **diatoms,** which are another important component of phytoplankton in the sea and in fresh water. Some diatoms occur in soil as well. These tiny autotrophs contain chlorophylls *a* and *c,* and their accessory photosynthetic pigment, **fucoxanthin,** gives them a yellow-brown color. Diatoms store food as the polysaccharide leucosin or as oil. Some forms have a gliding motion, but most are nonmotile. Other related organisms may have flagella.

The intricate skeletons of these organisms are impregnated with silica (SiO_2) and manganese, and often take the form of two overlapping **valves** that fit together like a round, flat box and its cover. Because the skeletons are highly resistant to decomposition, they accumulate in huge quantities and produce beds of **diatomaceous earth.** This material is an excellent abrasive and is widely used in industrial products such as fine polishes and toothpaste. A few of these delicate microorganisms are shown in Figure 15–3.

■ Examine a wet mount or prepared slide of diatoms.
■ If you are observing living cells, identify a cell that is pointed at either end.

Does the cell show any movement? ______________

■ Make a wet mount of diatomaceous earth and examine it for round, pillbox-shaped diatoms. Such diatoms are marine and therefore found only in the ocean.

Was your sample of diatomaceous earth of marine origin? ______________________________

Why is the fact that some diatoms store reserve food as oil very important to us?

____

____

Heterotrophic Protists in the Termite

There are four major groups of heterotrophic protists:

1. **Phylum Zoomastigina:** Protists that have one or more whiplike flagella, each of which is equivalent to a very long cilium. *Trichomonas* is a flagellated protist that is responsible for human vaginal infections and, in African countries, *Trypanosoma* causes sleeping sickness. (See Text Figure 20–17.)
2. **Phylum Sarcodina:** Protists that move and feed by means of pseudopods like *Amoeba.*
3. **Phylum Ciliophora:** Protists that have cilia, like *Paramecium.*
4. **Phylum Sporozoa:** Parasitic protists that may have evolved from flagellates and sarcodines and that are no longer motile. Malaria is caused by a sporozoan. (See Text Figure 20–19.)

Examples of the first three types can be seen in any environment that favors the growth of protozoans, such as the internal organs or intestine of a higher animal. Sporozoans are generally parasites of mammals.

Termites need a healthy population of protists in their hindgut in order to survive on their normal diet of wood, which is very rich in cellulose. Certain fla-

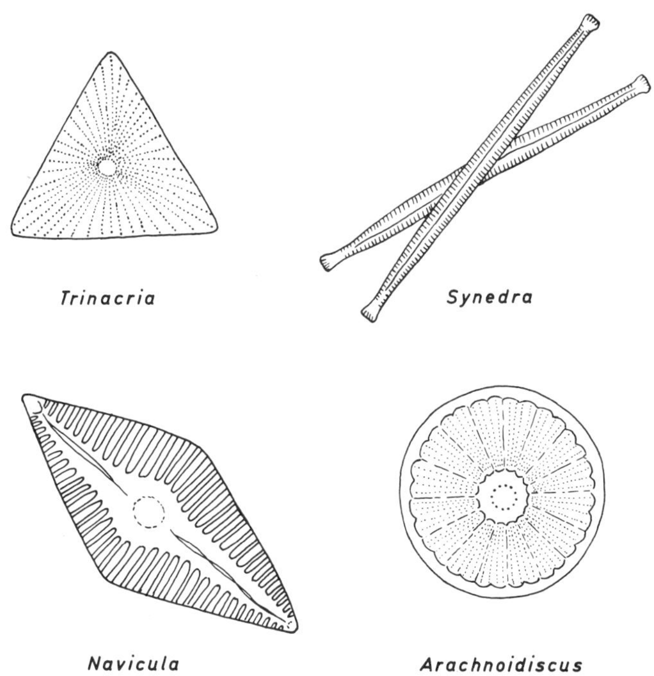

Figure 15–3. Diatoms. These members of Phylum Chrysophyta are photosynthetic and are the most important group of marine phytoplankton.

gellates produce the enzyme **cellulase,** which hydrolyzes the wood fragments ingested by the termite. (See Text Figure 20–9.) The flagellates in exchange release some of their nutrients into the gut of the termite. The termite absorbs and lives on these nutrients. The termite and the flagellate are **mutualistic symbionts.** Young termites must become infected with the flagellates at an early age or they will starve to death. Because termites shed their hindgut and foregut lining and content at each molt, termites that are isolated after molting will not become reinfected and will also starve.

Some of the common symbionts of termites are shown in Figure 15–4. Try to identify one or more of them in your prepared slide or termite wet mount.

- View the intestinal symbionts of a termite in a prepared slide under low, then high power in your microscope or prepare your own wet mount as follows.
- Place a termite in a drop of physiological saline (0.85%) on a slide, and press on the abdomen with the forceps to hold it in place.
- Pull off the head with another forceps; most of the intestine will come out with the head. Separate the free part of the intestine from the head and abdomen.
- Discard the rest of the termite, and tease apart the intestine with needles or insect pins.
- Add a coverslip and observe the intestinal flora under low power in your microscope.
- Add a drop of Protoslo and/or a drop of methylene blue stain and observe under high power.

Which are the most prevalent: amoebalike organisms, flagellates, or ciliates?

Which of the forms shown in Figure 15–4 do you see in your slide?

When you are finished with your work, return the prepared slides, clean up your work place, and store your microscope properly.

REFERENCES AND FURTHER READING

Barnes, R.D. *Invertebrate Zoology,* 5th ed. Philadelphia: Saunders College Publishing, 1987.

Buchsbaum, R. *Animals without Backbones,* rev. 2nd ed. Chicago: University of Chicago Press, 1975. [2nd ed. 1948]

Davis, B.D. *Microbiology.* New York: Harper and Row, 1980.

Jahn, T., and F. Jahn. *How to Know the Protozoa,* 2nd ed. Dubuque, IA: W.C. Brown, 1979.

Kudo, R.P. *Protozoology,* 5th ed. Springfield, IL: Charles C. Thomas, 1977.

Nester, E.W. *et al. Microbiology,* 3rd ed. Philadelphia: Saunders College Publishing, 1983.

Journal Articles

The September, 1981 issue of *Scientific American* is devoted to Microbiology.

Abraham, E.P. "The beta-lactam antibiotics." *Scientific American,* 76, June 1981.

Adler, J. "The sensing of chemicals by bacteria." *Scientific American,* 40, April 1976 (#1337).

Barghoorn, Elso S. "The oldest fossils." *Scientific American,* 30, May 1971 (#895).

Berg, H. "How bacteria swim." *Scientific American,* 36, August 1975 (no offprint).

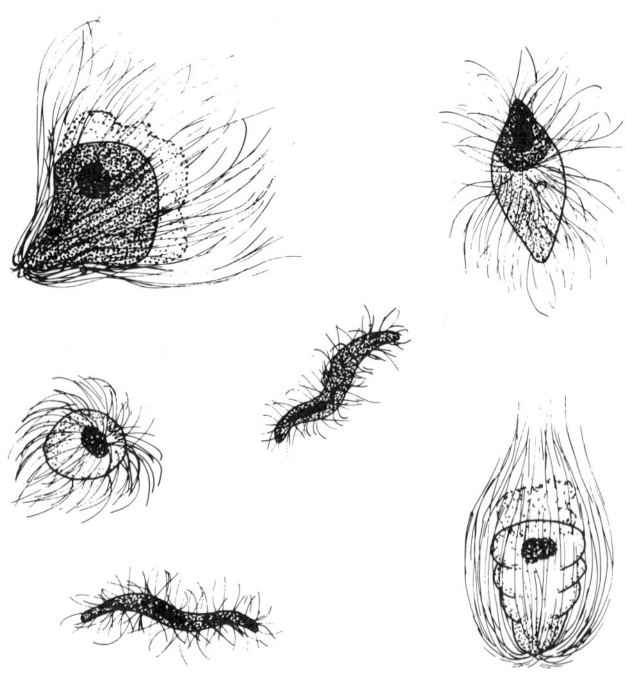

Figure 15–4. Termite gut flagellates. These symbionts must digest the cellulose that the termite ingests or it will starve. A few typical types are shown, including several species of *Trichonympha.* The large pear-shaped individual is *T. collaris,* the diamond-shaped species is *T. agilis,* and the round type is *T. sphaerica.*

Braude, A.I. "Bacterial endotoxins." *Scientific American,* 36, March 1964 (no offprint).

Brill, Winston J. "Biological nitrogen fixation." *Scientific American,* 68, March 1977 (#922).

Cairns-Smith, A.G. "The first organisms." *Scientific American,* 74, June 1985.

Costerton, J.W., G.G. Geesey, and K.-J. Cheng. "How bacteria stick." *Scientific American,* 86, January 1978 (#1379).

Delwiche, C.C. "The nitrogen cycle." *Scientific American,* September 1970.

Donelson, J.E., and M.J. Turner. "How the trypanosome changes its coat." *Scientific American,* 32, February 1985.

Fox, G.E. *et al.* "The phylogeny of prokaryotes." *Science* 209:457, 25 July, 1980.

Fraser, David W., and Joseph E. McDade. "Legionellosis." *Scientific American,* 82, October 1979 (#1447).

Gilbert, Walter, and Lydia Villa-Komaroff. "Useful proteins from recombinant bacteria." *Scientific American,* 74, April 1980 (#1466).

Godson, G. Nigel. "Molecular approaches to malaria vaccines." *Scientific American,* 32, May 1985.

Groves, David I., John S.R. Dunlop, and Roger Buick. "An early habitat of life." *Scientific American,* 64, October 1981.

Jannasch, Holger W., and Carl O. Wirsen. "Microbial life in the deep sea." *Scientific American,* 42, June 1977 (#926).

Kosikowski, Frank V. "Cheese." *Scientific American,* 88, May 1985.

Mandel, I.D. "Dental caries." *American Scientist* 67:680, 1979.

McCosker, John E. "Flashlight fishes." *Scientific American,* 106, March 1977 (#693).

Ourisson, Guy, Pierre Albrecht, and Michel Rohmer. "The microbial origin of fossil fuels." *Scientific American,* 44, August 1984.

Prestwich, Glenn D. "The chemical defenses of termites." *Scientific American,* 78, August 1983.

Schopf, J. William. "The evolution of the earliest cells." *Scientific American,* 110, September 1978 (#1402).

Vidal, Gonzalo. "The oldest eukaryotic cells. *Scientific American,* 48, February 1984.

Walsby, A.E. "The gas vacuoles of blue-green algae." *Scientific American,* 90, August 1977 (#1367).

Woese, Carl R. "Archaebacteria." *Scientific American,* 98, June 1981.

Student Name ______________________ **Date** ______________

TABLE 15–1. Cultures of Bacterial Samples

Nutrient Agar	Source of Inoculum	% of Plate	Number of Colonies Counted	Total Number of Colonies
Plate 1	AIR			
Plate 2	COUGH			
Plate 3	SUPPRESSED COUGH			

TABLE 15–2. Cultures of Oral Bacteria

A. DATA

Observation	Score
No color change	–
Color changing to yellow at top	+
Top half of tube yellow	++
Top $\frac{3}{4}$ of tube yellow	+++
Entire tube yellow	++++

Tube	Source of Inoculum	1 Day	2 Days	3 Days	4 Days
1	Before brushing				
2	After brushing				
3	Control				

B. INTERPRETATION

1 Day	2 Days	3 Days	4 Days	Susceptibility to caries
+	++	+++	++++	Marked
–	+	++	+++	Moderate
–	+/–	+	++	Average
–	–	+/–	+	Slight
–	–	–	+/–	Very low

Which pattern matches your observations most closely for Tube 1? ______________

Did brushing change your observations for Tube 2?

____

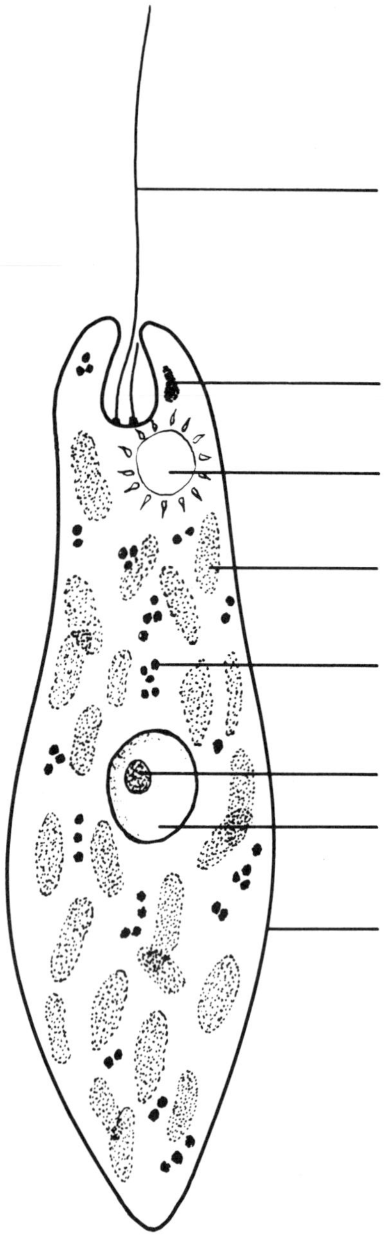

Figure 15–5. ***Euglena.*** Label the indicated structures.

TOPIC 16

Fungi and the Lower Plants

OBJECTIVES

When you have completed this topic, you should be able to:

1. List the plantlike characteristics of fungi.
2. List the characteristics that set fungi apart from the members of the plant kingdom.
3. Name three divisions of true fungi and an example from each division.
4. Describe the two main types of nutrition found in the fungi.
5. List the three divisions of algae, and give the characteristics of each division.
6. Define alternation of generations in plants, and give an example of an alga in which the sporophyte is dominant, another in which the gametophyte is dominant, and one in which neither is dominant.
7. Name the processes giving rise to plant spores, plant gametes, and plant zygotes.
8. Define: dikaryon, mycelium, coenocyte, hyphae, fruiting body, gill, basidium, ascus, zygospore, conidiospore, sporangium.
9. Define: holdfast, stipe, blade, air bladder, zoospore.
10. State some commercially important uses of fungi and algae.

TEXT REFERENCES

Chapter 20 (Q–W); Chapter 21 (A–D).

KEY TOPICS

alternation of generations
zygomycetes (Zygomycota)
ascomycetes (Ascomycota)
basidiomycetes (Basidiomycota)
slime molds
lichens
red algae (Rhodophyta)
brown algae (Phaeophyta)
green algae (Chlorophyta)

INTRODUCTION

Algae and fungi are grouped together because they both have relatively simple bodies. Their cells show little differentiation, and the egg cells are not protected by a layer of sterile jacket cells. The algae are all photosynthetic and the fungi are absorptive heterotrophs.

Most members of these groups are multicellular, but unicellular stages are involved in their life histories. The generalized life history for algae is shown in Figure 16–1 and includes both multicellular and unicellular stages. The multicellular stage may be haploid (N) or diploid (2N). Sometimes a multicellular haploid stage is followed by a multicellular diploid stage and then by multicellular haploid, and so on. In this case, the alga is said to show **alternation of generations** because the haploid and diploid stages alternate with each other.

Plants and fungi, unlike animals, produce **spores** when their cells undergo meiosis. If the haploid organism that grows from the germinating spore produces gametes, the gametes arise by differentiation, not by meiosis, because *the cells are already haploid.* Fusion of haploid cells or nuclei produces the diploid stage of the next generation. In algae and higher plants, the diploid plant that produces spores is called the **sporophyte** (= spore plant) and the haploid plant that produces gametes is called the **gametophyte** (= gamete plant).

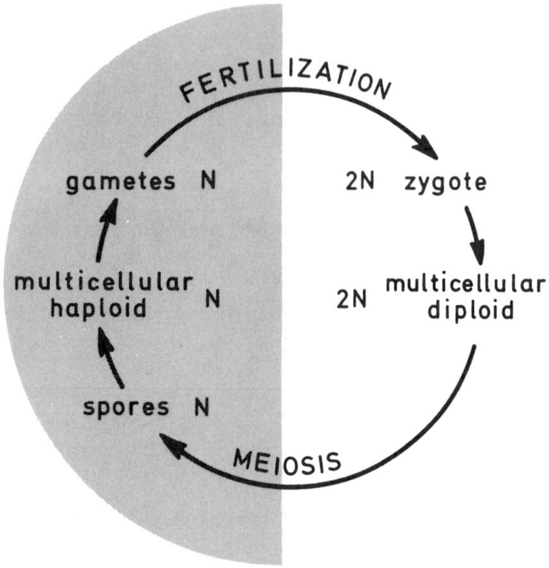

Figure 16–1. Generalized life history. Algal life histories may include one or both multicellular stages.

THE FUNGI (KINGDOM FUNGI)

Fungi are heterotrophic organisms that may have many plantlike features. They are nonmotile and their rigid cell walls contain **cellulose** or the related polysaccharide **chitin.** Because they can neither carry out photosynthesis nor ingest food, they must obtain nutrients by **absorption** through their cell membranes. Often they secrete hydrolytic enzymes into the food source to break it down sufficiently for absorption. If they feed on living cytoplasm, they are **parasites,** and if they feed on dead organisms or detritus, they are **saprobes.** Such fungi, the decomposers, are vital in recycling the dead bodies of organisms so that the nutrients in them can be used by living plants and animals.

True fungi always have **cell walls** and grow as a tangled mass of threads known as a **mycelium.** They can reproduce by **spores** and are classified on the basis of their structures for sexual reproduction. You will study examples of true fungi from the zygomycetes, ascomycetes, and basidiomycetes.

Rhizopus (Division Zygomycota)

The black bread mold *Rhizopus nigrans* forms a mat, or **mycelium,** of threadlike **hyphae.** The nuclei are haploid and the fungus lives in a terrestrial environment.

- Obtain a Petri dish containing a vegetatively (asexually) reproducing mycelium of *Rhizopus.*
- Examine the mold under the dissecting microscope without opening the dish.
- Note the **asexual spores,** which are tiny black spheres inside the **sporangia** that grow at the ends of certain erect hyphae known as **sporangiophores.** These asexual spores are formed by mitosis. Their production and dispersal is a form of vegetative reproduction and they will give rise to new haploid mycelia genetically identical with the parent. *Rhizopus* sporangia are shown in Figure 16–2.
- Make a wet mount of mycelium with sporangia, and examine it in the microscope under low, then high power.
- Note the cluster of rootlike hyphae at the base of the hyphae bearing the sporangium. These hyphae anchor the sporangium and are called **rhizoids.**
- Note the horizontally growing hyphae called **stolons** (see Text Figure 20–26).

A fungus without cross walls is said to form a **coenocyte,** in which all of the nuclei are in a common cytoplasm. If there are cross walls (septa), the fungus is **septate.**

Is the mycelium of *Rhizopus* coenocytic or septate?

____

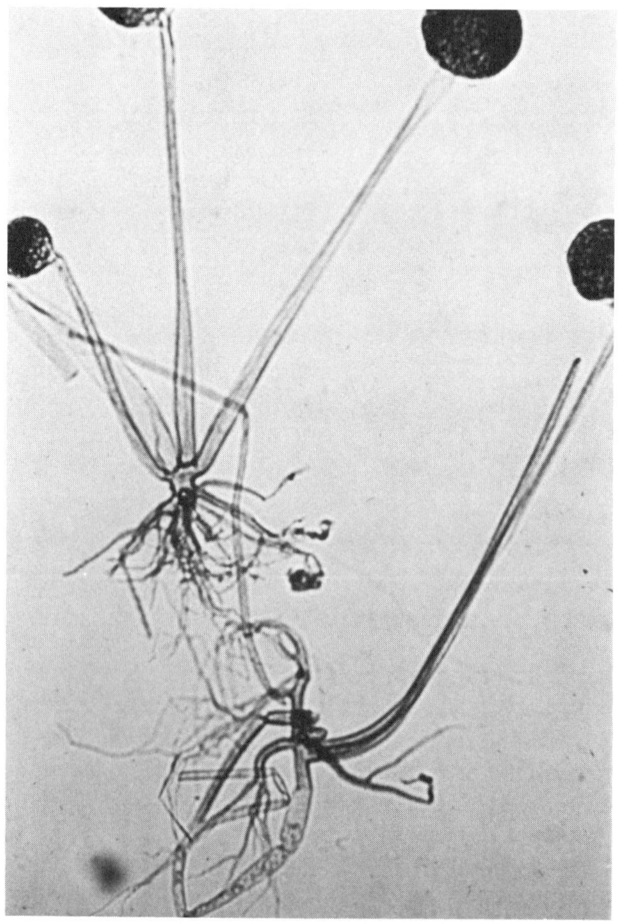

Figure 16–2. Asexual reproduction in *Rhizopus.* The hyphae and stolons of the mycelium grow over and into the food source. Sporangia on specialized hyphae produce and release the asexual spores by which *Rhizopus* usually reproduces. (Photo by Carolina Biological Supply Company.)

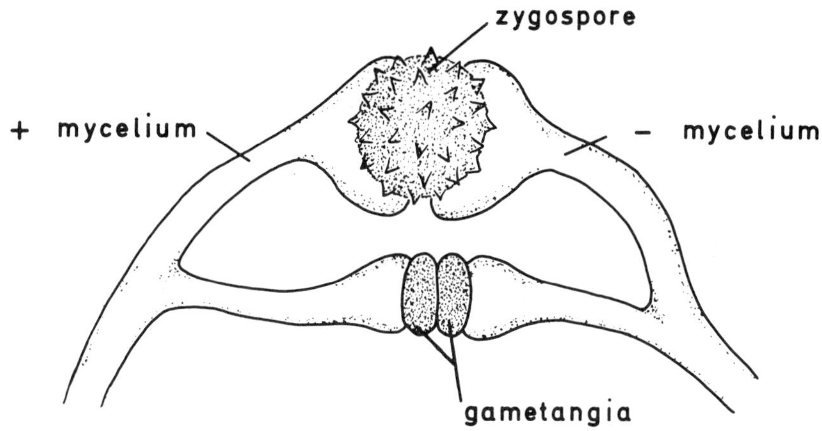

Figure 16–3. Sexual reproduction in *Rhizopus*. The diploid zygospore will undergo meiosis to produce a new haploid mycelium.

- Obtain another culture of *Rhizopus* in which two mating types (+ and −) of mycelia have fused and undergone sexual reproduction.
- Note the dark line in the mycelium where the + and − mating types fused. The dark color is due to the formation of **zygospores.**
- Make a wet mount of a bit of mycelium from the black region, or view a prepared slide, and look for cross bridges between the hyphae.
- Examine such a cross bridge for the presence of a dark sphere, the **zygospore.** (See Figure 16–3.)

When the two types of mycelia come into contact, the bridge is formed and the cell wall breaks down so that the cytoplasms of the two sides are in contact. A nucleus from each side moves into the bridge, and the haploid nuclei fuse to form a diploid nucleus. The zygote nucleus divides several times, and then undergoes meiosis to form a sporangium containing haploid spores. These spores, unlike the ones produced asexually, will be genetically different from either parent.

Yeast (Division Ascomycota)

A yeast such as brewer's yeast (*Saccharomyces cerevisiae*), is a unicellular ascomycete. It has an unusual method of reproducing asexually by **budding** off. (Some other fungi are dimorphic and can grow in a yeastlike form under favorable environmental conditions, such as in sugary solutions. At other times they grow as a mycelium.) Yeast also reproduces sexually by forming a saclike **ascus** containing spores produced by meiosis. The life history of a yeast is shown in Figure 16–4.

- Make a wet mount of a drop of actively growing yeast culture, and observe the cells in the microscope under high power.
- Note the cells that are producing smaller buds or short chains of buds.

Yeasts are important in the baking and brewing industries and are used to produce important industrial chemicals.

What by-product of yeast fermentation is important in baking? ______________________

What by-product is important in brewing?

____

Which of these by-products is produced in both aerobic and anaerobic respiration? ______________

Cup Fungi (Division Ascomycota)

Cup fungi such as *Peziza*, and mushrooms, are the fungi with which we are the most familiar because they are the most conspicuous and are sometimes edible. The large part of the cup fungus that we see is formed of masses of hyphae fused together in the **fruiting body** (ascocarp) formed during sexual reproduction. Each fruiting body is supported by an underground mycelium that is much larger than itself. The hyphae of the mycelium are divided into cells by cross walls, so these fungi are septate. Part of the fruiting body is called a **dikaryon** (dikaryon means "two nuclei") because each of its cells has two haploid nuclei. The cells in this part are different from diploid cells and are designated N + N rather than 2N. During sexual reproduction, some of the cells of the fruiting body undergo nuclear fusion to produce true diploid (2N) nuclei. Meiosis, which follows, produces haploid (N) **ascospores.** The large size of the fruiting body helps disperse the spores. The spores germinate to produce mycelia, and different mating types (+ and −) of mycelia interact to form a new dikaryon and fruiting body.

- Obtain and examine a preserved specimen of a cup fungus such as *Peziza*. The cup shape of the fruiting body allows rain to collect and splash around in-

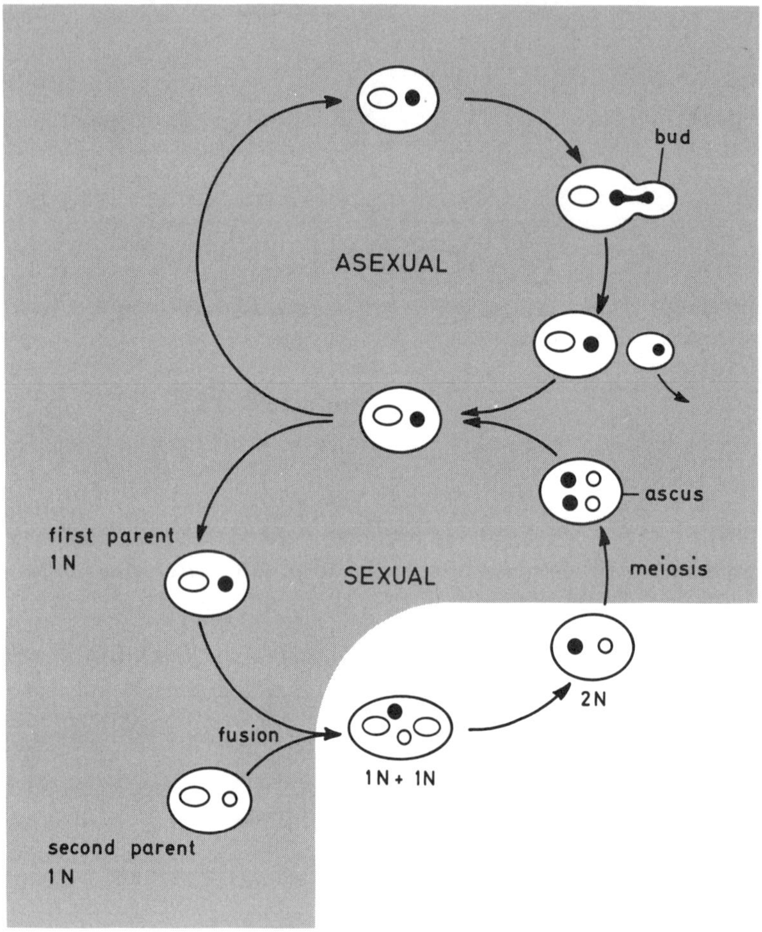

Figure 16–4. The life history of a yeast, *Saccharomyces cerevisiae.* Baker's yeast will reproduce in dough by budding. The cells that fuse in sexual reproduction are of two different mating types.

side where the ascospores are produced, and aids in splashing out the ascospores.

- Examine a cross section of a cup fungus under high power.
- Locate an **ascus.**

How many spores does an individual ascus contain?

____

Cup fungi can also reproduce asexually by forming chains of **conidiospores** at the tips of the hyphae (see Text Figure 20–27b). These tiny, airborne spores are the result of mitosis rather than meiosis. They pervade the entire biosphere, and germinate at every opportunity.

Ascomycetes are biologically interesting because some of them have become obligate symbionts with algae to form **lichens.** (Lichens will be considered later in this exercise.) Ascomycetes such as morels and truffles are gourmet delicacies, and the mold *Penicillium* is the source of the antibiotic penicillin. Other ascomycetes such as smuts and rusts are serious agricultural pests.

Mushroom (Division Basidiomycota)

The common edible mushroom *Agaricus campestris* is the fruiting body (basidiocarp) of a basidiomycete. As in the cup fungus, it is formed by the fusion of masses of hyphae. In sexual reproduction, underground mycelia of two mating types fuse and form a mycelium with two nuclei per cell. This mycelium is an N + N dikaryon. As in the ascomycetes, the mycelia are septate.

- Obtain a common mushroom, and make a cross section of it with a razor blade.
- Note the **stalk** and the **cap** that bears the **gills** on its undersurface.
- Examine the gills with the dissecting microscope.

The reproductive cells, the **basidia,** are located on the surface of the gills. Each basidium undergoes nuclear fusion to form a fused diploid nucleus that immediately enters meiosis. The haploid daughter nuclei form **basidiospores,** which are released from the surface of the gills. Each spore will germinate to form a new mycelium. Sexual reproduction in basidiomycetes is described in Figure 16–5.

- Make a wet mount of a small piece of gill, or examine a prepared slide through the cap of a mature mushroom such as *Coprinus.*
- Note the small, dark basidiospores on the edges of the gills. Each basidium produces four basidiospores.

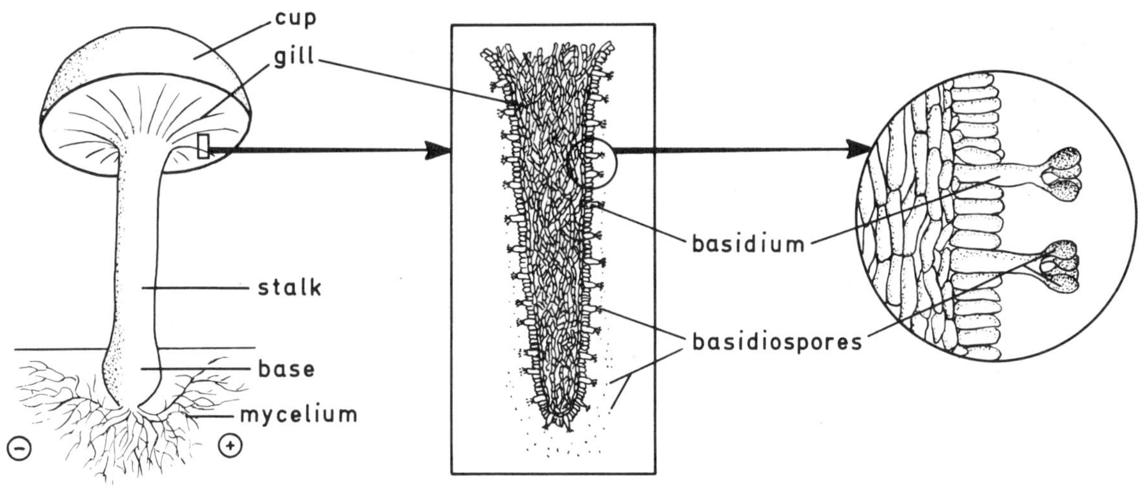

Figure 16–5. Basidiomycete reproduction. The mushroom is the fruiting body formed when + and − mycelia meet and fuse. Fusion of haploid nuclei, followed by meiosis, takes place in the basidium.

Which cells in the life history of a basidiomycete can undergo meiosis? ____________________

What is the difference between a diploid organism and a dikaryon?

____

____

Slime Molds (Divisions Acrasiomycophyta and Myxomycophyta)

Slime molds are grouped separately from true fungi because they have more animal-like features than other fungi. The **cellular slime molds** (Division Acrasiomycophyta) have amoeboid cells that can aggregate to form fruiting bodies. Each fruiting body produces asexual spores that disperse and germinate into more amoebae.

- Examine the demonstration culture of a cellular slime mold such as *Dictyostelium* in the dissecting microscope.
- Locate a sluglike **pseudoplasmodium** (grex), an aggregate of many amoebae that creeps over the agar leaving a slimy trail.
- Locate a **fruiting body** (sorocarp), which is supported by a thin stalk. When mature, it will burst to release the spores.

Acellular slime molds (Division Myxomycophyta) have the form of a gigantic coenocyte called a **plasmodium** in which hundreds of nuclei share a common cytoplasm (see Color Plate 1 and Text Figure 20–23).

- Obtain a Petri dish containing a culture of an acellular slime mold such as *Physarum,* and place it on the stage of your compound microscope.
- Examine the plasmodium under low power (100×).
- Note the "veins" of streaming cytoplasm with many nuclei.

If conditions become unfavorable, the plasmodium will dry up and turn into a resting stage known as a **sclerotium.** Pieces of dried sclerotium can be used to start new cultures of a plasmodium. *Physarum* can also reproduce by sporangia that form on its surface and contain haploid spores. Two types of haploid cells (amoebae and swarm cells) emerge from the spores and fuse to produce new plasmodia.

Lichens (Division Mycophycophyta)

These symbiotic organisms are usually classified by the type of fungus that they contain. The photosynthetic algal component may be a cynobacterium or a green alga. The hyphae of the fungus protect and support the alga, and the photosynthetic ability of the alga provides nourishment, so the lichen can flourish where neither the fungus nor the alga could live by itself. This relationship would be an example of **mutualism.** In some cases the fungus benefits more, so that the relationship is **commensal** or even controlled **parasitism** by the fungus. Lichens are pioneers that can colonize harsh and unfavorable environments where other organisms cannot survive, and they contribute to the breakdown of rocks in the early stages of soil formation. Species of lichen such as reindeer "moss" are widespread in the tundra region where they are an important source of food for herbivores.

There are three basic types of lichens (see Text Figure 20–32). **Foliose** lichens have a leaflike cell body

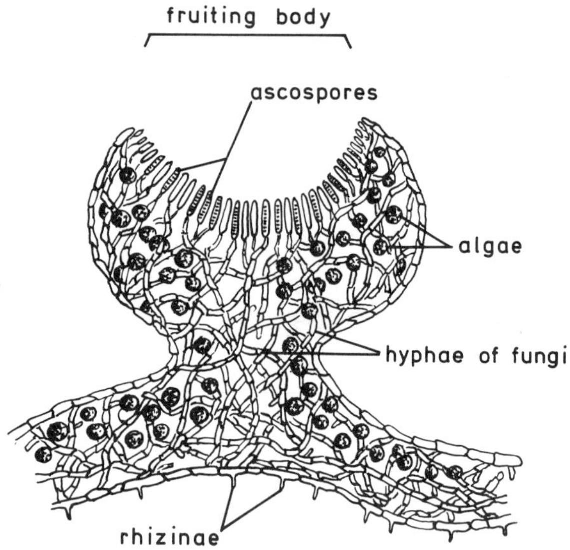

Figure 16–6. Lichen in cross section. This lichen shows the fruiting body of its ascomycete fungus. The lichen itself will reproduce by fragmentation.

and are the type most commonly seen. **Crustose** lichens live as a very thin coating that is tightly bound to its substrate. **Fruticose** lichens have a fruiting body that is a holdover from the fruiting body of the ascomycete fungus that is one of its symbiotic partners. The internal structure of a lichen is shown in Figure 16–6.

- Examine the specimens of foliose, crustose, and fruticose lichens on demonstration.
- Note that the color of lichens is highly variable; the pigments are thought to provide protection from harsh sunlight.
- Examine a prepared slide showing a lichen in cross section. Alternatively, with your instructor's permission, tease apart a small bit of lichen, make a wet mount, and examine it in the microscope.
- Note the hyphae and the algal cells.

Lichens often reproduce by **fragmentation,** in which a piece of the organism simply breaks off and is passively moved to a new location, where it grows into another lichen. Some lichens reproduce by releasing fungal spores.

THE LOWER PLANTS

Almost all plants are photosynthetic and so require only light and inorganic material for their nutrition. They must be able to grow rapidly to reach light. They do not move about, so muscle tissue is absent, and they rely on strong, cellulose cell walls for support and protection. The three plant divisions discussed below as well as the protist autotrophs are described by the term "algae." Green algae (Division Chlorophyta) are most closely related to the ancestors of land plants and have some examples that are intermediate between the single-celled algae and multicellular algae. They differ from red and brown algae in their photosynthetic pigments and the form in which they store food. These are the conservative traits important in the classification of algae.

Red Algae (Division Rhodophyta)

These marine plants may be closely related to the cyanobacteria. Both groups contain only chlorophyll *a* and the accessory pigments **phycocyanin** and **phycoerythrin,** a red pigment found in no other organisms. The photosynthetic membranes are unstacked in both groups and the chloroplasts of red algae may have arisen from cyanobacteria symbiotic within their ancient precursors. Carbohydrate is stored in the cytoplasm as a modified type of starch. The body of a multicellular red alga is usually filamentous or leaflike, and it is attached to the surface by a **holdfast.** There is no flagellated form in the life history, which often includes both multicellular haploid and diploid stages. (See Text Figures 21–2 and 21–3.)

- Examine the specimen of red alga, such as *Chondrus,* on demonstration.

What is the body form of the demonstration alga?

____

Commercially important substances such as carrageenan and agar are obtained from species of red algae, and these algae may be important in building up and maintaining the integrity of coral reefs. Red algae can live at greater depths than any other algae.

If red light is absorbed more strongly by seawater than light of any other color, why can red algae live at the greatest depths?

____

____

Brown Algae (Division Phaeophyta)

These algae contain chlorophylls *a* and *c* and the carotenoid, **fucoxanthin,** and they may be related to the golden algae (relatives of the diatoms in Phylum Chrysophyta), which also contain these pigments. Brown algae store carbohydrate as **laminarin,** and the golden algae store it as a modified laminarin (leucosin).

Examine the specimen of a brown alga such as *Fucus* on demonstration.

- Locate the **holdfast,** the **stipe** (stalk), and the **blade** ("leaf").

Does your specimen contain **air bladders** filled with gas? ____________

Why would air bladders be important?

____

■ Make a cut across one of the blades to see its internal structure.

Are all parts of the blade photosynthetic?

____

In algae of substantial size, the air bladders float the photosynthetic blade near the surface where it can obtain more light. The outer surface of the blade is covered by a gelatinous coating to prevent it from drying out when it becomes exposed to air.

■ Examine the tips of the blades for swollen sacs called **receptacles;** Eggs and sperm are produced within the receptacles (see Text Figure 21–5).

The life history of *Fucus* is similar to the life cycle of animals. The plant itself is diploid, and cells within the receptacles undergo meiosis to produce eggs and flagellated sperm, thereby skipping the multicellular haploid stage. The gametes fuse to form zygotes, and then each zygote grows into a new diploid alga.

The giant brown algae, the kelps, are the source of alginate, a thickening agent important in making ice cream, toothpaste, and many other products. Brown algae have also been used for fertilizer and livestock feed.

Green Algae (Division Chlorophyta)

Green algae contain unicellular and multicellular organisms and are the algae thought to be most closely related to higher plants. Both groups contain chlorophylls *a* and *b* and carotenoids, and store carbohydrate as starch. The body of a green alga may have a holdfast, but its structure is usually simpler than that of a brown alga.

Green algae show three important trends.

1. **Unicellular forms** → **colonial** forms → **multicellular** forms with division of labor among cells.
2. **Isogamy,** where gametes are of equal size → **anisogamy,** where one gamete is larger → **oogamy,** where the egg is much larger and nonmotile.
3. Development of **alternation of generations** with both multicellular haploid and multicellular diploid stages in the life history.

Some of the unicellular and multicellular motile green algae clearly form a related series. For example, the individual cells of *Gonium* or *Pandorina* are almost identical to those of the unicellular alga *Chlamydomonas*. These colonial organisms may have arisen from unicellular ancestors whose cells failed to separate after mitosis. *Gonium* in turn may have given rise to more elaborate colonies such as that of *Volvox,* in which there are hundreds of cells and some division of labor.

The development of oogamy has the obvious advantage that a nonmotile egg can be much larger than one that has to swim, and so can contain food reserves for the early development of the zygote. If the zygote divides by mitosis into a multicellular diploid plant before meiosis, it can produce may more spores than if it undergoes meiosis directly. In certain green algae fertilization and development of the multicellular diploid stage occurs with the body of the female parent so that the embryo can be protected and nourished. It is thought that algae having this strategy were so successful that they were the immediate ancestors of higher land plants.

■ Make a wet mount of *Chlamydomonas,* and view it under high power in your microscope.

■ Note the two **flagella,** the **chloroplast,** and the spot of red pigment (eyespot). The chloroplast contains a **pyrenoid** where it stores starch.

How is *Chlamydomonas* different from *Euglena?* (*Euglena* is described in Topic 15 and illustrated in Figure 15–5.)

____

____

In asexual reproduction, *Chlamydomonas* (haploid) produces four or more progeny per parent by mitotic division. During sexual reproduction, which is isogamous, each parent cell produces many small gametes of equal size, again by mitosis. Gametes from two mating types fuse to form a diploid zygote which may become dormant to survive the harsh conditions that brought about sexual reproduction. Eventually the dormant zygospore will become active and undergo meiosis to produce spores directly without going through a multicellular diploid stage. These spores swim about and develop into the cells of the next generation. Note that the zygote is the only diploid stage in this life history.

■ Make a wet mount of *Gonium* and examine it under low power.

How many cells are there, on the average, in a colony? _______________

Do the cells of *Gonium* appear any different from those of *Chlamydomonas?* _______________

The gametes produced by *Gonium* in sexual reproduction are flagellated but are different in size (anisogamy). Otherwise the life history is like that of *Chlamydomonas.*

Volvox is a motile colony of hundreds of haploid cells with some cell differentiation.

■ Place a drop of *Volvox* culture on a slide without a coverslip, and view it under your dissecting microscope.

What structures propel the colony about?

____

- Add a coverslip and view an immobilized colony under high power. If necessary, suck away some of the water with a piece of tissue, or use Protoslo.
- Note the spacing of the cells. They are connected by thin threads of cytoplasm that allow the swimming movements of the colony to be coordinated.

Refer to the life history of *Volvox* illustrated in Figure 16–7.

Volvox produces egg and sperm cells, and the sperm must swim to the nonmotile egg (oogamy). The diploid zygote undergoes meiosis to form a new haploid colony. In asexual reproduction, vegetative cells are released inside the mother colony, where they develop into new daughter colonies. The daughters are released when the old colony breaks open. This is a good example of altruism among plants.

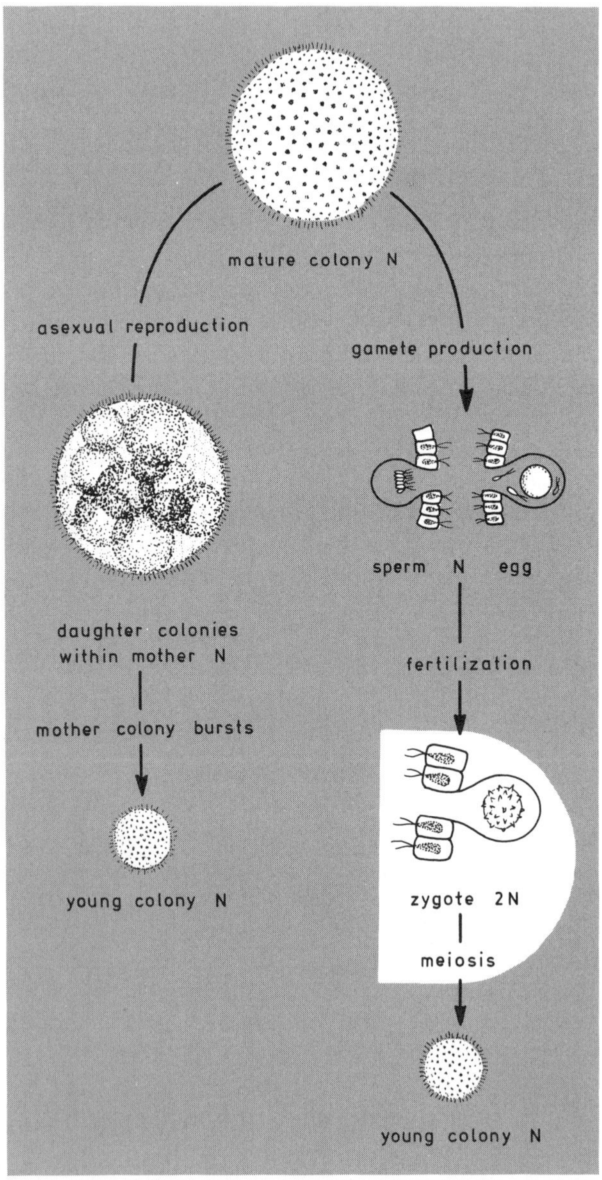

Figure 16–7. The life history of *Volvox*. The zygote is the only diploid stage of this colonial green alga.

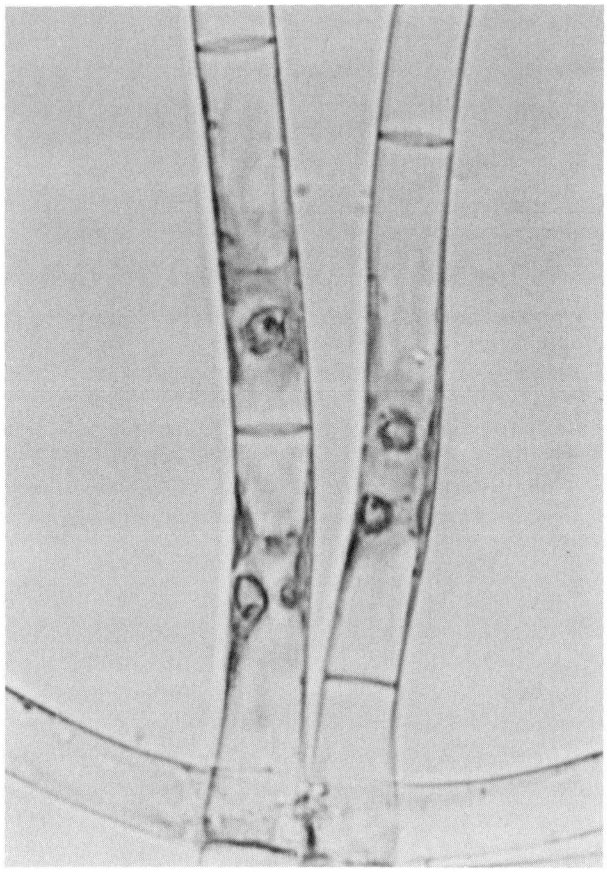

Figure 16–8. *Ulothrix*. This filamentous green alga is haploid and, like *Volvox*, lacks a multicellular diploid stage. Its body resembles the haploid gametophyte stage of some higher plants. (Photo by Carolina Biological Supply Company.)

- Locate a colony of *Volvox* that contains small colonies produced by vegetative reproduction.

The filamentous algae also have unicellular and multicellular forms, but usually they are nonmotile. An example is *Ulothrix*, which has a life history similar to that of *Volvox* in that it has a multicellular haploid stage, in this case the filament (see Text Figure 26–10). *Ulothrix* filaments are shown in Figure 16–8.

- Make a wet mount of living *Ulothrix*, or examine a prepared slide of this organism with your microscope.

What is the shape of the individual cells?

____

Are all of the cells alike? ______________

- Look for the holdfast at one end of the filament. (It may be broken off.)

What is the shape of the single chloroplast?

____

Do you see storage organs for carbohydrate (pyrenoids) in the chloroplast? ______________

Ulothrix can reproduce asexually by forming flagellated **zoospores** (see Text Figure 17–5). It can also form tiny gametes (isogamous), which are smaller than the zoospores, and fuse to form **zygospores.** The zygospores undergo meiosis to form meiotic zoospores, each of which will grow into a new filament.

Another green alga, *Ulva,* has a multicellular diploid stage that is identical in appearance to its multicellular haploid stage. Its history follows the pattern shown in Figure 16–1 (see Text Figures 21–8 and 21–6c). The motile female gamete is slightly larger than the motile male gamete, so reproduction is anisogamous.

■ Examine a specimen of *Ulva* (sea lettuce).

What is the form of the body?

____

Is there a stalk, as in the brown algae?

____

Into which stage do the motile spores grow?

____

Which stage develops from the zygote?

____

Which stage is present in *Ulva* but is absent in a brown alga such as *Fucus?*

____

Although plant life histories may at first seem very confusing, they must all conform to the general pattern given in Figure 16–1. Meiosis always takes place in a diploid cell (or diploid nucleus, as in fungi) and must produce haploid cells, which are almost always spores. It may be helpful to learn an example of each type of life cycle:

1. Neither multicellular haploid nor multicellular diploid stage present:

   ____

2. Multicellular haploid stage present:

   ____

3. Both multicellular haploid and multicellular diploid stages present:

   ____

4. Multicellular diploid stage present:

   ____

Which type of life cycle corresponds to that of humans and other animals? ______________

Return all demonstration materials and store your microscopes properly before leaving the laboratory.

REFERENCES AND FURTHER READING

Alexander, M. *Microbial Ecology.* New York: John Wiley and Sons, 1971.

Alexopoulos, C. J., and H. C. Bold. *Algae and Fungi.* New York: Macmillan, 1967.

Alexopoulos, C. J., and C. W. Mims. *Introductory Mycology,* 3rd ed. New York: John Wiley and Sons, 1979.

Balbach, M., and L.C. Bliss. *A Laboratory Manual for General Botany,* 6th ed. Philadelphia: Saunders College Publishing, 1982.

Banks, H. *Evolution and Plants of the Past.* Belmont, CA: Wadsworth, 1970.

Bold, H.C., C.J. Alexopoulos, and T. Delevorya. *Morphology of Plants and Fungi,* 4th ed. New York: Harper & Row, 1980.

Bold, H.C., and C.L. Hundell. *The Plant Kingdom,* 4th ed. Englewood Cliffs, NJ: Prentice-Hall, 1977.

Bold, H.C., and M.J. Wynne. *Introduction to the Algae.* Englewood Cliffs, NJ: Prentice-Hall, 1978.

Cooke, R.C. *Fungi, Man and His Environment.* London: Longman, 1977.

Darley, W.M. *Algal Biology: A Physiological Approach.* Boston: Blackwell Scientific Publications, 1982.

Dawson, E.Y. *Marine Botany: An Introduction.* New York: Holt, Rinehart and Winston, 1966.

Hale, M.E. *How to Know the Lichens.* Dubuque, IA: W.C. Brown, 1979.

Jensen, W.A., and F.B. Salisbury. *Botany: An Ecological Approach,* 2nd ed. Belmont, CA: Wadsworth, 1984.

Miller, O.K. *Mushrooms of North America.* New York: Dutton, 1977.

Prescott, G.W. *How to Know the Freshwater Algae,* 3rd ed. Dubuque, IA: W.C. Brown, 1978.

Smith, Alexander, and Nancy Weber. *The Mushroom Hunter's Field Guide.* Ann Arbor: University of Michigan Press, 1980.

Taylor, W.R. *Marine Algae of the Northeastern Coast of North America,* 2nd ed. Ann Arbor: University of Michigan Press, 1957.

Journal Articles

Abraham, E.P. "The beta-lactam antibiotics." *Scientific American,* 76, June 1981.

Bonner, J.T. "Chemical signals of social amoebae." *Scientific American,* 114, April 1983.

Gibor, A. "Acetabularia: A useful giant cell." *Scientific American,* 118, November 1966 (#1057).

Graham, L.E. "The origin of the life cycle of land plants." *American Scientist* 73: 178, March–April 1985.

Lamb, I.M. "Lichens." *Scientific American,* 144, October 1959 (#111).

Litten, W. "The most poisonous mushrooms." *Scientific American,* 90, March 1975.

Matossian, Mary K. "Ergot and the Salem witchcraft affair. *American Scientist* 70(4): 358, July 1982.

Niederhauser, J.S., and W.C. Cobb. "The late blight of potatoes." *Scientific American,* 100, May 1959 (#109).

Preston, R.D. "Plants without cellulose." *Scientific American,* 102, June 1968 (#1110).

Strobel, Gary A., and Gerald N. Lanier. "Dutch elm disease." *Scientific American,* 56, August 1981.

TOPIC 17

Higher Plants

OBJECTIVES

When you have completed this topic, you should be able to:

1. Define embryophyte and list the characteristics of these plants that link them to the green algae.
2. Describe the problems faced by plants that moved onto the land.
3. Describe the adaptations of land plants to each problem listed in Objective 2.
4. Give the adaptations of bryophytes that allowed them to live on land.
5. List the problems of terrestrial life that bryophytes have not solved.
6. Describe the life history that applies to club mosses, horsetails, and ferns.
7. Compare the horsetails and ferns with respect to the following features: size of presently living species, stem, leaves, roots, vegetative propagation, type of gametophyte, and type of sporangium.
8. Given a plant specimen or description, place it correctly in one of these groups: Bryophyta, Lycophyta, Sphenophyta, Pterophyta, Coniferophyta, or Anthophyta.
9. Name the embryophyte group(s) in which the gametophyte is the dominant generation.
10. Describe the gametophyte of a fern and a conifer.
11. Give reasons why gymnosperms persisted, even though angiosperms became dominant.

TEXT REFERENCES

Chapter 21 (E–I).

adaptation to life on land
mosses (Bryophyta)
club mosses (Lycophyta)
horsetails (Sphenophyta)
ferns (Pterophyta)
gymnosperms
conifers (Coniferophyta)
angiosperms (Anthophyta)

INTRODUCTION

Although some algae and fungi can eke out a living on land, their relatively undifferentiated bodies never allowed them to develop adaptations for a truly terrestrial life. Most members of the plant groups that you will be studying in this exercise live exclusively on land, taking advantage of its plentiful sunlight, CO_2 and oxygen, and absence of predators and competition. They have been able to cope in some fashion with the resulting problems of obtaining water, supporting their bodies in air, and providing a mechanism for fertilization. Land plants are thought to have arisen from green algae because both groups contain chlorophylls *a* and *b* and carotenoid accessory pigments, and they usually store their food reserves in the form of starch. An alga that developed some kind of waterproofing and rootlike structures to draw water up from the soil would have been able to live out of water on the shore. This ancestral plant would also have had to nourish and protect the zygote and developing embryo from drying out, and it might have done this by retaining the embryo within its body. Such a plant may have given rise to the more highly adapted land plants.

All of these plants show alternation of generations. Sometimes there is a large photosynthetic gametophyte (N) and a tiny sporophyte (2N) that can produce only relatively few spores for dispersal of the species. In other cases the sporophyte is a large photosynthetic plant that propagates the species through dispersal of many spores or seeds. The trend from simpler plants to more highly evolved plants is for the dominant gametophyte to become smaller and less conspicuous and for the sporophyte to become large and dominant.

Reproduction in many land plants still involves a flagellated sperm that must swim to the egg through a film of water. These plants are restricted to a moist environment for fertilization to occur. All of the land plants, however, do have the same important adaptation for reproduction on land: The zygote is held within the female reproductive organ where it is surrounded by a layer of sterile jacket cells. These cells protect the embryo while it starts to develop into a new sporophyte within the female reproductive organ. Plants with this feature during reproduction are called **embryophytes.**

MOSSES (DIVISION BRYOPHYTA, CLASS MUSCI)

Bryophytes (mosses and liverworts) are different from the other landplants in that they lack vascular tissue. Because true roots, stems, and leaves must have vascular tissue, the organs of a moss are described as "rootlike," "stemlike," and "leaflike." As a result, mosses are limited in several important ways. They cannot grow large in size because they have no strong supporting tissue to keep their bodies erect. They must live in damp places because their rootlike rhizoids cannot penetrate deep into the soil to extract the moisture they need. On the other hand, they are able to colonize bare rock because their rhizoids are so small. Finally, their photosynthetic parts above ground and the nonphotosynthetic parts below ground must be close together because there is no vascular tissue for the transport of nutrients and minerals. The gametophyte is the dominant stage in the moss life history.

- Obtain a single moss plant (gametophyte), and examine it under the dissecting microscope.
- Note the "**stem,**" "**leaves,**" and **rhizoids.**

Is any part of the moss plant nonphotosynthetic? If so, which part?

____

____

What are the functions of the rhizoids?

____

____

____

- Make a wet mount of a "leaf."

Are there guard cells or stomata? ____________

- Use a razor blade to make a very thin section of the "leaf" or obtain a prepared slide and view it in the compound microscope.

Do you see xylem, phloem, or large air spaces in the "leaf"? If so, which ones?

____

Is there a cuticle? ____________

During reproduction the flagellated sperm produced by the male moss plant (gametophyte) swim to the female gametophyte through a film of water and fertilize the egg. The zygote grows into a sporophyte and remains attached at the tip of the female plant (see Figure 17–1). Usually the sporophyte is nonphotosynthetic and obtains its nutrition from the female gametophyte. The spores are formed in a capsule at the end of a long stalk and are released from as high up as the plant can possibly reach (3–4 cm).

Why should the spores be released as high as possible?

____

____

____

- Obtain a moss plant that has a sporophyte growing on it, if living material is available. If not, obtain a prepared slide of a moss sporophyte.
- Note the long **stalk** and the **capsule.**

Is either part of the sporophyte photosynthetic?

____

- Mount the capsule in a drop of water on a slide, and remove the tip of it with teasing needles. Observe in the dissecting microscope.
- Try to look into the capsule. You will see a circle of toothlike projections that help disperse the spores.
- Tease apart the capsule to locate the spores.

Are the spores haploid or diploid? ____________

Is the sporophyte haploid or diploid?

____

What process gave rise to the spores?

____

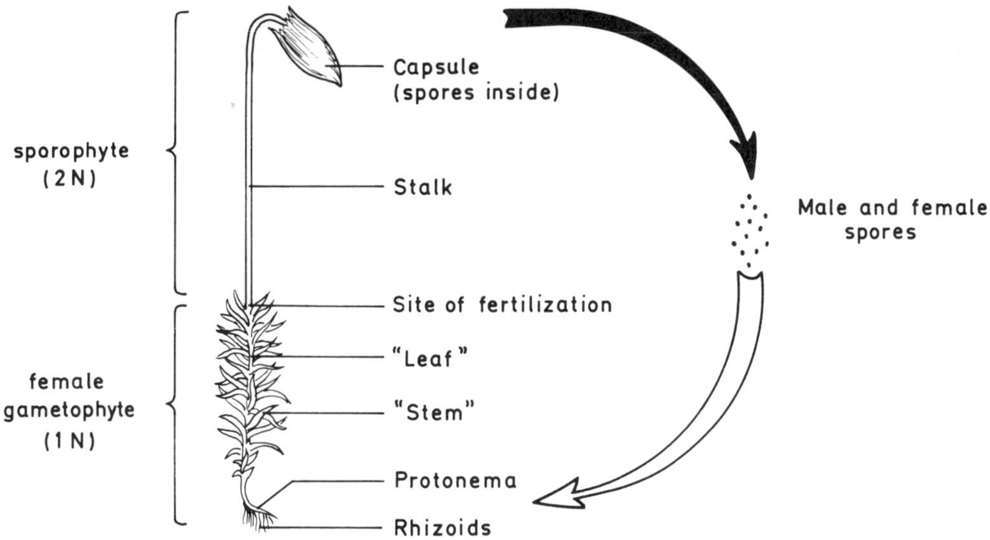

Figure 17–1. Sexual reproduction in a moss. The diploid sporophyte is parasitic on the haploid female gametophyte. Circle the names of all haploid parts.

The spore germinates into a threadlike **protonema,** which eventually forms the buds of the new gametophyte. The protonema is somewhat like a filamentous green alga.

■ Label the location of meiosis in the female moss plant shown in Figure 17–1. Circle the names of all haploid parts shown in the figure.

Mosses cope with the dryness of land by living in a damp environment and retaining as much water as possible. The mat of many moss plants crowded close together acts like a sponge to hold water and cut down evaporation. Some mosses even have special cells that are adapted to absorb water when it is available. Sphagnum moss, for example, has large, reinforced cells with pores in their cell walls for taking up water.

■ Obtain a 100-mL beaker and a weighed 1- to 5-g sample of dry sphagnum moss (note exact weight). Your instructor may ask you to weigh your own sample.

WEIGHT: ______________g

■ Fill the beaker about two-thirds full with water, and make a mark at the surface of the water with a wax pencil.
■ Immerse the moss in the beaker, and allow it to soak up as much water as possible.
■ Remove the moss with a forceps.
■ Fill a 100-mL graduated cylinder with water to exactly the 100-mL mark, and add water from it to your beaker until the water level again reaches the pencil mark.
■ Note how much water was needed to refill the beaker. This amount is equivalent to the amount removed by the moss.

VOLUME: ______________mL (g)

■ Calculate how many times its weight of water the moss can absorb:

$$\text{absorption} = \frac{\text{mL (g) of water removed}}{\text{dry weight of moss (g)}}$$

$$= \frac{\text{VOLUME}}{\text{WEIGHT}} = \underline{\qquad\qquad} =$$

Sphagnum moss is widely used by commercial greenhouses and home gardeners to hold moisture around the roots of plants, especially during shipping.

CLUB MOSSES (DIVISION LYCOPHYTA)

The remaining plants to be studied in this laboratory all belong to Division Tracheophyta and have true **vascular tissue.** This system of **xylem** and **phloem** conducts water, minerals, and the products of photosynthesis from one part of the plant to another. The organs of the plant are considered **true roots** and **true leaves** because they contain vascular tissue. Xylem also forms **wood** to support the plant body in tracheophytes that have woody tissue. The presence of vascular tissue had tremendous implications for the evolution of this group: The plant could grow tall and the nonphotosynthetic root system could become extensive. Plants lacking vascular tissue, such as bryophytes, are small in size, because there is no way for water to move from deep in the soil to the leaves, or for the leaves to nourish distant roots. The photosynthetic and nonphotosynthetic parts of a nonvascular plant must be quite close together.

Club mosses and horsetails are fern allies because all these plants have a similar life cycle (see Figure 17–2). They have a large sporophyte with true vascular tissue, and the spores grow into tiny gametophytes that are independent of the parent sporophyte. There is a flagellated sperm and the zygote grows into a new sporophyte. Although present-day ferns and their allies do not have woody tissue, in former times they did and reached the proportion of trees. Some species of tree ferns have survived to the present.

- Examine a living or preserved club moss such as *Lycopodium*, and note the leaves, stems, and roots.

What is the form of the leaf?

____

What is the value of this adaptation?

____

- Cut through the tip of one of the branchlets.

Is the interior hollow or solid? ____________

The leaf remains green through the winter, so it is **evergreen,** as are some ferns, most gymnosperms, and some angiosperms. Possession of evergreen leaves is therefore not a good characteristic for classifying plants.

The club moss reproduces vegetatively by its underground stem, the **rhizome,** which sends up shoots to form new photosynthetic stems.

- Examine the rhizome of your specimen.

Does it have leaves? ____________

During sexual reproduction cones called **strobili** form at the tips of the branches, and the spores are produced within them in **sporangia.** (See Figure 17–6 at the end of this topic.) The sporangium always forms on the surface of a leaf called a **sporophyll** (= spore leaf).

- Examine the strobilus, if your specimen has one. It consists of a cluster of sporophylls, which produce and disperse the spores. After the spores germinate, growth of the tiny gametophytes and fertilization go on underground. The zygote grows up into a new sporophyte.
- Label the parts of the club moss shown in Figure 17–6.

Although ancient club mosses included very large trees, the present species are all small. It is illegal to collect them in most areas because they are so few in number and do not reproduce rapidly. Sexual reproduction in club mosses is so slow that most propagation is vegetative, by the spreading out of the rhizomes.

HORSETAILS (DIVISION SPHENOPHYTA)

Fossil horsetails grew as large as trees, but there is only a single genus of small plants that survives today. (See Text Figure 21–15.)

- Examine a living or preserved specimen of a horsetail such as *Equisetum hyemale* or *Equisetum arvense.* (In *E. arvense* the green shoots that look like pine needles are the branches. Look for the tiny leaves on these branches.)

Describe the **leaves:**

____

____

Where do they occur?

____

Is the **stem** photosynthetic? ____________

- Note the ribs along the sides of the hollow stem, and the joints or **nodes.**

The stem continues underground as a rhizome. The stalks die in the fall, but the **rhizome** survives the winter and sends new shoots up in the spring. The horsetail spreads vegetatively as the rhizome grows and sends up more shoots. True roots grow from the rhizome.

If your specimen is *E. arvense,* sexual reproduction involves separate nonphotosynthetic **fertile shoots,** which bear conelike **strobili** at the tips and are different in structure from the green **sterile shoots.** (See Figure 17–6.) The spores are formed within the sporangia, which are within the strobili. If your specimen is *E. hyemale,* the strobili are found at the ends of the photosynthetic shoots. Winglike **elaters** on the spores aid in their dispersal by the wind after they are shed from the sporangium.

- Examine a strobilus. The sporangia contained within it form the spores.

The spores germinate into nonphotosynthetic gametophytes that grow underground. After fertilization, the zygote grows into another sporophyte.

- Label the parts of the horsetail shown in Figure 17–6. If you studied *E. hyemale,* label the parts described above on the left-hand shoot only.

Although this group is not very important today, during the time of the first settlers and the pioneers, horsetails were called scouring rushes and were used as an abrasive for cleaning pots and for polishing. The stems of the horsetails contain grains of silica (the material of which sand is made), which are good for scouring.

FERNS (DIVISION PTEROPHYTA)

Although you would never name a child "Horsetail" or "Liverwort," you might choose the name "Fern" in honor of these beautiful woodland plants. The life history of a fern is shown in Figure 17–2. It is similar to those of the fern allies (club mosses and horsetails). The fern gametophyte, however, is an independent, photosynthetic plant that produces both eggs and sperm.

- Obtain a fern plant (sporophyte) and study its structure.
- Note the deeply lobed leaves, called **fronds.**

Do the fronds have many veins? ______________

Ferns, gymnosperms, and angiosperms are grouped together as having **megaphylls** (= large leaf) because of their large, many-veined leaves. These leaves evolved in a different way from the **microphylls** (= tiny leaf) found in fern allies (see Text Figure 21–14).

- Note the **petiole** (stalk) of the frond. It attaches the frond to the underground stem.
- Follow the petiole underground to where it joins the horizontally growing stem, the **rhizome.**

Fronds of most ferns do not survive the winter, but the rhizome does. It sends up new fronds in the spring in the curled up form known as **fiddleheads.** Each season the rhizome grows longer and sends up new fronds to spread the fern by vegetative propagation.

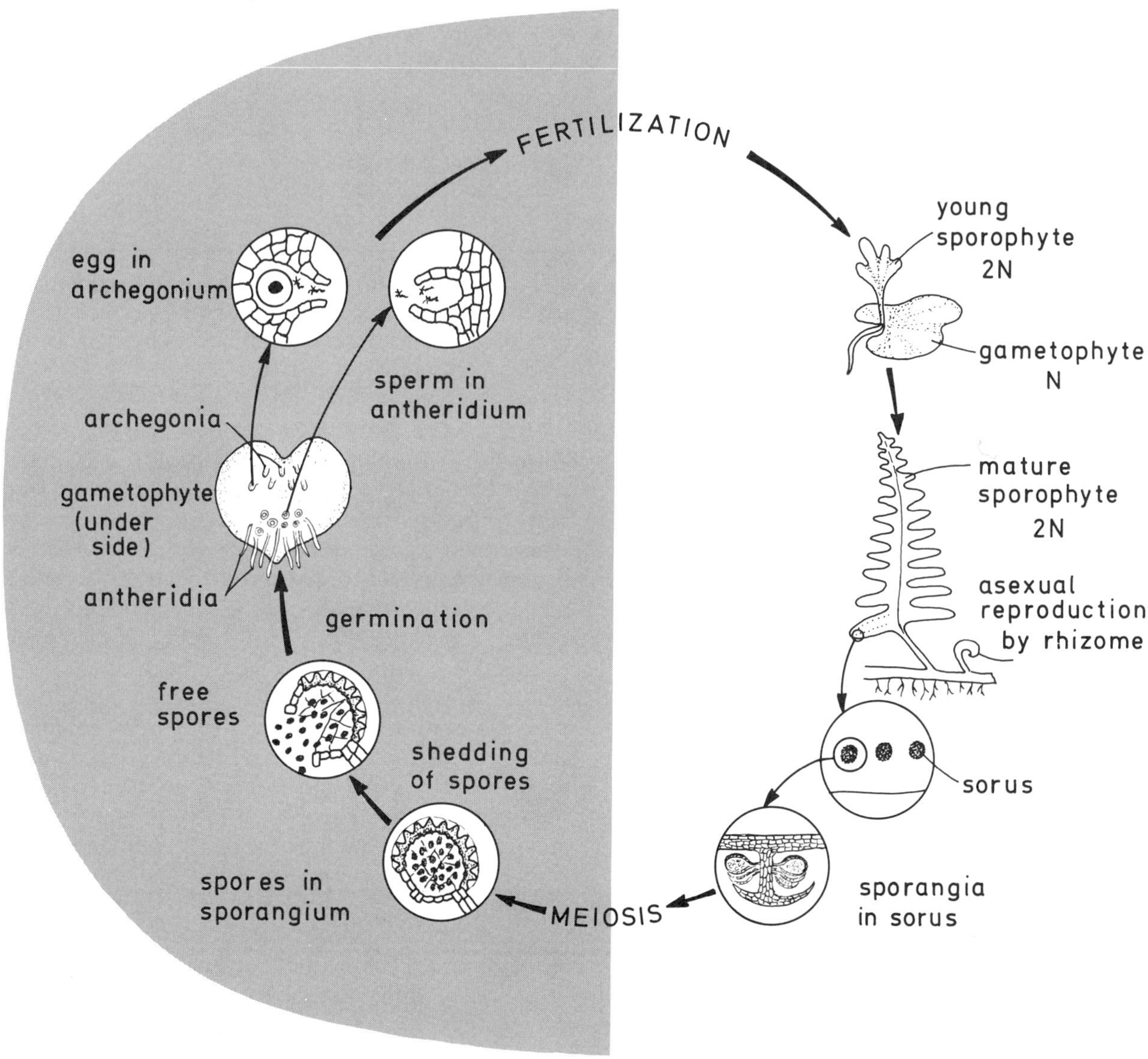

Figure 17–2. Life history of a fern. The sporophyte is dominant and the small gametophyte (valentine plant) is independent and photosynthetic.

- Examine a prepared cross section of a fern stem and note the arrangement of the vascular tissue.

Certain leaves called **sporophylls** bear the sporangia in which the spores are produced. Sometimes there are separate, fertile fronds, but more often the ordinary fronds take on the reproductive function.

- Examine the undersides of your fronds for the presence of **sori,** which are little brown or colored dots. A sorus is a cluster of sporangia.
- Examine a prepared slide showing the sorus of a fern in cross section.
- Locate a **sporangium,** which is a sphere-shaped capsule suspended from a stalk.
- Note the heavy, segmented ring around most of the sphere. This is the **annulus,** which is sensitive to moisture and helps release the spores by snapping the capsule shut suddenly.
- Note the **spores** within the capsule of the sporangium.
- If your frond has sori, scrape the surface of one into a drop of water, make a wet mount, and examine the sporangia in the microscope.
- Locate the spores that are visible through the transparent walls, and find the annulus, which has thick cell walls.
- Find the place in the annulus that is not thick-walled. This is where the capsule will break open to release the spores.

After the sporangium is mature, the cells of the annulus dry out, and the annulus straightens out to form a rod, thereby breaking open the capsule. At a certain point the annulus becomes so strained that it snaps back and flings the spores out of the capsule. The spores are thus actively spread out away from competition with the parent plant. (See Text Figure 21–6.)

Each spore grows into a small plant called a **prothallus,** which is heart-shaped. (It is sometimes called a "valentine plant.") It bears both an **archegonium** (female), and **antheridium** (male) and produces both eggs and sperm. The prothallus is anchored onto the substrate by **rhizoids.** The sex organs and rhizoids are all located on the undersurface. The sperm must swim over to the eggs, which are fertilized and retained within the archegonium. (See Text Figure 21–10.)

- Examine the living prothallus of a fern if living material is available; otherwise look at a prepared slide.
- Locate the **archegonia,** which are vase-shaped. They may not be present if you are looking at a living prothallus.
- Locate the globe-shaped **antheridia.**

Why is it advantageous to have the sex organs located beneath the plant?

____

____

The young sporophyte grows right out of the archegonium. The leaves grow upward from the apical notch of the prothallus and the rhizome, and the roots grow downward.

- Examine living or preserved young fern sporophytes.
- Note that the prothallus persists even after the sporophyte is well established.

Ferns have survived in spite of competition with the seed plants, and they are plentiful and widespread today. Most ferns are small, but there are some tree ferns still in existence.

GYMNOSPERMS AND ANGIOSPERMS

Seed plants are the **gymnosperms** (= naked seed) and the **angiosperms** (= hidden seed), the flowering plants (Division Anthophyta). Gymnosperms include the cycads, which look like palm trees but bear cones (Division Cycadophyta), *Ephedra* and relatives (Division Gnetophyta), the ginkgo tree (Division Ginkgophyta), and the familiar conifers (Division Coniferophyta). Most of these plants have mastered the problems posed by reproducing on land by doing away with flagellated sperm. Instead, male spores are retained within the parent and develop into immature male gametophytes called **pollen grains.** These are not susceptible to drying out and can be spread about in large numbers by the wind. In addition, the young sporophyte embryo does not have to develop into a mature sporophyte at once, but can survive as a **dormant seed** until the conditions become favorable because it is enclosed with its food supply inside a tough seed coat. The production of pollen and seeds allowed these plants to reproduce in the absence of free water and was an important factor in their evolutionary success.

Gymnosperms and angiosperms have **woody tissue** that supports the stem of the plant so that it can reach great size. A large, taller plant has an advantage in obtaining sunlight for photosynthesis and in dispersing its seeds. These plants also have **hormones** that control the growth and development of the plant organs. They were able to become highly adapted to the environment and to change in response to the seasons by producing seeds and becoming dormant at

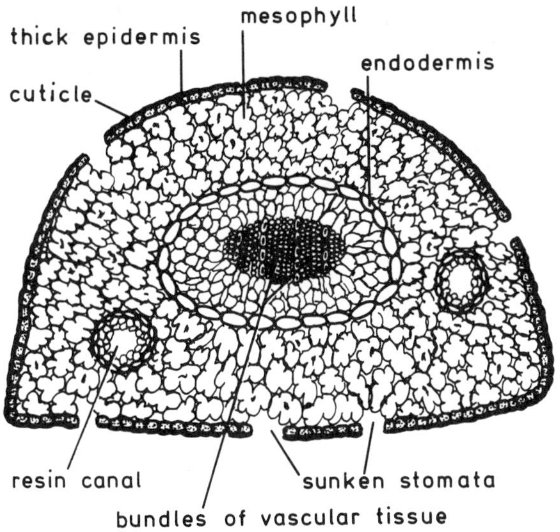

Figure 17–3. Pine leaf in cross section. This evergreen leaf has a reduced surface area and a thick cuticle to conserve moisture.

the right time because of the control of these processes by hormones.

Gymnosperms have continued to survive in spite of the enormous success of angiosperms because they are better adapted to poor, rocky soils and extreme cold. Their **evergreen** leaves allow them to continue photosynthesis all year round, even under conditions of extreme cold. Many land plants, in particular conifers, have evolved a close symbiotic relationship with fungi called **mycorrhizae** that are associated with their roots. These fungi greatly increase the surface area for absorption and facilitate the uptake of minerals from poor soil. They also protect the plant from toxic minerals released from soil in the wake of acid rain.

- Obtain a branch of a conifer, or observe one in a field demonstration.
- Note the needlelike **leaves** and the **bark** on the branches and trunk.
- Note that the leaves are **evergreen** and do not drop off in the fall as do the leaves of **deciduous** trees. (A few conifers are deciduous: the larch, or tamarack, for example.)
- Examine a prepared slide of a cross section of a leaf (needle). Compare what you see with Figure 17–3.

What is the advantage of evergreen leaves?

____

Why are evergreen leaves usually needlelike or covered with a thick protective cuticle?

____

____

The life history of a pine is shown in Text Figure 21–19. Sexual reproduction in conifers is restricted to the **sporophylls** that make up the **cone.** There are separate male and female mother spores (heterospory) that develop into the male and female gametophytes within separate male and female cones (see Figure 17–4). The gametophytes are nonphotosynthetic and are nourished by the parent sporophyte. Thus they can live wherever the sporophyte lives. The pollen (male gametophyte) must travel from the male cone to the female cone; then fertilization and early development occur within the female gametophyte. Gymnosperm seeds are naked in that they are located on the surface of the sporophyll rather than within a fruit (ovary) as are angiosperm seeds.

- Locate some developing cones which are green and soft. The larger ones are the female **seed cones.** The brownish male **pollen cones** are smaller and usually occur in clusters; they are present for only a short time.
- Examine a male pollen cone. Each sporophyll or scale bears a pollen sac, also called the **microsporangium** because meiosis takes place there.

Pollen is formed within the pollen sac, and each grain can develop into a tiny, parasitic male gametophyte. During the formation of a pollen grain, a cell undergoes meiosis to form four microspores. Each microspore grows into a pollen grain containing several cells by means of mitosis. One cell nucleus becomes the reproductive nucleus (sperm nucleus), and one gives rise to the pollen tube (tube nucleus), which allows the sperm nucleus to reach the egg. When the mature pollen grain is released from the pollen cone, a few of the grains will reach the scales of a female cone and land on a mature ovule (immature seed). This is the process of **pollination.**

- Examine a prepared slide of pollen from a conifer.
- Note the winglike projections of the mature grains that help them to travel further in the wind.

What are the advantages of a pollen grain over a flagellated sperm cell?

____

____

- Examine an older **seed cone;** remove one scale.
- Look for two **ovules** containing **megasporangia** on its upper surface. If the ovules have already matured into seeds, they may have been shed, leaving two hollow impressions where they once grew.

Within each megasporangium, meiosis followed by mitosis gives rise to the cells that make up the parasitic female gametophyte, including several egg cells. When pollen lands on the sticky surface of the ovule,

A

B

Figure 17–4. Cones of red pine (*Pinus resinosa*). (A) Female. A seed cone contains "naked" seeds with thin, papery "wings" that aid in their dispersal by the wind. Two pale, rounded seeds can be seen near the base of a scale (sporophyll) at the lower left. (B) Male. The small pollen cones disintegrate soon after they release their masses of yellow, dustlike pollen in early summer. The entire male gametophyte (pollen grain), not just a sperm cell, travels from the male to the female cone. (From Arms & Camp BIOLOGY 3/e)

the pollen tube grows through an opening in the ovule called the **micropyle** and on through the tissues of the female gametophyte until it reaches an egg cell. The sperm nucleus enters the egg in the process of **fertilization,** and the new embryo starts to develop within the megasporangium. A diagram of fertilization is shown in Figure 17–5. The seed of the pine can be understood as an extended pine family in which the young sporophyte is contained within the tissue of its parental female gametophyte, and the gametophyte in turn is contained within the reproductive organs of the grandparental sporophyte. Although several eggs may be fertilized within the female gametophyte, one embryo will win out and be released in the mature seed. The seed contains stored food as well as the embryo, has a tough seed coat, and is carried by the wind.

The biology of angiosperms will be studied in more detail later (Topics 30 and 31). Reproduction in angiosperms is similar to that in gymnosperms. **Pollen** from the stamen lands on the pistil or carpel(s) and a **pollen tube** grows down to the ovule at its base. The **female gametophyte** within the ovule consists of only a few cells, including a haploid **egg cell** and a special diploid cell. This cell is also fertilized and forms the **endosperm,** a source of nutrients that is incorpo-

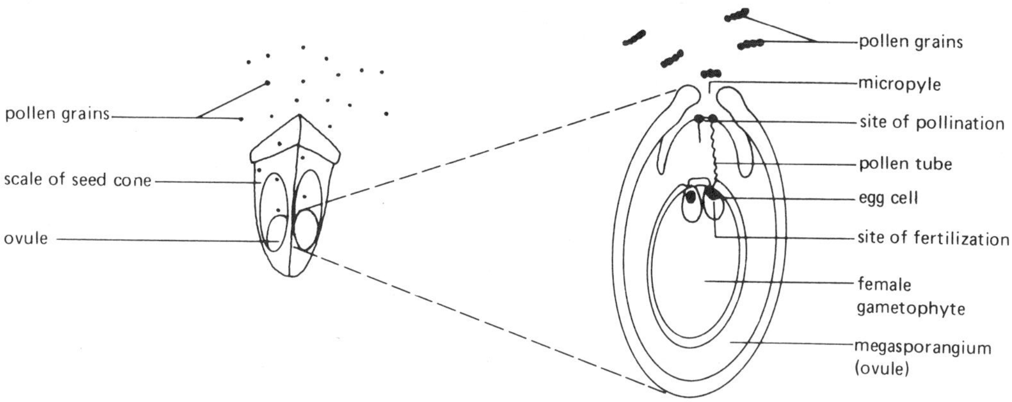

Figure 17–5. Fertilization in a pine cone. Pollination occurs when the pollen grains reach the surface of the ovule and the pollen tube begins to grow. Fertilization occurs when the sperm nucleus from the pollen tube unites with the egg nucleus within the female gametophyte. These two events may be separated by a long period of time.

rated into the seed. **Double fertilization** of both the egg cell and the endosperm cell is unique to angiosperms. (See Figure 31–7.) The outer layer of the ovule forms the **seed coat** to protect the embryo and endosperm. Finally, the base of the pistil containing the ovules—namely, the ovary—develops into a **fruit.** The fruit may be fleshy, as in a peach, or hard, as in a bean pod. The fruit serves to protect or disperse the seeds.

- For your own review, complete Table 17–1 at the end of this topic by checking each feature that applies to the groups you have studied. List the characteristics of all embryophytes in the space below the table.

Before leaving the laboratory, be sure to return all demonstration materials and store your microscopes properly.

REFERENCES AND FURTHER READING

Banks, H. *Evolution and Plants of the Past.* Belmont, CA: Wadsworth, 1970.

Bold, H.C., C.J. Alexopoulos, and T. Delevorya. *Morphology of Plants and Fungi,* 4th ed. New York: Harper and Row, 1980.

Bold, H.C., and C.L. Hundell. *The Plant Kingdom,* 4th ed. Englewood Cliffs, NJ: Prentice-Hall, 1977.

Jensen, W.A., and F.B. Salisbury. *Botany: An Ecological Approach,* 2nd ed. Belmont, CA: Wadsworth, 1984.

Raven, P.H., R.F. Evert, and S.E. Eichhorn. *Biology of Plants,* 4th ed. New York: Worth, 1986.

Ray, P.H., T.A. Steeves, and S.A. Fultz. *Botany,* 5th ed. Philadelphia: Saunders College Publishing, 1983.

Thomas, Barry. *The Evolution of Plants and Flowers.* New York: St. Martin's Press, 1981.

Weier, T.E., C.R. Stocking, and M.G. Barbour. *Botany: An Introduction to Plant Biology,* 6th ed. New York: John Wiley and Sons, 1982.

Journal Articles

Altschul, Siri von Reis. "Exploring the herbarium." *Scientific American,* 96, May 1977 (#1359).

Echlin, Patrick. "Pollen." *Scientific American,* 80, April 1968 (#1105).

Graham, L.E. "The origin of the life cycle of land plants." *American Scientist* 73: 178, March–April, 1985.

Pettitt, John, Sophie Ducker, and Bruce Knox. "Submarine pollination." *Scientific American,* 134, March 1981.

Regal, P.J. "Ecology and evolution of flowering plant dominance." *Science* 196: 622, 6 May 1977.

Student Name ______________________________ **Date** ______________

TABLE 17–1. Characteristics of the Embryophyte Groups

Characteristics	Musci (Moss)	Lycopsida (Club Moss)	Sphenopsida (Horsetail)	Filicineae (Fern)	Coniferopsida (Conifer)
Cuticle					
True roots, stems, leaves					
Flagellated sperm					
Rhizoids in gametophyte					
Vascular tissue					
Rhizomes					
Separate male and female gametophyte					
Parasitic or nonphotosynthetic sporophyte					
Photosynthetic gametophyte and sporophyte					
Parasitic or nonphotosynthetic gametophyte					
Airborne spores					
Pollen					
Embryo protected within parent plant					
Embryo protected by seed coat					
Seed enclosed within fruit					
Woody vascular tissue					

Characteristics common to all embryophytes:

____

____

____

____

____

CLUB MOSS

_________phyte

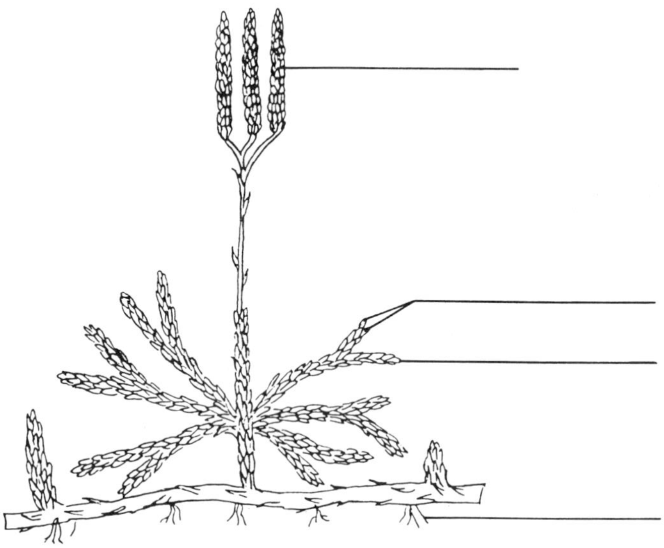

CLUB MOSS

_________phyte

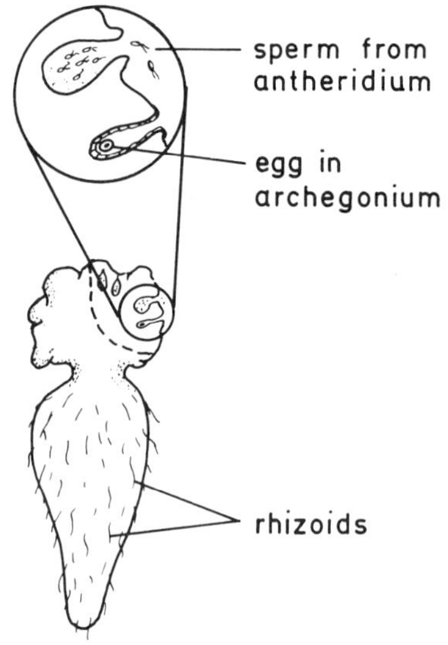

HORSETAIL

_______phyte

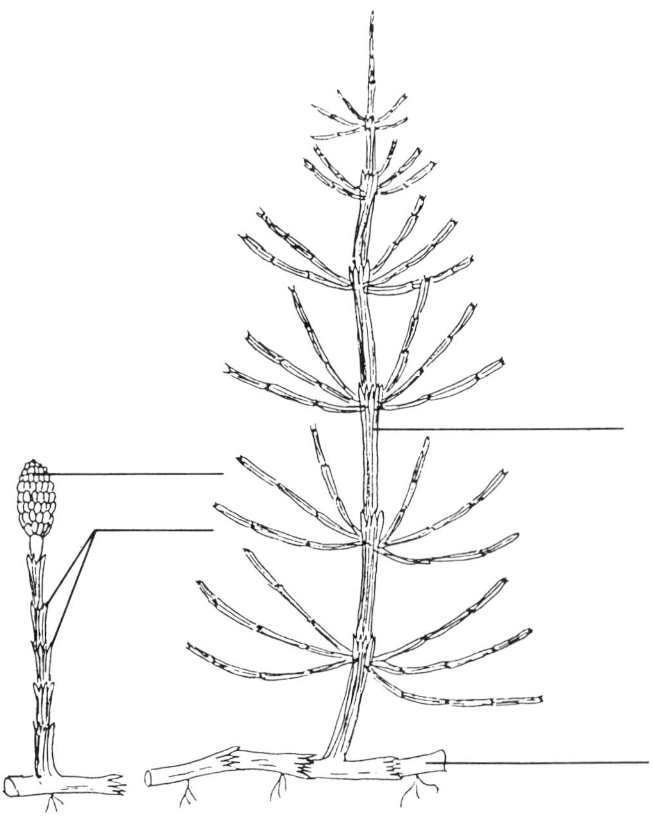

HORSETAIL

_______phyte

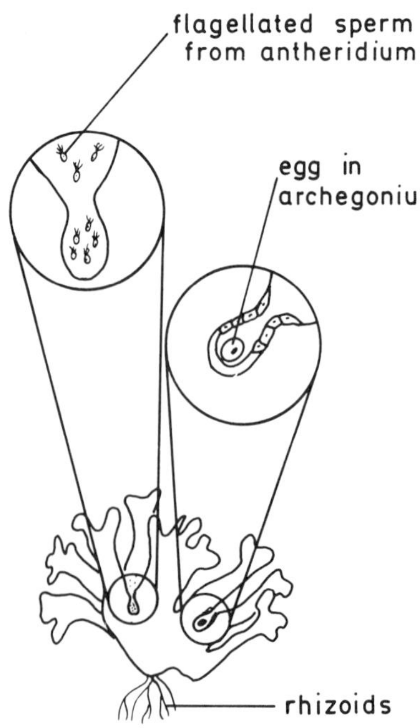

Figure 17–6. Stages in the life history of a club moss and a horsetail. Label each stage as a sporophyte or gametophyte under its name, and label the parts of each plant marked with label lines.

TOPIC 18

Lower Invertebrates

OBJECTIVES

When you have completed this topic, you should be able to:

1. Describe the advantages of multicellularity, larger size, radial symmetry, bilateral symmetry, the mobile stage of life history, the sessile stage of life history, the sense organs, cephalization, and the nervous system.
2. Explain or use the following terms correctly: invertebrate, larva, plankton, filter feeder, hermaphrodite, gonad, alternation of generations.
3. Define coelom, coelomate, and acoelomate, and state the advantage that a coelom gives to an organism possessing it.
4. Give the distinguishing features of members of the phyla Porifera, Cnidaria, Platyhelminthes, Nematoda, and Rotifera.
5. Describe the following structures and give the phylum or phyla in which each of them is found: tentacles, nematocysts, choanocytes, gastrovascular cavity, flame cells, pseudocoelom, polyp, medusa, and gonads.

TEXT REFERENCES

Chapter 22 (A–H).

KEY TOPICS

body symmetry
coelom
alternation of generations
larva
cnidarians (Cnidaria)
flatworms (Platyhelminthes)
nematodes (Nematoda)
rotifers (Rotifera)

INTRODUCTION

Topics 18–21 cover diversity within Kingdom Animalia. Animals are multicellular heterotrophs and during sexual reproduction produce gametes in multicellular gonads. About 95% of animal species lack a backbone and are called **invertebrates.** Vertebrates, animals with backbones, are a subphylum within Phylum Chordata. Many invertebrates are still marine and their body forms and lifestyles reflect their early origin in the primitive seas. They are generally not good swimmers and so are either **planktonic** and drift about or are **sessile** and attach themselves to one fixed spot. The sessile forms must produce a large number of gametes for reproduction to succeed, and they disperse the species by means of a **motile larval stage** in the life history.

As competition between unicellular organisms became intense, larger and more complex organisms had a selective advantage, and multicellular organisms such as the animals came into existence. **Multicellularity** allowed differentiation of cells that performed different specific functions. Different types of cells became organized into **tissues** and the tissues into **organs,** specialized structures adapted to carry out a certain function or process. For example, a complete digestive tract, in which there is a one-way passage of food from mouth to anus, can have specialized parts and is much more effective than a simple bag.

In organisms with more sophisticated organs and organ systems, the body plan changed to accommodate the new complexity. In nematodes, rotifers, and more complex organisms, there is a space between the gut and body wall called the **coelom.** This space allows the gut and body to move independently and provides a protective space in which the internal organs and a circulatory system can be contained. When the coelom is not fully lined by mesoderm, as in nematodes and rotifers, it is a **pseudocoelom.** In annelids, molluscs, and the other organisms that we will be studying in subsequent exercises, the coelom is a true coelom because it is fully lined by mesodermal tissue.

Compare the body plans shown in Figure 18–1 and be sure you can relate each type to the phyla in which it is found.

In the descriptions of organisms to follow, we will be using certain terms that you need to understand:

Symmetry: The body can be split at least once to form halves that are mirror images.

Radial Symmetry: Many planes of symmetry around a central axis, as in a sea star.

Bilateral Symmetry: One plane of symmetry, as in a fish.

Sagittal or **Longitudinal Section:** The body is cut from head to tail, dividing the fish into right and left halves.

Frontal Section: Body is cut in half from head to tail, separating the fish's back from its belly.

Cross Section: Body is cut in half from top to bottom to separate the head end of the fish from the tail end.

Anterior: Toward the head or front.

Posterior: Toward the tail or rear.

Dorsal: Toward the back or top.

Ventral: Toward the belly or bottom.

Sessile animals commonly show **radial symmetry** so that they can receive food and detect predators arriving from any direction. Motile animals will often have **bilateral symmetry,** which adapts them for locomotion in a three-dimensional world.

PHYLUM PORIFERA: *SCYPHA* (CLASS CALCISPONGIAE)

The household sponges with which everyone is familiar are the synthetic imitations of real sponges, which have internal skeletons made of **spongin.** Other types of sponges have skeletons of **calcium carbonate** or **silica** and are classified on that basis. The basic structure of a sponge is a mass of cells organized around a central cavity called the **spongocoel.** Water enters through small pores (porifera means "pore bearer") and exits through the **osculum,** which is usually raised above the surface of the body. (See Color Plate 3.) Lining the interior of the sponge or of the canals leading into the spongocoel are special flagellated cells called **choanocytes** or **collar cells.** These are very similar to flagellated protozoans that exist as single cells or as simple colonies. Such protozoans and sponges may have had a shared ancestor.

Sponges are **hermaphroditic,** so an individual can form both eggs and sperm during sexual reproduction. Since adult sponges are sessile, hermaphroditism makes sexual reproduction possible even for isolated individuals. The **flagellated larva** leaves the spongocoel of the parent, swims to a new location, and attaches to grow into a new sessile adult. Such a mobile part of the life history allows the species to spread to new locations.

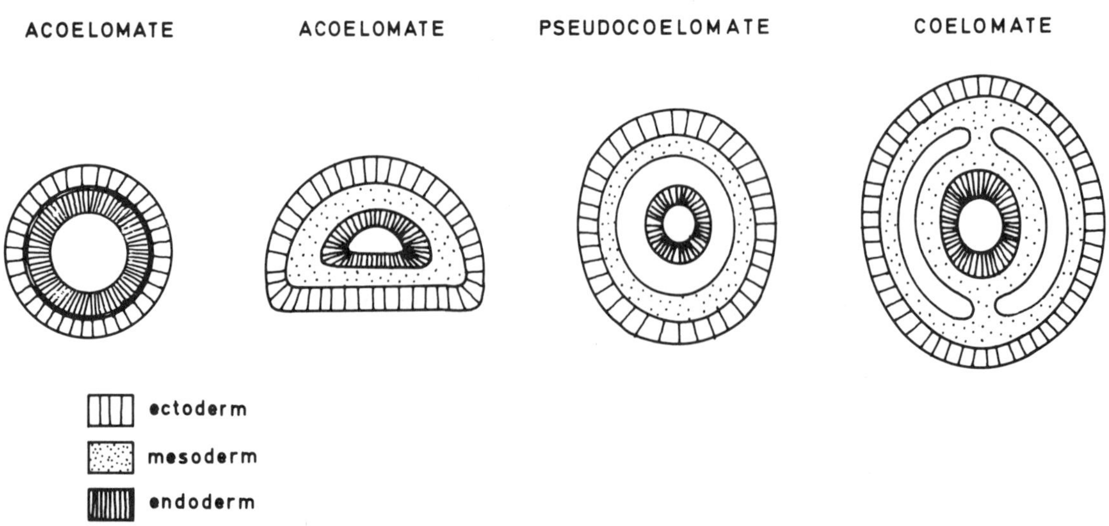

Figure 18–1. Body plans. Cnidarians are acoelomates with two cell layers and a mesoglea. Flatworms are acoelomates with three cell layers including a true mesoderm. Pseudocoelomates such as the nematodes and rotifers have a body cavity, but it is between the inner and middle layers of cells. In true coelomates, the body cavity is formed within the middle cell layer and the organs are protected and supported by this layer.

- Study a specimen of a sponge such as *Scypha,* which has been cut in longitudinal section, after placing it in a culture dish partly filled with water. Use your dissecting microscope when appropriate.
- Locate the **osculum,** the **base,** and the calcium carbonate **spicules.** The calcium carbonate composition of the spicules places this particular sponge in **Class Calcispongiae.** Sponges may reproduce asexually by budding off new individuals near the base.
- Note the hollow **spongocoel** from which water exits through the osculum. When sponges reproduce sexually, eggs and sperm are produced and fertilization occurs within the spongocoel.
- Find the **ostia** (singular ostium) or pores on the surface of the sponge through which water enters the **incurrent canals** within the body wall.
- Label these structures in Figure 18–6 at the end of this topic.
- Obtain a prepared slide that shows the structure of the body wall of *Scypha* and view it in the microscope under low, then high power.
- Note that the incurrent canals lead into the **radial canals,** which in turn lead to the spongocoel. The choanocytes line these canals but not the interior of the spongocoel.
- Label the location of the choanocytes and indicate the direction in which water will flow into and out of the sponge in Figure 18–6.

The sponge is a **filter feeder** and is organized to maintain a steady flow of water through its body for food collection, gas exchange, and excretion. The animal feeds on tiny food particles that the choanocytes trap and ingest by phagocytosis. Since the choanocyte is of such central importance to the life of the sponge, it is reasonable that there may be an evolutionary relationship between sponges and certain protozoans that strongly resemble choanocytes. Sponges apparently elaborated on the capabilities of this one cell type, but did not develop other types of specialized cells. They seem to be unrelated to the other phyla of multicellular animals and represent an offshoot of evolution that ended without reaching the tissue level of organization.

PHYLUM CNIDARIA: *HYDRA* (CLASS HYDROZOA)

Cnidarians are predatory carnivores. (See Color Plate 3.) They have a saclike body consisting of two cell layers, epidermis and gastrodermis, surrounding a **gastrovascular cavity** with a single opening. In the **polyp** form, the organism is more or less sessile, and the opening to the cavity is directed upward. In the motile **medusa** form, the opening is underneath (see Text Figure 22–8). Both forms usually have fingerlike extensions of the body called **tentacles** with specialized stinging capsules called **nematocysts.** There is much cellular specialization in the form of reproductive, sensory, secretory, and primitive nerve cells. Movement is accomplished by the contraction of muscle fibers coordinated by a network of nerve cells.

Hydra is an example of a polyp that has lost the medusa stage (common in freshwater cnidarians). Individuals are either male or female, and during sexual reproduction, reproductive cells in the **gonads** form eggs or sperm. After fertilization, the zygote will eventually develop into a small version of the adult. *Hydra* can also reproduce by budding off a small individual complete with tentacles from the side of its body (see Color Plate 2). The internal structure of *Hydra* is shown in Figure 18–2.

- Examine a living *Hydra* using the dissecting microscope.
- Locate the **basal disc** by which it is attached, the cylindrical **body,** and the crown of **tentacles.**

What kind of symmetry does *Hydra* show?

____

What advantages does this type of symmetry have for a sessile animal?

____

____

- If brine shrimp are available, add a drop of *washed* brine shrimp to your *Hydra* and observe its eating behavior. (Salt water will quickly kill the *Hydra* by upsetting its water balance.)

Is the shape of the *Hydra* different after ingestion?

____

Is digestion of the food intracellular or extracellular?

____

What happens to the wastes after the edible parts of the brine shrimp have been absorbed?

____

____

____

____

The tentacles are studded with specialized cells called **cnidocytes.** Each such cell produces a threadlike **nematocyst,** which stings and entangles prey so that it can be brought to the mouth and ingested.

- Examine a prepared slide of nematocysts in the microscope under high power.

The nematocyst consists of a tiny thread coiled up inside a capsule. It is fired in response to a nervous impulse when a brine shrimp touches the cnidocil trigger. The thread turns inside out and then entangles

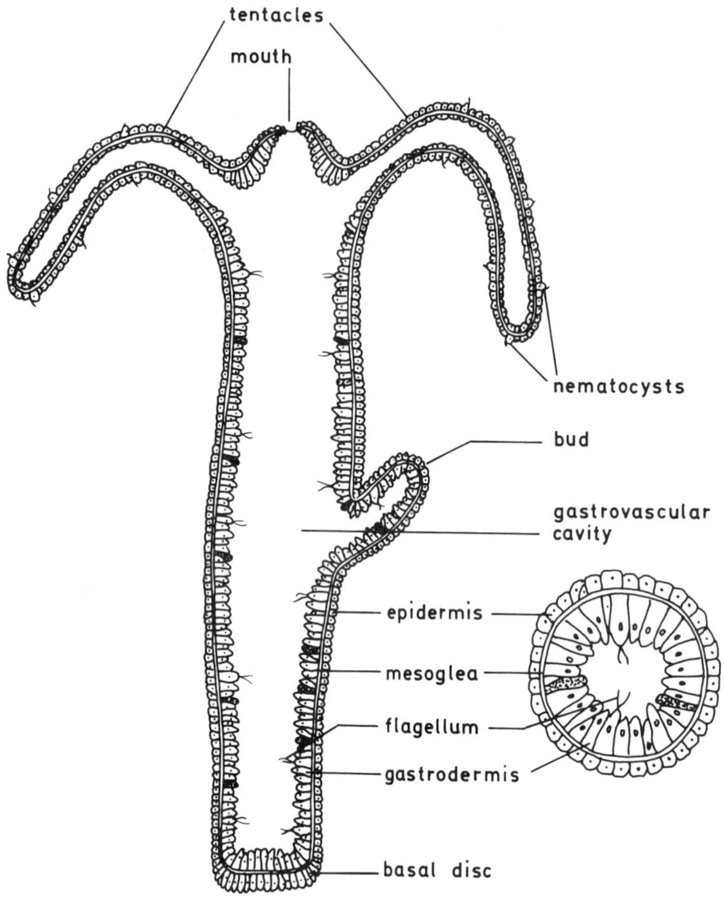

Figure 18–2. ***Hydra.*** Prey is captured by the stinging or entangling nematocysts and stuffed into the mouth by the tentacles. Movement occurs by the contraction of muscle fibers in the mesoglea.

or paralyzes the prey organism (see Text Figure 22–10). Toxins in the nematocysts of cnidarians such as fire coral and the Portuguese man-of-war can be harmful or even fatal to humans.

Other cnidarians, such as the jellyfish *Gonionemus,* have both a medusa stage and a polyp stage in their life histories. These organisms show **alternation of generations:** The medusa gives rise to the polyp, the polyp then gives rise to the medusa, and so forth.

PHYLUM PLATYHELMINTHES

Flatworms have three body layers, **ectoderm, mesoderm,** and **endoderm,** but there is no body cavity, so they are **acoelomate.** As in cnidarians, the gastrovascular cavity has a single opening. Flatworms have true organs and organ systems. Motile flatworms show **cephalization** and **bilateral symmetry.** These features are adaptations for locomotion and for detection and response to new stimuli as the animal moves along. (See Color Plate 3.)

Class Turbellaria: Planaria

- Look at the outline of a planarian in Figure 18–7 at the end of this topic and note the shape of the anterior (front) and posterior (rear) ends of this animal. The interior anatomy is shown in Text Figure 22–12.
- Examine a living planarian such as *Dugesia,* or look at a prepared slide in the microscope.
- Locate the anterior end, with its earlike **auricles** and pigmented **eyespots** for light perception. This concentration of sense organs in a head region at the front of a motile organism is what is meant by **cephalization.**

Cilia on the underside of the worm are used in its locomotion. The **pharynx,** mouth (also serves as anus), and genital pore are also located on the underside.

- Turn the worm upside down to look for these structures.
- Feed your planarian with a bit of meat. Try to observe the extension of its pharynx as it feeds.

If you are looking at a prepared slide, how many branches does the gastrovascular cavity have?

____

Is this a complete digestive system? ____________

The **excretory system** consists of highly branched interconnecting tubes which are lined with ciliated **flame cells** that move the contents along to the excretory pores. Its major function is probably water balance.

The nervous system consists of nerve cells grouped in nerve cords along each side of the worm and concentrated in the ganglia (two masses near the eyespots), sometimes called the "brain." This "ladder" type of nervous system reflects the cephalization and bilateral symmetry obvious from the outside.

An individual worm is **hermaphroditic:** It produces both eggs and sperm but does not fertilize itself. Two individuals mate by transferring sperm to each other during copulation. Fertilization is internal, and eggs are laid in cocoons or jelly. The most common method of reproduction is asexual reproduction by transverse fission in which the animal splits in half behind the pharynx, and each half regenerates the missing parts.

Class Cestoda: Tapeworm

The members of this class are highly specialized **endoparasites.** The adult tapeworm lives in the intestine of a vertebrate and is firmly attached by its **scolex,** which contains suckers and hooks. The rest of the ribbonlike body is a series of segments called **proglottids.** Each proglottid contains male and female reproductive organs, nerves, and excretory canals. There is no digestive system.

How does the tapeworm obtain food?

____

____

____

- Examine a preserved whole tapeworm such as the dog tapeworm, *Dipylidium caninum,* and locate the scolex.

Immature proglottids are found at this end, and the ripe proglottids stuffed with eggs are broken off from the opposite end and leave the body in the feces. The structure of a mature proglottid is shown in Figure 18–3.

- View a mature proglottid in the microscope, and identify the **ovaries,** which are centrally located; the **uterus** between the ovaries; and the **vaginas,** which lead out through the genital pores at the sides of the proglottid.
- The **testes** fill most of the proglottid. Sperm pass out through the vas deferens to the genital pore.
- Find the **excretory canals,** which run along the sides and posterior of the proglottid.
- Find the **longitudinal nerves,** which pass down the sides of the proglottid.

The proglottids of a tapeworm fertilize themselves or an adjacent proglottid. In an infected animal, competition will lead to the success of one individual tapeworm and the rest will be eliminated.

PHYLUM NEMATODA: *ASCARIS*

Roundworms may be the most abundant animals on earth. They occur by the millions in soil where they damage plants and crops and may reach a large size, as in the giant roundworm, which parasitizes humans. The **complete digestive tract** of the roundworm is a tube from mouth to anus within the tube of the body wall ("tube within a tube" body plan). The space between the inner and outer tubes is called a **pseudocoelom** because it is not completely lined by mesoderm. There is a protective **cuticle** covering the outside of the worm's epidermis that prevents parasitic roundworms from being digested by their host. There are separate male and female sexes and fertilization is internal. The mesoderm contains longitudinal muscles, so the worms can move by a whiplike

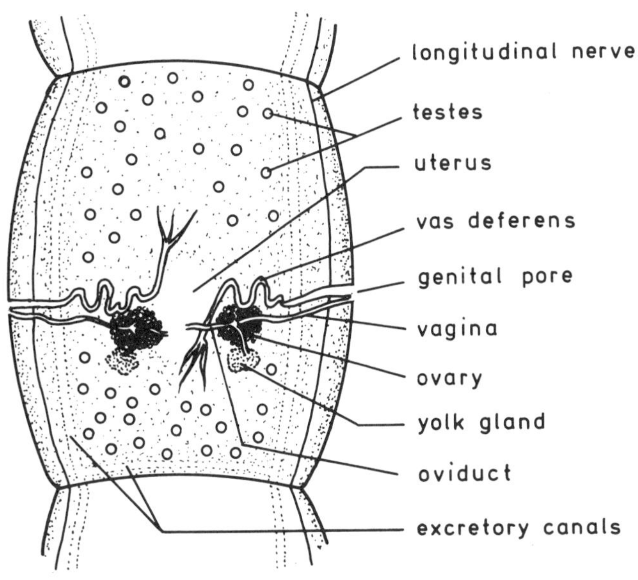

Figure 18–3. Mature proglottid of the dog tapeworm, *Dipylidium caninum.* Copulation between two proglottids results in internal fertilization of eggs. Fertilized eggs completely fill the ripe proglottid, which breaks off from the tapeworm and eventually bursts to release the eggs.

thrashing of the body, but they cannot creep like an earthworm. There are dorsal and ventral nerve cords but no respiratory or circulatory system.

External Anatomy

- Examine preserved specimens of male and female roundworms such as *Ascaris.*
- Identify the male, which has a distinctly curved posterior end with copulatory spicules.
- The opposite anterior end has a **mouth** surrounded by three **lips.**
- Locate the **anus,** which is a slit across the posterior end of the worm on its ventral side.
- Find the female **genital pore** on the ventral side of the worm about one third of the way from the head.

Internal Anatomy

The internal organs of *Ascaris* are shown in Figure 18–4.

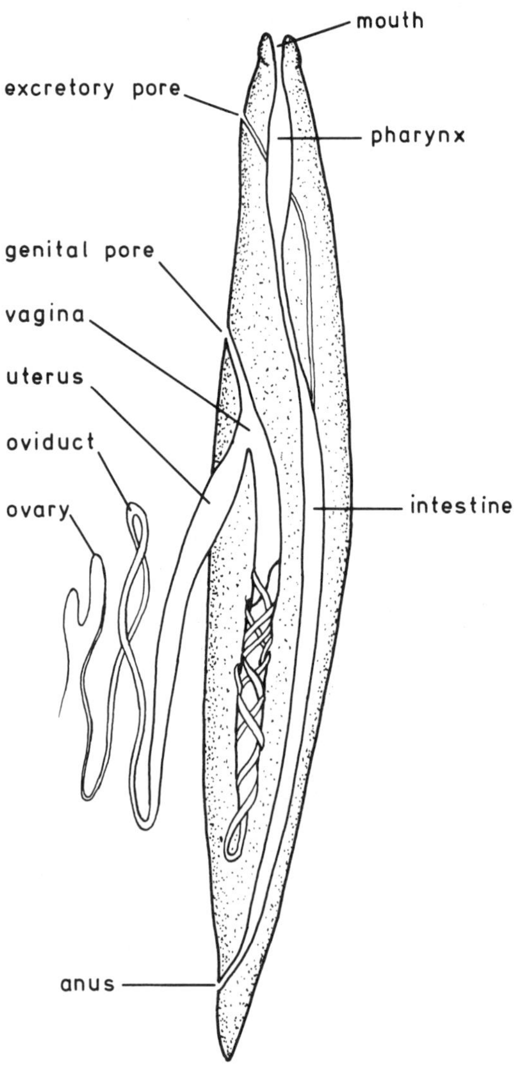

- Make a longitudinal slit along the dorsal side of the female worm, and pin back the sides.
- Cover the worm with some water to minimize drying.
- Find the **intestine,** which runs through the body from mouth to anus.
- In the middle of the intestine, locate the coils of the female reproductive system. The **ovaries** are threadlike and connect to the **oviducts,** which in turn lead into the larger **uteri.** The uteri join just before exiting through the genital pore.

These structures can be seen more easily in cross section. (See Figure 18–4.)

- Examine a cross section of *Ascaris* in the microscope, and find the large, flattened **intestine.**
- Identify the small, round structures that are sections through the tubes of the **oviduct** and the larger **uterus,** which is filled with eggs; the **ovary** is solid.
- Locate the dorsal and ventral **nerve cords.**
- Identify the **cuticle** surrounding the worm and the longitudinal **muscle cells,** which are the large cells surrounding the body cavity.
- The space just under the cuticle in either side of the worm is the **lateral line** in which the **excretory canals** are located.

In the male *Ascaris* a threadlike testis leads into ducts that empty into the cloaca at the posterior of the digestive tract. As you already learned in Topic 10, fertilization is internal, and the eggs have passed

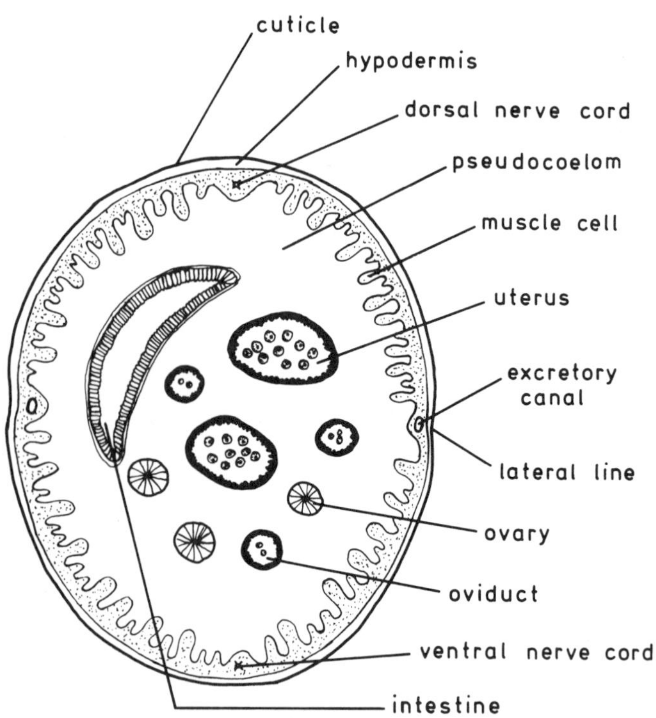

Figure 18–4. A female roundworm, *Ascaris*. There is no respiratory or circulatory system.

through the early stages of development before they leave the body of the female.

Anatomy of a Nematode Cyst

A parasite has a great advantage if its eggs or larvae are well protected and can survive adverse conditions until they can reach a new host. In *Trichinella spiralis*, the parasitic roundworm that causes the disease trichinosis, fertilized eggs develop into larvae within the reproductive tract of the female worm and are then deposited within the host's lymphatic system. The larvae are carried throughout the body and those that invade skeletal muscle will form cysts and survive. The cysts have a low metabolic rate and a protective wall. If the host is a pig, the meat might be ingested without being properly cooked, and the cyst walls would be broken down in the digestive tract of the new human host. The young worms quickly mature and produce hundreds of offspring. These larvae enter the lymphatic system, burrow into the host's skeletal muscle, and again encyst.

- Examine a prepared slide that shows the cysts formed in skeletal muscle by young trichina worms.

Do the encysted worms harm their host?

What are two ways in which the disease trichinosis could be controlled?

1. ______________________________

____

2. ______________________________

____

PHYLUM ROTIFERA

Rotifers are small, pseudocoelomate animals that move about actively and can ingest small particles of food. The structure of a typical rotifer is shown in Figure 18–5. The beating of circles of cilia in the **corona** gives the phylum its name (rotifer means "wheel bearer"). These cilia draw food into the mouth. Food is then ground up in the mastax and passes on through the stomach, intestine, cloaca, and out the anus. The corona is also used in locomotion.

Rotifers have an excretory system with flame cells and a bladder. The body is covered by a protective cuticle. Muscle bands can contract to change the shape of the body.

- Examine a living rotifer or a prepared slide.
- Locate the **corona** and **sensory antenna** at the anterior end.

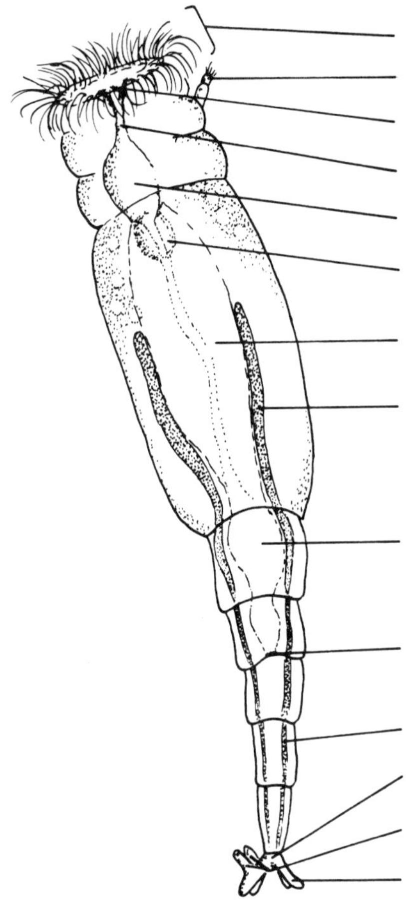

Figure 18–5. A rotifer, *Philodina*. This type of rotifer is always female and reproduces by parthenogenesis.

- Find the **toes** and **spurs** at the posterior end. A rotifer may attach to the substrate by its toes.
- Locate the **mastax.** It is part of the one-way, complete digestive tract.

How could you decide that a rotifer is not a giant protozoan?

____

____

____

In certain species of rotifers there are no males, and the unfertilized eggs develop into new female rotifers. The eggs may develop in a brood pouch that is located within the body of their mother. When males are present and are involved in reproduction, they are usually small and degenerate and fertilize the female by the crude process of **hypodermic impregnation:** the male stabs the female in any part of her body with his penis and releases the sperm which have to swim through the body to the eggs in order to fertilize them.

Rotifers can withstand drying and will lose most of their body water to enter a desiccated form that can survive for years. When water again becomes available, the animal will rehydrate and become active shortly afterward.

When you are finished with this exercise, return your materials and store your microscopes properly.

REFERENCES AND FURTHER READING

Barnes, R.D. *Invertebrate Zoology,* 5th ed. Philadelphia: Saunders College Publishing, 1987.

Barth, R.H., and R.E. Broshears. *The Invertebrate World.* Philadelphia: Saunders College Publishing, 1982.

Buchsbaum, R. *Animals Without Backbones,* rev 2nd ed. Chicago: University of Chicago Press, 1975.

Buchsbaum, R., and L.J. Milne. *The Lower Animals: Living Invertebrates of the World.* Garden City, NY: Doubleday, 1962.

Pennack, R.W. *Freshwater Invertebrates of the United States,* 2nd ed. New York: Wiley, 1978.

Russell-Hunter, W.D. *A Life of Invertebrates.* New York: Macmillan, 1979.

Storer, T.I. *et al. General Zoology,* 6th ed. New York: McGraw-Hill, 1979.

Villee, Claude A., W.F. Walker, and R.D. Barnes. *General Zoology,* 6th ed. Philadelphia: Saunders College Publishing, 1984.

Journal Articles

Blonston, Gary. "To build a worm." *Science* 84, 63, March 1984.

Gierer, A. "Hydra as a model for the development of biological form." *Scientific American,* 44, December 1974 (#1309).

Gilbert, John J. "Developmental polymorphism in the rotifer *Asplanchnia sieboldi.*" *American Scientist* 68(5): 636, November 1980.

Goreau, T.F., N.I. Goreau, and T.J. Goreau. "Corals and coral reefs." *Scientific American,* 124, August 1979 (#1434).

Jackson, Jeremy B.C., and Terence P. Hughes. "Adaptive strategies of coral reef invertebrates." *American Scientist* 73(3): 265, May–June 1985.

Koehl, M.A.R. "The interaction of moving water and sessile organisms." *Scientific American,* 124, December 1982.

Mansour, T.E. "Chemotherapy of parasitic worms: New biochemical strategies." *Science* 205: 462, 1979.

Moore, J. "Parasites that change the behavior of their host." *Scientific American,* 108, May 1984.

Savory, T.H. "Hidden lives." *Scientific American,* 108, July 1968 (#1112).

Student Name ______________________ **Date** ______________

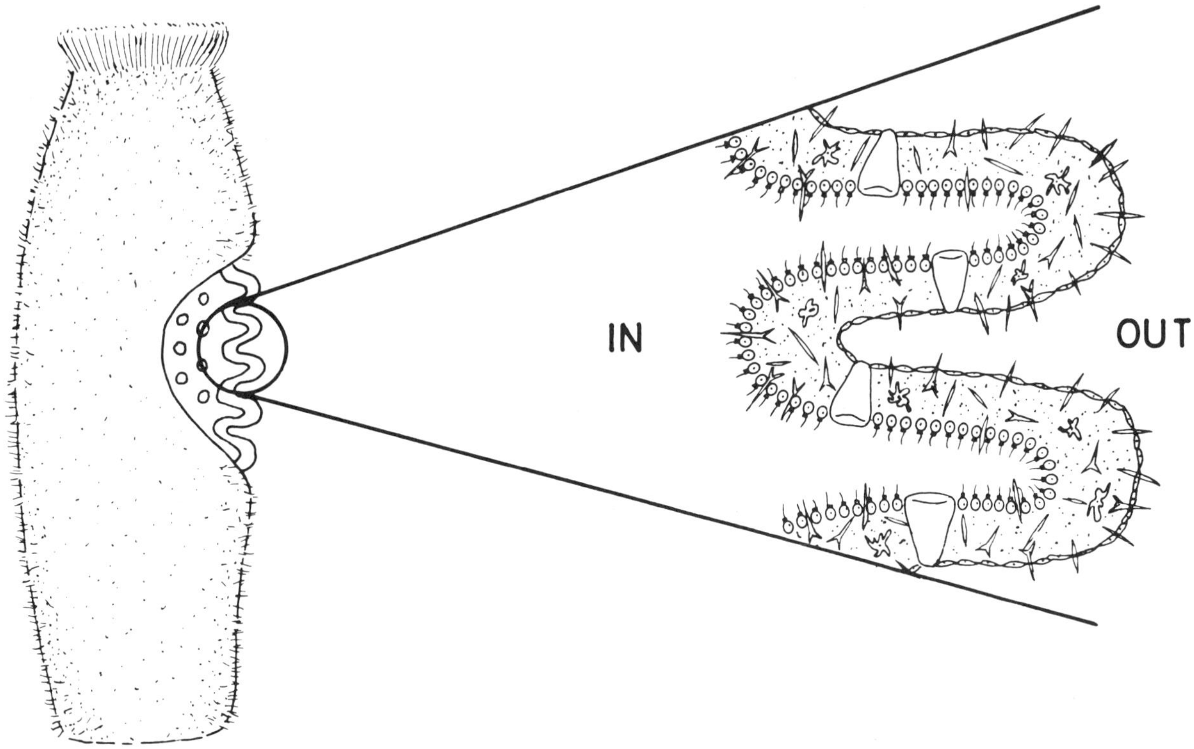

Figure 18–6. ***Scypha.*** This sponge is of the syconoid type and has calcareous spicules. Label the osculum, base, spongocoel, spicules, and the location of the choanocytes. Indicate with arrows how water enters and exists from the sponge.

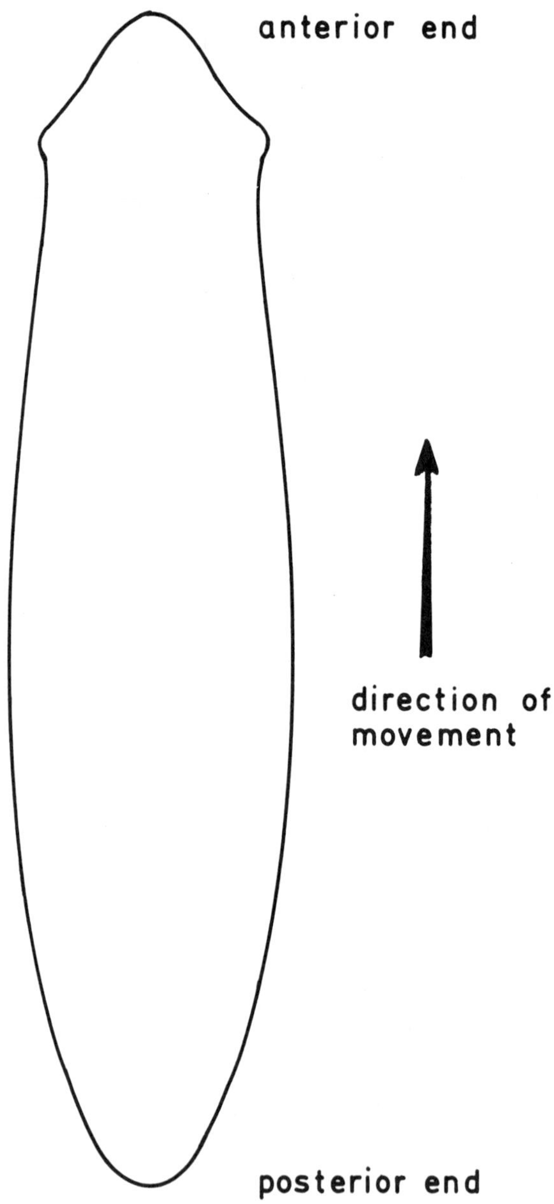

Figure 18–7. A planarian. Draw in and label the parts of the digestive system, nervous system, and sense organs of the flatworm.

TOPIC 19

Higher Invertebrates: Annelids and Molluscs

OBJECTIVES

When you have completed this topic, you should be able to:

1. Define coelom, and enumerate the advantages to an organism of having one over not having one.
2. List the features of protostomes that distinguish them from deuterostomes.
3. List the major features of an annelid and of a mollusc.
4. Describe the adaptations of the earthworm that allow it to live on land.
5. Describe the advantages and disadvantages of the exoskeleton in a mollusc.
6. Explain the evidence that annelids and molluscs share a common ancestor.

TEXT REFERENCES Chapter 22 (I, J).

KEY TOPICS

coelom
protostome
deuterostome
annelid
mollusc

INTRODUCTION

Annelids and molluscs, as well as the animals that we will study in later exercises, have a true **coelom** that arises by splitting of the middle layer of cells in the body, the mesoderm (see Figure 18–1). The mesoderm surrounds all of the organs that develop within the coelom, protecting and supporting them, and it can give rise to layers of muscle so that the gut and other organs can move and function independently from the body wall. The space within the coelom makes room for a specialized digestive tract, an independently active circulatory system, other internal organs, and large gonads for seasonal reproduction. In some organisms such as the earthworm the fluid in the coelom acts as a **hydrostatic skeleton** on which the muscles of the body wall can exert pressure during locomotion. In many coelomates, the coelomic fluid itself helps in circulating nutrients and in gas exchange.

Annelids, molluscs, arthropods and various other phyla share some important features of embryological development and so are grouped together as **protostomes.** For example, the primitive opening in the embryo, the blastopore, develops into the mouth in these animals (proto = "first", stoma = "mouth"). The protostomes show so much variation, however, that many of these groups are probably only distantly related. A second group of organisms, the **deuterostomes,** shares different features. The blastopore in these animals becomes the anus, and the mouth is formed in a quite different way. Deuterostomes are thought to be evolutionarily related to each other and include the echinoderms (sea stars, sea urchins) and the chordates (tunicates, vertebrates, ourselves).

These are some important differences between the two groups, but there are many exceptions:

Protostomes

1. Blastopore becomes mouth.
2. Spiral cleavage of embryo.

3. Determinate cleavage (fixed fate).
4. Coelom formed by mesoderm splitting.

Deuterostomes
1. Blastopore becomes anus.
2. Radial cleavage of embryo.
3. Indeterminate cleavage (fate can be altered).
4. Coelom formed by mesoderm pouches.

A common theme during the evolution of the annelids and molluscs was the appearance of **metamerism** or serial repetition of body parts. Annelids such as the earthworm are obviously segmented. Molluscs are thought to have arisen from an annelid ancestor, but the current forms have tended to lose the segmentation that was the outstanding feature of the annelid ancestor. (A hypothetical unsegmented ancestral mollusc is shown in Text Figure 22–15). A recently discovered primitive mollusc, *Neopilina,* does show metamerism and supports the notion of a common ancestor with the annelids. Another important connection is the **trochophore larva** produced by both annelids and molluscs during reproduction.

PHYLUM ANNELIDA

Annelids have **metamerically segmented** wormlike bodies, and most are divided into segments internally by partitions called **septa.** In polychaetes and oligochaetes the body contains circular and longitudinal muscles that squeeze the fluid within each segment to lengthen or shorten the worm during its creeping movements (see Text Figure 31–13). Thus the coelomic fluid acts as a **hydrostatic skeleton** and there is no hard exoskeleton as there is in molluscs. The **closed circulatory system** contains pigmented blood for improved gas exchange, and some annelids have **gills.** The **complete digestive tract** processes food in one direction from mouth to anus and has specialized parts. The nervous system shows **cephalization** with sense organs and masses of nerve cells called ganglia at the anterior end. The **nephridia** are true excretory organs that can actively absorb and excrete substances to regulate the composition of the coelomic fluid.

Class Oligochaetae: Earthworm

Many oligochaetes such as the earthworm have taken up a terrestrial life, but have not completely adapted to land: they secrete a thick mucus to keep their skin moist and must remain within their damp burrows in the ground unless the surface is soaking wet. They eat soil and decaying vegetation and have a well-developed complete digestive system to process their vegetarian diet. Earthworms occur in the soil in large numbers and are extremely important in maintaining its loose, aerated condition.

The earthworm's head is much reduced to adapt to its burrowing way of life. As an oligochaete (= few bristles), it has only four pairs of setae on each segment for traction instead of the parapodia paddles found in polychaetes. (See Color Plate 3.) Gas exchange takes place over the entire moist body surface and is aided by an efficient closed circulatory system in which the blood is moved by little pumps ("hearts").

The individual earthworm is hermaphroditic, producing both eggs and sperm, but does not fertilize itself. Rather, during copulation eggs and sperm are shed into a protective cocoon of mucus where external fertilization and development take place (see Text Figure 17–7). Earthworms, in contrast to polychaetes, do not have a trochophore larva. Instead, a young worm hatches directly from the cocoon.

EXTERNAL ANATOMY AND BEHAVIOR

- Examine a live or preserved earthworm and find the **clitellum,** an enlarged section near the anterior end that secretes mucus during copulation.
- Locate the **mouth** under the protruding **prostomium.**
- Counting back from the prostomium, find the ninth segment, and look for openings of the **seminal receptacles** between the ninth and tenth and the eleventh and twelfth segments on the worm's ventral surface. Sperm are stored here when worms copulate.
- The **oviducts** open on the fourteenth segment to release the eggs.
- **Seminal vesicles** open on the fifteenth segment to release sperm.
- Look at one segment closely, and locate the **setae** on the lower part of the side of the worm. They are made of chitin and help give the worm traction during locomotion.
- Find the **excretory pores** that occur on the sides of most of the segments. Each pore is the opening of one **nephridium,** or excretory organ.

Earthworms do have a very important adaptation to terrestrial life: a special type of external fertilization. When two worms mate, they overlap ventral surfaces so that the anterior tip of one worm is a little past the clitellum of the other and vice versa (see Text Figure 29–7). Sperm are released by both worms and have to migrate over many segments in mucus-covered grooves to reach the seminal receptacles, so the worms must remain joined for quite a long time. After mating, the mucus secreted by the clitellum passes over the head of the worm and picks up eggs from the oviduct openings and sperm from the sperm receptacles as it moves. The cocoon is deposited in the soil and a young worm hatches from it eventually.

If live earthworms are available, investigate locomotion and other behavioral responses.

- Place the worm on a moist paper towel in a dissecting pan. Study its locomotion under the dissecting microscope.

Referring to Figure 19–2, which muscles contract to cause elongation of a segment? ______________

Which muscles contract to cause shortening of a segment? ______________

How does the worm respond to the bright light of your microscope?

____

____

- Touch the front end of the earthworm gently with a probe. What is the response? ______________
- Hold the probe still in the path of the moving earthworm. How does the worm respond to this foreign object?

____

____

- Dip your probe in 5% acetic acid. (This is similar to vinegar and not acidic enough to injure the worm.) Move the probe toward the front end of the worm until you get a response.

Describe the response. ______________________________

____

How do you think the response is important to the earthworm? ______________________________

____

INTERNAL ANATOMY

Use Figures 19–1 and 19–2 as guides to the internal structure of the earthworm. If you need to review basic anatomical terms such as "dorsal" and "cross section," refer to Topic 18, Introduction.

- Open the worm from the darker **dorsal side,** and keep it covered with water while you work.
- Notice the internal segmentation that corresponds to the external segmentation. Identify the internal **septum** between adjacent segments.
- Locate the **dorsal blood vessel,** which carries blood from posterior to anterior. The circulatory system has hemoglobin and the vessels have one-way valves.
- Locate the five pairs of pumping blood vessels ("**hearts**") that connect the dorsal and ventral blood vessels.

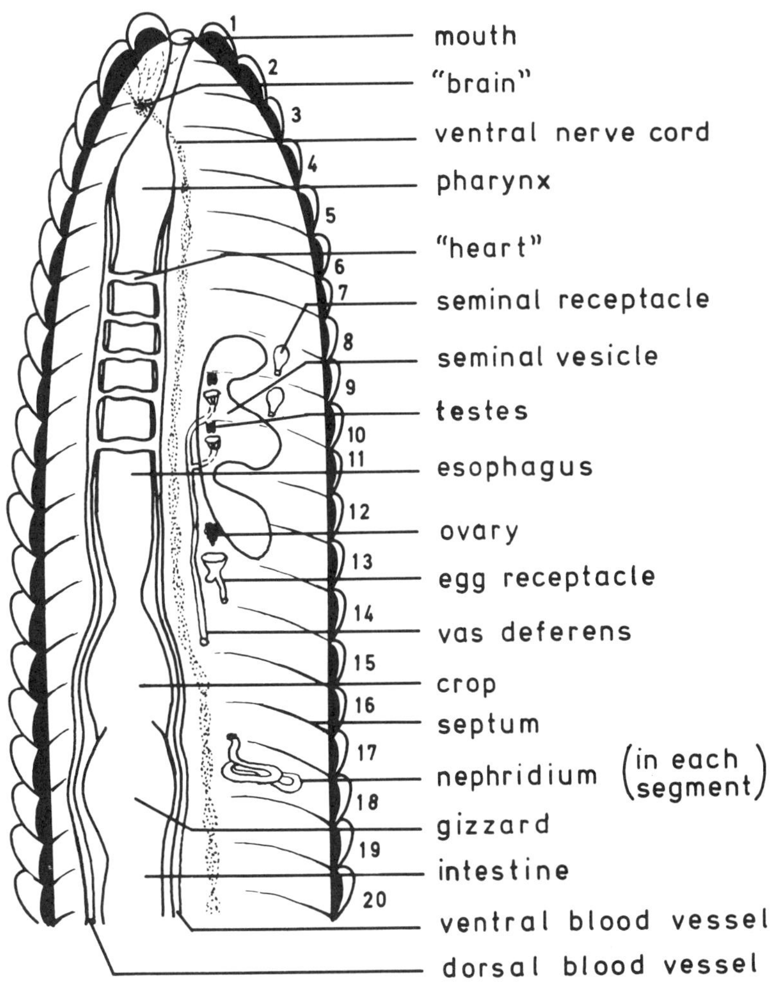

Figure 19–1. Internal structure of the earthworm, *Lumbricus*. The dorsal side is toward the left. Sense organs are lacking, but there is a well developed nervous system, closed circulatory system, complete digestive tract, and excretory organs.

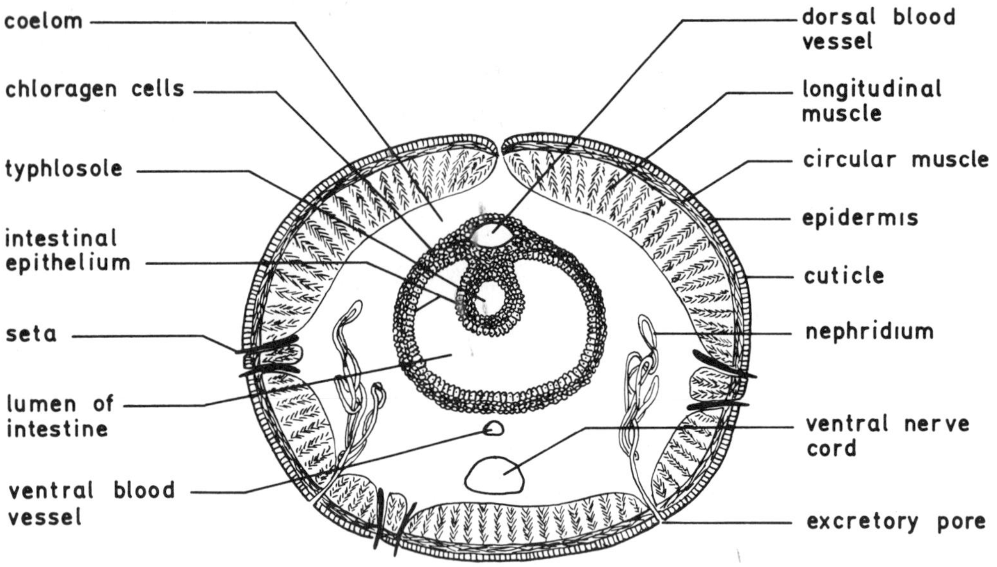

Figure 19–2. Cross section of the earthworm, *Lumbricus*. The surface area of the intestine is doubled by the typhlosole, and the spacious, fluid-filled coelom acts as a hydrostatic skeleton in locomotion.

- Find the **ventral blood vessel,** and the subneural blood vessel beneath the ventral nerve cord. Blood in these vessels flows from anterior to posterior.
- Follow the large digestive tube back from the mouth as far as the clitellum, identifying the muscular **pharynx,** the **crop** (about segment 16), the **gizzard** (about segment 19), and the **intestine** (all the rest).
- Within each segment find the paired **nephridia.** Each of these excretory organs is a coiled tube beginning with a nephrostome and leading to an **excretory pore.** The beginning of the tube is in the anteriorly adjacent segment. Fluid enters by ciliary action, and the tube excretes waste into the fluid and reabsorbs needed substances as it passes out of the body.
- Locate the saclike **seminal receptacles** in segments 9 and 10.
- Sperm made in the tiny testes are stored in the multi-lobed **seminal vesicles** until copulation. Then they pass out through the **vas deferens.**
- **Ovaries** are located in segment 13, and the eggs are stored in the adjacent **egg receptacles** before being released through the **oviducts.**

The earthworm does not have eyes or ears, but the epidermis contains light-sensitive cells, and sensory bristles that are sensitive to vibration. The paired ventral nerve cords have fast-conducting giant neurons that allow the worm to retreat rapidly in a startle response if it senses danger. In the head region, the nerve cords connect to swellings of nerve cells called ganglia by tracts of fibers that pass on either side of the digestive tract, just in front of the pharynx.

- Locate the **ventral nerve cords,** beneath the ventral blood vessels.
- Locate the whitish paired **ganglia** at the anterior end of the ventral nerve cord; these represent the earthworm's "brain." They are just dorsal to the digestive tract, and just anterior to the pharynx.
- Obtain and study a prepared slide of the cross section of an earthworm; compare your slide with Figure 19–2.
- Locate the intestinal epithelium that forms the gut.
- Locate the large space it surrounds, which is the **lumen** or interior of the intestine; the infolding at its top is the **typhlosole,** which increases the surface area available for absorption.
- Notice the large coelom, which is the fluid-filled space between the muscular body wall and intestine, and which contains the body organs. In the earthworm the coelom acts as a hydrostatic skeleton.
- Locate the blood vessels and the ventral nerve cord.
- Surrounding the gut are **chloragen cells,** which aid in digestion and may represent primitive liverlike cells.
- Find the coils of nephridium within the coelom. Depending on the section, you will see only parts of these paired organs, which are found in each segment.
- Surrounding the coelom is a thick layer of longitudinal muscle. It functions to shorten the worm.
- A thinner layer of circular muscle surrounds the longitudinal muscle. It contracts to lengthen the worm.
- Locate the **cuticle** on the outside of the worm and the **hypodermis** that secretes it.

PHYLUM MOLLUSCA

Molluscs are mostly marine animals such as chitons, snails, the octopus, and shellfish (see Color Plate 3). They have these major body parts: a muscular **head-foot,** a **visceral mass,** and a hard protective **shell** secreted by the **mantle,** which lies over the soft body to create a mantle cavity. Except for bivalves, molluscs have a rasplike feeding organ, the **radula.** There is a **complete digestive tract** and commonly an **open circulatory system.** The respiratory system usually has **gills** and the excretory system has **nephridia.** Molluscs are true coelomates with a body cavity surrounded by mesoderm. The sexes are usually separate and the life history may include a **trochophore larva.**

Class Bivalvia: Clam or Mussel

As in many sessile animals, the bivalves (clams, mussels, oysters, and scallops) have developed protection, in this case a pair of heavy, calcareous shells (valves) that are secreted by the mantle. They are hinged and can be very tightly clamped together to enclose the soft body and protect it from predators. In those forms that are subjected to periodic drying when the tide falls, the closed shell prevents the loss of moisture. Some bivalves, such as the scallop, can swim after a fashion by closing and opening their relatively lightweight valves. The muscle involved in swimming is the part of the scallop used for human consumption.

These animals show bilateral symmetry and, if you orient one with the hinge at the top and the "points" at the top left, as shown in Figure 19–3, you will be viewing the left side. The gills are enclosed in a chamber between the mantle and the visceral mass, which is sealed off from the rest of the viscera. Water passes in and out of the gill chamber through siphons, which are modified parts of the mantle. The internal organs are bathed in the fluids of an open circulatory system that is sufficient for the needs of this slow-moving animal. Pigmented blood is pumped by a two-chambered heart.

Would you expect an active, predatory mollusc such as a squid to have an open or closed circulatory system? _______________

Bivalves are filter feeders, so there is no radula. They extract edible particles from the water circulating through the gill chamber. Labial palps aid in moving the food trapped in mucus into the mouth. There is a complete digestive tract, a true nephridium, and a foot for digging or locomotion. The nervous system is greatly reduced and there is almost no cephalization.

- Orient the clam or mussel as shown in Figure 19–3. Focus on the large adductor muscles that close the shell. These are shown in Figure 19–5 and must be cut to open the shell.
- Insert a tool into the space between the lower edges of the shells, and pry the shells apart enough to insert something to act as a wedge.
- Use a scalpel to cut through the muscles without damaging the internal organs, especially in the region near the hinge. Separate the shells to expose the mantle.
- Refer to Figures 19–4 and 19–5 to identify the internal organs.
- Identify the **adductor muscles** and the muscular **foot.**
- Remove the **mantle,** which covers the **gills.**
- Carefully remove the left gill. Slit the bottom of the foot in order to see the organs within it.

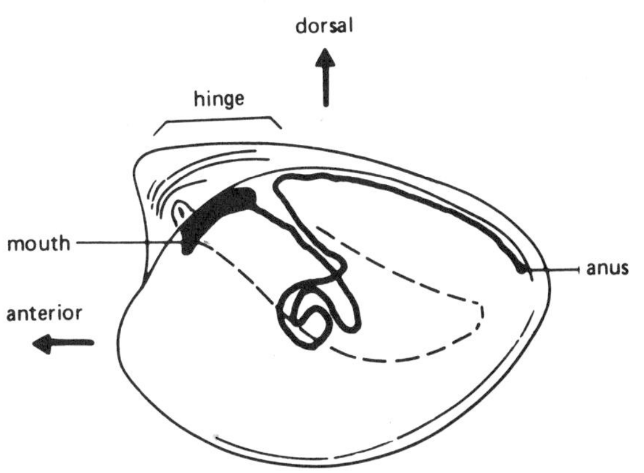

Figure 19–3. A bivalve mollusc. If a clam or mussel is held in this orientation, with the hinge on top and the "points" aiming to the left, the anterior will be to the left and the dorsal part of the animal on top. The approximate location of the digestive tract is shown in black, and the position of the gills is marked by the dashed line.

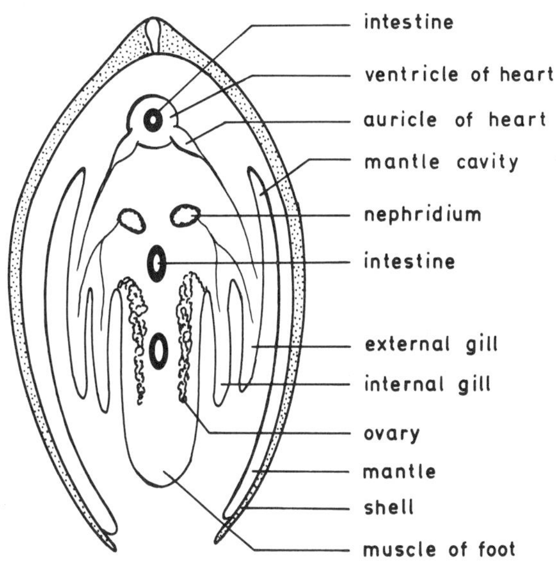

Figure 19–4. Cross section of a bivalve. Water circulates within the mantle cavity, bathing the gills for gas exchange.

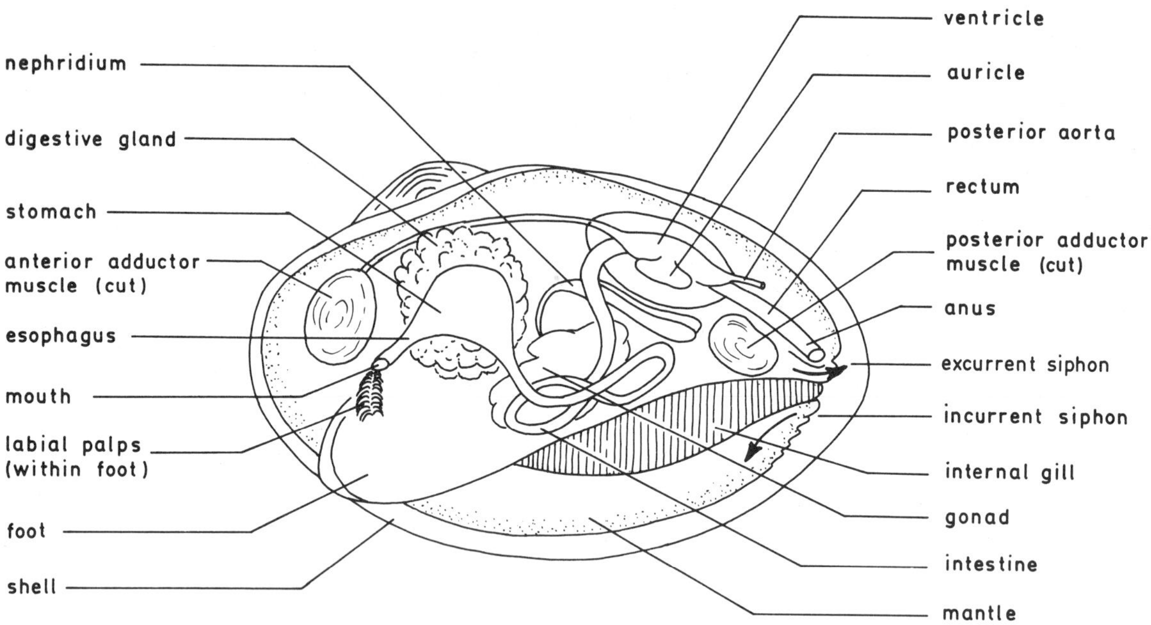

Figure 19–5. Internal structure of a bivalve. The left gill and mantle have been removed to show the internal organs.

- Trace the digestive tube from the mouth through the **gonad** (yellowish) and **heart** to the anus.
- Locate the **digestive gland** and the **nephridium** (brown or green).
- Pass a probe into the **siphons** at the animal's posterior to see where water circulates. Water enters the mantle cavity through the incurrent siphon, circulates through the filaments of the gills, and leaves through the excurrent siphon. The siphons are necessary to take in water free from sediment. The water moves by ciliary action, and particles in the water are the only food of the bivalve.

Because bivalves are filter feeders, any poisonous substances in their food particles will become concentrated in their bodies. During the season that toxic algae are prevalent, it is unwise to risk eating shellfish, and the gathering of shellfish for food is usually prohibited by law.

Most bivalves are either male or female, and they carry out reproduction by shedding a large number of gametes into the water where external fertilization takes place. Sometimes the young are brooded in a specialized part of the gill. The embryos develop into motile trophophore larvae.

When you are finished with your work, dispose of your specimens, clean your dissecting equipment, and store your microscopes properly.

REFERENCES AND FURTHER READING

See Topic 18 for general references on invertebrates.

Barrington, E.J. *Invertebrate Structure and Function,* 2nd ed. New York: Halstead, 1979.

Lane, F.W. *Kingdom of the Octopus.* New York: Sheridan House, 1965.

Morton, J.E. *Molluscs,* 4th ed. London: Hutchinson, 1979.

Journal Articles

Boycott, Brian B. "Learning in the octopus." *Scientific American,* 42, March 1965 (#1006).

Feder, Howard M. "Escape responses in marine invertebrates." *Scientific American,* 92, July 1972 (#1254).

Gosline, J.M., and M.E. DeMont. "Jet-propelled swimming in squids." *Scientific American,* 74, January 1985.

Jones, Douglas S. "Sclerochoronology: Reading the record of molluscan shells." *American Scientist* 71(4): 384, 1983.

Nicholls, John G., and David Van Essen. "The nervous system of the leech." *Scientific American,* 38, January 1974 (#1287).

Reid, R.G.B., and R.R. Bernard. "Gutless bivalves." *Science* 208:609, May 1980.

Roper, Clyde F.E., and Kenneth J. Boss. "The giant squid." *Scientific American,* 96, April 1982.

Stanley, Steven M. "Mass extinctions in the ocean." *Scientific American,* 64, June 1984.

Ward, Peter, Lewis Greenwald, and Olive E. Greenwald. "The buoyancy of the chambered nautilus." *Scientific American,* 190, October 1980.

Willows, A.O.D. "Giant brain cells in mollusks." *Scientific American,* 68, February 1971 (#1212).

TOPIC 20

Higher Invertebrates: Arthropods

OBJECTIVES

When you have completed this topic, you should be able to:

1. List the features of members of the phylum Arthropoda.
2. Describe adaptations of terrestrial arthropods that allow them to live on land.
3. List three factors that contributed to the success of insects.
4. Define metamorphosis, and explain the difference between ametabolous, paurometabolous, hemimetabolous, and holometabolous metamorphosis.
5. Define the following arthropod structures and give the class(es) in which they are found: trachea, gill, book lung, spiracle, hemocoel, green gland, swimmerets, spinnerets, Malpighian tubules, antennule, antenna, labium, mandible, simple eye, compound eye, cephalothorax, carapace, chelicerae, pedipalps.

TEXT REFERENCES

Chapter 22 (K).

KEY TOPICS

coelom
arachnids (Class Arachnida)
crustaceans (Class Crustacea)
insects (Class Insecta)

INTRODUCTION

The majority of animal species are arthropods, and the biomass of insects alone outweighs that of human beings. Arthropods and annelids share a common ancestry, but the arthropods have gone on to fill every possible habitat on land and in the sea. Their success is mainly due to the development of a hard **exoskeleton** of protein and chitin, a type of polysaccharide. This skeleton is light compared with the heavy shell of molluscs, and it is jointed, allowing flexibility of the body and appendages (arthropod = "jointed foot"). This versatile exoskeleton protects the animal's body and prevents terrestrial arthropods from drying out. It also serves as an attachment point for muscles so that well-developed locomotion is possible.

The presence of the exoskeleton also creates certain problems, however. Because the skeleton can't grow with the animal, it has to be shed periodically during a **molt,** and the temporarily unprotected animal is vulnerable to attack during and immediately after the molting process. The skin cannot be used for detecting stimuli, so antennae and sensory bristles are used instead.

Although arthropods are coelomate, the **coelom** is lost during development. The body cavity that you will see has a different origin. The body usually shows some signs of **segmentation,** and there is pronounced **cephalization,** with the mouth, nervous system, and sense organs concentrated in the region of the head. The paired, **jointed appendages** may be modified for other purposes in addition to locomotion. The ancestral arthropod is

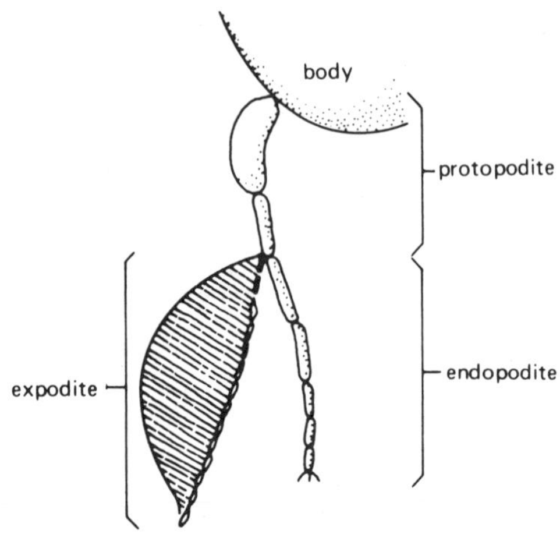

Figure 20–1. The basic arthropod appendage. In most arthropods, one branch or the other of the basic two-branched appendage is lost, but sometimes both branches are retained, especially in the crustaceans.

thought to have had many segments with a pair of appendages attached to each, somewhat like the fossil trilobites. The original ancestral appendage was probably branched, as is shown in Figure 20–1. Usually one branch or the other was lost, but sometimes both were retained. Modified appendages are used for chewing, biting, stinging, transferring sperm, chemoreception, and touch perception.

The sexes are separate, and arthropods usually copulate and have internal fertilization. Most arthropods care for the eggs in some way, either by depositing them in or on a food source, or carrying them around on their bodies until the young hatch. The larvae are often very different from the adults and must undergo **metamorphosis,** in which they change into the adult form in a short period of time.

All arthropods have a **complete digestive tract,** a double **ventral nerve cord,** an **open circulatory system** consisting of a body cavity or hemocoel filled with hemolymph (blood) pumped by a dorsal heart, and an excretory organ.

CLASS ARACHNIDA

Spiders, ticks, mites, and scorpions are the major groups of arachnids. Besides four pairs of **walking legs,** they all have **chelicerae,** appendages for piercing and sucking on their prey, and **pedipalps,** which are sensory organs and help hold the prey during feeding. Pedipalps are modified for sperm transfer in males and they are the prominent pincers of scorpions. Most arachnids are carnivorous predators or external parasites. The unsavory reputation of this group is well deserved, as far as humans are concerned: venomous spider bites, scorpion stings, lice, scabies caused by mites, and Rocky Mountain spotted fever spread by ticks. There is even a mite that lives at the base of human eyelashes and sometimes causes severe irritation of the eyelids.

- Examine the variety of preserved arachnids on demonstration. What obvious feature is shared by all arachnids? ______________________________

Do arachnids have antennae? _______________

- Note the division of the body into the **cephalothorax** and the larger **abdomen.**

Are the legs attached to the cephalothorax or to the abdomen? _______________

The Spider: External Anatomy

Spider anatomy is shown in Figure 20–2. These predatory arachnids usually have **poison glands,** which open near the tips of the chelicerae. Once the prey is bitten, it will be quickly paralyzed by the poison. The spider will secrete digestive enzymes into the prey tissue and then suck up the resulting broth. While spider bites are unpleasant, only a few species can inflict serious or fatal bites on humans.

- Examine a preserved spider in the dissecting microscope.
- Identify the cephalothorax with its **chelicerae** and **pedipalps.**
- Examine the **eyes:** Usually there are 4–8, and they are connected to the brain.

How many eyes does your specimen have?

____

The cephalothorax also contains the poison glands and a powerful **sucking stomach** with which the spider ingests its liquid diet.

The abdomen contains **book lungs** for gas exchange, gonads, digestive glands, and **silk glands.** Spiders spin their webs from a protein that is extruded in liquid form from spinnerets at the tip of the abdomen and polymerizes into silk. The cocoons, which hold the eggs and developing young, are made from the same silk.

- Look for the spinnerets on your spider specimen.

Manipulation of silk for a variety of purposes probably accounts for the great diversity and success of spiders.

CLASS CRUSTACEA

Crustaceans are mostly marine and are predators or scavengers. They include a wide variety of animals: shrimp, lobsters, hermit crabs, brine shrimp, cope-

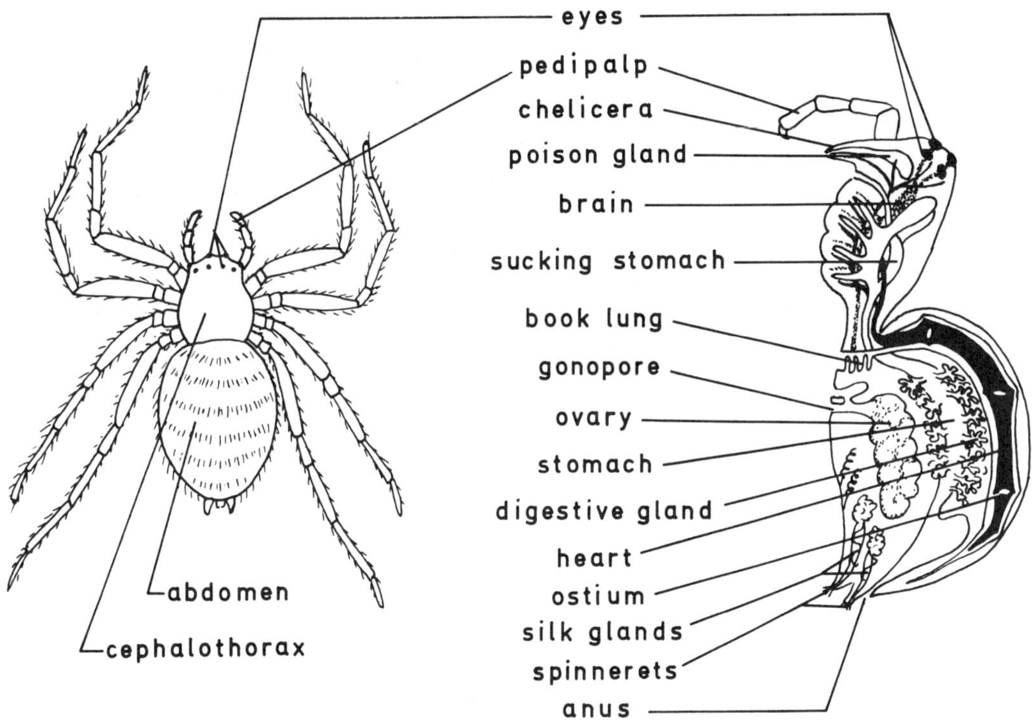

Figure 20–2. Anatomy of a spider. The piercing mouth parts and sucking stomach obtain the liquid diet. Silk is spun from the silk glands by spinnerets in the abdomen.

pods, barnacles, beach fleas, and wood lice (sow bugs). Copepods are tiny crustaceans that are an important link in the marine food chain and are the krill on which some whales feed. Along with centipedes, millipedes, and some insects, crustaceans have true jaws or **mandibles.** They also have two pairs of accessory feeding appendages, the **maxillae,** two pairs of **antennae,** and their appendages are often branched. In crustaceans such as the lobster, the shell is reinforced with calcium carbonate. Pigments in the shell give this animal protective coloration. The loss of one of these heat-sensitive pigments gives cooked lobsters their bright red color by revealing another pigment, which is heat stable.

The Crayfish: External Anatomy

■ Examine a preserved crayfish such as *Procambarus,* and use Figure 20–3 to locate the following structures: **antennae, antennules, rostrum, compound eyes, chelipeds** (pincers), the **carapace** which covers the **cephalothorax,** the segmented **abdomen** ("tail"), **walking legs, telson,** and **uropods.**

How many pairs of walking legs does the crayfish have in addition to the chelipeds? ______________

■ On the ventral side locate the **swimmerets** on the abdomen, and examine the first pair. They are

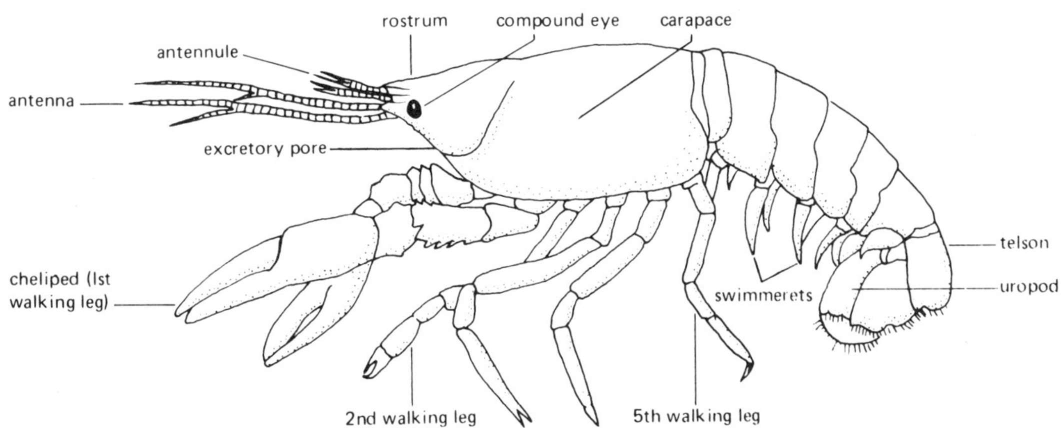

Figure 20–3. External structure of a male crayfish. The sperm ducts open at the base of the fifth pair of walking legs, and the first pair of swimmerets are modified for transferring sperm during copulation. In the female, the oviducts exit at the base of the third pair of walking legs.

modified for the transfer of sperm in those mature males that are ready to mate.

Is your crayfish a mature male? ___________

- Also on the ventral side, locate the **mouth,** with the **mandibles** or true jaws, five pairs of specialized appendages between the mandibles and chelipeds, and the **anus.**

The mandibles tear the food, and the other appendages detect and hold the food during feeding. One pair is specialized to move water through the chamber around the gills.

- Look for the excretory pore at the base of the antenna; it connects to the green gland that you will see later.

Internal Anatomy: The Crayfish

Use Figure 20–4 as your guide to the internal structure of the crayfish.

- Carefully dissect away the dorsal half of the carapace to expose the internal organs of the cephalothorax.
- Locate the diamond-shaped **heart** of the open circulatory system. The heart sends hemolymph to the spaces around the organs through several arteries. After passing through the gills, the hemolymph drains back into the heart through the openings in it (ostia).
- Remove the heart in order to see the **gonads** (ovaries or testes). Sperm exit at the base of the fifth pair of walking legs, and eggs leave the oviducts at the base of the third pair of walking legs.
- Trace the digestive system from the mouth through the **esophagus** to the **cardiac stomach** where food is ground up, then to the intestine leading to the anus.
- Note the **extensor** (straightens) and **flexor** (curls up) muscles which fill the abdomen ("tail") of the crayfish.
- Find the **green gland** located just behind each antenna. This excretory organ opens through the excretory pore that you identified before.
- Note the large **liver,** or digestive gland, beneath the first part of the intestine.
- Remove the internal organs to expose the double **ventral nerve cord,** which lies along the bottom of the body cavity.
- Locate a pair of nerves leading from the ventral nerve cord around the esophagus to the large ganglion, or **"brain"** in the head. Short nerves connect eyes, antennae, and antennules to the brain.

The nervous system is the main internal part of the crayfish that still shows segmentation: Each segment

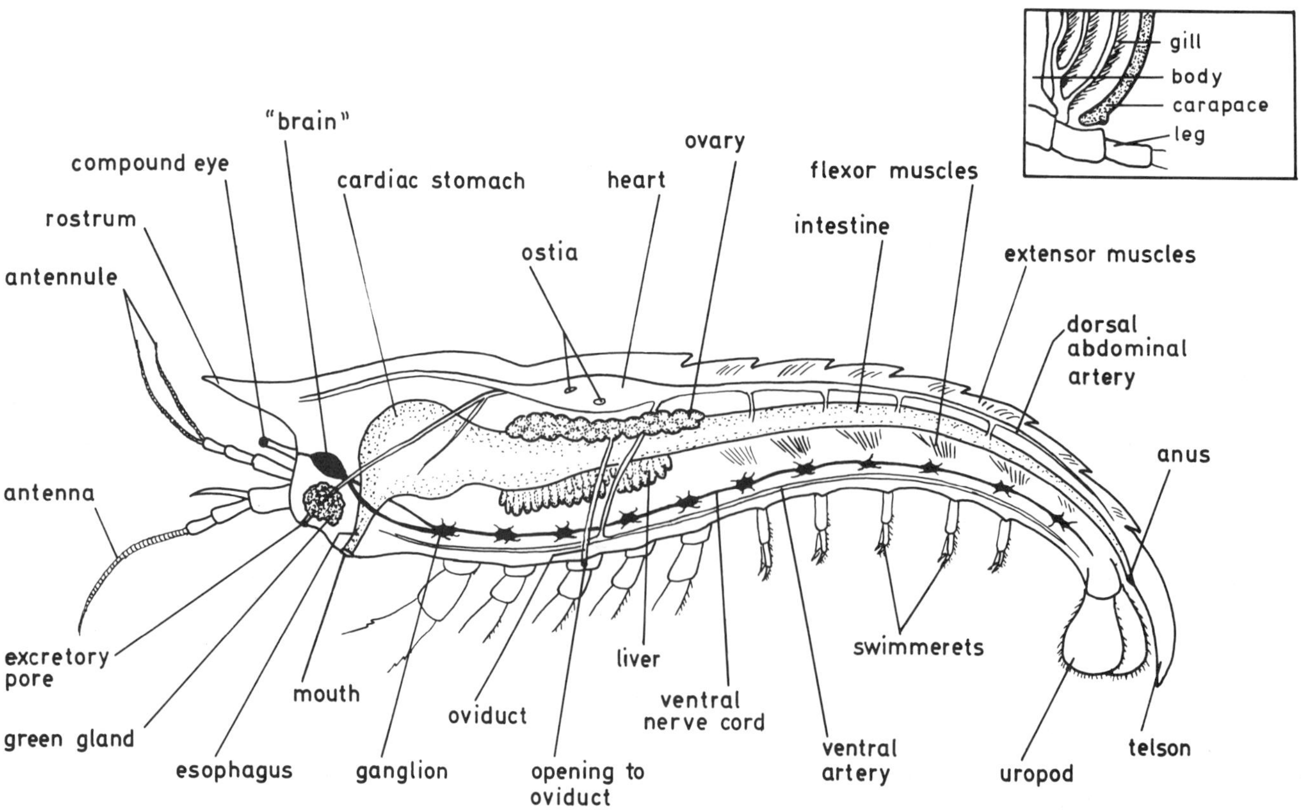

Figure 20–4. Internal anatomy of a female crayfish. In the male, the testes lead into paired sperm ducts, which open at the base of the fifth pair of legs.

of the animal has a ganglion that controls most of the functions in that segment.

- Find a **ganglion,** or swelling, along the ventral nerve cord. It controls the organs in its own segment but is also subject to influence by messages from adjacent ganglia and the brain.

The gills of the crayfish are located outside the body cavity in a space between the body wall and the carapace.

- Remove a segment of the carapace on one side of the crayfish to expose the **gills.**
- Detach one of the walking legs, and note the small gill that is attached to the leg.

The leg and gill are a modified form of the branched appendage shown in Figure 20–1. The leg represents the endopodite, the gill represents the exopodite, and the base of the leg is the protopodite. Branched appendages are common in crustaceans.

Male and female crustaceans copulate, fertilization is internal, and the fertilized eggs are attached to the swimmerets of the female. She carries them around, aerating them with her swimmerets, until they hatch into young that resemble the parents. In some marine crustaceans, the eggs hatch into a nauplius larva that bears three pairs of appendages.

CLASS INSECTA

Insects are the most successful invertebrate group and the only group besides vertebrates to inhabit all terrestrial environments. All insects are terrestrial or have evolved from terrestrial ancestors. They have come to occupy habitats in which they don't directly compete with vertebrates and have thus remained small in size. They survive on land by tolerating fluctuations in the salt and water content of the body, and they have developed a unique **tracheal system** for respiration. The insect body consists of three parts: head, thorax, and abdomen. The head has antennae, **compound eyes** (unique to crustaceans and insects), and modified mouth parts. The thorax always bears three pairs of walking legs and one or two pairs of wings, if wings are present. As in the other arthropods, the body cavity is filled with hemolymph, which is pumped through the **open circulatory system** by the heart. Excretion is performed by **Malpighian tubules,** which empty into the gut. Most insects copulate and lay eggs adapted to land. The young go through a series of molts before becoming sexually mature adults. Some insects have larval stages that are very different from the adult in structure, diet, and habitat.

The Grasshopper: External Anatomy

Grasshoppers are common herbivorous insects found in dry grasslands and open spaces. Use Figure 20–5 to identify the parts of its body.

- Examine a preserved grasshopper, and note the large jumping legs and the huge compound eyes. The compound eyes of insects consist of many simple visual units called ommatidia adjacent to each

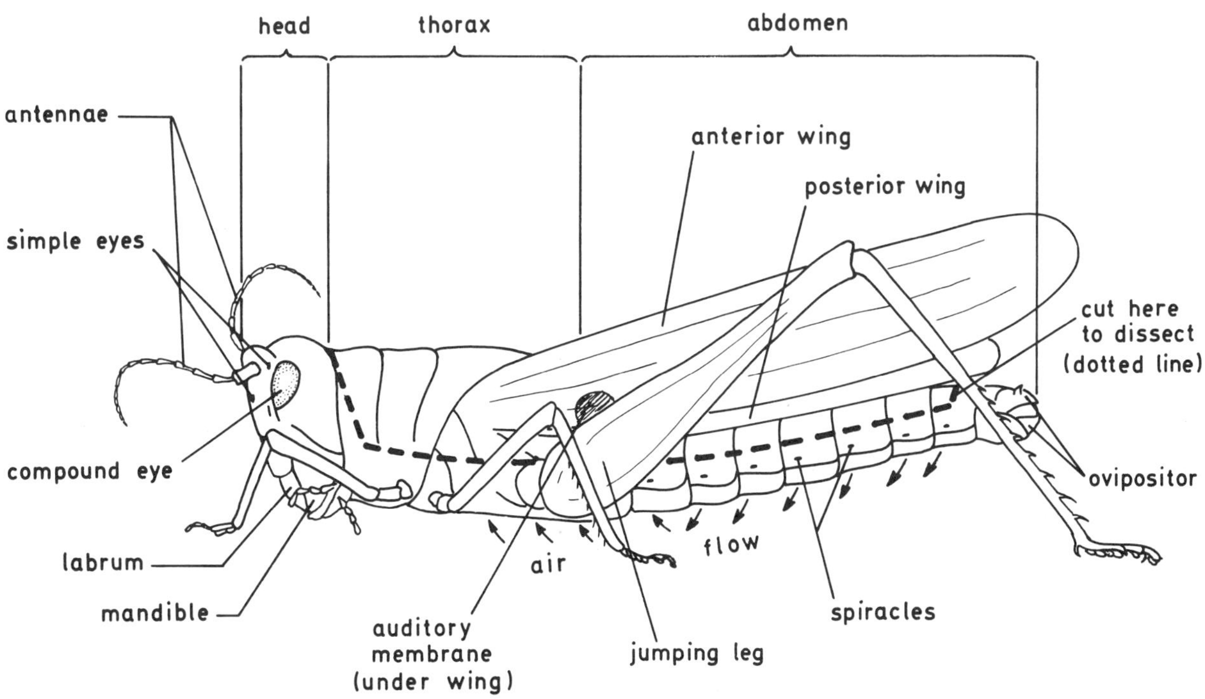

Figure 20–5. External structure of a female grasshopper. The male lacks the ovipositor and has a blunt abdomen. Open the exoskeleton by cutting along the dotted lines.

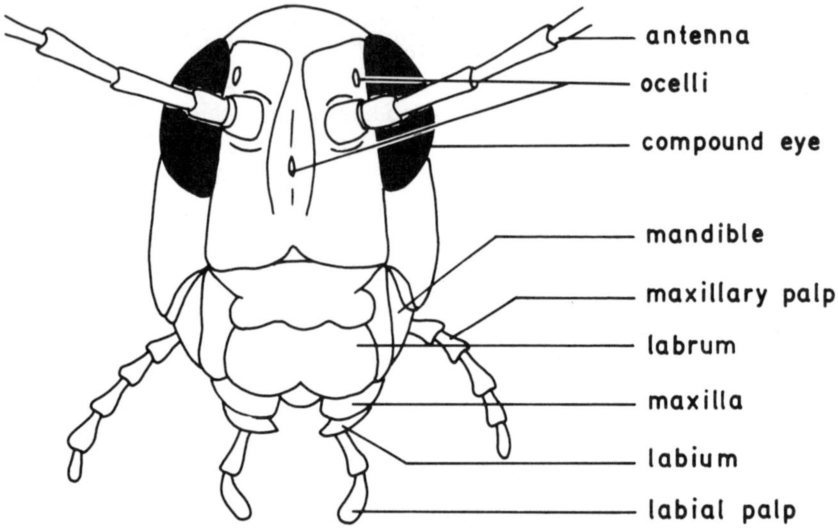

Figure 20–6. Head-on view of a grasshopper. The mouth is located below the labrum, and the tongue-like hypopharynx (not visible) is between the mandibles and labia.

other. These individual units detect degrees of light and dark and combine to form a mosaic. The combination of the pattern of light and dark on the whole array of visual units can be interpreted by the grasshopper's brain as an image. The image on a TV screen is formed by light and dark spots in a similar way. The resolution of the image is not very good compared with that of the vertebrate eye, but the insect eye can follow a rapidly changing image much better than we can. Insects have color vision and can see ultraviolet wavelengths invisible to us.

- Find the three **simple eyes** (ocelli) on the top of the head. They detect quickly moving objects.

Are the **antennae** jointed? ____________

- Examine the mouth parts shown in Figures 20–6 and 20–7.
- Find the **labrum** (upper lip); **labia** (lower lips, usually fused); **mandibles** (true jaws), which surround the mouth; and **maxillae** (second jaws).
- Within the mouth parts is the **hypopharynx** ("tongue"). In other insects the hypopharynx may be modified for piercing or sucking.
- The two pairs of appendages below the mouth (maxillary and labial palps) have sensory receptors for "tasting."

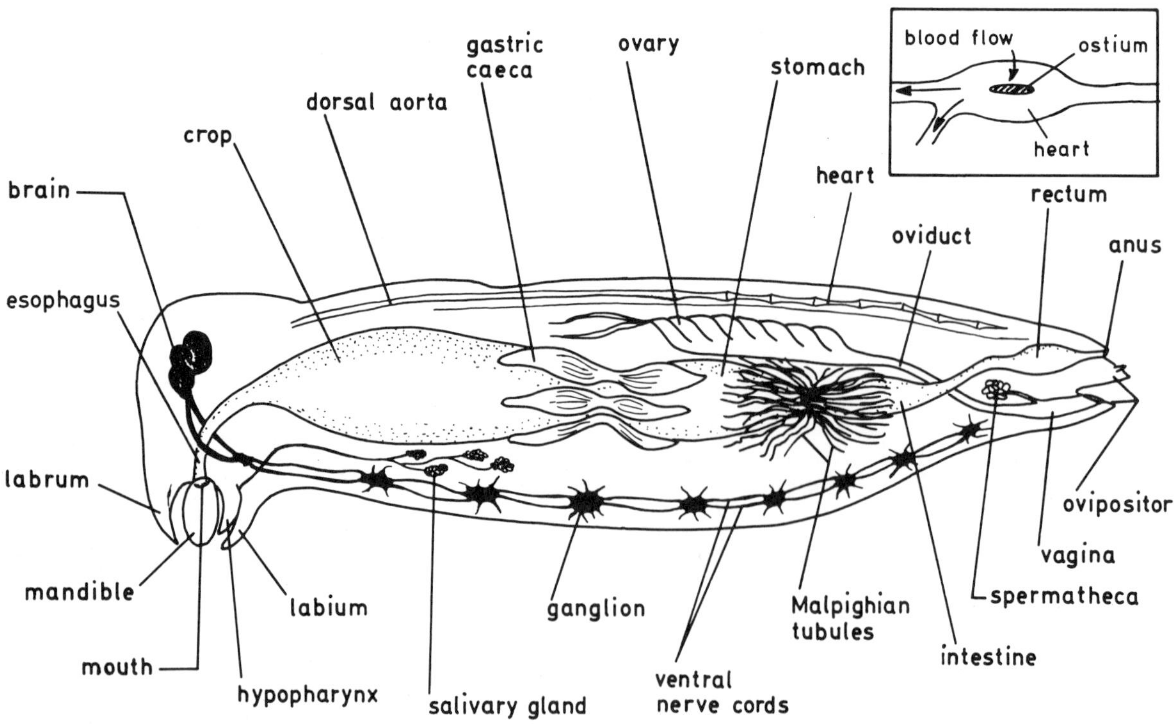

Figure 20–7. The internal organs of a female grasshopper. In the male, the testes lead into seminal vesicles and then to the penis. There is no ovipositor.

- The segmented **thorax** bears the three pairs of **legs** and two pairs of **wings.**
- Stretch out the wings, and note the difference between the protective anterior wings and the posterior flight wings. They are given support by stiff **veins.**
- The abdomen is blunt in males and in females has a pointed **ovipositor** at the posterior end for laying eggs.

Is your grasshopper male or female?

____

- Locate the **spiracles** along the sides of the thorax and abdomen. They connect to the **tracheal system** through which insects carry on gas exchange. The three anterior pairs of spiracles take in air, and the posterior ones let it pass out. The spiracles can be closed, if necessary, to prevent too much water loss under dry conditions.
- Just above the jumping legs find the **auditory membranes.** These allow the grasshopper to detect sound and communicate with other grasshoppers, especially those of the opposite sex.

The Grasshopper: Internal Anatomy

The internal organs of the grasshopper are shown in Figure 20–7.

- Remove the wings of the grasshopper.
- Cut along the dotted line as shown in Figure 20–5, and remove the dorsal part of the exoskeleton.
- Study the internal part of the exoskeleton that you have cut off, and find the muscles that attach to the wings; these are the **insect flight muscles,** which are rich in mitochondria and perform more efficiently than any vertebrate muscle.
- Note the muscles along the body wall that move the legs. In an animal with an exoskeleton, muscles at the joints work in pairs to extend (straighten) or flex (bend) the joint.
- Note the body cavity, or **hemocoel,** which in life is filled with hemolymph. There is no respiratory pigment because respiration is carried out by the tracheal system.

What, then, is the function of insect hemolymph?

____

____

- Look for the **dorsal aorta** and the **heart.** Hemolymph entering the heart through small openings (ostia) is pumped forward and enters the hemocoel near the brain.
- Locate the **esophagus** through which food passes from the **mouth** to the large **crop** (foregut).
- Find the outpocketings of the gut, called gastric caeca, at the posterior end of the crop.
- The large **stomach** (midgut) leads into the **intestine** (hindgut).
- The **Malpighian tubules** excrete nitrogenous wastes, mainly **uric acid,** into the gut at the point where these sections of the gut join.
- Locate the **rectum.** It absorbs water efficiently, resulting in almost solid feces that are emptied through the **anus.**

The sex organs are located in the posterior part of the abdomen.

- In the male, locate the **testes,** which produce sperm that are stored in the **seminal vesicles** and eventually deposited in the female by the **penis.**
- In the female, the sperm are deposited in the **vagina** and stored in the **spermatheca.** Eggs produced in the large **ovaries** are fertilized by the sperm stored within the body of the female and are deposited with the ovipositor.
- Remove the digestive organs to see the double **ventral nerve cord.** At its anterior end, nerves pass around the esophagus to the **ganglion** of the brain.

Does the ventral nerve cord have ganglia in each segment, as in the crayfish? ______________

Insect Metamorphosis

Metamorphosis differs in the various groups of insects: In some insects the young that hatches from the egg is similar to the adult, but in others it is completely different. There are four types of development.

1. **Ametabolous** (e.g. silverfish): The young are tiny versions of the wingless adult.
2. **Paurometabolous** (e.g. grasshopper): The young **nymphs** are similar to the adults, but lack wings and are sexually immature. (See Text Figure 10–17.)
3. **Hemimetabolous** (e.g. dragonfly): The young **naiads** are aquatic larvae with gills and are unlike the adults.
4. **Holometabolous** (e.g. *Drosophila*): The worm-like **larva** passes through the resting stage of the **pupa** before emerging as a mature adult.

- Turn back to the LIFE HISTORY section of Topic 11, and review the life history of *Drosophila,* a fly that belongs to the holometabolous order Diptera.
- If you have not already done so, examine a culture bottle of *Drosophila* and identify the **egg, larva, pupa,** and **adult.**
- Note the white, oval eggs. Each one has two tiny extensions and a waterproof covering to protect it in a terrestrial environment.

- The largest larvae (maggots) are the ones that are about to pupate, and the smallest ones are newly hatched from the egg. Each stage of larval development is called an **instar,** and the stages are separated by **molting** (*Drosophila* has three instars).
- Note the dark-colored pupae, which contain flies about to emerge as adults.
- Try to find newly hatched adults whose wings are not yet fully extended. They need to dry in the air and stiffen before they can be used for flight.

When you are finished with this laboratory, dispose of your specimens as directed by your instructor, clean your dissecting tools before putting them away, and store your microscopes properly.

- Answer the review questions at the end of this topic.

REFERENCES AND FURTHER READING

See Topic 18 for general references on invertebrates.

Borror, D.J., D.M. DeLong, and C.A. Triplehorn. *An Introduction to the Study of Insects,* 5th ed. Philadelphia: Saunders College Publishing, 1981.

Chapman, R.F. *Insects: Structure and Functions,* 2nd ed. New York: Elsevier, 1976.

Evans, H.E. *Life on a Little-Known Planet.* Chicago: University of Chicago Press, 1984.

Gertsch, W.J. *American Spiders,* 2nd ed. New York: van Nostrand Reinhold, 1979.

Schmitt, W.L. *Crustaceans.* Ann Arbor: University of Michigan Press, 1965.

Wilson, E.O. *Insect Societies.* New York: Academic Press, 1971.

Journal Articles

Batra, Suzanne W.T. "Solitary bees." *Scientific American,* 120, February 1984.

Brownell, Philip H. "Prey detection by the sand scorpion." *Scientific American,* 86, December 1984.

Burgess, J.W. "Social spiders." *Scientific American,* 100, March 1976.

Cameron, James N. "Molting in the blue crab." *Scientific American,* 76, May 1985.

Dyer, Fred C., and James L. Gould. "Honeybee navigation." *American Scientist* 71(6): 587, 1983.

Forster, Lyn. "Vision and prey-catching strategies in jumping spiders." *American Scientist* 70(2): 165, March 1982.

Heinrich, Bernd. "The regulation of temperature in the honeybee swarm." *Scientific American,* 146, June 1981.

Hinton, H.E. "Insect eggshells." *Scientific American,* 84, August 1970 (#1187).

Hölldobler, Bert, and E.O. Wilson. "The evolution of communal nest-weaving in ants." *American Scientist,* 71(5): 490, September 1983.

Horridge, G. Adrian. "The compound eye of insects." *Scientific American,* 108, July 1977 (#1364).

Jackson, Robert R. "A web-building jumping spider." *Scientific American,* 106, September 1985.

Jones, Jack Colvard. "The feeding behavior of mosquitoes." *Scientific American,* 138, June 1978 (#1392).

LaBarbara, Michael, and Steven Vogel. "The design of fluid transport systems in organisms." *American Scientist* 70(1): 54, January 1982.

Merritt, R.W., and J.B. Wallace. "Filter-feeding insects." *Scientific American,* 132, April 1981.

Rothschild, Miriam, Y. Schlein, K. Parker, C. Neville, and S. Sternberg. "The flying leap of the flea." *Scientific American,* 92, November 1973 (#1284).

Ryker, Lee C. "Acoustic and chemical signals in the life cycle of a beetle." *Scientific American,* 112, June 1984.

Seeley, Thomas D. "The ecology of temperate and tropical honeybee societies." *American Scientist* 71(3): 264, 1983.

Wicksten, Mary K. "Decorator crabs." *Scientific American,* 146, February 1980 (#1462).

Wilson, E.O. "Slavery in ants." *Scientific American,* 32, June 1975 (#1323).

Student Name ______________________________ **Date** ____________________

REVIEW

Peripatus is a small, velvety animal that is particularly interesting because it combines several important features of arthropods and some features of annelids (see Figure 20–8).

State whether each of the following characteristics of *Peripatus* is also found among the annelids, arthropods, both or neither. Then decide in which group you would place *Peripatus* and defend your answer.

wormlike body ______________________________

thin, soft cuticle ______________________________

unjointed legs ______________________________

claws on legs ______________________________

serially duplicated organs ______________________________

jaws derived from appendages ______________________________

open circulatory system ______________________________

heart ______________________________

coiled excretory tubes with external pores ______________________________

carnivorous nutrition ______________________________

double ventral nerve cord ______________________________

respiratory pores all over body ______________________________

I suggest placing *Peripatus* in phylum ____________________. My rationale for this choice is ______

____

____

____

____

(Most biologists place *Peripatus* in its own phylum, Onychophora = claw bearer.)

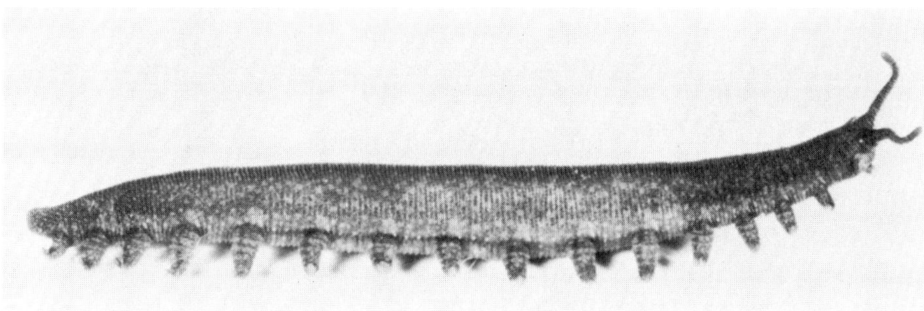

Figure 20–8. *Peripatus*. This rare, nocturnal animal lives in tropical rain forests and may be on the way to extinction. (Photo by Carolina Biological Supply Company)

TOPIC 21

Vertebrate Anatomy and Evolution

OBJECTIVES

When you have completed this topic, you should be able to:

1. List the three important characteristics of chordates.
2. Explain why a deuterostome invertebrate larva was the most likely ancestor of the vertebrates.
3. List characteristics that would permit you to distinguish members of the following vertebrate classes:
   Chondrichthyes, Osteichthyes, Amphibia, Aves and Mammalia.
4. Name the class of any vertebrate presented to you.
5. Define and give examples of analogous structures, homologous structures, and vestigial structures.
6. Describe the anatomy of one representative vertebrate in detail, and be able to describe the major features that are similar to and different from those of the other classes of vertebrates.
7. Identify any of the structures in the review list at the end of this topic in any vertebrate in which it occurs.
8. List and explain the problems encountered by water-dwelling vertebrates in adapting to life on land.
9. Explain the adaptations of a representative member of each of the terrestrial classes of vertebrates that allow it to live on land successfully.

TEXT REFERENCES Chapter 22 (L–O).

KEY TOPICS

deuterostome characteristics
chordate characteristics
vertebrate characteristics
analogous structures
homologous structures
vestigial structures
cartilaginous fish (Chondrichthyes)
bony fish (Osteichthyes)
amphibians (Amphibia)
reptiles (Reptilia)
birds (Aves)
mammals (Mammalia)
transition to land

INTRODUCTION

The remaining two major phyla, Echinodermata and Chordata, include most of the **deuterostomes.** Echinoderm structure and embryology are described in Topic 9. Although the adult deuterostome animals are very diverse, they share some conservative characteristics such as embryological development. In contrast to the protostomes, they are similar in having some or all of these important features of development:

1. The blastopore becomes the anus and the mouth develops in another way.

2. Cleavage of the egg is radial.
3. Cleavage is indeterminate.
4. Mesoderm is formed by pouches from the gut.
5. The coelom is formed from mesodermal pouches.

Because of these similarities, deuterstomes are thought to represent phyla that had a common ancestor and so are evolutionarily related. The details are not clear, but echinoderms and chordates seem to be more closely related than the other phyla. A likely candidate for their common ancestor is an ancient relative of the acorn worm (Phylum Hemichordata) because this animal has pharyngeal gill slits like a chordate and its larva is similar to that of an echinoderm. A simplified diagram of deuterostome relationships is shown in Figure 21–1.

Here we will study some vertebrates, members of Subphylum Vertebrata in Phylum Chordata. Along with the other Chordates (urochordates and cephalochordates), vertebrates have at some point in their life history:

1. a dorsal notochord
2. a dorsal, hollow nerve cord
3. pharyngeal gill slits.

A typical chordate is a coelomate with three germ layers. The body is a tube within a tube, segmented, and bilaterally symmetrical. There is an endoskeleton, segmented blocks of muscle and a post anal tail (see Text Figure 22–30). Blood flows in a closed circulatory system forward toward a ventral heart, then backwards in a dorsal vessel. When these features are compared with those of an annelid or arthropod, the enormous divergence of the protostome and deuterostome branches of the animal kingdom become apparent.

The first chordates were probably filter feeders and the adults of the simpler chordates of today are certainly very different from the typical chordate described above. Their motile larvae, however, do fit that picture. For example, the ammocoete larva of the lamprey (a jawless vertebrate), is a small, motile, filter feeder with a notochord that also has muscular gills for more efficient filter-feeding. It is thought that chordates probably arose from motile larvae like these that developed sexual maturity, a process called **neoteny.** This is known to have occurred in some tunicates (see Color Plate 3). The stiff notochord meant that these early chordates could use their blocks of muscles for effective swim-

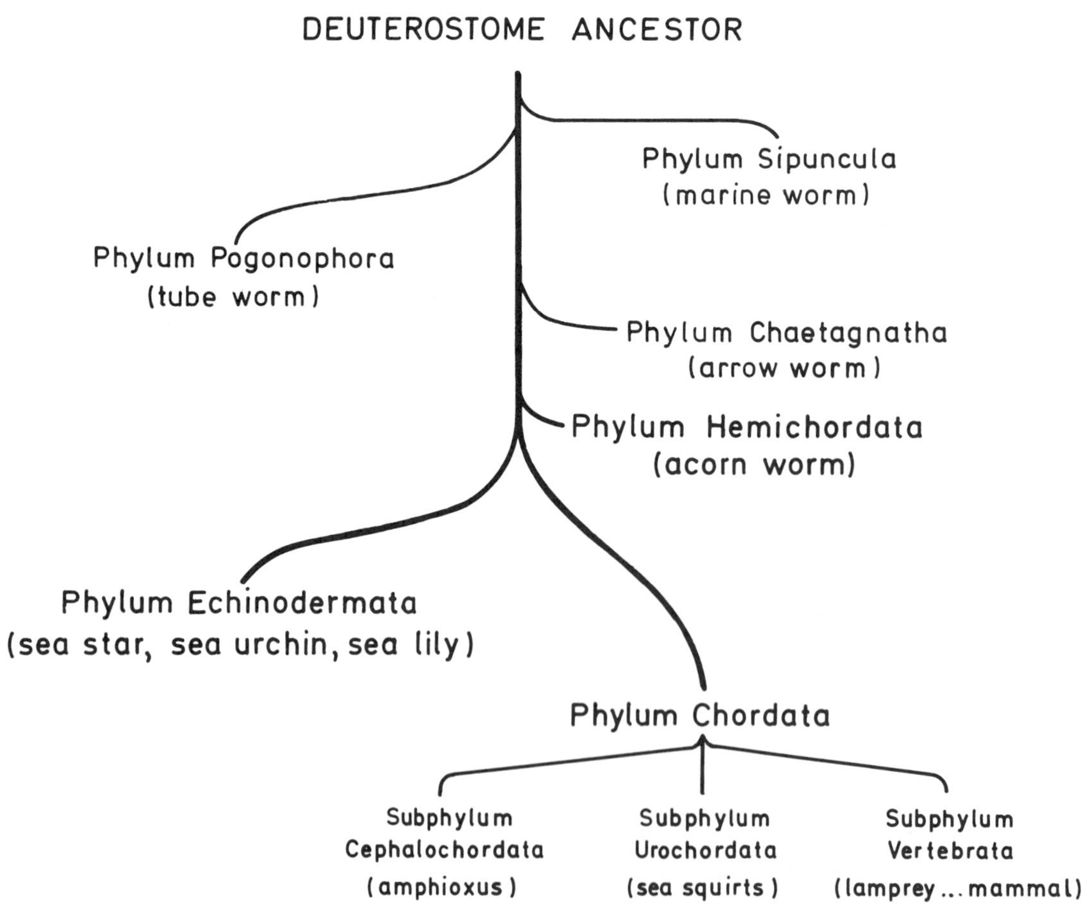

Figure 21–1. The deuterostome relatives of vertebrates. These phyla share certain features of early embryology and so are considered to be related. Echinoderms and chordates are most closely related.

ming. Cephalization and modification of the filter-feeding apparatus allowed them to find and ingest prey and so switch to a carnivorous way of life. Such a small, actively swimming carnivore would have had the great advantage of being able to seek out food at the water's surface, and so could invade the rich planktonic zone not available to the bottom-dwelling forms. Development of this lifestyle was the key to the evolutionary success of vertebrates.

These are some of the important ways in which the typical chordate body plan has been modified in the vertebrates.

1. A vertebral column replaces the embryonic notochord. It surrounds and protects the nerve cord and serves as the backbone in locomotion.
2. The dorsal hollow nerve cord develops a brain with three primary vesicles dividing it into forebrain, midbrain, and hindbrain, with many cranial nerves.
3. The pharyngeal gill slits that are present in the embryo may be modified for other purposes, such as the Eustachian tubes.
4. Along with the marked cephalization of the nervous system and sense organs, a skull or cranium develops to surround and protect these structures in the head region.
5. There are red blood cells (erythrocytes), which contain hemoglobin.
6. The ventral heart has 2, 3, or 4 chambers.
7. There is a pair of nephric kidneys that develop from mesoderm.
8. There is a solid liver.
9. There is a pancreas or pancreatic tissue.
10. There are specific endocrine organs.
11. The eye has a lens and a retina.
12. There are olfactory organs, which are usually paired.
13. There are one to three semicircular canals forming part of the inner ear.
14. Muscles attach to the endoskeleton for efficient locomotion, often by means of two pairs of appendages.

These are the classes of Vertebrata that we will be studying in this exercise:

Class Chondrichthyes (cartilaginous fishes: for example, sharks, skates, and rays)

Class Osteichthyes (bony fishes: for example, perch, mackerel)

Class Amphibia (amphibians: for example, frogs, toads, newts, and salamanders)

Class Aves (birds: for example, robin, osprey, penguin, duck, ostrich)

Class Mammalia (mammals: for example, cows, horses, whales, rabbits, humans)

In contrast to the lamprey and its relatives in Superclass Agnatha (= without jaws), all of the other vertebrates have true jaws and are placed in Superclass Gnathostoma (= jaw mouth). The advent of true biting jaws about 400 million years ago at the end of the Silurian period was a revolutionary development in vertebrate evolution and made vertebrates competitive with the large predaceous invertebrates of that time.

Before beginning your examination of vertebrate diversity, take a few moments to focus on the vertebra, a small but important structure that has greatly contributed to the evolutionary success of this group. Bony vertebrae replace the notochord in some, but not all vertebrates.

■ Examine the **vertebra** on demonstration, and identify the parts shown in Figure 21–2.

The **neural spine** (dorsal spinous process) is one of the set of bumps you feel if you run your hand down a cat's back. The spinal cord runs through the large opening, known appropriately as the neural canal. It is protected by the bony neural arch. Lateral displacement of adjacent vertebrae would put pressure on the spinal cord, resulting in severe pain and possibly permanent damage.

The **centrum** marks the former position of the embryonic notochord, which has been surrounded and mostly replaced by bone (see Text Figure 31–1). A spongy **intervertebral disc,** a remnant of the notochord, lies between the centra of adjacent vertebrae and allows some flexibility of the backbone. A "slipped" or ruptured disc will change the alignment of the vertebrae and put pressure on the spinal cord.

In this laboratory, you will learn something about the vertebrates by observing the demonstra-

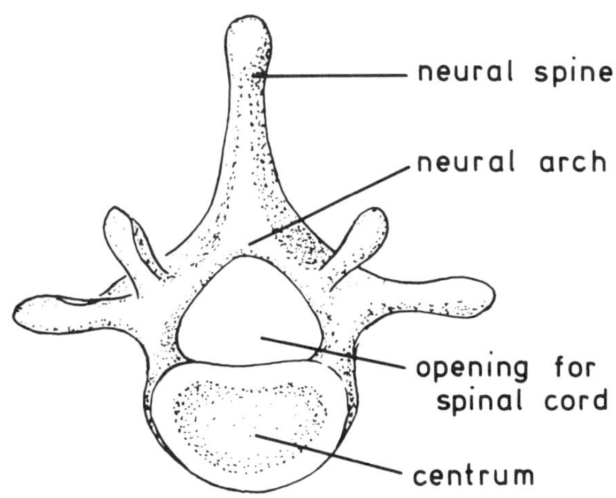

Figure 21–2. A vertebra. In the vertebral column the spinal cord would pass through the large opening and there would be soft discs between adjacent vertebrae to cushion the centra.

tions, dissecting a representative of one of the vertebrate classes, and studying the dissections made by other students. At the end of this exercise there is a list of important terms and structures that you should be able to identify in any vertebrate in which they occur.

When you study the anatomy of the various vertebrates, you will often notice structures that appear similar, such as the leg of a frog and that of a cat. If the structures develop from similar rudiments in the embryo and have a similar internal structure and development, they are considered **homologous** and point to an evolutionary relationship between the animals. Homologous structures may be modified for purposes as different as flight (the wing of a bat) and swimming (the fin of a whale).

Structures that appear externally similar and are used for similar purposes but have very different internal structures and development are **analogous.** The wings of butterflies and birds are analogous limbs used for flight, but their structure and development are distinct, and their presence does not imply a close evolutionary relationship.

Sometimes a structure seems to have no function, but is homologous to a similar structure in a related animal that does have a function. Such a structure that has lost its ancestral function is said to be **vestigial.** A good example is the human appendix, which is homologous with the caecum of other mammals.

Your instructor will provide dissection guides to help you understand the structure of your specimen. If you are going to work with a preserved specimen, rinse off the preservative at the sink before examining the external features. Dissect from the ventral side, first removing the skin by separating it and peeling it back from the abdominal wall. Once the abdomen is open, use your fingers and blunt probe for most of the dissection. You can study much of the internal anatomy by moving organs aside so that you can look underneath. If it is necessary to remove a structure, do it carefully without damaging adjacent organs.

When working with preserved specimens, *YOU SHOULD NOT WEAR CONTACT LENSES.* Use eyeglasses or safety glasses, if you have them, and protect your hands with liberal amounts of hand cream before beginning work.

Read the general procedures section in Topic 22 before beginning work on your specimen.

CLASS CHONDRICHTHYES

These fish have cartilaginous skeletons, although their ancestors had bony skeletons, as do the present-day bony fish in Class Osteichthyes. Many cartilaginous fish are predatory carnivores, but some are bottom feeders and the largest shark is the filter-feeding whale shark. Sharks appear in the fossil record during the Devonian, over 350 million years ago, when they evolved from primitive bony fish.

A typical shark is well adapted to a highly motile, carnivorous way of life. The most anterior of the gill arches have been moved forward to form jaws used for predation. The shark's powerful musculature and vertebral column make it an excellent swimmer and allow it to pursue and catch prey. Although the shark has paired fins, they are used for maneuverability, and the thrust for swimming comes from sinusoidal movements of the shark's body and tail.

- Study the preserved dogfish shark on demonstration.

Where are the **jaws** located in relation to the head?

____

Why do you suppose the belly is light and the back is dark?

____

____

- Feel the skin of the shark. The roughness is caused by **placoid scales,** or dermal denticles, which arise from the dermis of the skin.

Modified denticles form the teeth, which are homologous with the teeth of other vertebrates. They can easily be replaced as they become worn or lost. The internal structure of a single denticle is shown in Figure 21–3.

- Remove a section of skin on the side of the shark's body to see the segmented musculature underneath. The skin acts as an external tendon for the muscles when the shark swims fast.

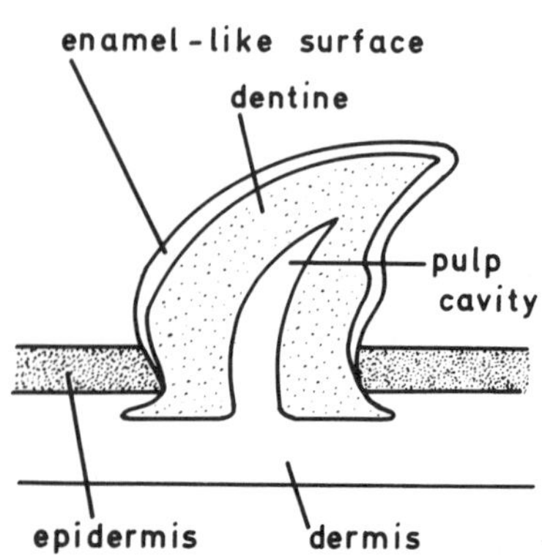

Figure 21–3. The jaw and tooth of a shark. As teeth become worn or are lost, they are replaced by new ones from the rows on the inner sides of the jaws.

What other sign of segmentation do you see in the shark's external anatomy?

____

- Behind the eye, locate the **spiracle** through which some water may enter the pharynx.

How does water leave the pharynx?

____

____

How many pairs of fins does the shark have?

____

The significance of paired fins is that they were modified and used as rudimentary limbs in the ancestors of terrestrial vertebrates (and in some contemporary fish), which could thereby crawl around on land as well as swim.

- The **pectoral fins,** just behind the gill slits, are important in balance and maneuvering.
- If your specimen is a male, the posterior **pelvic fins** will be modified into claspers that transfer sperm during copulation.

Is your shark male or female? ___________

- The **eye** has a lens and retina with rods and cones so the shark can apparently see color; its vision is not especially keen. Instead of an eyelid, it has a white membrane that covers the eye when the shark strikes its prey.
- Along the side of the shark is a light stripe marking the position of the **lateral line organ,** which is sensitive to pressure changes in the water. Part of the system in the head region (ampullae of Lorenzini) is modified to sense electrical fields. This sense allows the shark to detect hidden prey. Because strong electrical fields repel sharks, this sense also allows humans to discourage sharks from visiting beaches by setting up artificial electric fields in the water.
- Locate the **nares** (nostrils) anterior to the mouth. Within them are the **olfactory sacs,** which contain chemoreceptors that allow sharks to sense chemicals in the water. These are blind sacs that do not connect to the oral cavity.

Within the skull the shark has an inner ear with **semicircular canals.** There are three canals, in the three perpendicular planes. The inner ear detects movement of the head and pressure changes and allows the shark to keep track of its position. All vertebrates have a similar organ of equilibrium.

- Open the shark's abdomen by cutting around the cloacal area and then toward the head. Cut through the pectoral girdle (skeleton between pectoral fins), and pin back the flaps of the body wall

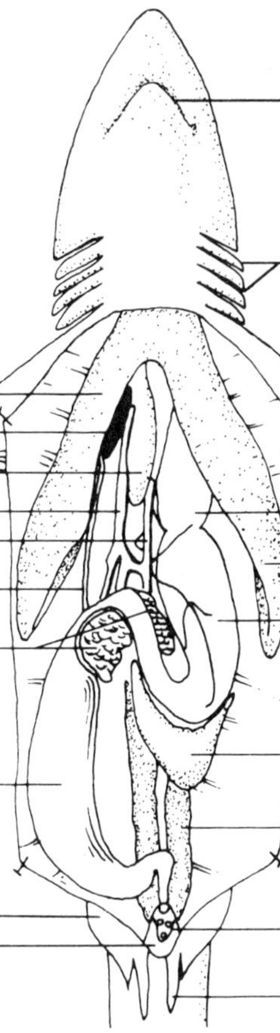

Figure 21–4. Internal anatomy of a male shark.

so that the dissection looks like the shark shown in Figure 21–4.

- Locate the following internal organs, most of which appear in Figure 21–4:

Esophagus. Slit open to see the papillae inside.

Stomach (J-shaped). Open the stomach to determine the shark's last meal. Notice the rugae (folds) inside).

Small Intestine. Short, as in many carnivores. Open it to see the **spiral valve** inside.

What is the function of the shark's spiral valve?

____

____

Rectum. Empties into the cloaca.

Rectal Gland. Excretes sodium chloride for osmoregulation.

Pancreas. Two lobes.

Liver. This organ is composed of three large oil-filled lobes and is critical in maintaining the buoyancy of open ocean sharks, which must constantly swim or sink, so to say, due to lack of sufficient ventilation of the gills.

Gall Bladder. A greenish organ located underneath the liver.

Spleen. Triangular shape, caudal (toward the tail) to the stomach.

Kidneys. Long, bandlike organs that extend along almost the entire length of the body cavity under the peritoneal membrane. They drain through ducts (archinephric ducts) into the cloaca. Most sharks are marine and must conserve water, but the kidneys are not efficient and remove a lot of water as they remove urea from the blood. Sharks osmoregulate by keeping the level of urea in the blood and tissue high, so that water actually tends to enter the body.

Testes or **ovaries.** Located near the anterior ends of the kidneys. In the female, eggs must migrate, helped along by ciliary action, forward to enter the openings of the oviducts. They pass down into the uterus, which will be enlarged if the shark is pregnant. In the male shark, sperm from the testes pass through common urogenital ducts to the cloaca. The **claspers** are used to transfer sperm into the female's reproductive tract during copulation; therefore, fertilization is internal. In dogfish sharks, the young develop in the female's body and then hatch as small sharks. This system is called **ovoviviparous** development because the young are nourished by a yolk sac, rather than by the mother. Development lasts up to two years, and many young are born at the same time. In some sharks, the developing young are in intimate contact with and obtain nourishment from the mother's circulatory system, so that development is **viviparous.** (See Text Figure 1–7.)

Cloaca. Common opening of the digestive and urogenital systems. The **urinary papilla** within the cloaca has the opening of the urogenital system.

Heart. Four parts, two chambers (see Text Figure 25–5). Blood from the body drains into the **sinus venosus** and then enters the heart's **atrium.** The atrium leads into the muscular **ventricle,** which then forces the blood out through the **conus arteriosus.** The conus arteriosus is also muscular and has internal valves. There is a single circulation because there is only one pathway through the single atrium and ventricle (see Text Figure 25–6). Deoxygenated blood is pumped by the heart to the gills in the **ventral aorta** and **afferent branchial arteries** (af- = to, toward). It is reoxygenated as it passes through the gill capillaries. It then leaves in the **efferent branchial arteries** (ef- = from, away), enters the **dorsal aorta,** and is distributed to the body under relatively low pressure.

Respiration in many sharks partly depends on the shark's continuously swimming through the water so that fresh water enters the pharynx from the mouth and spiracles, then passes over the gill capillaries and out through the **gill slits.** The muscular action of the floor of the mouth and in the branchial region helps suck in and expel the water, and in some sharks provides sufficient gill ventilation without swimming.

- The shark has six gill slits on either side of the pharyngeal region, counting the ones modified to form the spiracles. Gill slits are present at some stage in all vertebrate embryos, a holdover from filter-feeding ancestors.
- Notice the **gill rays,** which support and protect the delicate gill filaments within the gill slits, and the **gill rakers,** which prevent food from passing out through the gill slits.

The brain of the shark is protected by a cartilaginous braincase. Your instructor may ask you to open the braincase and study the anatomy of the shark's brain. The shark's existence is mainly concerned with sensations of smell, sight, and equilibrium with little provision for decision making. This animal is truly an "eating machine" and as such has been highly successful and has not changed much since it first evolved.

CLASS OSTEICHTHYES

Like sharks, many bony fish are fast-swimming carnivores, but they have also evolved into a great variety of other types. The first true bony fish evolved from jawless vertebrates at the end of the Silurian period, over 400 million years ago, and fish were abundant during the Devonian period which followed. They have continued to evolve extensively to the present day, and modern ray-finned fish, such as the perch, first appeared late in the Cretaceous period, about 65 million years ago.

- Study the preserved perch on demonstration. Handle the specimen carefully because the spines are sharp.

Is the body as flexible as that of the shark?

____

- Remove a few scales, make a wet mount, and examine them under low power in the microscope. These are **ctenoid scales** and provide the fish with a tough but flexible covering. All the scales are covered by a layer of very thin skin.
- Locate the growth rings on the exposed surface of the scale. They can be used to determine the approximate age of the fish because they are further apart in spring and summer and closely spaced in fall and winter.
- Note the spiny dorsal **fins.**

How many sets of *PAIRED* fins does the fish have?

____

These fins (pectoral above and pelvic below) are homologous with the fins of the shark and with the limbs of land animals. During swimming, most of the propelling force comes from the tail and caudal fin, with the other fins used for balance and direction control.

- Find the **anus** in front of the anal fin. It serves only as the opening of the digestive tract, unlike the cloaca of the shark, which is also the urogenital opening.
- The **operculum** covers the gill slits, and water enters through the mouth.

Where is the **mouth** located in relation to the head?

____

How is this an advantage?

____

____

- Examine the inside of the mouth.

Do you find teeth or a tongue? ______________

In bony fish that have biting teeth, the teeth are conical, all alike (homodont), and set into the jaw bones. These are considered primitive characteristics. Some fish have teeth on the roof of the mouth and in the throat to help hold and swallow prey.

Is there an eyelid to cover the **eye**? ______________

- Find the pores of the **lateral line organ** along the side of the fish. This organ senses pressure change in the water, as in the shark.
- The **olfactory sacs** within the **nares** (nostrils) are located anterior to the eyes. They contain chemoreceptors used in olfaction and are blind sacs that do not connect with the oral cavity.

Within the bony skull is the inner ear with three perpendicular **semicircular canals.** The inner ear is an organ of equilibrium used to detect the fish's position.

Some of the internal organs of the perch are shown in Figure 21–5.

- Open the abdominal cavity of the fish by filleting away the muscles along one side, and locate as many organs as you can.

Liver. A reddish organ in the perch.

Esophagus. Leads to the stomach.

Stomach. Has three pouches called pyloric caeca; empties into the duodenum, the first part of the small intestine.

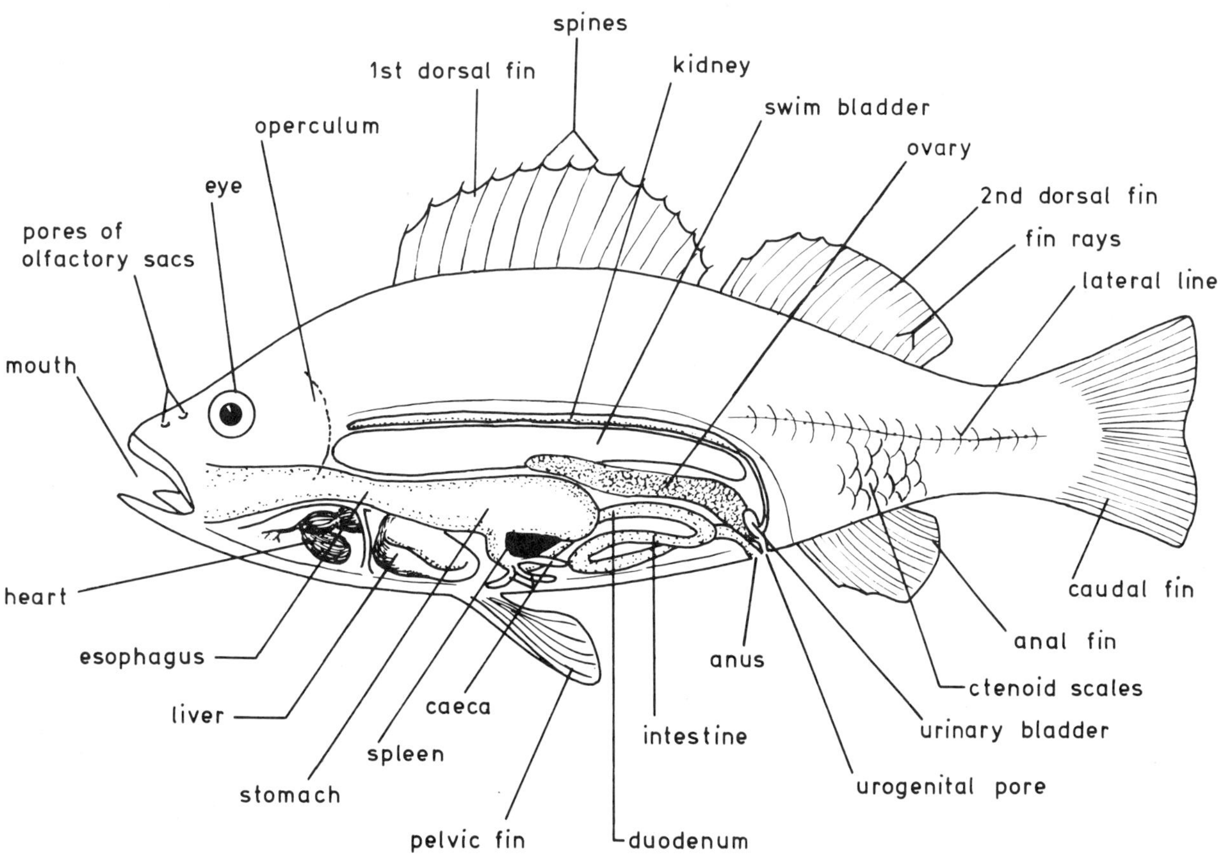

Figure 21–5. Internal anatomy of a female perch.

Intestine. There is no spiral valve, so the intestine is longer than in the shark. It exits from the body via the anus.

Spleen. Located next to the stomach. The spleen may not be obvious in your perch.

Swim Bladder. This whitish, translucent organ fills the dorsal part of the body cavity. It is derived from the pharynx during development and makes the fish buoyant. As a result, the fish can remain stationary without moving its fins. Gases are secreted into the swim bladder by the red organ or rete mirabile. In some fish the swim bladder is used as a rudimentary lung and it may have been modified for air-breathing in the ancestors of amphibians.

Kidneys. Blackish or brownish organs that are diffuse and are located dorsal to the swim bladder, beneath the peritoneal membrane. They help this freshwater fish retain salts and excrete a copious, dilute urine.

Urinary Bladder. Located at the very tail end of the body cavity (look hard, behind intestine). Copious, dilute urine exits through the **urogenital pore** just posterior to the anus.

Testes or **Ovaries.** The gonads fill much of the posterior part of the body cavity below the swim bladder. The testes are white and elongated and the ovary may be filled with eggs. Gametes (eggs and sperm) are shed to the exterior through the urogenital pore.

Most fish reproduce by external fertilization and a great many small eggs are shed. Some fish brood or care for the eggs and young. Many don't, however, and in this case there is often very high mortality among their offspring. This type of reproduction in which the eggs develop externally is **oviparous.**

Heart. This four-part organ has two chambers: one **atrium** and one **ventricle.** Blood enters the heart from the body through the **sinus venosus** and leaves through the muscular **bulbous arteriosus.** The **ventra aorta** and **afferent branchial arteries** take blood from the heart to the gills where it is reoxygenated in the gill capillaries. It leaves in the **efferent branchial arteries,** enters the **dorsal aorta,** and is sent to the rest of the body. Fish have basically the same pattern of circulation as sharks (see Text Figures 25–5 and 25–6).

- Remove the bony **operculum** that covers the **gill slits.**

What is the function of the operculum?

____

How many gill slits does the perch have?

____

- Notice the **gill rays,** which support and protect the delicate gill filaments (easier to see on the perch skeleton), and the **gill rakers,** which prevent food from passing out through the gill slits.

Water enters the mouth, passes from the pharynx through the gills, aerating the capillaries in the gill filaments so that gas exchange can occur, and finally exits through the operculum. If the fish is not moving, it can pump water through the gills by opening and closing its mouth and using muscles of the pharynx to force water out through the gills. The gills are also important in water balance, because they absorb salts that are constantly being lost in the urine of a freshwater fish.

The brain of the perch is small and deals mainly with smell, sight, and equilibrium. Olfactory nerves leading from the olfactory sacs to the olfactory bulbs are prominent.

- Examine the teleost skeleton on demonstration.
- Notice that the flexible fin rays interact with the fish's skeleton but are not directly attached to it. They make the fish highly maneuverable. The hardened, sharp spine rays of the perch are for protection; in some fish the spines also have poison glands.

How is the skull different from your own?

____

____

CLASS AMPHIBIA

Amphibia, Reptilia, Aves, and Mammalia are grouped together as the four classes of **tetrapods.** Those animals that walk on all four limbs are **quadrupeds** and those that walk on two limbs are **bipeds.** The amphibian ancestors of tetrapods evolved from fishlike animals that were aquatic, but present-day tetrapods have for the most part taken up a permanent existence on the land and become terrestrial. The first tetrapods to leave the water benefited from the advantages of life on land: little competition, absence of predators, abundant plant and insect food, and a high concentration of O_2. They used lungs rather than gills for respiration and were somewhat similar to modern lungfish. According to the fossil record, these amphibian ancestors evolved over 300 million years ago, and amphibians were abundant during the Carboniferous period.

If you have ever watched a fish out of water suffocate and quickly die, you realize that the evolution from an aquatic to a terrestrial existence must have been gradual. Many adaptations to the problems posed by living on land arose during the transition: support for the body; a mechanism for respiration with air; modification of the skin, gut, and kidneys to prevent dehydration; sense organs that function in air; and protection of the eggs and developing young. The amphibians did not solve all of these problems, and so are not fully adapted to terrestrial life.

Frogs and toads belong to the Anurans, a group of amphibians that have no tail and that have a skeleton adapted for jumping. The pectoral and pelvic girdles are fused to the backbone, which is very short.

- Examine the frog skeleton on demonstration.

Are the vertebrae also fused? ____________

- If live frogs are available, study the behavior of the intact animal.
- Observe its stance and its jumping and swimming motions.

The ancestral tetrapod limb had five similar digits, as in the relatively unspecialized human leg. This limb form was greatly modified for special purposes in many vertebrates, which thus have more highly evolved limbs than we do.

Are the frog **limbs** attached to the sides or to the ventral side of the trunk? ____________

How many **digits** ("fingers") are there on the forelimbs? ____________

On the hindlimbs? ____________

Are the digits longer on the forelimbs or on the hindlimbs? ____________

Do the digits have claws? ____________

What adaptation for swimming do you see?

____

If the frog is a male, there may be an enlarged pad on the side of the inner digit of its forelimb. It helps the male grip the female (amplexus) during mating (see Text Figure 28–2).

Is the skin of the living frog dry or moist?

____

It is used for respiration when the frog is warm and active, and secretes a protective mucus that minimizes water loss. Sometimes the mucus contains a poison to repel predators.

- Notice the **cloaca** at the posterior end of the abdomen into which the digestive and urogenital systems open.
- Study the head region: The **eyes** are very large and adapted to spotting insects, which are the staple of the frog's diet.
- Find the **nictitating membrane,** which is a transparent eyelid that can close over the eye to keep it clean without blocking the frog's vision. It is analogous to the blinking membrane over the shark eye.

Are opaque upper and lower eyelids present?

____

- Locate the external eardrum or **tympanic membrane.** The ear is derived from the lateral line organ, which is still present in aquatic and larval amphibians.

If you studying a live frog, pith or anesthetize it according to the directions given by your instructor. Pithing instructions are given in Topic 25.

- Open the mouth of the pithed, anesthetized, or preserved frog, and find the sticky tongue with which it traps insects. It is a true tongue.

Is the frog's tongue attached like yours?

____

Pass a hair or some soft probe into one of the **external nares** (nostrils) to see the connection with the oral cavity. The openings inside are the **internal nares.**

- Locate the **vomerine teeth,** which are in the roof of the mouth. They hold the prey before it is swallowed by the frog.

Are there other teeth present on the upper jaw?

____

On the lower jaw? ____________

- Cut the corners of the mouth so that you can open it all the way.
- Find the large opening into the digestive tract and the small opening to the respiratory tract. The latter is called the **glottis,** and it is located on the floor of the pharynx.
- Look for the openings at the posterior corners of the mouth that lead into the **vocal sacs** in a male frog.
- Find the openings alongside the entrance to the esophagus that lead into the **Eustachian tubes,** which connect with the middle ear.
- Open the abdomen of the frog as follows: First cut the skin of the abdominal wall *TO ONE SIDE OF THE MIDLINE* from the pelvic girdle to the throat region. Then cut the abdominal and chest muscles and the pectoral girdle. Continue the cuts to each side at the top and bottom of the incision. Then fold back the flaps so that your frog looks like the dissection shown in Figure 21–6. This figure shows most of the organs in the frog's abdominal cavity.
- If your specimen is a female, the abdomen may be filled with masses of dark-colored eggs. Remove them carefully so that you can see the other organs.
- Find as many of these organs as you can.

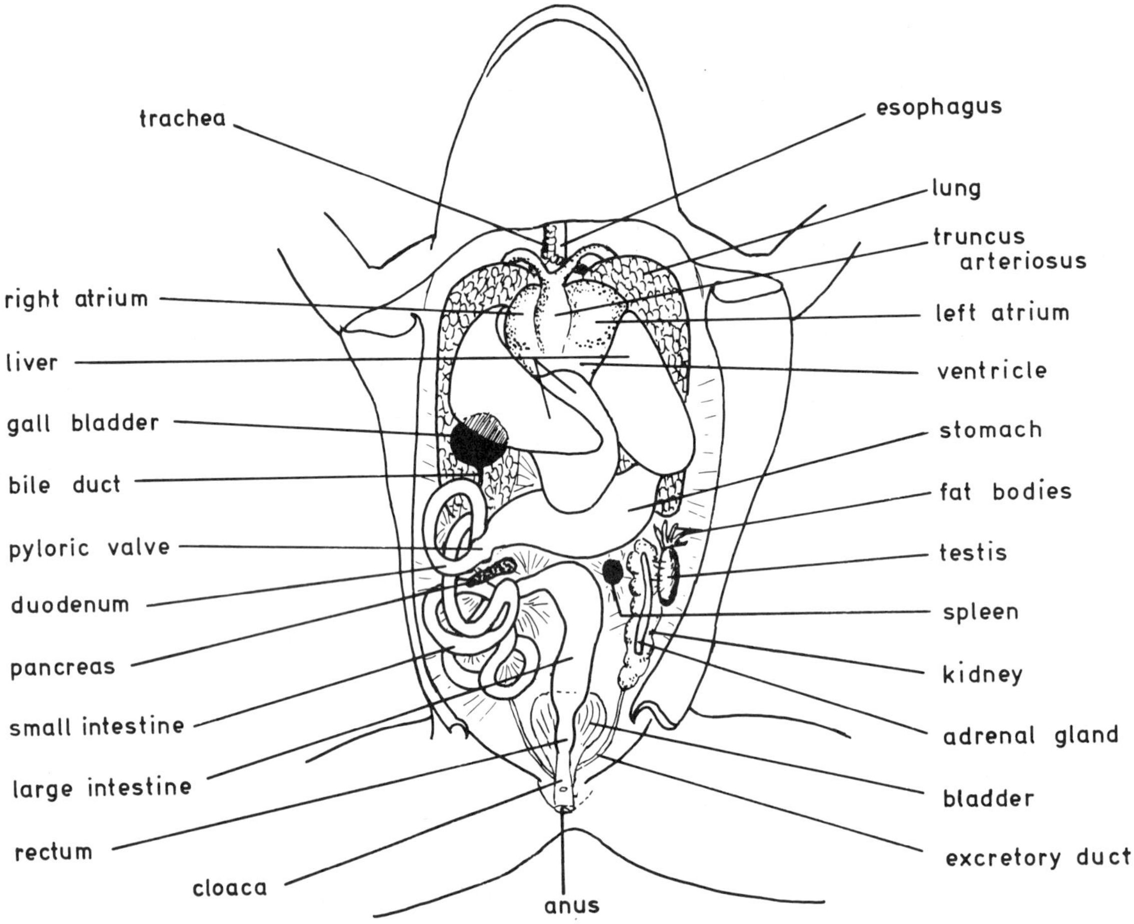

Figure 21–6. Internal anatomy of a male frog. Move the larger organs aside in order to find those that lie deeper within the body cavity.

Liver. Large with three lobes.

Gall Bladder. Bile from the liver is stored here and passes through the bile duct to the duodenum.

Esophagus: Leads to the stomach.

Stomach. Pyloric valve separates the stomach from the duodenum of the small intestine.

Small Intestine. Relatively short because the frog is carnivorous.

Large Intestine.

Rectum. Opens into the cloaca.

Fat Bodies. Yellow structures with fingerlike projections.

Pancreas. Light-colored lobed organ located near the joining of the stomach and small intestine.

Spleen. Deep red round organ.

Kidney. The kidneys are flattened, reddish organs located in the dorsal part of the abdominal cavity, and extending about half its length. They can form concentrated urine.

Adrenal Glands. Yellow bandlike organs on the ventral surface of the kidneys. These endocrine glands are hard to see.

Archinephric Ducts. Carry both urine and sperm in the male. Refer to a more detailed diagram in a separate dissection guide provided by your instructor before attempting to locate them. Urine drains through these ducts from the kidneys to the cloaca, where it is stored in the bladder.

Bladder. A thin, translucent sac in the ventral part of the posterior abdomen that opens into the cloaca. Water is actively reabsorbed from urine by the bladder.

Testes. Light yellow cylindrical organs located near the anterior ends of the kidneys. Sperm travel through **sperm ducts** into the kidneys, and then through the archinephric ducts of the kidneys to the seminal vesicles from which they are released through the cloaca.

Ovaries. Located near the anterior end of the kidneys and attached to the body wall.

Oviducts. Long white coils. The enlarged **ovisac** at the posterior end holds the eggs prior to spawning.

During mating, the male mounts on the back of the female, a position called **amplexus,** and eggs and sperm are simultaneously released into the water. External fertilization follows amplexus, and the fertilized eggs develop into herbivorous larvae called tadpoles. The aquatic tadpoles undergo **metamorphosis** into the terrestrial, carnivorous adults. (See Text Figure 10–16.)

- If you have a live specimen, observe the beating heart. Locate the **right atrium, left atrium,** and **ventricle** of the three-chambered heart. The large branching vessel is the main artery, the **truncus arteriosus.**
- Lift the ventricle to see the **posterior vena cava** and **anterior vena cavae,** which lead into the thin-walled **sinus venosus.**

In frogs there are two circuits of the circulatory system, the **systemic circuit** to the body and the **pulmonary circuit** to the lungs, but they are only partially separated. Deoxygenated blood from the body returns to the right atrium, and oxygenated blood from the lungs returns to the left atrium. The blood enters the ventricle and is pumped out to the body and lungs. The spongy internal structure of the ventricle and a spiral valve in the truncus arteriosus helps keep the oxygenated blood from the lungs and deoxygenated blood from the body separate. Some of the blood returning from the skin will be rich in oxygen, however, so there is mixing of oxygenated and deoxygenated blood in the ventricle to some extent. Thus the efficiency of the circulatory system is not as good as it would be if the circuits for oxygenated and deoxygenated blood were separated more completely. In reptiles, birds, and mammals, the skin is no longer used for respiration and in crocodilian reptiles, birds, and mammals the two circuits are completely separate.

Be sure you understand how a frog heart differs from a fish heart and from your own mammalian heart.

- Locate the small **lungs** by raising the liver. Both the lungs and the moist skin serve as respiratory organs to oxygenate the blood when the frog is warm or active. The skin is sufficient when the frog is cool and inactive.

Does the frog have a diaphragm, a partition separating the heart and lungs from the other organs?

____

The frog's brain has olfactory, visual, and equilibrium functions, but there are now cerebral hemispheres as well. The cerebral hemispheres help coordinate the activities of the other parts of the brain and allow the frog to take advantage of more complex behavior patterns.

CLASS AVES

Reptiles, birds, and mammals are classed together as the **amniotes** because they produce an **amniotic egg** (cleidoic egg) with a protective shell (see Text Figure 22–4a). The amniotic egg is fertilized internally and in some reptiles and all birds a shelled egg is laid. During its **oviparous** development, the young bird is surrounded by several protective membranes and a sac of amniotic fluid. Amniotes do not depend on water for respiration or reproduction and so are the only truly terrestrial vertebrates.

Birds and mammals are the only two classes of vertebrates that are **endothermic** and generate body heat by their rapid metabolism, in contrast to other vertebrates that use behavioral tricks to absorb heat from the environment. Endothermic animals closely regulate their high body temperature by physiological and anatomical adaptations. As a result they can live where reptiles and amphibians cannot, and they can remain active during very cold weather. Their respiration and circulation efficiency must be high because they are such active animals and must continuously meet the energetic demands of thermoregulation.

Birds are the only vertebrates that have **feathers,** possibly derived from the scales of their reptilian ancestors. This important development took place during the Jurassic period over 150 million years ago, during the time of *Archeopteryx,* a fossil bird that has feathers, but also many reptilian characteristics (see Text Figure 15–6 and 22–44). Birds are so closely related to certain fossil reptiles, that they are considered to be dinosaurs that have survived to the present time.

- Look at the feathers on demonstration.
- Study a contour feather in the dissecting microscope, and identify its parts:

Quill. The hard pointed end that attaches to the bird's skin.

Superior Umbilicus. A small hole in the quill through which blood vessels pass.

Shaft. The hollow rod that supports the feather.

Barbs. The flexible filamentlike structures attached to the shaft. Notice how there is a system of tiny hooks called **barbules** along each barb. These zip up with the ridges of adjacent barbs, like velcro, to hold the barbs into position and make the feather surface smooth (see Text Figure 22–45a). The **down** at the base of the contour feather lacks these barbules.

If there are down and pin feathers also on demonstration, describe how a down feather differs from a pin feather.

____

____

What are the two main functions of the feathers?

____

____

- Examine a preserved pigeon or other bird. There is only a tiny **tail,** and the forelimbs have become the **wings** of this bipedal animal. A special gland near the tail secretes oil for preening the feathers.
- The jaws in birds move the horny **bill** or beak, in which are the **external nares.**

Does the bird have a **tongue**? _____________

Are teeth present? _____________

Is there a **nictitating membrane** over the eye?

____

Are there **eyelids?** _____________

The flexible neck allows great mobility of the head, and the brain is fairly large, filling much of the skull. Birds have very good color vision.

- Examine the bird's feet, which retain **scales** like those of reptiles.

How many digits are present on each foot?

____

Are there claws as in reptiles? _____________

One adaptation of birds for flight is that parts of the skeleton are fused for strength and many of the bones are hollow so that weight is reduced as much as possible (see Text Figure 31–20).

- Examine the internal structure of the bird skeleton on demonstration, and notice the internal struts that give the bones added strength. Some of the internal spaces in the bones are connected to the respiratory system.
- Notice the enlarged breastbone, the **sternum.** In birds that fly, the powerful pectoral muscles that move the wings are attached to the keel of the sternum.
- Locate the **cloaca,** which is the common exit for the urinary, genital, and digestive tracts. The urine is mainly uric acid, a nontoxic waste product adapted to the closed system of a shelled amniotic egg. Birds have no bladder, another adaptation for flight. Urine and feces are eliminated together in the bird's excrement, which in large deposits is called **guano** and makes excellent fertilizer.

The internal organs of the bird are shown in Figure 21–7.

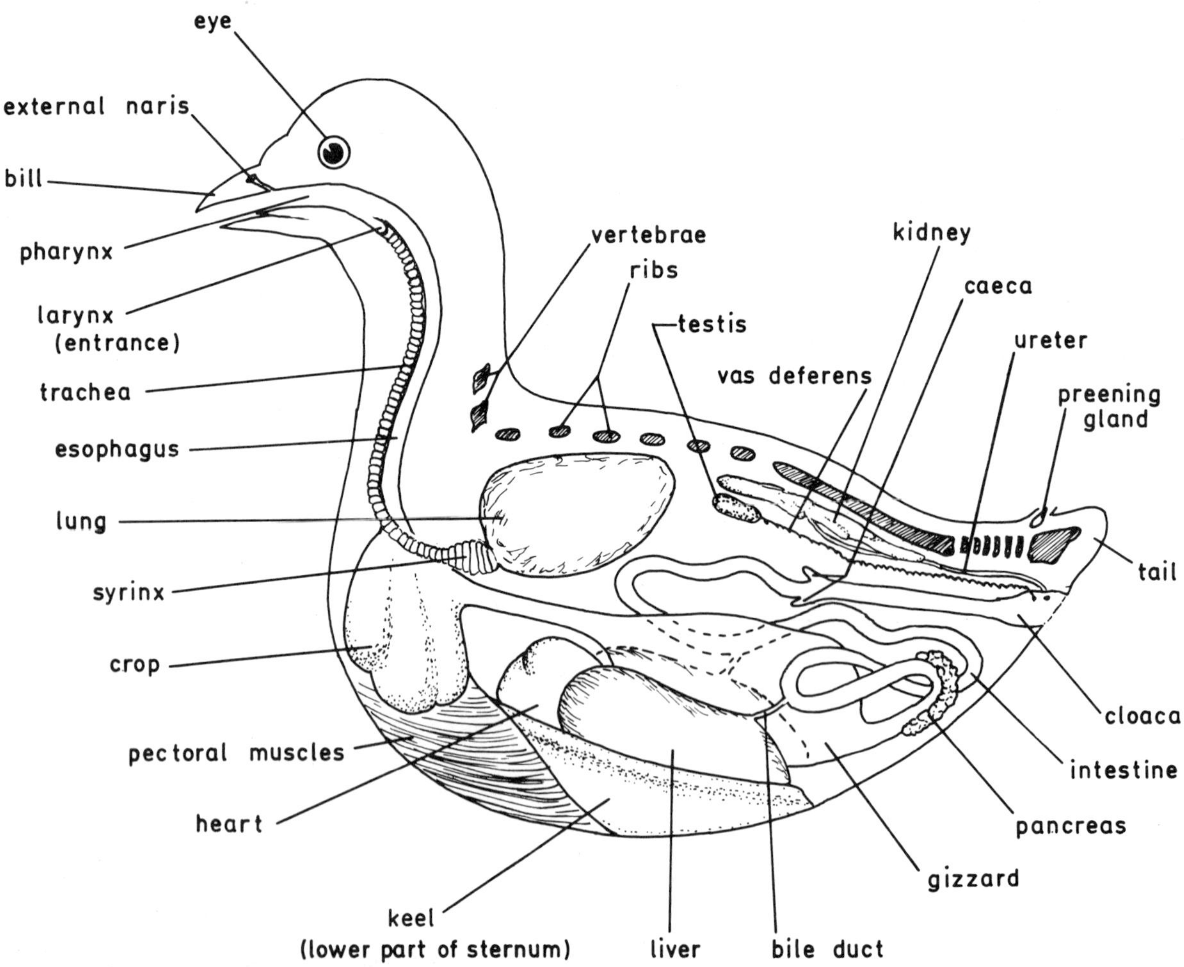

Figure 21–7. Internal anatomy of a male pigeon. There is no gall bladder or urinary bladder. The lungs connect to several air sacs within the abdominal and thoracic cavities and within the major bones (not shown).

- Remove the skin and feathers over the region just below the neck, and expose the **crop,** which is an outgrowth of the esophagus. It is a storage sac for the food items that are generally swallowed whole.
- Cut open the crop to see what it contains.

In pigeons, the crop of both males and females produces "crop milk" during the breeding season. It is regurgitated and used to feed the young birds.

- Open the abdomen of the bird, first removing the skin and feathers, and locate the following structures, most of which are shown in Figure 21–7.

Liver. Empties into the small intestine directly through the bile ducts. There is no necessity for a gall bladder, because feeding and digestion are continuous.

Gizzard. Derived from the first part of the stomach and used to grind up the food, which is often hard kernels of grain. It serves the same function as our own grinding teeth. Open the gizzard and see if any small pebbles are inside.

Small Intestine. Food from the gizzard passes through the stomachlike proventriculus, to this part of the digestive system.

Large Intestine. Two small vestigial caeca are attached near the end of the large intestine.

Pancreas. Located in the folds of the small intestine.

Kidney. Three-lobed organ against the dorsal wall of the body cavity. Bird kidneys produce a concentrated urine containing uric acid.

Ureter. Leads into the cloaca. There is no bladder.

Testes. Located at the anterior end of the kidney; small unless the bird is preparing to breed.

Deferent Duct. This very small tube leads from the testis to the cloaca. There is no penis in the pigeon nor in most birds. (A copulatory organ is present in the duck and ostrich groups.) The cloacal areas of the male and female are simply pressed tightly together during mating.

Ovary. Only the left ovary is present; small unless the bird is breeding.

Oviduct. Again, only the left one is present. If the bird was breeding, there will be eggs in various stages of production within the oviduct. As the soft, fertilized egg passes along, albumin and membranes are added. The hard calcareous shell is added just before the egg is laid and is usually not yet present. It is necessary so that the eggs won't collapse when the parent birds sit on them during incubation. All birds lay eggs and development is always **oviparous.**

- Locate the **heart,** which is underneath the anterior portion of the sternum. You will have to cut away some of the bone. The four fully separate chambers are the **right atrium, left atrium, right ventricle,** and **left ventricle.** Birds require a very efficient circulatory system, and the systemic and pulmonary circuits are completely separated (see Text Figures 25–5 and 25–6). The main artery, the aorta, has retained only the right branch, which becomes the **dorsal aorta.**
- Trace the path of a red blood cell from the right atrium to the dorsal aorta, using Figure 24–1 as your guide.
- Remove one of the wings, and cut through the sternum and through the ribs near the vertebral column to expose the underlying organs.
- Locate the compact **lungs** in the most anterior and dorsal part of the abdominal cavity, pressed against the dorsal body wall.

The lungs of the bird are unique in that they are connected to a system of **air sacs**. You will not be able to identify air sacs in your dissection. During ventilation, air is drawn through the lungs and into the air sacs, so the lungs do not expand much. During expiration, the air leaves the sacs and enters the lungs. Finally the air leaves after completing a circuit through the more-or-less one-way system. Oxygenation of the blood takes place as the air flows through tiny air capillaries within the lungs. Because this respiratory system uses continuous flow, it is very efficient and meets the stringent demands of the bird's circulatory system during flight. The air sacs are also important in cooling the body of the bird during strenuous flight.

The vocal sounds made by a bird are produced in a special organ, the **syrinx,** which is located at the posterior end of the trachea where it divides into the two bronchi. The **larynx** at the entrance to the trachea is not involved in vocalization in birds.

- Remove one of the lungs and locate the **syrinx** of the pigeon.

The brain of the pigeon has greatly enlarged **cerebral hemispheres,** which permit the complex behavior patterns characteristic of birds such as courtship, pair bonding, mating, and migration. The **cerebellum** is well developed for the coordination required in flight.

CLASS MAMMALIA

The first mammals evolved from reptilian stock during the Triassic period, over 200 million years ago. They were small insect-eaters and were probably active at night when their reptilian competitors were not. This was possible because, like the birds that evolved later, they are **endothermic** and maintain a high body temperature. Mammals have **hair** and sweat glands, important in thermoregulation. Their mammary glands produce **milk,** which is usually released through the **nipples** to nourish the young after birth. The young are generally retained within the body of the female during their **viviparous** development so that they are kept warm, but the mother can move about. They are nourished through the blood vessels of the **placenta.** This organ allows the fetal and maternal circulations to exchange nutrients and wastes during development. Characteristics of the reptilian ancestors of mammals are sometimes retained, however: Certain mammals lay eggs, some have scaly tails, and many have claws on their hands and feet.

The major orders of mammals are described in your text on pp. 431–435. Because the variation among the different orders is so great, it is difficult to pick one that is "typical." In this exercise you will concentrate on the features of the rat in Order Rodentia. Rats resemble the ancestral mammal in many ways, such as small size and rapid reproduction, and the rodents are the most successful of all mammals other than humans. As you study the rat, try to keep in mind that the organs and systems that you ar studying may be greatly modified in other mammals.

- If you are studying live rats, observe their behavior while they are being anesthetized by an overdose of barbituate. Notice the sniffing behavior and the gradual loss of coordination. Barbituates and alcohol have a similar effect on the cerebellum, and the rat's behavior is similar to that of a person who is very drunk.
- In the fully anesthetized or preserved rat, note the hair and nipples that characterize all mammals. The skin also has sweat glands and subcutaneous fat for temperature regulation, and it secretes chemical signals important in social behavior.
- Note the elongated snout with **vibrissae** (whiskers) used for tactile sensing.

Why is it advantageous to have the vibrissae as far forward as possible?

____

- Locate the **lips** and **external nares** (nostrils).

Does the **eye** have a nictitating membrane?

____

- Find the external ears, or **pinnae** (singular: pinna). The **eardrums** are in a protected position within the skull.
- Notice the **scales** on the rat's tail.

Are there scales on the legs, as in the bird?

____

Are there **claws** on the digits of the hands and feet?

____

- Locate the **anus,** the posterior opening of the digestive tract.
- If your specimen is a male, locate the **penis** in the fur-covered **prepuce** or foreskin.

Is the penis directed anteriorly or posteriorly?

____

- Anterior to the anus in the male is the **scrotum** containing the **testes.** The testes are located outside the abdominal cavity because they must be cooler than body temperature to produce viable sperm.
- If your specimen is a female, locate the **vagina** anterior to the anus and the **urinary opening** anterior to the vagina. The urinary and genital systems have separate openings as in human females.
- Open the rat's mouth and investigate the attachment of the **tongue.**
- Locate the **glottis,** which opens into the trachea (leading to the lungs), and its protective covering, the **epiglottis.** It keeps food out of the trachea when the rat swallows.
- On the roof of the mouth distinguish between the anterior **hard palate** and the posterior **soft palate.** The openings of the nasal passages are covered by the soft palate.
- Locate the large **incisors,** which are the gnawing teeth characteristic of rodents. These teeth are constantly being worn down so they must grow at a fast rate.

The teeth of mammals show a lot of variation and have become specialized to serve different purposes. **Incisors** are front teeth adapted for cutting or cropping the food as in humans and deer or for gnawing as in rodents. The **canines** are further back in the jaw and are for biting or holding prey and for tearing flesh as in a carnivore such as a wolf or a lion. In herbivores they are greatly reduced and also serve for cropping. In the rear of the jaw are the **premolars** and **molars,** which have a flattened surface and are used to crush and grind the food. Molars are especially prominent in herbivores, which must spend a great deal of time chewing food before it can be digested properly.

- Examine the teeth of the skulls on demonstration and compare them to those of the rat.
- Sometime during this lab, also examine and compare the mammalian skeletons on demonstration.
- Cut open the skin of the rat over the abdomen from the pelvic girdle to the ribs, and cut sideways to form two flaps. Cut through the abdominal muscles to expose the internal organs.

The anatomy of the internal organs is shown in Figure 21–8. As you study your specimen, try to identify as many of the following organs as possible.

Esophagus. Leads to the stomach.

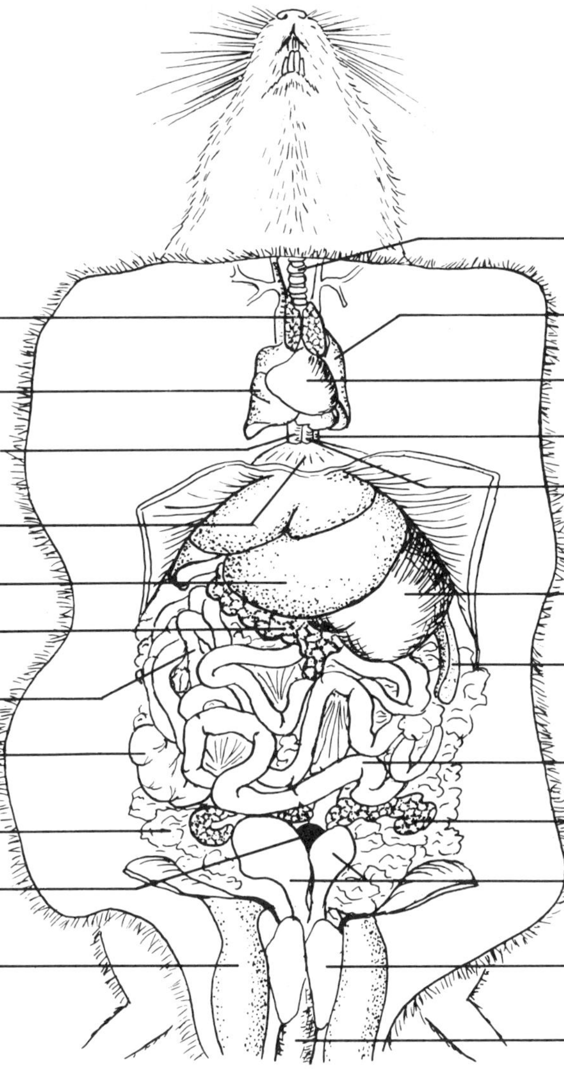

Figure 21–8. Internal anatomy of a male rat. Move the intestine aside to find the organs of the urogenital system.

Stomach. Separated from the small intestine by the pyloric sphincter, a muscular valve.

Small Intestine. Tightly coiled and very long compared to the size of the animal. A long intestine is characteristic of herbivores and omnivores like the rat.

Caecum. Enlarged sac at the beginning of the large intestine very important in digestion in herbivores such as the horse, and homologous to the vestigial human appendix.

Large Intestine. Ends in the muscular **rectum,** which probably contains some fecal pellets.

Liver. There is no gall bladder in the rat.

Pancreas. Located in the loop of the small intestine.

Spleen. Dark red organ that is located next to the stomach.

Kidneys. Compact organs in the dorsal part of the abdominal cavity. Urine travels through the **ureter** to the bladder where it is stored. Urine leaves the body through the **urethra,** which also carries sperm in the male. You will have to dissect below the intestine and abdominal membranes to see these structures.

Adrenal Glands. Small oval glands at the anterior end of the kidneys that are very difficult to see.

Testes. Carefully cut open the scrotal sacs and the tough membrane that surrounds the testis. Sperm are stored in the highly convoluted tubules of the **epididymis,** on the surface of the testis, and are carried in the tube called the **vas deferens** to the urethra.

Seminal Vesicles and **Prostate.** These glands add fluids to the semen before it passes into the urethra.

Ovaries. Pea-sized organs attached to the body wall.

Oviducts. Tightly coiled tiny tubes leading from the ovaries to the horns of the uterus.

Uterus. Consists of two tubes, or **horns,** in which the embryos develop. The horns join together at the posterior end and enter the **vagina.**

If your specimen is female, is she pregnant?

____

- Cut through the sternum (breastbone) and ribs to open the thorax and expose the **trachea;** it is kept open by its cartilaginous rings.
- Locate the muscular **diaphragm,** which is a sheet separating the abdomen from the thorax. When it contracts, air is sucked into the lungs.
- Remove the glandular material anterior to and covering the heart. This is the **thymus gland,** which is important during the development of the immune system.

The structure of the heart and circulatory system is like that of other mammals (see Text Figures 25–5 and 25–6). As in birds, the systemic and pulmonary circuits are completely separate because the chambers of the four-chambered heart are completely separate.

- Turn to Figure 24–1 and identify the four chambers of the heart: **right atrium, left atrium, right ventricle,** and **left ventricle.**
- Find the **pulmonary artery,** which is the most ventral vessel at the anterior end of the heart and carries blood from the right ventricle to the lungs. Blood returns to the left atrium of the heart in the **pulmonary vein.**
- Find the **aorta,** which is dorsal to the pulmonary artery. It carries blood from the left ventricle to the body. The aorta is the main artery of the system, and it developed from the left systemic arch that you saw in the frog.
- Lift the ventricles of the heart to see the **posterior vena cava,** which carries blood from the body to the heart. It joins with the **anterior vena cava,** which brings blood from the head and forelimbs, to enter the right atrium.
- Find the dark **thyroid gland** covering the trachea in the neck region.
- Locate the **right** and **left lungs.**

Unlike those of the bird, the lungs are closed sacs and are filled by the sucking action of rib and diaphragm muscles. Within the lungs, gas exchange takes place across the thin membranes of the tiny air sacs (alveoli) located at the ends of the finely branching air tubes. Air exits the same way it came in, so flow is two-way.

The brain of the rat has relatively large cerebral hemispheres. The enlarged olfactory bulbs show that smell is a very important sense in the rat. Vision is poor and unimportant. The visual functions are performed by the rear parts of the cerebral hemispheres.

CLEANUP

When you are finished with your dissection, store or dispose of your specimen as directed. Wash and dry your dissecting equipment and wipe your work area clean for the next student.

REVIEW

Review what you have learned in this exercise by making sure that you have some understanding of each of the following structures or terms. If you are not sure of any, refer back to the text of this exercise. You should be able to locate the structures in boldface in any vertebrate in which they occur.

amnion
amniotic egg (cleidoic egg)
anterior
anterior vena cava
anus
aorta
aquatic
artery
atrium
auditory
biped
bladder (urinary)
branchial arteries (afferent, efferent)
bulbous arteriosus
caecum
carnivore
cerebellum
cerebrum
claspers
claws
cloaca
colon, or **large intestine**
conus arteriosus
crop
ctenoid scale
dermis
diaphragm
digits
dorsal
ear
endothermic
epidermis
esophagus
Eustachian tubes
external fertilization
eye
fins
gall bladder
gametes
gill slits
gizzard
glottis
gut
heart
herbivore
homologous structures
internal fertilization
intestine (large or small)
jaws
kidney
larva
larynx
lateral line
liver
lungs
milk
mucus
nares, (singular: naris) or **nostrils**
nictitating membrane
nipples
notochord
olfactory (lobes, bulbs, sacs)
omnivorous
optic
oral cavity
ovary
oviduct
oviparous
ovoviviparous
palate (hard, soft)
pancreas
pectoral girdle
penis
pharyngeal gill slits

pharynx
pinna
placenta
placoid scale
posterior
posterior vena cava
pulmonary artery, vein
quadruped
rectal gland
renal
scales (of epidermis)
sinus venosus
skull or **cranium**
spiracle
spiral valve
spleen
stomach
swim bladder
syrinx
tadpole
teeth (incisors, canines, premolars, molars)
terrestrial
testis
tetrapod
tongue
trachea
tympanic membrane or **eardrum**
ureter
urethra
uterus
vagina
vein
ventral
ventricle
vertebra
vestigial structures
vibrissa
viviparous
vomerine teeth

REFERENCES AND FURTHER READING

Alexander, R.M. *The Chordates.* New York: Cambridge University Press, 1975.

Attenborough, D. *Life on Earth: A Natural History.* London: BBC/Collins, 1979.

Barrington, E.J.W. *The Biology of the Hemichordata and Protochordata.* Edinburgh: Oliver and Boyd, Ltd., 1965.

Bellairs, A., and R. Carrington. *The Life of Reptiles.* New York: Universe Books, 1970.

Blair, W.F. *et al. Vertebrates of the United States,* 2nd ed. New York: McGraw-Hill, 1968.

Bone, Q. *The Origin of Chordates,* 2nd ed. Burlington, NC: Carolina Biological Supply Co., 1979. [#45-9618]

Colbert, E.H. *Evolution of the Vertebrates,* 3rd ed. New York: John Wiley and Sons, 1980.

Gilbert, P.W., R.F. Mathewson, and D.P. Rall. *Sharks, Skates and Rays.* Baltimore: Johns Hopkins, 1967.

Marshall, N.B. *The Life of Fishes.* Cleveland: The World Publishing Co., 1966.

McFarland, William N., F. Harvey Pough, Tom J. Cade, and John B. Heiser. *Vertebrate Life.* New York: Macmillan, 1979.

Nichols, D. *Echinoderms,* 4th ed. London: Hutchinson, 1969.

Noble, G.K. *The Biology of the Amphibia.* New York: Dover Publications, 1954.

Romer, A.S., and T.S. Parsons. *The Vertebrate Body,* 6th ed. Philadelphia: Saunders College Publishing, 1986.

Walker, W.F. *Vertebrate Dissection,* 7th ed. Philadelphia: Saunders College Publishing, 1986.

Welty, J.C., and Baptista, L. *The Life of Birds,* 4th ed. Philadelphia: Saunders College Publishing, 1988.

Young, J.Z. *The Life of Vertebrates,* 3rd ed. New York: Oxford University Press, 1981.

Young, J.Z., and M.J. Hobbs. *The Life of Mammals,* 2nd ed. Oxford: Clarendon Press, 1975

The following dissection guides are published by W.C. Brown Co., Dubuque, IA:

Ashley, L.M. *Laboratory Anatomy of the Shark,* 4th ed. 1983.

Ashley, L.M. *Laboratory Anatomy of the Turtle,* 1962.

Chaisson, R.B. *Laboratory Anatomy of the Necturus,* 3rd ed. 1976.

Chaisson, R.B. *Laboratory Anatomy of the Perch,* 3rd ed. 1980.

Chaisson, R.B. *Laboratory Anatomy of the Pigeon,* 2nd ed. 1972.

Chaisson, R.B. *Laboratory Anatomy of the White Rat,* 4th ed. 1980.

Odlaug, Theron O. *Laboratory Anatomy of the Fetal Pig,* 6th ed. 1980.

Oldham, J.C., and H.M. Smith. *Laboratory Anatomy of the Iguana.* 1975.

Robinson, M.C. *Laboratory Anatomy of the Domestic Chicken.* 1970.

Underhill, R.A. *Laboratory Anatomy of the Frog,* 4th ed. 1980.

Journal Articles

Alldredge, A. "Appendicularians." *Scientific American,* 94, July 1976.

Archibald, J. David, and William A. Clemens. "Late cretaceous extinctions." *American Scientist* 70(4): 377, 1982.

Baker, MaryAnn. "A brain-cooling system in mammals." *Scientific American,* 130, May 1979 (#1428).

Buffetaut, Eric. "The evolution of the crocodilians." *Scientific American,* 130, October 1979 (#1449).

Carey, Francis G. "Fishes with warm bodies." *Scientific American,* 36, February 1973 (#1266).

Greenwalt, Crawford H. "How birds sing." *Scientific American,* 126, November 1969 (#1162).

Gruber, S.H., and A.A. Myrberg, Jr. "Approaches to the study of the behavior of sharks." *American Zoologist* 17: 471, 1977.

Hartline, Peter H., and Eric A. Newman. "The infrared 'vision' of snakes." *Scientific American,* 116, March 1982.

Heller, H. Craig, Larry I. Crawshaw, and Harold T. Hammel. "The thermostat of vertebrate animals." *Scientific American,* 102, August 1978 (#1398).

Jensen, D. "The hagfish." *Scientific American,* 82, February 1966 (#1035).

Johansen, Kjell. "Air-breathing fishes." *Scientific American,* 102, October 1968 (#1125).

Mossman, D.J., and W.A.S. Sarjeant. "The footprints of extinct animals." *Scientific American,* 74, January 1983.

Myers, C.W., and J.W. Daly. "Dart-poison frogs." *Scientific American,* 120, February 1983.

Ostrom, John H. "Bird flight: How did it begin?" *American Scientist* 67: 461, 1979.

Provine, Robert R. "Wing-flapping during development and evolution." *American Scientist* 72: 448, September 1984.

Russell, Dale A. "The mass extinctions of the late mesozoic." *Scientific American,* 58, January 1982.

Schmidt-Nielsen, Knut. "How birds breathe." *Scientific American,* 72, December 1971 (#1238).

Wann, Langston Jr. "Pterosaurs." *Scientific American,* 122, February 1981.

TOPIC 22

Animal Nutrition and Digestion

OBJECTIVES When you have completed this topic, you should be able to:

1. Describe how a mammal ingests and breaks up its food mechanically.
2. Name the three major classes of food molecules that are macronutrients.
3. Describe in general terms what happens to each class of food chemically.
4. Name the parts of the mammalian digestive tract in order, and describe the function of each part in the digestion of each of the macronutrients.
5. List those organs that secrete digestive enzymes, and give the substrates and products of digestion for these enzymes.
6. Describe where in the gut absorption takes place, what substances are absorbed in each place, and what happens to the food after it passes out of the gut lumen into the body.
7. Explain the role of the liver in digestion.
8. Distinguish between: ingestion/digestion; excretion/egestion; bile/enzyme; villus/microvillus; pepsinogen/pepsin; chyme/chylomicron, peristalsis/peritoneum.
9. Describe what happens to each class of food molecules if it is eaten in excess of the body's needs.
10. Explain how the digestive system of a mammalian omnivore (human, pig) differs from that of an herbivore (cow) or carnivore (wolf).
11. Describe the role of symbiotic bacteria in the gut of yourself and of other mammals.

TEXT REFERENCES Chapter 23

KEY TOPICS

carbohydrate digestion	digestive enzymes	pancreas
protein digestion	salivary glands	omnivore adaptations
lipid digestion	liver	gut symbionts
mammalian digestive tract		

INTRODUCTION

In this and the following labs you will put together what you are learning about animal structure and function by examining a representative mammal, the fetal pig, in some detail. You will be looking at the digestive system, circulatory system, respiratory system, excretory system, endocrine system, and reproductive system, and you should try to understand how these systems fit together into an integrated organism.

Why a pig?

It turns out that pigs closely resemble humans, anatomically and physiologically. Pigs, like ourselves, are omnivores, so their digestive tracts are similar to ours. Their circulatory systems can undergo atherosclerosis, like ours, so they are used as models in research on the function and diseases of the circulatory system. Special "minipigs" have been developed for laboratory work, because 600-lb porkers are not very practical. Last, but not least, these pigs are available to us as a by-product of the meat packing industry. When sows are slaughtered, the fetal pigs are discarded, except those used for biological study (or dog food). Thus the animals are not killed unnecessarily just to obtain the fetal pigs.

The fetal pig develops for 112–115 days before it is mature enough to be born. It will then be about 330 mm long from snout to rump. The pigs usually used for dissection are roughly 102–107 days old.

Since the pig is so like the human, you will most likely find the dissection very interesting once you get underway. It will be a little messy, but this may be your best or only chance to learn firsthand how your body functions in health and disease. In the end, we hope what you learn will be worth the trouble of making a careful dissection.

The fetal pigs have been prepared for dissection by being placed in a preservative such as formalin. (These preservatives are strong and irritating to the hands. Hand cream applied to the hands before and after dissection can be helpful.) In addition, these pigs have had their circulatory systems injected with colored latex to facilitate blood vessel identification. Veins should appear blue and arteries red, although exceptions occur.

You should not wear contact lenses during dissection because the fumes from the preservatives will irritate your eyes.

Your dissection must be planned and executed *carefully*. Thus it is especially important to read the instructions beforehand. Dissection is a procedure to help you learn the anatomy and structural relationships of the animal. If you do not follow the directions, you may remove, misplace, or lose parts before you have a chance to observe them. Remember that you are going to be using this animal for several laboratory periods, and parts must be preserved *intact* for future study. There are two good rules to keep in mind: Do as little cutting as possible, and *never* remove a part unless told to do so. You may be asked to work on your pig together with your lab partner. In that case, each person should do part of the actual dissection to get a feel for the procedures involved. Work slowly at first, and do not hesitate to ask your instructor for advice on how to proceed.

Much of the work of dissection is clearing away connective tissue to expose the organs and free them up so they can be moved. Do not expect all the organs shown in the diagrams to be immediately visible without some probing underneath on your part.

Dissecting Tools

For dissection you will need:

Razor Blade (single edge) or scalpel to make the initial cuts through the skin. Use these sparingly. Most students tend to be heavy-handed with them and cut too deeply.

Scissors to use for most of the cutting. Use scissors instead of the scalpel whenever possible. Use heavy bone scissors for cartilage and bone. Oil your scissors at the end of each lab to prevent corrosion.

Forceps to use for lifting and moving structures.

Blunt Probe to use for lifting parts, separating tissues, and probing ducts.

Fingers can lift, separate, and expose organs far more efficiently than any of the above instruments. Fingers are your best dissecting tools.

Anatomical Terms

You need to know a few adjectives used to denote orientation and direction in anatomical work. Be sure you understand the following and refer back to this list whenever necessary:

Left. The pig's left.

Right. The pig's right.

Anterior. Toward the head.

Posterior. Toward the tail.

Cranial. Toward the head (in humans, the term "superior" is used).

Caudal. Toward the tail (in humans, "inferior").

Dorsal. Toward the back (in humans, "posterior").

Ventral. Toward the belly (in humans, "anterior").

Lateral. Toward the side.

Medial. Toward the midline.

Proximal. Describes the part of a structure nearest a point of reference.

Distal. Describes the part of a structure that is farthest removed from a point of reference.

Pelvic. The hip region.

Pectoral. The shoulder region.

The terms **left** and **right** should be self-evident, but unfortunately these terms lead to more confusion than any of the other terms. It is important that you remember that left and right always refer to the *pig's* left or right, not your own!

Many more terms are given in the following directions than you may be expected to learn.

Your instructor will give you some indication as to how much detail you will be expected to master, depending on the amount of laboratory time spent in dissection. To a certain extent, you will be expected to use your own judgment in deciding what is important. As a rough guide, you should be familiar with the location and function of those structures that are also mentioned in your text.

Care of Your Specimen

Although your specimen is preserved in embalming fluid, it will stay useful for dissection only if it is kept from drying out. It must be kept in a sealed plastic bag containing a little fluid at all times when you are not dissecting. During the laboratory period, you should add a little water or dissecting fluid from time to time to keep the tissues moist and flexible.

Cover the pig with wet paper towels whenever you are not actually dissecting.

When you are ready to stop the dissection until the next period, make sure the organs are moist before you seal the specimen up in its plastic bag.

EXTERNAL ANATOMY

- At the sink, remove the pig from its plastic bag and rinse it off under the tap. (Save the bag.) Do not be alarmed at the "drowned rat" appearance of your specimen. Remember that mammals, including humans, are born dripping wet, since they spend the fetal period in a watery environment.

The body of a mammal is divided into a number of regions. The **head** is separated from the **trunk** by the movable neck. The trunk is generally divided into the anterior **thorax** and posterior **abdomen** and **pelvis.** Most mammals have a **tail.**

- Examine the head and locate the **eyes,** the external ears, or **pinnae** (singular: pinna), the nostrils, and the mouth.
- Note the paired legs and their various joints.
- Compare the pig forelimb with your own arm, and try to figure out the corresponding parts.
- Do the same for the hindlimb.
- Pick up the pig and manipulate the joints.

Do they move the same way yours do?

____

- Notice the toes.

How many are there on each foot? ______________

How many toes does the pig actually walk on?

____

The pig walks on the tip of his toenails, a method of locomotion known as "unguligrade."

How does this differ from the human method of walking?

____

____

____

____

____

____

- Observe the **umbilical cord** by which the fetal pig was attached to the placenta during development within the uterus.
- Observe the cut end of the cord, and notice the cut ends of the blood vessels that carried blood containing dissolved nutrients and wastes to and from the placenta of the mother. The human umbilical cord is normally cut at birth, and shrivels up, so all that remains of it is the **navel.**
- On the ventral surface of the thorax and abdomen find the two rows of **mammary glands.** These glands are one of the distinctive features of mammals. The presence of **hair** is a second distinctive characteristic of all mammals.

Your next task will be to determine the sex of your pig.

- Raise the tail and observe the **anus,** the posterior end of the digestive tract.
- If your specimen is a female, there will be a second opening just ventral to the anus, **the urogenital opening.** This is the exit for both the reproductive and excretory systems. Protruding from the urogenital opening in the female is a small, fleshy **urogenital papilla.**
- If your specimen is a male, there is no second opening near the anus. Instead, locate the small **urogenital opening** just posterior to the umbilical cord. The duct leading to it runs forward from between the legs in a long muscular tube, the **penis,** which can be felt under the skin.
- If your specimen is male, find the scrotal sacs in the area ventral to the anus. The scrotal sacs are not very apparent in the fetal pig. Around the time of birth, the testes eventually descend into them.

Is your pig male or female? ______________

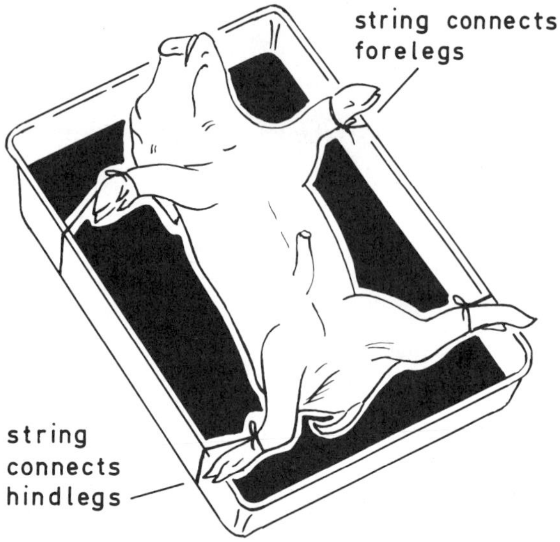

Figure 22–1. Fetal pig specimen. Stretch the legs as far apart as possible and tie them securely. Remove the pig for storage by slipping the strings out from under the dissecting pan.

ANATOMY OF THE DIGESTIVE SYSTEM

- Place your pig in the dissecting tray so that its ventral side is up.
- Cut two pieces of string about 2 ft long.
- Tie a piece of string securely around the ankle of each of the right legs.
- Run the two pieces of string under the dissecting tray, and tie them tightly to the ankles of the left legs, thus holding the pig securely in place in the dissecting tray, as shown in Figure 22–1.

If you stretch the legs as far apart as possible at this point, it will be a lot easier to locate the inner organs during your dissection.

The digestive system of the fetal pig consists of the **mouth, oral cavity, pharynx, esophagus, stomach, small intestine,** and **large intestine.** The **salivary glands, liver,** and **pancreas** are accessory organs. (See Figure 22–5.) As you study the anatomy of the digestive system, you should review the functions of each of these structures. The one-way flow of food from mouth to anus of this **complete digestive tract** allows greatly specialized sections that can act independently of each other. It also allows the animal to continue to eat while previously eaten food is still being digested.

The Mouth and Pharynx

The mouth opens into the oral cavity.

- Use bone scissors to cut through the corners of the mouth to the back of the jaw, so that the mouth opens widely. (See Figure 22–2.)
- Note the **hard palate,** which makes up the anterior part of the roof of the mouth. It is made of bone and covered with folds of mucous membrane.

The hard palate separates the oral cavity from the nasal cavities.

- Note that the membranes continue posteriorly, forming the **soft palate,** which does not contain bone.

In humans, there is a fleshy protuberance from the soft palate called the **uvula.**

- Locate the uvula *on your neighbor,* not the pig.
- Notice the sensory papillae on the surface of the muscular **tongue.** Taste buds within the papillae are responsible for the sensations associated with food ingestion: salty, sweet, sour, and bitter.

How is the tongue attached?

____

Ducts from the **salivary glands** enter the rear of the oral cavity and produce **salivary amylase,** which breaks down starch enzymatically during mastication.

What are the products of starch digestion?

____

What substance other than starch is required for the reaction?

____

What is this type of reaction called?

____

Have any **teeth** erupted from the gums in your specimen? ______________

- Observe the various types of teeth on demonstration (see Text Figures 22–47 and 23–12).
- Cut into the gum of the pig along one side and expose one or more of the developing teeth.

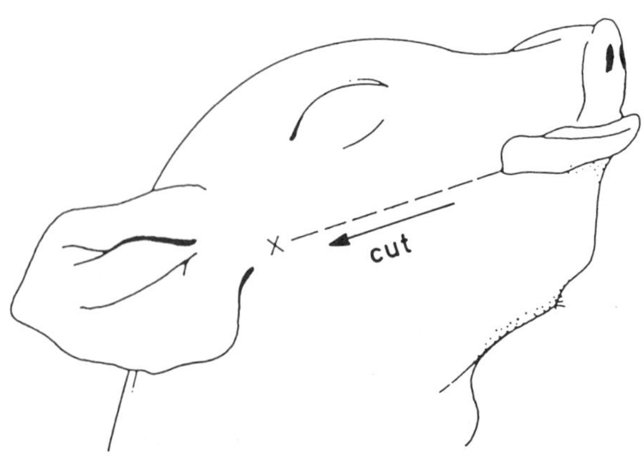

Figure 22–2. Cutting the jaw. Use heavy scissors to cut through the jaw cartilage so that the mouth can be fully opened.

- Identify a tooth as an **incisor** (for biting, gnawing), **canine** (for stabbing, tearing), or **molar** (for crushing, grinding). Canines are important for carnivores and incisors and molars for herbivores. Omnivores tend to use all three types of teeth.

Describe some digestive and nondigestive uses of teeth:

____

____

____

____

- In the region of the lower jaw, remove the skin just in front of the ear and locate the salivary glands. They are the color and consistency of used chewing gum and quite different from the stringy muscles.

What two functions does saliva have in the digestive process?

____

____

At the base of the tongue, the oral cavity ends and the **pharynx** begins. The pharynx is the common passageway for the digestive and respiratory tracts because both air and food must pass through this structure. You may have to extend the cuts you made earlier to open the mouth wide enough to see the pharynx and associated structures. A sore throat is really a sore pharynx or pharyngitis.

- At the posterior end of the tongue find a trough-shaped fold of tissue, the **epiglottis.** You may have to reach deep into the pharynx with your probe to locate the epiglottis and pull it out into view. Dorsal to this is an opening, the **glottis,** which leads into the **larynx,** part of the respiratory system. (See Figure 22–3.)

During swallowing the epiglottis forms a kind of lid over the glottis and deflects the food away from the glottis. When the epiglottis fails to do this, the food enters the glottis and choking results.

- Dorsal to the glottis find the second, wider opening into the **esophagus.**

The esophagus is a collapsible muscular tube connecting the pharynx to the stomach. Food enters the esophagus by the reflex swallowing of the pharynx and passes through the esophagus and gut by waves of muscular contractions called **peristalsis.**

- Pass the end of your probe into the glottis and then into the esophagus. Be certain that you can differentiate them. Notice that air entering through the nose must cross the pathway of food during swallowing.
- Relate Figure 22–3 to your pig to be sure you understand the anatomy of the pharynx.

At this time you will not follow the course of the esophagus as it proceeds through the thorax. It will be easier to study it when you dissect the respiratory system.

The esophagus marks the point at which food has been more or less irreversibly **ingested** or taken into the body by the actions of the teeth, tongue, and pharynx. Next the food will be **digested,** or broken down chemically by hydrolysis into small molecules that can be absorbed into the body proper. Some ingested material never becomes part of the body at all; it remains within the **lumen** (interior) of the gut and will be **egested** as waste.

The Abdominal Cavity

The abdominal region is that area, posterior to the ribs, in which the ventral body wall has no bony sup-

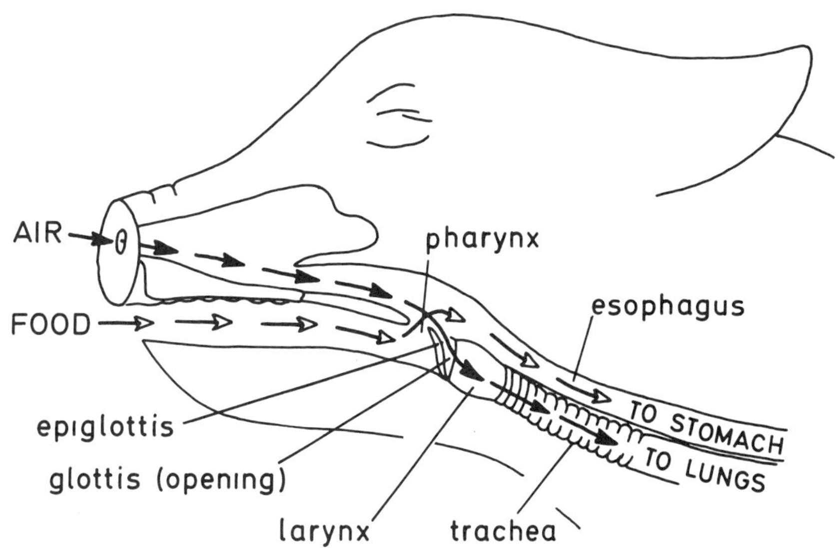

Figure 22–3. The pharynx in action. Air and food passages cross here, and there is some risk that food may enter or block the glottis, causing suffocation.

port. The body wall in this region encloses a large **peritoneal cavity.** Most of the digestive, excretory, and reproductive organs are found within the peritoneal cavity, together with some organs belonging to other systems. Each of the three systems mentioned above has one or more ducts that pass to the exterior through a ring of bones called the **pelvic girdle.**

- Look at Figure 22–4 and study the cuts that you are going to make to open the abdominal cavity.
- In opening the peritoneal cavity, use the umbilical cord as a landmark, and be careful to leave a strip of tissue caudal to the umbilical cord and to avoid cutting structures that lie directly in the midline.
- With the scalpel or razor blade make a shallow cut (1) from a point about 5 mm anterior to the umbilical cord forward to the most anterior pair of mammary glands.
- Cut around the umbilical cord on both sides (2), and continue the two parallel cuts posteriorly, keeping them *outside* the mammary glands.
- Deepen these cuts carefully until you reach the body cavity. You will have cut deeply enough when you see the dark-colored abdominal organs underneath the transparent membrane called the **peritoneum.**

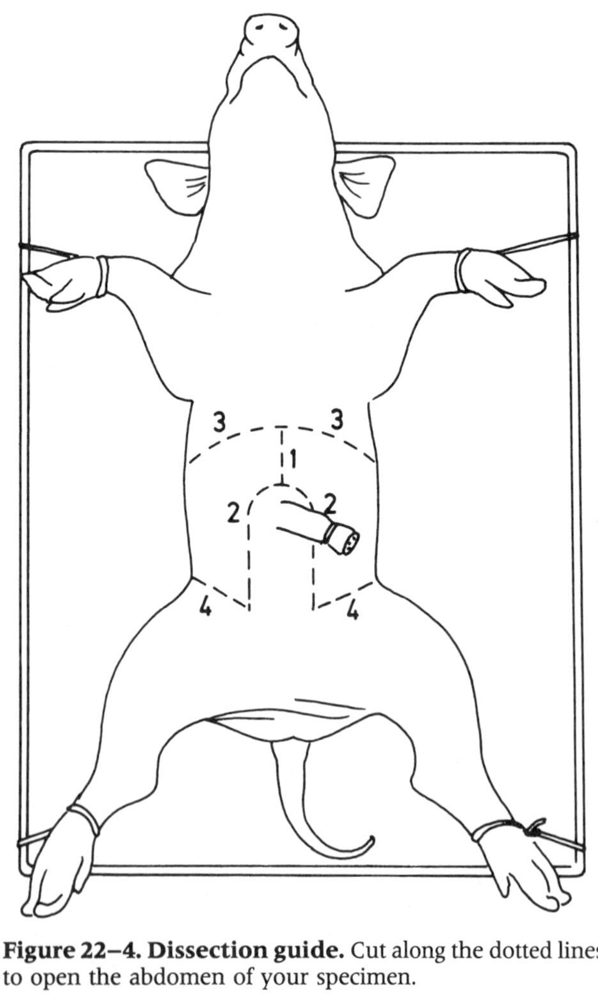

Figure 22–4. Dissection guide. Cut along the dotted lines to open the abdomen of your specimen.

- Locate the posterior end of the sternum, or breastbone, and cut laterally on each side following the posterior edge of the rib cage (3).
- Make similar cuts just anterior to the hind legs (4).
- Now fold out the flaps of the body wall, exposing the abdominal cavity. Its anterior end is bounded by a transverse partition, the **diaphragm.**
- Pull up and back on the umbilical cord and you will see the **umbilical vein** going to the liver. This vein carries food and oxygen to the fetus from the placenta.
- Tie the ends of a piece of thread or string to the umbilical vein in two places to form a loop. The vein may now be cut between the ends of the string. The string will let you find the vein easily later on in the dissection.
- If the body cavity contains much fluid, flush it out with water. During the dissection, paper tissues can be used to soak up any remaining fluid in the body cavity.
- Examine the lining of the abdominal cavity. This shiny membrane is called the **peritoneum.** The part of the membrane next to the body wall is called the **parietal peritoneum,** while that covering the surfaces of the internal organs is the **visceral peritoneum.** The thin, transparent sheets of tissue that connect the two are called **mesenteries.** The mesenteries suspend and support the visceral organs.
- Examine the large, dark brown **liver** located just posterior to the diaphragm.
- Notice that the liver consists of a number of lobes.
- Lift up the posterior edge of the liver and find the cut end of the umbilical vein anterior to the right side of the liver. (It should be attached to your string.)
- Just below the entrance of the umbilical vein into the liver you should see the **gall bladder,** a greenish sac embedded in the surface of the liver.

The liver secretes **bile,** a mixture of **bile salts** (lipids similar to cholesterol), and excreted wastes. (**Excretion** refers to the removal from the body of substances that have entered and been a part of body cells.) Bile is temporarily stored in the gall bladder and at the appropriate time enters the small intestine where it acts to **emulsify** fats. This breaks the insoluble globs of fat into tiny droplets so that lipid-digesting enzymes have a greatly increased surface area on which to act.

Once food molecules have been digested and absorbed into the blood stream, the liver plays an important role in storing carbohydrate as glycogen and in metabolizing amino acids and lipids.

What are some other nondigestive functions of the liver?

____

____

____

____

Does the liver secrete any digestive enzymes?

____

Describe the composition and location of gallstones:

____

Use Figure 22–5 as your guide in locating the other abdominal organs.

- With the liver still elevated, identify the **stomach** on the left side of the abdominal cavity.
- Locate the point near the midline where the esophagus enters the stomach. This is the **cardiac** portion of the stomach. Food from the stomach leaves through the **pyloric** end of the stomach. The flow of food is governed by a circular sphincter muscle called the **pyloric valve.** It releases partly digested food called **chyme** from the stomach into the small intestine.
- Squeeze the pyloric valve and note how muscular it is.
- Cut the stomach open lengthwise, and notice the ridges (rugae) on the inside. They serve to increase the internal surface area (see Text Figure 32–7).
- Observe the internal entrance of the esophagus and the pyloric valve, which marks the exit into the small intestine.

Are these valves normally open or closed?

____

- Notice that the stomach is not empty even though the fetus receives its nourishment from its mother through the umbilical vein. The fetus constantly drinks in the amniotic fluid surrounding it. The contents of the stomach consist of epithelial cells, hair, and other substances from the amniotic fluid.

Cells of the gastric glands, located in the lining of the stomach, secrete **hydrochloric acid** and **pepsinogen.** Pepsinogen is an inactive precursor of the enzyme **pepsin,** which hydrolyzes proteins to polypeptides. HCl maintains the acidic conditions in the stomach (pH 1–2) required for pepsin to act. The stomach is protected from digesting itself because the mucosa is alkaline and the cells are tightly pressed together sideways so that acid cannot penetrate the tissue. When the protection fails, a sore called a peptic

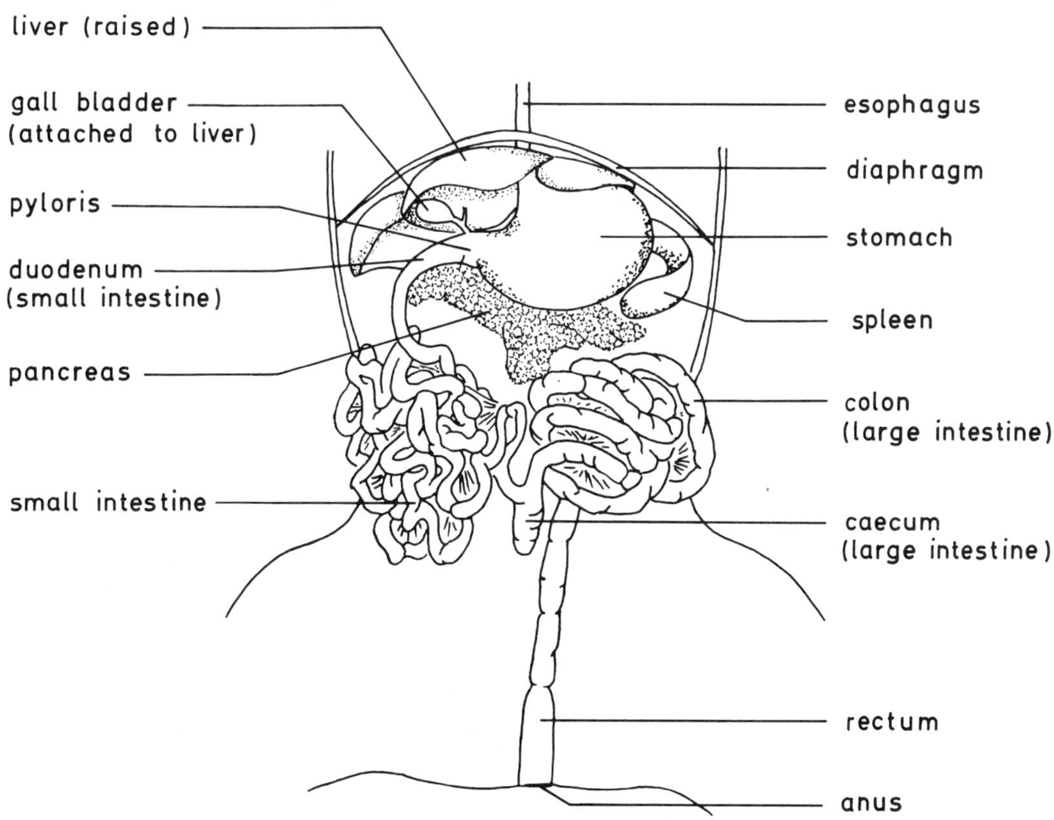

Figure 22–5. Abdominal organs. This diagram shows the approximate location of organs within the abdominal cavity. You will have to lift the liver to see the gall bladder underneath and move aside some of the other organs to see all of them.

ulcer results. Stomach secretions are regulated by the hormone **gastrin** from the stomach's own mucosa.

Does any absorption of food molecules take place from the stomach? ____________

How would your own feeding habits have to be modified if your stomach were removed?

____

____

____

- Note the tongue-shaped **spleen,** which is attached to the stomach by a mesentery. The spleen is part of the circulatory system. In humans it acts as a fine filter to remove debris and possibly harmful bacteria from the blood.

What color is the spleen? ____________

- Identify the anterior end of the **small intestine** leading from the stomach; it is the C-shaped **duodenum.** This is a very important part of the small intestine because the ducts from the liver, gall bladder, and pancreas enter here. Those from the liver and gall bladder join as the **common bile duct,** which can be seen as a white cord leading from the gall bladder to the duodenum. The **bile** produced by the liver and stored in the gall bladder enters through this duct.

What is the function of bile in digestion?

____

What is bile made of?

____

- If you lift the stomach and press the intestinal mass posteriorly, you will expose the light-colored glandular **pancreas.** (Glandular tissue has the appearance of used chewing gum.) The structural relationship between the duodenum and the pancreas has whimsically been called "the romance of the intestines" because the pancreas lies in the "arms" of the duodenum! The pancreas is connected to the duodenum by two tiny ducts that are too small to be easily seen. The juice secreted by the pancreas enters the small intestine through these ducts. No digestion takes place within the pancreas itself.

The pancreas is also an endocrine gland and secretes the hormones **insulin** and **glucagon,** which together regulate blood glucose.

What is the major target organ for these hormones?

____

All major food classes (carbohydrate, protein, fat) and nucleic acid can be digested by enzymes from the pancreas. These enzymes, as is pepsin in the stomach, are secreted in an inactive form. **Enteropeptidase,** an enzyme secreted by the small intestine, activates the protein-digesting enzyme **trypsin** and trypsin in turn activates the other enzymes. A neutral pH is required for activity, so pancreatic juice includes **sodium bicarbonate** to neutralize the acidic chyme entering from the stomach. If secretion of sodium bicarbonate is not sufficient, acid could attack the intestinal lining to cause a duodenal ulcer.

What are the substrates and products of these pancreatic enzymes?

Enzyme	Substrate(s)	Product(s)
trypsin		
chymotrypsin		
carboxypeptidase		
aminopeptidase		
lipase		
amylase		
RNase		
DNAse		

The small intestine itself, in addition to enteropeptidase, secretes aminopeptidase to hydrolyze small polypeptides and alkaline phosphatase to hydrolyze phosphate-containing molecules. The enzymes needed to digest the disaccharides maltose, sucrose, and lactose are produced here as well. The small intestine also secretes three important hormones that regulate the digestive system. **Secretin** stimulates the liver and pancreas to release their secretions while suppressing stomach acid secretion. **Cholecystekinin** (CCK) stimulates the gall bladder and pancreas and acts on hunger centers in the brain to regulate food intake. **Gastric inhibitory peptide** suppresses peristalsis and secretion of stomach acid when there is no food to digest.

- Trace the rest of the small intestine posteriorly from the duodenum. Notice the great length of the small intestine crowded into the small space. Intestine length tends to be short in carnivores but long in herbivores and omnivores because plant material is harder to digest and larger amounts must be eaten.
- Cut the small intestine next to the pyloris and next to the entrance to the large intestine. Dissect it free.
- Measure its length and compare it with the pig's snout-to-rump length.

Humans, like pigs and rats, are omnivores and have a small intestine of intermediate length. If this length is comparable in pigs and humans, how long would the small intestine be in a person whose sitting height is 1 m?

	Pig	Human
Height (length)	cm	1 m
Small intestine	cm	m

- Cut off a section of small intestine.
- Slit it open lengthwise and examine it with a hand lens or under the dissecting microscope.

The velvetlike appearance is due to small projections called **villi,** which greatly increase the interior surface area. The cells on the surface of the villi are in turn covered with a brush border of **microvilli** to further increase the surface area available for absorption. (See Text Figure 32–10.) The villi contain capillaries and lymphatic vessels that take up the end products of digestion and circulate them to the rest of the body, especially the liver. Amino acids and sugars enter the blood directly by diffusion or active transport. Amino acids are either used, or immediately metabolized for energy, since little protein can be stored by the body. Carbohydrates can be stored as glycogen in the liver or muscles. Partly digested fats enter the **lacteals** (blind ends of the lymphatic vessels) as tiny droplets called **chylomicrons.** They travel through the lymph and blood to the liver for further digestion. Fats may be metabolized for energy or stored as body fat. The small intestine also absorbs **vitamins,** organic molecules needed by the body in small amounts.

- Push the intestine to the right side, and relocate the point at which the small intestine enters the **large intestine.** It enters from the side, making a blind pouch called the **caecum.**

The caecum in plant-eating animals is usually large (up to 35 cm long in rabbits) and contains many microorganisms that produce **cellulase,** an enzyme that can digest cellulose. The caecum is of little importance in the pig and unimportant in humans where it is reduced to the vestigial **appendix.**

- The main portion of the large intestine is called the **colon.** It forms a tight double coil and then goes to the dorsal abdominal wall and continues posteriorly along the midline. This type of coiling is uncommon in mammals. The human colon is shaped like an upside down U.

What substances are absorbed by the colon?

____

Does the colon secrete digestive enzymes?

____

- Remove a short segment of the colon and cut it open.

How does the inner surface compare with that of the small intestine?

____

The colon contains a large number of bacteria that are constantly being lost in the feces but that are also very important in human nutrition. They synthesize **vitamin K,** which is absorbed from the gut lumen and plays a vital role in blood clotting. Because they lack gut bacteria, newborn babies may be deficient in vitamin K and therefore especially susceptible to bleeding from small wounds.

Later in your dissection you will follow the colon through the pelvic girdle to the **rectum** and **anus** where feces formed from waste and bacteria exit from the body. The large intestine both egests and excretes since some wastes are undigested food material but others (e.g. bile pigments) have been excreted into its lumen.

The **transit time** of food through the digestive tract can be speeded up if the diet is high in bulk and fiber. This is desirable, since our intestinal bacteria constantly produce **carcinogens** such as nitrites that may lead to cancer of the colon.

HISTOLOGY OF THE INTESTINE

The intestinal wall is made up of four different layers. Review its structure by looking at a cross section of the intestine under low power of the compound microscope. (Refer to Figure 5–3.) The central opening, or **lumen,** has the villi projecting into it. (See Text Figures 4B and 23–9.)

- Examine one **villus.** Covering the villus is a layer of epithelial cells called the **mucosa.** Under the mucosa is an area of connective tissue, the **submucosa.**
- Identify the blood vessels running through the submucosa. The digested food must move from the intestinal lumen, through the mucosa and submucosa, into these blood vessels.
- Locate the **lacteals** or blind ends of the lymphatic vessels.

What kind of food molecules are absorbed by the lymphatic vessels? ____________

What are the tiny droplets of this food called?

____

Where do they go for further processing?

____

- Outside the submucosa, locate the **mucularis,** which consists of two layers of muscle: an inner circular layer and an outer longitudinal layer. This arrangement allows the peristaltic contractions of the intestine to move the food through the digestive tract.

- An adventitia of connective tissue covers the muscularis. The outermost tissue is a single cell layer, the **visceral peritoneum.**

Why does the intestine have villi?

____

What is the function of the goblet cells of the mucosa?

____

How does the shape of the intestine change when the inner layer of muscles contracts?

____

____

How does the shape of the intestine change when the outer layer of muscles contracts?

____

____

What tissue supports the blood vessels leading to and from the intestine? _______________

Reminder: When you are finished with your dissection for today, close up the abdomen of your pig and return it and any scraps of tissue to its plastic bag. Seal the bag, label it with your name, and store it properly. Clean your dissecting tools, oil the scissors to prevent rusting, and leave your workplace clean for the next student.

REFERENCES AND FURTHER READING

Ganong, W.F. *Review of Medical Physiology,* 11th ed. Los Altos, CA: Lange, 1983.

Gordon, M.S. *Animal Physiology: Principles and Adaptations,* 4th ed. New York: Macmillan, 1982.

Grollman, S. *The Human Body. Its Structure and Physiology,* 4th ed. New York: Macmillan, 1978.

Guthrie, H. *Introductory Nutrition.* St. Louis: Mosby, 1979.

Guyton, A.C. *Textbook of Medical Physiology,* 7th ed. Philadelphia: W.B. Saunders, 1985.

Morton, John. *Guts: The Form and Function of the Digestive System.* New York: St. Martin's Press, 1967.

Prosser, C.L. *Comparative Animal Physiology,* 3rd ed. Philadelphia: W.B. Saunders, 1973.

Solomon, E.P., and P.W. Davis. *Human Anatomy and Physiology.* Philadelphia: Saunders College Publishing, 1983.

Journal Articles

Bennett, William, and Joel Gurin. "Do diets really work?" *Science 82,* 42, April 1982.

Brown, M.S., and J.L. Goldstein. "How LDL receptors influence cholesterol and atherosclerosis." *Scientific American,* 52, November 1984.

Davenport, H.W. "Why the stomach does not digest itself." *Scientific American,* 86, January 1972 (#1240).

Deutsch, J.A., W.G. Young, and T.J. Kalogeris. "The stomach signals satiety." *Science* 201: 165, 14 July 1978.

Fernstrom, John D., and Richard J. Wurtman. "Nutrition and the brain." *Scientific American,* 84, February 1974 (#1291).

Geller, Stephen A. "Autopsy." *Scientific American,* 124, March 1983.

Harlan, J.R. "The plants and animals that nourish man." *Scientific American,* 88, September 1976.

Kappas, Attallah, and Alvito P. Alvares. "How the liver metabolizes foreign substances." *Scientific American,* 22, June 1975 (#1322).

Kretchmer, N. "Lactose and lactase." *Scientific American,* 70, October 1972 (#1259).

Lieber, Charles S. "The metabolism of alcohol." *Scientific American,* 25, March 1976 (#1336).

Moog, Florence. "The lining of the small intestine." *Scientific American,* 154, November 1981.

Scrimshaw, Nevin S., and Lance Taylor. "Food." *Scientific American,* September 1980 (#734).

Scrimshaw, Nevin S., and Vernon R. Young. "The requirements of human nutrition." *Scientific American,* 50, September 1976.

TOPIC 23

Gas Exchange in Animals

OBJECTIVES

When you have completed this topic, you should be able to:

1. Distinguish between ventilation (breathing), gas exchange, and cellular respiration.
2. Give two examples of groups of organisms that use the general body surface for gas exchange.
3. Name some organisms that do not need special structures for gas exchange.
4. Cite one group of organisms that uses internal gills and one group that uses external gills for gas exchange.
5. Explain a tracheal breathing system, and state the group of organisms in which it is found.
6. Give the advantages and disadvantages of using air compared with water as a respiratory medium.
7. Describe the structures through which air must travel from the outside to the lungs of a mammal.
8. Explain the function of: the Eustachian tubes, turbinate bones, epiglottis, tracheal cartilage rings, alveoli, and diaphragm.
9. Explain negative pressure breathing in a mammal, describing what happens during inspiration and expiration.
10. Explain positive pressure breathing in a frog.
11. Describe how the rate of ventilation of human lungs is controlled.

TEXT REFERENCES Chapter 24 (A–D, G)

KEY TOPICS

respiratory surface	tracheal system
ventilation	positive and negative pressure breathing
gas exchange	control of human ventilation
gills	

INTRODUCTION

Cells require oxygen as the final electron acceptor in the electron transport system. CO_2 is produced in the citric acid cycle. Thus all aerobic cells need to obtain oxygen and dispose of CO_2.

Unicellular organisms can obtain enough oxygen by simple diffusion and can dispose of their CO_2 and wastes in the same way. All larger and thicker organisms utilize a respiratory system of some sort to ensure that every one of their cells has an adequate oxygen supply. The respiratory system has a large, moist **respiratory surface** across which gases can diffuse, and a means of circulating the **respiratory medium,** air or water, next to the respiratory surface. This is the process of **ventilation** or, in the case of air, **breathing.** Animals whose respiratory system is in the form of lungs, gills, and sometimes the body surface, have a circulatory system intimately associated with the respiratory system. The circulatory system carries oxygen from the respiratory surface and distributes it throughout the body. Those organisms with a tracheal system, on the other hand, distribute air directly to all body cells and do not use the circulatory system for gas exchange. Distribution of gases from the respiratory surface to and from the body's cells is **gas exchange.** Metabolism of food molecules inside cells uses up O_2 and produces CO_2 in **cellular respiration.**

GENERAL BODY SURFACE

Frogs and many invertebrates use the skin of the **body surface** as a respiratory surface.

How is this surface "ventilated"?

____

____

____

The frog uses its lungs as well as its skin for respiration.

Name some invertebrates that lack respiratory organs and use only the skin:

____

____

____

GILLS

Gills consist of feathery outpocketings that greatly increase the area of the respiratory surface. In some aquatic animals they are external and unprotected, but usually they are protected in some way.

Necturus (Axolotl)

- Examine the specimen on demonstration. In a living animal the exposed gills are pink.

Why are gills pink or red?

____

Perch

The gills are protected by a cover called the **operculum,** and water circulates from the mouth through the spaces between the gills and out through the opening of the operculum. The pattern of circulation through the gills of a fish is shown in Figure 24–6 in your text. Flow of water is opposite or **countercurrent** to that of the blood, so that the greatest possible amount of oxygen diffuses into the blood. The gills themselves are so large that a **gill skeleton** is necessary to give support to the delicate filaments.

- Examine the gill skeleton of a fish. The spiny anterior protrusions are called **gill rakers** and help keep large food particles away from the delicate gill filaments. The filaments themselves are supported by **gill rays.** If the fish is not moving, water can be pumped out through the gills by muscular action of the pharynx.

Water contains much less oxygen than air, and the fish must expend considerable energy to ventilate its gills. Being aquatic, however, it does not lose water across its respiratory surface.

TRACHEAL SYSTEM

Insects, centipedes, and millipedes obtain oxygen from air rather than from water. Their respiratory system, the **tracheal system,** is quite different from that of the lungs found in vertebrates: It does *not* depend on the circulatory system. Air enters the body through small openings called **spiracles,** which lead into the **tracheal ducts.** The ducts divide repeatedly into **tracheoles** and finally end in tiny, fluid-filled tubes where gases enter and leave the body tissues. The tiny tracheoles extend everywhere in the body, so that no cell is too far away from its air supply.

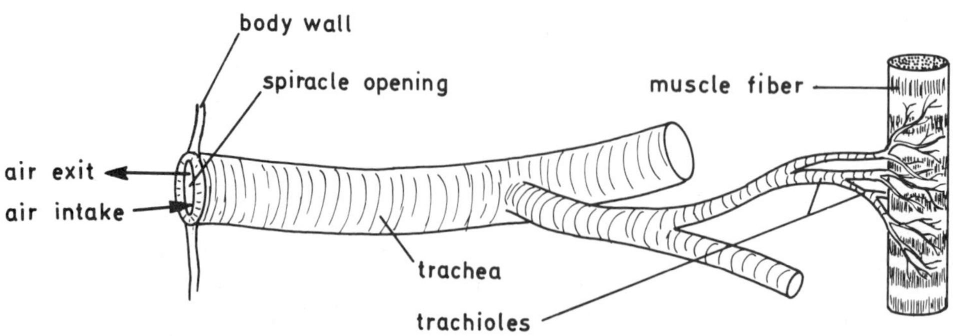

Figure 23–1. Trachea of an insect. Movement of body muscles causes the air within the trachea to circulate. Gases enter the body tissues through the fine, fluid-filled endings of the trachea.

Figure 23–1 shows how air entering the spiracles reaches a muscle cell. The circulation of air is speeded up in larger, active insects by contraction of the abdominal muscles and flight muscles. Water is lost as air circulates through the tracheal system, so drying out is a serious problem for insects.

- Examine a specimen of a grasshopper or a cockroach. Locate the spiracles along each side of your specimen. The location of grasshopper spiracles is shown in Figure 20–5.

Why does a film of oil quickly kill an insect?

____

____

Why is a strong solution of a nontoxic detergent a good insecticide?

____

____

- Remove the dorsal part of the exoskeleton by making cuts along its edges.
- Locate a **trachea,** the tube leading from a spiracle into the insect's body.
- Remove a bit of muscle, and tease it apart in a drop of water or Ringer's solution. Observe it under the compound microscope to see the fine tracheal tubes that carry oxygen to all of the muscle cells.

Why does the insect need a circulatory system if it isn't used for gas exchange?

____

____

Dispose of your specimen properly and clean your slide and dissecting tools before continuing.

THE RESPIRATORY SYSTEM OF THE FETAL PIG

In mammals, the lining of the **alveoli** within the **lungs** is the respiratory surface where gases can be exchanged (see Color Plate 4). Air is circulated to the alveoli through the **nose, pharynx, larynx, trachea,** and **bronchi** during ventilation of the lungs. Air breathers such as the pig trade the advantage of an abundant supply of oxygen in the air for the disadvantage of continuous water loss from the large, moist surface of the alveoli.

Nose

- Examine the external openings (**external nares**) of your pig's snout. Here air enters the nasal passages.
- Cut across the nose about 1 cm from the tip (see Figure 23–3) with your razor blade or scalpel, and open the cut, or look at the demonstration. Inside you see the curved **turbinate bones,** which help increase the surface area of the passages.

Why is a large surface area needed?

____

____

____

What is the role of the hairs, cilia, and mucous lining of the nasal passages?

____

____

____

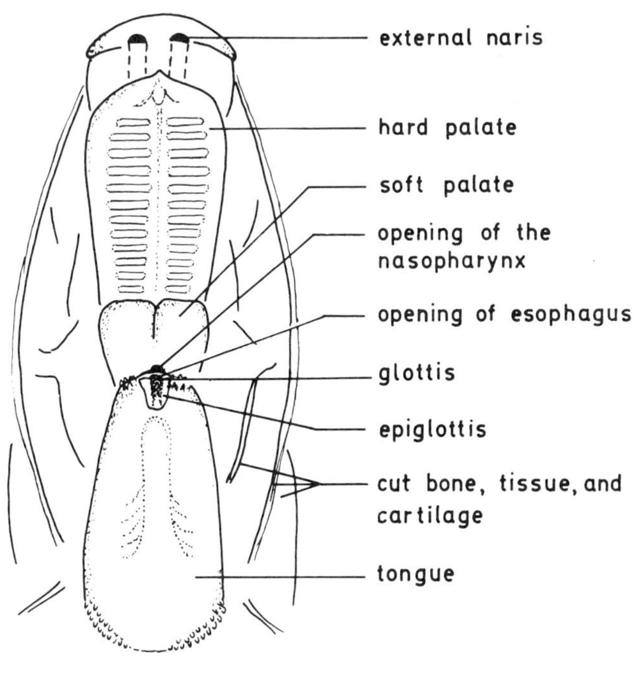

Figure 23–2. Inside the fetal pig's mouth. Cut the jaws open far enough to see the oral cavity and the oropharynx. The glottis is the opening within the folds of the epiglottis. The soft palate must be cut to find the openings of the Eustachian tubes and nasal passages.

Do comparable structures play a role in human nostrils? _______________

Pharynx

- Open the pig's mouth and review the location of the **glottis** and **epiglottis.** (See Figure 23–2.) They are located deep in the throat or oropharynx.

Air enters the glottis, the opening to the lower respiratory tract, from the opening of the nasopharynx (pharynx above the soft palate). If the epiglottis is not evident, you may have to cut more deeply and fully open the pig's mouth. Use your blunt probe to fish for it, if necessary.

If food accidentally enters or blocks the glottis, a person may choke and suffocate. Usually the food can be dislodged by using the **Heimlich maneuver.** This process involves pressing inward and upward on the abdomen just below the sternum in order to push the diaphragm upward and thus force food out of the glottis by air pressure.

- Slit the soft palate along its midline, and locate the opening of the **nasal passages.**

A pocket in this part of the pharynx has the openings of the **Eustachian tubes** that lead to the middle ear. They allow pressure within the middle ear to equilibrate with the pressure outside so that there isn't any strain on the eardrum. Nose and throat infections can spread to the middle ear through the Eustachian tube.

Larynx

- Touch the throat region of the pig to locate the hard, round larynx, or voice box.
- Make the necessary cuts to expose the throat region, making sure that the larynx is included within the exposed area (see Figure 23–3).
- Beneath the skin of the throat are muscles and glands. The muscle tissue is mostly in ribbons, with the fibers running lengthwise.
- Note the large masses of light-colored glandular tissue, the **thymus glands.**

These glands are part of the immune system, which produces antibodies as a defense against disease organisms. They are prominent early in development when the system is being established. Later in life the thymus decreases in size, and it is of little importance in the adult.

- Up near the lower jaws you may notice salivary glands under the skin.
- In the region of the larynx, use your probe or fingers to separate the muscles so that the **larynx** itself and the **trachea** posterior to it are exposed.

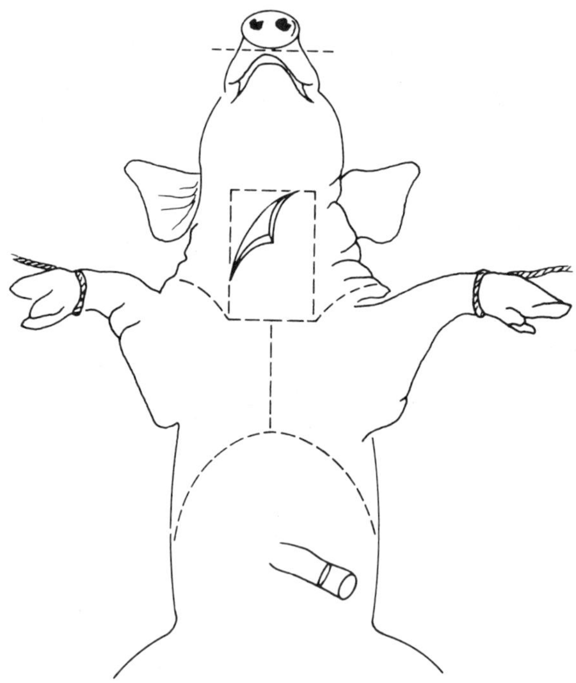

Figure 23–3. Dissection guide. Cut along the dotted lines to open the thorax and throat regions. Be careful not to damage underlying structures in the midline and shoulder regions.

Air passes from the pharynx into the glottis, through the larynx where the vocal cords are located, and through the trachea into the lungs themselves.

- Notice that the larynx and trachea are hard; the trachea contains rings of cartilage that give it support.

Do the cartilage rings completely encircle the trachea?

____

Why must the trachea be supported by cartilage?

____

____

____

The epithelial lining of the trachea secretes mucus to cleanse incoming air by trapping foreign particles. It is also covered with cilia, which constantly move the mucus toward the pharynx where it can be swallowed. (See Text Figure 24–8.)

- Ventral to the trachea near the joining of the neck and trunk, locate a smooth, brownish-colored gland, the **thyroid.** This endocrine gland secretes thyroxin, a hormone important in regulating the body's metabolism. Thyroxin controls the metabolic rate by signaling the mitochondria to respire faster or slower.
- Locate the **esophagus,** a collapsed flabby tube just dorsal to the trachea. To be sure that you locate the right structure, pass a probe into the opening of the

esophagus in the pharynx, and feel for it with your fingers beneath the trachea. Food passes from pharynx to stomach through the esophagus by peristalsis. The esophagus sometimes becomes irritated when food from the stomach backs up into it, resulting in "heartburn." Antacids relieve heartburn by neutralizing acid *in the esophagus,* not in the stomach.

- Following the dotted lines in Figure 23–3, open the thorax.

Cut carefully without damaging the underlying organs and blood vessels.

- Use your scissors to cut through the bone and cartilage in the center. Be careful that you do not cut into the heart and blood vessels just underneath.
- Locate the posterior boundary of the thoracic cavity by reaching through the cut with a finger and feeling for the **diaphragm,** the sheet of muscle and tendon that separates the thoracic cavity from the abdominal cavity.
- Then make two additional cuts, one to the left and one to the right, through the chest wall just anterior to the diaphragm. Follow the margin of the rib cage as indicated by the curved line in Figure 23–3.
- Cut toward the shoulders with care as far as necessary to fold back the thorax walls.

The thoracic cavity is divided into three parts by these membranes:

1. The **right pleural membrane** contains the right lung.
2. The **left pleural membrane** contains the left lung.
3. The **pericardial membrane** surrounds the heart.

Each is a double membrane, lining the organ and the body wall with a very thin space in between called the **pleural cavity** (around lungs) or the **pericardial cavity** (around heart). Inflammation of the pleural membranes, **pleurisy,** makes breathing painful and difficult.

- Remove the remains of these membranes so that the heart and lungs are fully exposed. Also remove and discard any thymus gland tissue that you may find in this region.

The structure of the lower respiratory system is shown in Figure 23–4.

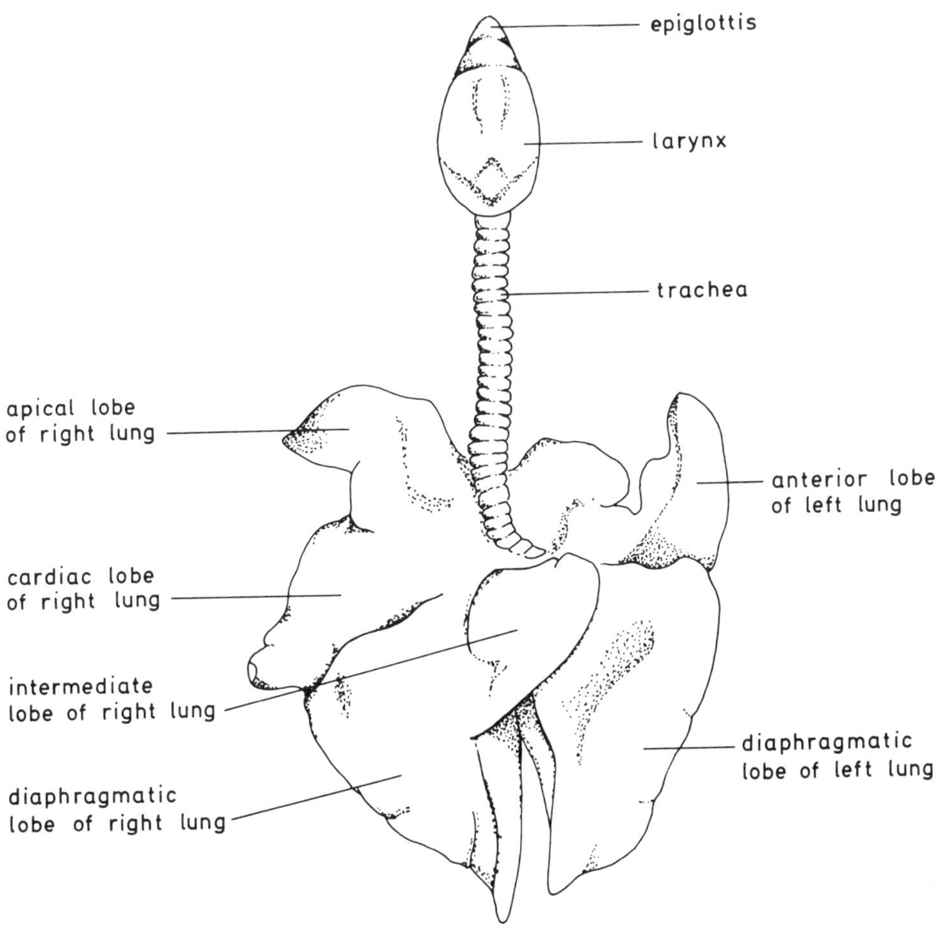

Figure 23–4. Lower respiratory tract of the fetal pig. The bronchi and bronchioles are surrounded by the spongy tissue of the lungs. In the fetal pig, the right lung has four lobes and the left has two (sometimes three).

Air from the trachea enters the **bronchi,** which divide into smaller and smaller **bronchioles.** Bronchitis is an inflammation of the bronchi and bronchioles that may follow a heavy cold.

Finally the air ends up in the microscopic, thin-walled moist sacs called **alveoli** (singular alveolus). (See Color Plate 4.) Alveoli are the real functional units of the lung. It is across the enormous surface area of their moist walls that gases actually enter and leave the body. Oxygen enters the blood capillaries by diffusion and facilitated diffusion, and CO_2 crosses from the capillaries into the alveoli. Because they are so delicate, alveoli must be coated with a surfactant to reduce surface tension so they can be easily inflated with each breath. In the disease emphysema, alveoli are destroyed and the surface area of the lung decreases.

- Examine the **lungs** of your pig. In addition to respiration, the lungs are important in oxidizing toxins that may be present in the blood.

Are the lungs equal in size? ____________

The right lung has four lobes and the left lung two or sometimes three. The lungs are filled with fluid in the fetal pig and are much more compact than they will be after they are inflated at birth. Fresh lung tissue is filled with air and pink in color due to its capillary beds. The point where the trachea divides into the right and left bronchi is hidden from view beneath the heart and blood vessels.

How does the fetal pig obtain oxygen and dispose of CO_2?

____

____

When you are finished with your dissection for today, store your fetal pig properly, clean your dissecting tools, and oil your scissors.

HOW LUNGS ARE VENTILATED

Vertebrate lungs can be ventilated in two ways. In mammals the lungs are enclosed in the airtight thoracic cavity: Enlargement of this cavity creates **negative pressure** outside the lungs so that the air rushes into them. Negative pressure is the same as suction. Air is pulled into the lungs. The frog, on the other hand, uses **positive pressure** from contraction of the oral cavity to "push" air into the lungs. Negative pressure inflation of the lungs is like sucking in on your bubble gum so that it forms a bubble inside your mouth. Positive pressure inflation is like blowing a regular bubble.

Negative Pressure Breathing

Inspiration in mammals is due to expansion of the chest cavity when the rib cage lifts and the diaphragm contracts. Muscles between the ribs and radial muscles in the diaphragm must contract for this to happen. The positions of the diaphragm and rib cage during inspiration and expiration are shown in Figure 23–5. Negative pressure breathing has the advantage that the animal need not use the muscles of the mouth and pharynx for breathing and so can eat and breathe at the same time. Also see Text Figure 24–10.

- Place one hand on your stomach just below your ribs to feel the effect of your diaphragm contracting, and place the other hand on your rib cage.
- Take a few normal breaths.

Does the rib cage move? ____________

Does the stomach move? ____________

- Try to breathe without moving the rib cage at all.
- Then try not to move your diaphragm.

Can you get enough air using only one set of muscles?

____

Breathing "from the diaphragm" is important in singing and in playing wind instruments.

Positive Pressure Breathing

The frog is a good animal to observe using positive pressure breathing because it is easy to see how it is done (see Text Figure 24–9).

- Observe the live frog on demonstration when it is out of the water.

What movement relating to breathing is most obvious?

____

A frog can open and close the nostrils and can close off the glottis leading to the lungs. The floor of the mouth can be lowered to draw in air and raised to force air into the lungs. Meanwhile, valves in the nostrils must be closed to prevent air from escaping through them while the air is under positive pressure. Air is expelled from the lungs due to their elasticity and to pressure from the abdominal organs.

- Complete Table 23–1 at the end of this topic to show what happens during inspiration in positive pressure breathing.

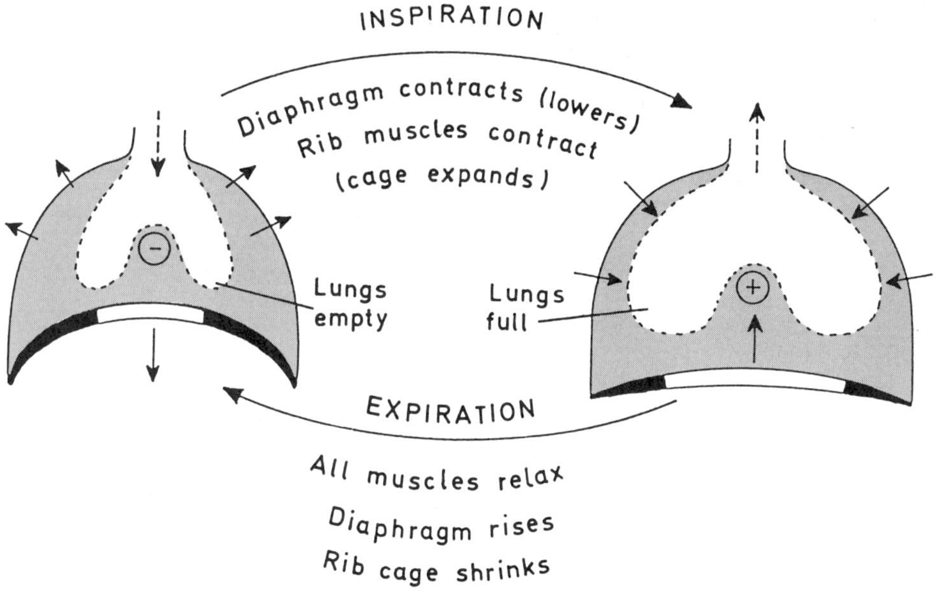

Figure 23–5. Negative pressure breathing. Contraction of rib muscles and diaphragm enlarges the chest cavity creating negative pressure (suction). Ribs rise and the diaphragm lowers while air rushes in during inhalation. Relaxation of muscles causes the ribs to lower and the diaphragm to rise during exhalation, when air is forced out of the lungs under positive pressure.

HOW THE RATE OF BREATHING IS CONTROLLED

The rate at which you breathe is automatically controlled so that there will be enough oxygen and so that CO_2 will be promptly removed. **Chemoreceptors** in the medulla (part of the spinal cord just below the brain) respond to the pH of the blood. Blood pH will decrease as the CO_2 level increases due to the formation of carbonic acid:

$$H_2O + CO_2 \rightarrow H_2CO_3 \rightarrow H^+ + HCO_3^-$$

As soon as the CO_2 level rises, the receptors detect carbonic acid and the **inspiratory center** in the medulla speeds up the breathing rate to get rid of the CO_2 (see Text Figure 24–13). *Breathing rate is normally determined by the level of CO_2* rather than that of oxygen, probably because CO_2 is easier for the body to detect. The inspiratory center shuts itself off by sending a message to an **expiratory center,** also in the medulla, via the pneumotaxic center in the pons of the brainstem. The expiratory center works by inhibiting the inspiratory center, allowing relaxation of all the muscles involved in inspiration. The inspiratory center is also inhibited by stretch receptors in the lungs that send signals over the vagus nerve to the expiratory center when the lungs are full of air. If the blood oxygen level ever becomes abnormally low, oxygen chemoreceptors in the aorta and carotid arteries stimulate the inspiratory center, but oxygen concentration does not play a role in normal breathing. The control of breathing is not perfect, and **apnea,** a spontaneous lapse in breathing may occur, particularly during sleep or under the influence of drugs.

In these experiments, stop at once if you begin to feel faint. If you have any medical problem with your heart or lungs, be a timekeeper, not a subject.

Work with a partner to make the following measurements:

- While sitting down and breathing normally, measure the number of breaths per minute three times. Record your results to the nearest half breath in Table 23–2 at the end of this topic.
- Hyperventilate by breathing as deeply and as fast as you can for 20 breaths. Then breathe normally.

Stop if you feel faint.

- Measure your breathing rate for the first minute after you resume breathing, and record it.
- Wait until you are breathing normally again before continuing.
- Breathe into a plastic bag for 2.5 minutes, and record your breathing rate for each half-minute interval. Calculate breaths per minute for each interval.

Stop if you feel faint. Do not exceed 3 min.

■ Answer the questions that follow Table 23–2.

When you are finished with your work, store or dispose of all specimens and materials, clean your dissecting tools, and store your microscope properly.

REFERENCES AND FURTHER READING

ALSO SEE TOPIC 22

Altman, P., and D. Dittmer, eds. *Respiration and Circulation.* Bethesda, MD: Federation of American Societies for Experimental Biology, 1970.

Hainsworth, F.R. *Animal Physiology.* Reading, MA: Addison-Wesley Publishing Company, 1981.

Heimlich, Henry J. *Dr. Heimlich's Home Guide to Emergency Medical Situations.* New York: Simon and Schuster, 1980.

Ruch, T.C., and H.D. Patton, eds. *Physiology and Biophysics,* Vol. 2, 20th ed. Philadelphia: W.B. Saunders, 1974.

Schmidt-Nielsen, Knut. *Animal Physiology: Adaptation and Environment,* 3rd ed. New York: Cambridge University Press, 1983.

Journal Articles

Avery, M.E., N.-S. Wang, and H.W. Taeusch, Jr. "The lung of the newborn infant." *Scientific American,* 74, April 1973 (no offprint).

Comroe, Julius H., Jr. "The lung." *Scientific American,* 56, February 1966 (#1034).

Feder, Martin E., and Warren W. Burggren. "Skin breathing in vertebrates." *Scientific American,* 106, November 1985.

Hock, R.J. "The physiology of high altitude." *Scientific American,* 52, February 1970 (#1168).

Hong, S.K., and H. Rahn. "The diving women of Korea and Japan." *Scientific American,* 34, May 1967 (#1072).

Johansen, K. "Air-breathing fishes." *Scientific American,* 102, October 1968 (#1125).

Jones, M.D. Jr. *et al.* "Oxygen delivery to the brain before and after birth." *Science* 216:324, 16 April 1982.

Kolata, G. "Cell biology yields clues to lung cancer." *Science* 218:38, 1 October 1982.

Kylstra, J.A. "Experiments in water breathing." *Scientific American,* 66, August 1968 (#1123).

Naeye, Richard L. "Sudden infant death." *Scientific American,* 56, April 1980 (#1467).

Student Name ______________________________ **Date** ____________________

TABLE 23–1. Positive Pressure Breathing in the Frog.

	Nostrils (Open/Closed)	Glottis (Open/Closed)	Mouth Floor (Raised/Lowered)
1. Air enters mouth.			
2. Air enters lungs.			

3. Finally air leaves the lungs under pressure from the stretched chest wall, and the breathing cycle is ready to start again.

TABLE 23–2. Rate of Human Breathing

Stop if you feel faint. Do not exceed 3 min breathing into bag.

Condition	Time Interval	Breaths	Breaths per Min
Normal	1st min		
	2nd min		
	3rd min		
After hyperventilation	1st min		
Breaths into bag	1st half min		
	2nd half min		
	3rd half min		
	4th half min		
	5th half min		
	6th half min		

Hyperventilation:

Did you feel an urge to breathe right after hyperventilating? ____________

Was your blood pH higher or lower? ____________

Was your blood CO_2 higher or lower? ____________

As a result, were your breaths faster or slower than normal? ____________

Does increased oxygen in your blood affect your breathing rate? ____________

Breathing into Bag:

Did your breathing become faster or slower after breathing into the bag? ____________

Why?

____

____

There was still oxygen left in the bag. What would have happened if the CO_2 had been removed from each breath that you exhaled?

____

____

TOPIC 24

Animal Transport System: Structure

OBJECTIVES When you have completed this topic, you should be able to:

1. Name four functions of the circulatory system.
2. Cite the function of each of the major parts of the mammalian circulatory system: heart, arteries, veins, lymph veins, and capillaries.
3. Describe the function of the placenta, umbilical arteries, and the umbilical vein.
4. Describe the path of blood flow through the four chambers of the heart and the five major vessels connecting to it.
5. Trace the path of a blood cell from the umbilical vein to the right atrium of the heart and back to the umbilical artery by the shortest route.
6. Name the parts of the circulatory system that have valves, and state the function of valves in regulating blood flow.
7. Describe the changes in structure and function that occur in these structures when the fetal pig is born: umbilical artery, umbilical vein, hepatic portal vein, ductus venosus, foramen ovale, pulmonary artery, pulmonary vein, ductus arteriosus.

TEXT REFERENCES Chapter 25 (B, C, E); Chapter 26 (A–E).

KEY TOPICS

closed circulatory system	arteries	capillaries
double circulation	veins	lymphatic system
mammalian heart		

INTRODUCTION

In a multicellular animal, the cells are bathed in an internal "sea" of **extracellular fluid.** If the cells are to survive, the pH, water, and salt content of this fluid must be closely regulated. It is also the medium by which food enters the cells and wastes are removed. The circulatory system of a mammal controls the composition of the extracellular fluid and maintains a favorable environment for the body cells in four major ways:

1. The transportation of nutrients, gases, and wastes.
2. The transportation of hormones.
3. The protection of the body by the immune system.
4. The regulation of body temperature.

The mammalian **heart** is a double pump that forces the blood alternately through the separate pulmonary (lung) and systemic (body) circuits of the double circulatory system. A circuit is the path that blood takes from the heart, through the blood vessels, and back to the heart. In **double circulation** blood is first sent to the lungs for oxygenation (pulmonary circuit) and then it returns to the heart to be pumped to the body under high pressure (systemic circuit). The **arteries** are the main vessels carrying blood *from the heart* to the rest of the body. The arteries end in the tiny **capillaries,** where exchange of nutrients, gases, wastes, and other substances with the tissues can occur. Finally the blood returns from the capillaries of the body *to the heart* in the large **veins.** Thus in each circuit blood travels as follows:

heart → arteries → capillaries → veins → heart

Since the blood is contained within vessels or the heart, this is a **closed circulatory system.** The blood and the capillaries are the physiologically active parts of the circulatory system. The main purpose of the heart and large arteries and veins is simply to move the blood quickly to the capillaries so that exchange can occur there. The internal structure of blood vessels is shown in Color Plate 4.

THE HEART

Study the diagram of the adult heart in Figure 24–1. This view is from the ventral side and shows the internal structure of the heart.

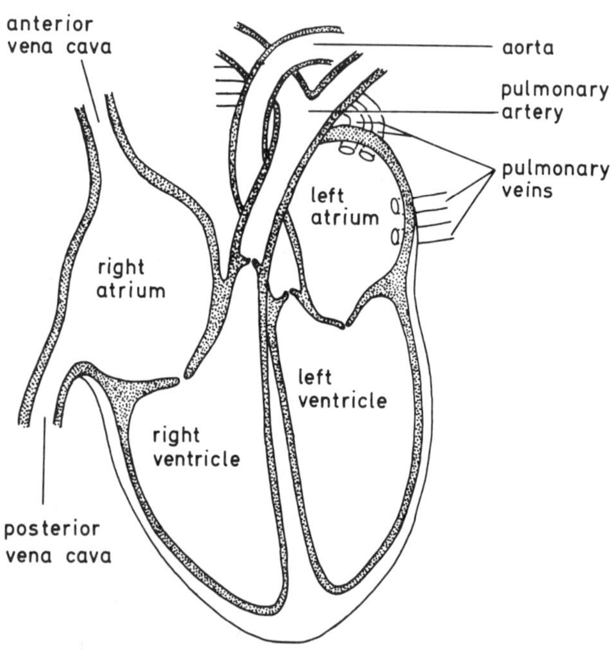

Figure 24–1. Adult heart. Blood from the anterior and posterior venae cavae flows into the right atrium, through the atrioventricular valve (tricuspid) into the right ventricle, and leaves through the pulmonary artery to the lungs. Blood then returns from the lungs in the pulmonary veins, enters the left atrium, passes through the atrioventricular valve (mitral) into the left ventricle, then leaves via the aorta.

Adult Heart

- Use the demonstration heart or a model of the heart to identify the major structures. The ventral side of the heart can be identified easily because it has a diagonal line of fat. The dorsal surface is usually flatter and has a short, straight line of fat.
- Find the ear-shaped **right atrium** and **left atrium** located on either side of the heart. The bulk of the heart consists of the **right** and **left ventricles.**
- Examine the cut across the base or apex of the heart.

Which ventricle has the greatest muscle mass?

____

The contraction of the powerful left ventricle creates a peak of high blood pressure called **systole.** When the ventricles relax, blood pressure drops to **diastole.**

- Open the cuts in the heart so that you can see these valves (also see Text Figure 25–16):
  1. The **tricuspid** valve is between the right atrium and right ventricle.
  2. The **bicuspid** or **mitral** valve is between the left atrium and left ventricle.
  3. A **semilunar** valve is at the base of the pulmonary artery. There is also a semilunar valve at the base of the aorta.
- Note the "heart strings" (chordae tendineae) that hold the flaps of the valves in place.

Normal heart sounds are due to the closing of the two sets of atrioventricular valves (1 and 2). A heart **murmur** occurs when the valves do not close completely, allowing blood to leak backwards. A severely defective heart valve, usually the mitral, can be replaced by open heart surgery.

- Note the two large arteries connecting to the heart. They are thick-walled and very muscular.

In which direction does blood flow in these arteries?

____

The more ventral artery is the **pulmonary artery,** which carries deoxygenated blood from the heart to the lungs. (Note that not all arteries carry oxygenated blood.)

To which chamber of the heart is it connected?

____

The larger, more dorsal vessel is the **aorta.** This is the largest artery in the body and contains blood under the highest pressure.

To which chamber of the heart is the aorta connected?

____

There are three major veins connected to the heart. Because they are thin-walled, they are collapsed in this empty heart.

In which direction does blood flow in each of these veins?

____

Is the blood flowing in these vessels under high pressure or under low pressure?

____

- Find the two large vessels connecting to the right side of the heart. They are the **anterior vena cava** and **posterior vena cava.**

To which chamber of the heart do the anterior and the posterior vena cava connect?

____

Finally, the **pulmonary veins** come together on the dorsal surface of the heart. They bring oxygenated blood from the functioning lungs back to the heart.

To which chamber do the pulmonary veins connect?

____

- Trace the flow of blood from the right atrium to the aorta.
- Draw arrows in Figure 24–1 to show the direction that blood flows in each vessel and chamber.

Fetal Pig Heart

- Locate the **pericardial cavity,** which contains the heart and is lined by the pericardial membrane.
- If you have not already done so, remove and discard remnants of the **thymus gland** from the throat and chest region of your fetal pig. This gland is important during the development of the immune system and is the site where the white blood cells called T lymphocytes mature.
- Carefully remove the pericardial membrane to expose the heart.
- Identify the major chambers and vessels of the heart, using Figure 24–2 as your guide; note the light-colored **ventricles** that form the posterior part of the heart and the dark-colored flaps that are the ventral parts of the **atria.** Larger portions of the atria are hidden from view.
- You can cut off the tip of the heart with a razor blade to see the difference in size of the muscles making up the right and left ventricle.

Which ventricle pumps blood out under greater pressure? ______________

Part of the blood entering the heart follows the pattern that you traced for the adult heart. In the fetal pig, however, there is an opening between the right and left atria called the **foramen ovale.** It allows some of the blood entering the right atrium to pass directly to the left atrium and out through the aorta, instead of first circulating to the fetal lungs.

- Push the heart posteriorly and clean off pieces of thymus tissue and membrane so that you can see clearly the vessels that arise from the heart.

1. The most conspicuous vessel is the thick-walled, light-colored **pulmonary artery.**
2. Beneath is the slightly larger **aorta,** which is also thick-walled and light-colored.
3. The pulmonary artery is connected to the aorta by the short, thick **ductus arteriosus.** In the fetal stage, most of the blood leaving the heart in the pulmonary artery by-passes the right and left branches of this artery and instead passes through the ductus arteriosus and into the aorta directly.

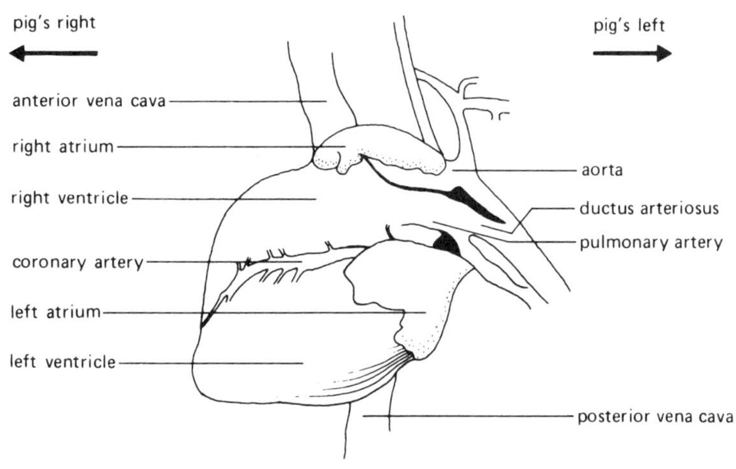

Figure 24–2. Fetal pig heart. Ventral view of its external structure. The thick white pulmonary artery is the most prominent vessel leaving the heart.

Why are the lungs by-passed before birth?

____

____

____

4. Returning to the anterior part of the heart, the large, thin-walled, dark or blue-colored vessel on the right is the anterior vena cava. It brings blood from the anterior part of the body to the right atrium of the heart. Lift the posterior tip of the heart to see the pulmonary veins, which join together on the dorsal surface of the heart and enter the left atrium.
5. On the right side of the heart is a large blue vessel, which is the posterior vena cava and which joins the anterior vena cava in entering the right atrium.

ARTERIES

Arteries are all the vessels carrying blood away from the heart. Remember that blood carried by arteries is sometimes rich in oxygen and sometimes poor in oxygen. If you have an injected pig, the arteries have been injected with pink latex so they are easily seen and resistant to breakage. The veins have been injected with blue latex so that you can tell the difference, but sometimes the injection results are not perfect and a certain vessel, the umbilical vein or artery, for example, may be the "wrong color." Ask your instructor for help if you have trouble locating a certain artery or vein. Use Figure 24–3 to identify your pig's arteries.

In cross section, arteries have thick walls that are muscular and have elastic connective tissue. Arteries lose their elasticity with age, leading to **arteriosclerosis.** In the most common form of this disease, **atherosclerosis,** their delicate endothelial lining becomes covered with lipid deposits. Narrow, hard arteries are likely to become plugged and then rupture because arterial blood is under very high pressure. When this happens in the brain, nerve cells are destroyed, resulting in a **stroke.**

Pulmonary Artery

- Locate the branches to the right and left lungs; lift the heart to see them.

What is the main vessel to which the pulmonary artery joins in the fetal pig?

____

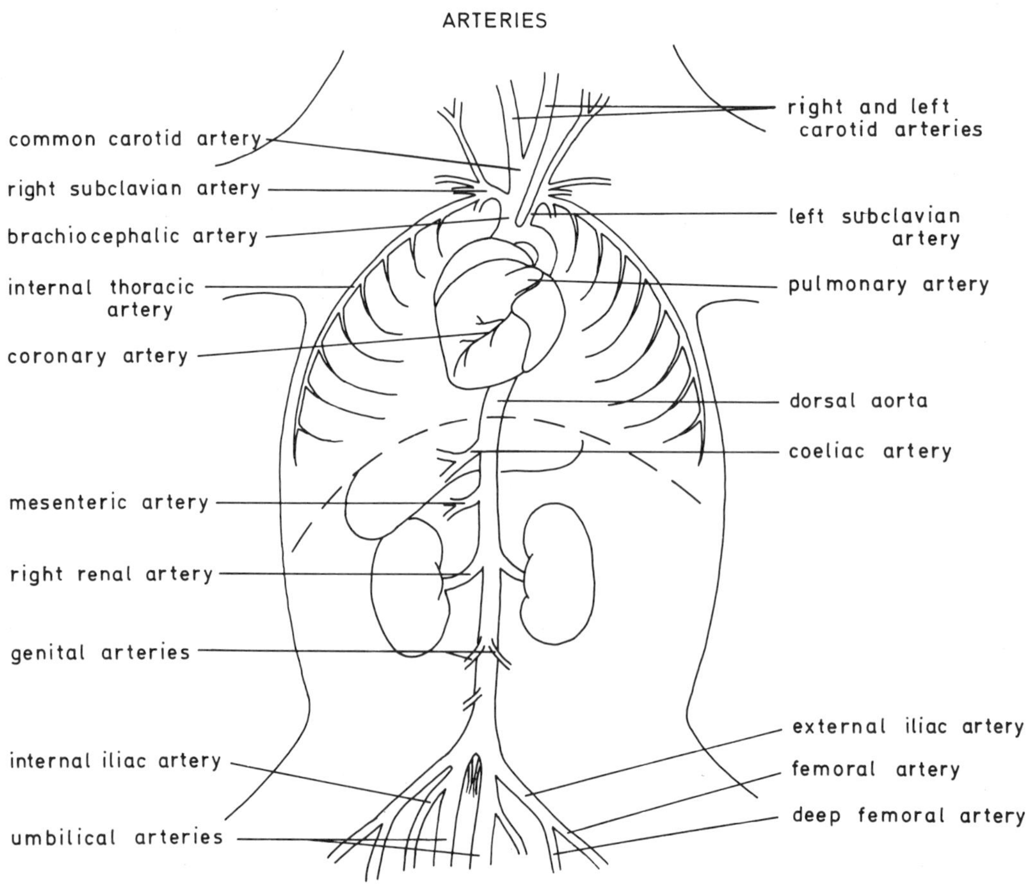

Figure 24–3. Arterial system of the fetal pig. In this view from the ventral side, the aorta arches dorsally, passes beneath the heart, and becomes the dorsal aorta deep within the thorax. The aorta divides to form the prominent umbilical arteries.

Aorta

- Note that this vessel arches as it leaves the heart, passes down in the thoracic cavity to the spine, and then runs posteriorly all the way to the umbilical arteries. It is the major artery in both the fetal and adult pig and contains blood under the highest pressure.
- Locate the following vessels that branch off from the aorta as it passes through the body (see Figure 24–3).

1. **Coronary Artery.** The first vessel to branch from the aorta. It is difficult to see the branch point even when you lift the atrium and look closely. The vessel itself is easy to see on the ventral surface of the heart. This is the vessel that becomes blocked during a heart attack. The blockage may be due to a blood clot (coronary thrombosis) or spasms of the artery (angina). As a result, the heart muscle gets insufficient oxygen and part of the heart muscle tissue may die. If the heart actually stops pumping, cardiopulmonary resuscitation (CPR) may keep the body oxygenated long enough for the damaged heart to be stimulated to resume functioning. If the damage to the heart is extensive, it cannot continue to function and the heart attack will be fatal.
2. **Brachiocephalic Artery.** A large branch that divides into the **common carotid artery** and the **right subclavian artery.** Trace the common carotid to its branch point into the right and left carotids, and trace one of the carotids into the throat region. This artery serves the head and neck. The right subclavian sends blood to the right forelimb. The **internal thoracic arteries** send blood to the chest and ribs.
3. **Left Subclavian Artery.** The branch that leaves the thorax and enters the left forelimb.

- Push the lungs to the pig's right, and locate the **dorsal aorta** deep in the thoracic cavity next to the vertebral column. Notice that it is still thick-walled and light-colored. Its walls are so thick that the color of the pink latex is not visible. Next to the aorta is the collapsed, flabby, tubular esophagus.
- Trace the aorta and esophagus posteriorly as they pass through the diaphragm and enter the abdominal cavity.

4. **Coeliac Artery.** Branches from the aorta just posterior to the diaphragm and serves the stomach, liver, spleen, and pancreas. Blood pools in the spleen, which is an important part of the circulatory system. It acts as a filter to remove bacteria and dead cells that are ingested by phagocytic white blood cells.
5. **Mesenteric Artery.** The branches of this artery send blood to the small intestine and to the anterior part of the large intestine.
6. **Renal Artery.** The right and left renal arteries send blood to the kidneys. Occasionally you might see a renal artery that is double instead of a single vessel.
7. **Genital Artery.** A pair of small genital arteries that branches next, to supply the testes in the male or the ovaries and uterus in the female.
8. **External Iliac Artery.** A pair of large arteries that supplies the hindlimbs. After each gives off a small branch, it branches into the **femoral artery** and **deep femoral artery.**
9. **Internal Iliac Artery.** The aorta ends at the spot where the external iliac branches off and it becomes the internal iliac artery for a short distance. The internal iliac in turn splits to form the umbilical arteries.
10. **Umbilical Arteries.** These large arteries arise from the internal iliac and carry blood from the fetus to the placenta. Blood in these vessels would be very low in oxygen and nutrients but high in CO_2 and wastes. (Remember that arteries sometimes carry oxygen-rich blood and sometimes oxygen-poor blood. There is no point in carrying oxygen-rich blood to the placenta where oxygenation of fetal blood is carried out.)

What major artery in your own body carries blood poor in oxygen?

____

You have traced the major arteries as an introduction to the complexity of the circulatory system anatomy. Each artery divides repeatedly into smaller **arterioles** and finally becomes a **capillary bed.** The capillaries join together into larger and larger **venules** and finally return blood to the large veins entering the heart.

Cover your pig with moist paper towels during the next part of the laboratory.

CAPILLARIES

The only purpose of the heart and all of the blood vessels is to provide a constant supply of fresh blood to the body's **capillaries.** The preserved capillaries in your fetal pig are much too small to be seen without special preparation for microscopy. Instead you should look at some living capillaries in which blood is actually circulating to get a feeling for the intense activity that goes on there. (See Text Figure 25–8.)

Here in the capillaries gases and molecules are exchanged between the blood and the **extracellular fluid** in which the tissues are bathed. Because the blood is still under fairly high pressure when it enters the capillaries, some of the noncellular fluid (plasma) is forced out into the intracellular spaces around the capillaries so the area for exchange is increased even more. Migratory white blood cells also leave the blood and enter the tissues. Blood loses most of its pressure as it passes through the capillaries, so blood leaving the capillaries in the venules and veins is under very low pressure. Excess extracellular fluid drains into the **lymphatic system** and is eventually returned to the circulatory system via large lymph vessels.

- Anesthetize a frog according to the directions given by your instructor.
- Always keep the animal wet with tap water because its moist skin is its major organ of respiration.
- Wrap a wet paper towel around the frog, leaving the feet exposed, and fasten the animal to a frog board with rubber bands.
- Spread the foot web over the hole in the frog board so that the toes are wide apart, and pin the foot down with fine insect pins, or use a toe clamp.
- Place the board on the stage of your microscope so that the hole in the board is lined up with the hole in the stage. Your setup should look similar to the one shown in Figure 24–4.
- Turn on the microscope light, and focus on the foot webbing using scan or low power. You will see lots of star-shaped dark cells: These are the melanocytes that can change shape and alter the frog's color. You will also see a network of fine arterioles and capillaries with rapidly moving blood cells. (If you do not see movement, ask your instructor for help immediately.)
- Taking care not to scratch the objective with the pins, switch to high power and observe the capillary bed more closely.

How wide is the smallest capillary compared with the width of a red blood cell (7–8 μm)?

____

- Observe the small capillaries.

Does the blood always flow in the same general direction?

____

Is the blood flowing in all of the capillaries or only in some?

____

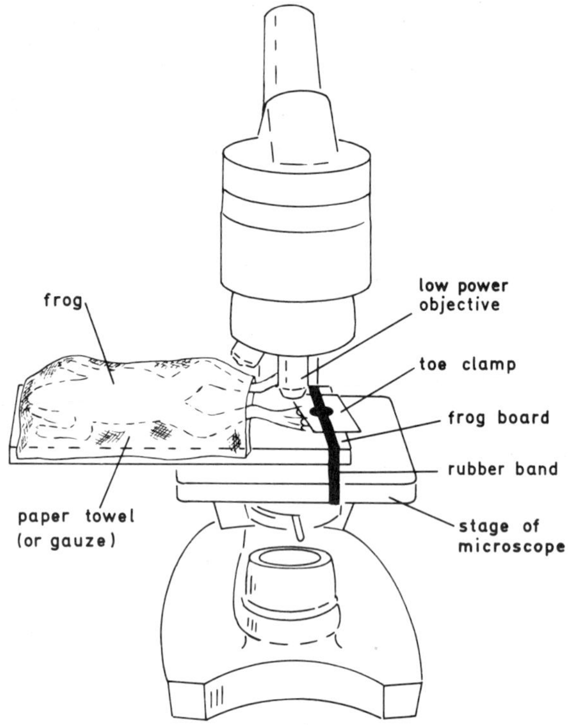

Figure 24–4. Frog capillary setup. Keep the frog and the delicate web between its toes moist.

Entrance of blood into a capillary is regulated by a precapillary sphincter muscle. Sphincters in the skin capillaries, for example, regulate body temperature by determining how much blood passes close to the cool skin surface.

- When you are finished, free the frog and return it to the recovery tank.

VEINS

Veins are thin-walled and contain much of the blood. Because the blood in them is under very low pressure, veins have **valves** so that blood can flow only toward the heart (see Text Figure 25–9). Movement of the blood is accomplished by pressure of skeletal muscles contracting against the veins. Blood will pool in the legs due to gravity if there is insufficient physical activity. Because veins are not as resilient as arteries, with time they may become stretched out and swollen. They are then called **varicose veins** (e.g. in the legs) or **hemorrhoids** when they occur in the rectum.

Lymph Veins

Lymphoid tissue is found in the lymph nodes,

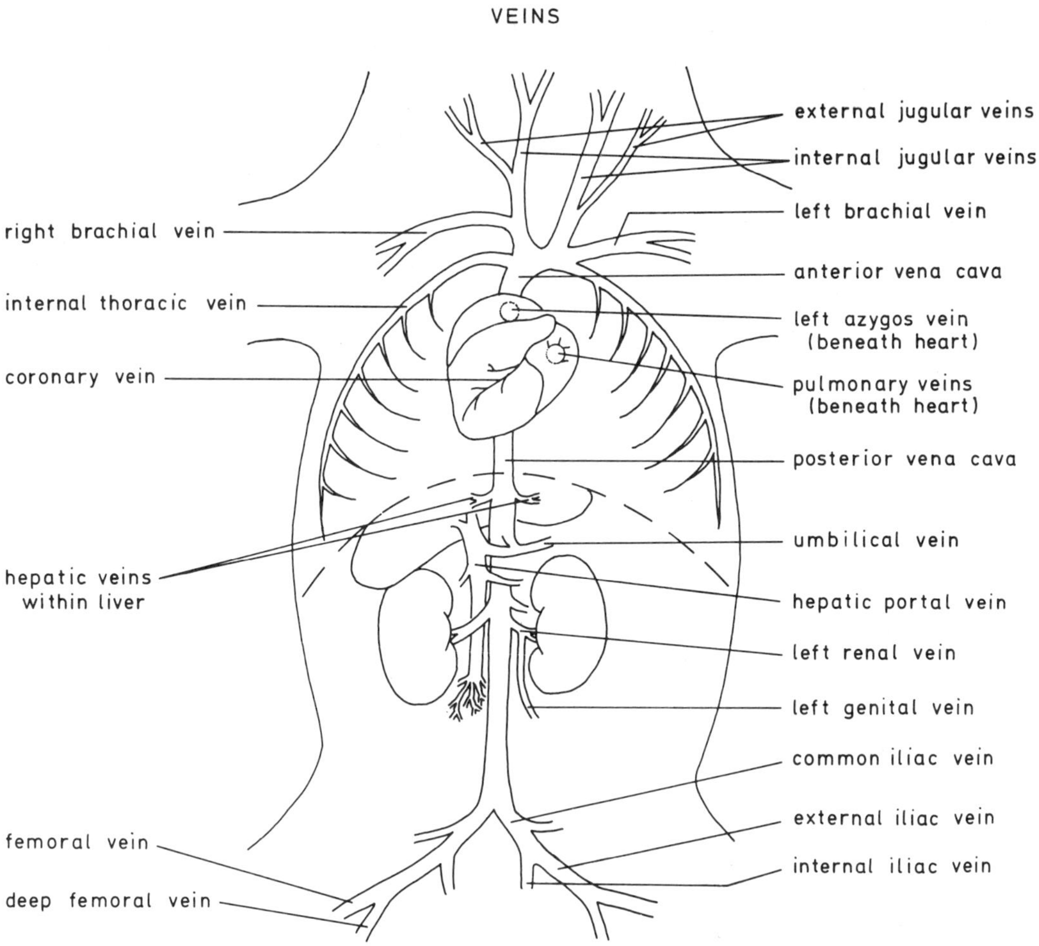

Figure 24–5. Venous system of the fetal pig. In this ventral view, the major veins leading to the heart are the venae cavae. The pulmonary veins are beneath the heart, and the hepatic veins are hidden within the liver.

spleen, liver, linings of the gut and respiratory tracts, tonsils, and adenoids. It is where most of the body's white blood cells are found and therefore is the basis of the **immune system.** The lymphatic system also includes a set of **lymph veins** that drain excess extracellular fluid called **lymph** from dead-end lymphatic capillaries and eventually empty back into veins of the circulatory system through the thoracic duct (left side) and the right lymph duct (right side). In this way plasma lost from the blood through the capillaries is returned from the tissues to the bloodstream. Along the way, lymph passes through **lymph nodes** where dead or foreign cells and debris are removed by phagocytic white blood cells (see Text Figure 35–6). Lymph transports protein and fat globules that are too large to pass through capillaries and enter the bloodstream. It can also temporarily take up excess water from the gut.

As in other veins, the flow of lymph is under low pressure created by muscular movement and it is in one direction because lymph veins have one-way **valves.**

Pulmonary Veins

Blood from the right and left lungs drains into the branches of the pulmonary veins. If the lungs are functioning, this blood will be rich in oxygen. You have already seen where the pulmonary veins join and enter the left atrium.

Do the pulmonary veins carry as much blood before birth as they do after birth? ______________

The **azygos vein** is also beneath the heart and drains blood from the chest wall into the right atrium. The rest of the venous system is shown in Figure 24–5.

Anterior Vena Cava

- Locate the vessels leading into the anterior vena cava:

1. **Coronary Vein.**

2. **Internal Jugular Vein.** Parallel to the common carotid, and drains the head and neck region.
3. **External Jugular Vein.** Joins the internal jugular close to the anterior vena cava.
4. **Brachial Veins.** The main vessels draining the forelimbs. They join with the external jugulars before entering the anterior vena cava. Lymph from the thoracic (left) and right lymph ducts enters the bloodstream here.
5. **Internal Thoracic Veins.** Carry blood from the chest and ribs.

Posterior Vena Cava

■ Locate the vessels draining the posterior portion of the body, which join the posterior vena cava and include:

1. **External Iliac Vein.** Drains the anterior part of the hindlimb via the **femoral veins.**
2. **Internal Iliac Vein.** Drains the posterior part of the hindlimb and joins the external iliacs to form the common iliac. The two common iliacs join to form the beginning of the posterior vena cava.
3. **Genital Veins.** Small veins that lead into the posterior vena cava and left renal vein.
4. **Renal Veins.** Drain blood from the right and left kidneys.
5. **Hepatic Veins.** Veins within the liver that drain into the posterior vena cava just below the diaphragm. (They are hidden within liver tissue and should not be confused with the hepatic portal vein described below.)
6. **Umbilical Vein.** When you originally opened the abdomen, you cut this vessel, which leads from the umbilical cord into the liver (use the string to find the cut ends); one of its branches leads into the hepatic portal system, and another branch, the ductus venosus, leads directly into the posterior vena cava.
7. **Ductus Venosus.** In the fetal pig, part of the blood in the umbilical vein branches off in this vessel and goes directly into the posterior vena cava. This allows some blood rich in oxygen and nutrients to be pumped directly out to the body without passing through a capillary bed (see Figure 24–6).

Does the umbilical vein carry blood rich or poor in oxygen? ______________

Why does the umbilical vein carry much of the blood from the placenta directly to the liver?

____

____

Hepatic Portal System

A portal system connects two capillary beds, in this case intestinal capillaries and liver capillaries. Blood flows as follows:

heart → arteries → intestinal capillaries
→ **hepatic portal vein** → liver → capillaries
→ veins → heart

This allows the liver to efficiently process nutrient-rich blood after digestion of a meal. The structure of the hepatic portal system is shown in Figure 24–6.

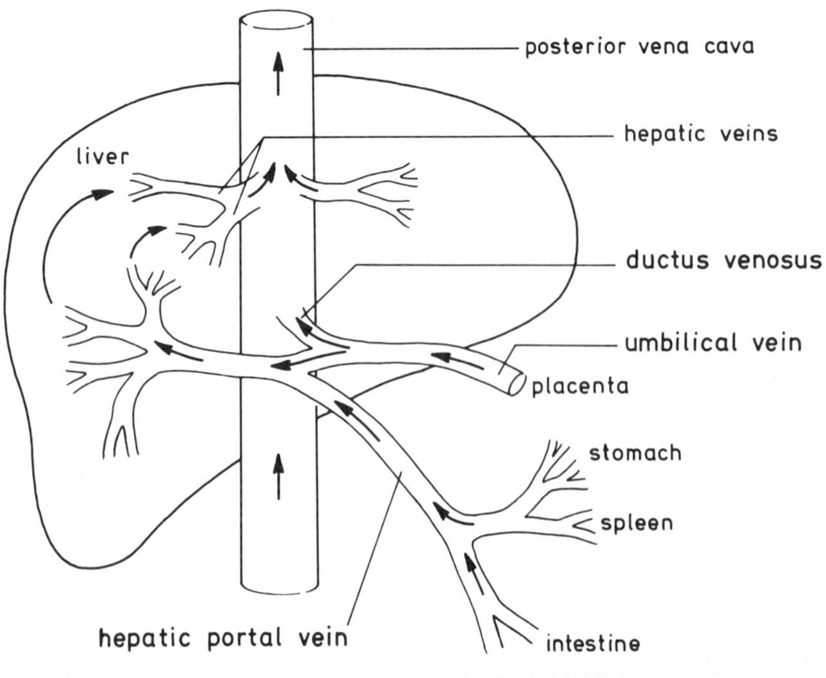

Figure 24–6. Hepatic portal system. The hepatic portal system drains from the intestine, stomach, and spleen through the hepatic portal vein into the capillary beds of the liver. In the fetal pig, where the placenta rather than the intestine provides nutrition, the umbilical vein empties partly into the portal vein and partly through the ductus venosus directly into the posterior vena cava.

■ Gently separate the liver and stomach from the small intestine and locate the **hepatic portal vein** draining blood from within the mesentery of the small intestine into the liver. It may not be injected with blue latex in your pig—if not, find it in another specimen. It is helpful to note that this vein does *NOT* end in the posterior vena cava, in contrast to other veins in the region.

Veins from the small intestine, large intestine, stomach, and spleen join to form the hepatic portal vein, which carries nutrient-rich blood to the liver after birth. During fetal development the gut has little function, so the hepatic portal system is not important in transport between gut and liver.

How do nutrients enter the fetal circulation?

____

____

____

____

TRANSFORMATION OF FETAL TO ADULT CIRCULATION AT BIRTH

At the time of birth the connection of the fetus with the placenta via the umbilical cord is broken. The umbilical vein no longer brings blood rich in oxygen and nutrients to the liver, and the umbilical artery does not carry spent blood to the placenta to pick up more oxygen and nutrients and exchange wastes with the maternal circulation.

The fetal circulation responds to these changes in the following way:

1. **Lungs.** When the lungs are ventilated, the pulmonary circuit becomes functional for the first time. Blood leaves the right ventricle through the pulmonary artery, exchanges oxygen and CO_2 in the capillaries of the lungs, and returns to the left atrium via the pulmonary veins. It is now extremely harmful to by-pass the pulmonary circuit. Therefore the foramen ovale seals shut and the ductus arteriosus closes and degenerates.
2. **Digestive System.** Nutrients enter the gut as the piglet suckles and are carried to the liver by the hepatic portal system which thereby replaces the umbilical vein. The ductus venosus no longer has any function, and both it and the umbilical vein degenerate.
3. **Kidneys.** The umbilical arteries and vein can no longer transport wastes to the placenta for removal from the fetal blood. Therefore the kidneys now have to fully assume that function. The renal arteries and renal veins become fully active.

When you are satisfied with your understanding of the circulatory system, return your pig to its proper storage place or container. Clean your dissecting pan and instruments, and oil your scissors.

REFERENCES AND FURTHER READING

ALSO SEE TOPIC 22

Singer, C.A. *A Short History of Anatomy and Physiology: From the Greeks to Harvey.* New York: Dover, 1957.

Journal Articles

Jarvik, Robert K. "The total artificial heart." *Scientific American,* 74, January 1981.

Johansen, Kaj. "Aneurysms." *Scientific American,* 110, July 1982.

Kaplan, Norman, M. "The control of hypertension: A therapeutic breakthrough." *American Scientist* 68: 537, Sept.–Oct., 1980.

Mayerson, H.S. "The lymphatic system." *Scientific American,* 80, June 1963 (#158).

Schmidt-Nielsen, Knut. "Countercurrent systems in animals." *Scientific American,* 118, May 1981.

Stallones, Reuel A. "The rise and fall of ischemic heart disease." *Scientific American,* 53, November 1980.

Winfree, Arthur T. "Sudden cardiac death: A problem in topology." *Scientific American,* 144, May 1983.

Wood, J.E. "The venous system." *Scientific American,* 86, January 1968 (#1093).

Zweifach, B.W. "The microcirculation of the blood." *Scientific American,* 54, January 1959 (#64).

TOPIC 25

Animal Transport System: Function

OBJECTIVES

When you have completed this topic, you should be able to:

1. List the major proteins present in blood plasma and give their functions.
2. List the major types of blood cells in human blood and give their functions.
3. Explain the function of hemoglobin and why it is found within cells rather than free in the blood.
4. Define blood pressure and demonstrate how to measure it.
5. Define pulse rate and describe how it is regulated by the nervous system.
6. Demonstrate how the beating of the frog heart is regulated by neurotransmitters.

TEXT REFERENCES Chapter 24 (E–F); Chapter 25 (C–E); Chapter 26 (B–F); Chapter 29 (B, C, G, H).

KEY TOPICS

amphibian circulatory system	blood composition	synapses
heart cycle	blood clotting	neurotransmitters
blood pressure	lymphatic system	vagus nerve
exercise	immune system	

INTRODUCTION

In Topic 24 you studied the anatomy of the circulatory system and heart from the point of view of transporting nutrients, wastes, and gases around the body. The circulatory system actually has other important functions. Review its functions by listing the four major ones below:

1. ______________________________

____

2. ______________________________

____

3. ______________________________

____

4. ______________________________

____

Several structures other than arteries, veins, capillaries, and the heart are important in the functioning of the circulatory system. The fluid that is lost in the capillaries drains back into the circulation through the **lymphatic system,** which also transport fats that are poorly soluble in water in its lymph vessels, and includes the lymphoid tissue and lymphocytes of the **immune system.** The **spleen** contributes white blood cells to the immune system and helps cleanse the blood of debris. The **bone marrow** forms red and white blood cells, and the **liver** and **thymus** contain cells of the immune system. The circulatory system does not act alone, but is closely involved with many other organs.

COMPOSITION OF THE BLOOD

Plasma

The liquid part of the blood, called the **plasma,** contains dissolved nutrients, wastes, salts, and the blood proteins, including albumin, globulins, fibrinogen, and lipoproteins. **Albumin** helps maintain the osmotic strength of the blood, and contributes to its buffering capacity. The **gamma globulins,** one group of immunoglobulins, are the enormous variety of antibodies circulating in the blood at all times. **Complement** proteins in the blood assist the immunoglobulins in destroying bacteria. **Fibrinogen** is the structural protein from which clots are formed. It is split by another blood protein, the enzyme thrombin, during the chain reaction of the clotting process. (See Text Figures 25–15 and 25–16.) Plasma from which fibrinogen has been removed is called **serum. Lipoprotein** binds to and transports insoluble lipids such as triglycerides and cholesterol.

Blood plasma is a very good buffer and maintains the pH of blood at around 7.4.

What ion determines the pH of a solution?

____

Why is it important that pH be kept relatively constant?

____

____

____

The following experiment demonstrates the buffering capacity of blood. Record your results in Table 25–1 at the end of this topic.

- Obtain two 100-mL beakers. Label them #1 and #2.
- Add 10 mL of 0.9% NaCl to beaker #1.
- Add 1 mL of blood plasma and 9 mL of 0.9% NaCl to beaker #2.

Avoid the risk of serum hepatitis: Do not pipet plasma by mouth, and avoid contamination if human plasma is used.

- Test the pH of each solution by wetting a glass rod in it, and touching a drop to a piece of pH paper; if pH sticks are used, place one stick into each beaker.
- If the pH values do not agree, adjust the pH of the solution in beaker #1 by adding 0.05 M HCl or 0.05 M NaOH one drop at a time. Stir the solution and test its pH after adding each drop.

If the pH of the solution in the beaker is too high, should you add HCl or NaOH? ______________

- When both solutions have the same pH, record the value in the table as **initial pH.**
- Use a pipet to restore the volume of the solution in beaker #1 to 10 mL. Discard the excess solution.
- Add five drops of 0.1% methyl orange indicator to each beaker.
- *While counting the drops,* add 0.1 M HCl to beaker #1, one drop at a time, stirring the solution with a glass rod, until there is a definite and *permanent* color change from orange to pink. (This change occurs at a certain pH.)
- Measure the pH of the solution and record the value under **final pH.**
- Repeat the procedure for beaker #2, counting the drops needed to reduce the pH to the same value as you did for beaker #1, so that *both solutions are the same color.*

Which solution was more resistant to a change in pH upon addition of acid? ______________

How does a buffer react chemically upon the addition of acid or base?

____

____

____

Formed Elements

The three major types of blood cells are the **red blood cells** (erythrocytes), the **white blood cells** (leukocytes), and the **platelets.** Some of these types are illustrated in Figure 25–1 and Color Plate 1. To view some of your own blood cells, use the following procedure, illustrated in Figure 25–2.

- Obtain two clean slides, alcohol, sterile cotton, and a sterile blood lancet in an unopened wrapper.

Avoid the risk of serum hepatitis: Never use a previously used lancet. Do not touch materials contaminated with another person's blood.

If you are unable or unwilling to have your finger pricked, work with a partner who doesn't mind, or view a prepared, stained slide.

- Massage a finger on your left hand if you are right-handed, and ask your partner to wipe it with alcohol and prick it for you or do it yourself.
- Drip a drop of blood onto one slide.
- Hold the second slide almost vertical and move it along until it just touches the drop of blood. The drop will spread out along the edge of the second slide.

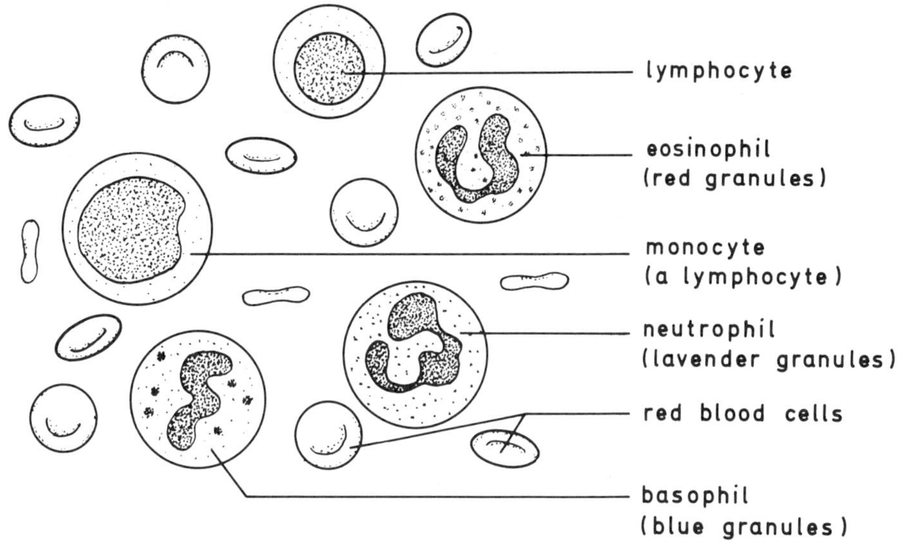

Figure 25–1. Human blood cells. White blood cells are very rare compared with red blood cells (1 : 700).

- Keep the second slide at an angle, and move it back to smear out the blood into a thin layer. Discard the slide that you used for spreading.
- Allow the smear to dry, and place the slide on a staining rack.
- Add enough drops of Wright's stain to cover the surface of the slide, and let it stand for 6 min.
- Add 10 drops of distilled water, and let it stand for 5 min.
- Finally rinse the slide with distilled water from a squeeze bottle to remove the dye. Rinse it with more distilled water at the sink, and stand it on end to dry.
- Examine the dry slide under high power with the compound microscope.

Most of the cells are the pink-staining red blood cells, which have the form of biconcave discs (donuts without the holes).

Do human red blood cells have nuclei?

____

Where are these cells formed?

____

What is the respiratory pigment contained within them?

____

Red blood cells have a limited life span (about 120 days) because they have no nuclei. Old red blood cells are phagocytized in the spleen, and their hemoglobin is degraded in the liver and leaves the circulation as the bile pigments. Much of the iron can be recycled. Insufficient iron intake is one cause of **anemia,** a deficiency of red blood cells in which their concentration falls below the normal level of about 5×10^9 cells per mL. **Polycythemia** is an abnormally large concentration.

Less than 1% of the cells are white blood cells. The larger ones are the **granulocytes** and have a grainy cytoplasm. Cells with light pink granules are the **neutrophils.** They phagocytize bacteria and dead body cells. Those with large deep red or orange granules are the **eosinophils,** which increase in number during allergic reactions or during parasitic infections. Cells with closely packed blue-violet granules are **basophils** (very rarely seen). They release histamine in response to allergens or tissue injury.

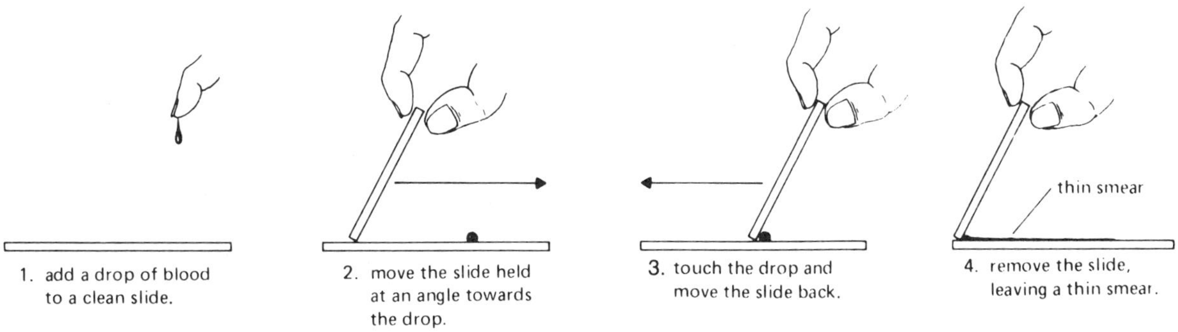

Figure 25–2. Blood smear preparation. After the smear has dried, stain the cells with Wright's stain.

Other white blood cells are **agranulocytes** and have a clear cytoplasm. Small cells with only a thin ring of cytoplasm are the **lymphocytes,** which are important in the response of the immune system. **Mononucleosis** is a disease, probably caused by a virus, that results in an increase in atypical, large lymphocytes. Normal cells with a larger amount of cytoplasm are **monocytes.** They migrate into the tissues where they mature into phagocytic **macrophages.**

- Locate 20 or 30 white blood cells, and note whether they are granulocytes (grainy cytoplasm) or agranulocytes (clear cytoplasm).
- Try to identify at least one example of each of the white blood cell types shown in Figure 25–1.

Blood is unique in that it is a liquid tissue and so requires the rapid repair mechanism of clotting whenever blood vessels are damaged. **Platelets (thrombocytes)** are very important in that they initiate the early stages of blood clotting by disintegrating at the site of injury. They are fragments of giant cells found in the bone marrow and are much smaller than white blood cells. Since they are very fragile, they were probably destroyed during the staining procedure.

Hemoglobin

The concentration of hemoglobin in red blood cells is very high. That the hemoglobin is packed into these cells is advantageous because otherwise the osmotic concentration of the blood would be extremely high. The environment within the cells can also be closely regulated to control the binding of oxygen to hemoglobin. It is not known why mammalian red blood cells, in contrast to those of other vertebrates, have a biconcave shape and lack nuclei.

HEMOGLOBIN CONTENT

Use the following procedure to measure the hemoglobin content of your blood.

- Obtain two spectrophotometer tubes and add 5 mL of distilled water to each one.
- Wipe your finger with alcohol and carefully prick your finger as you did before (or have your partner take a turn at being pricked).
- Draw up exactly 0.02 mL of blood in the capillary pipet provided for this purpose.
- Blow the blood into the water in one of the tubes, and rinse three times with water from the tube.
- Add 1 drop of concentrated ammonium hydroxide (NH_4OH) to the tube and to a second blank tube.
- Wait 10 min.
- Meanwhile, turn on a Spectronic 20 spectrophotometer to warm up, and set the wavelength at 545 nm.
- With the sample chamber empty, adjust the **absorbance** to infinity (left knob).
- Then use the blank tube to adjust the **absorbance** to zero (right knob).
- Determine the absorbance of the solution at 545 nm. Record this value in Table 25–2 at the end of this topic.
- Calculate the approximate concentration of hemoglobin in your blood by multiplying the absorbance at 545 nm by the factor 28.8. (Hemoglobin concentration is directly proportional to absorbance over a certain range.)

The normal range for hemoglobin concentration is 16 ± 2 g/100 mL for males and 14 ± 2 g for females. Indicate in the table whether your hemoglobin concentration is in the low, normal, or high range. **Anemia** is the condition of having an abnormally low concentration of hemoglobin and it can be due to a number of different causes, including insufficient iron intake.

SICKLE CELL HEMOGLOBIN

Persons with sickle cell anemia have hemoglobin S rather than normal hemoglobin. This hemoglobin contains beta chains with a single amino acid substitution in which glutamic acid is replaced by valine. As a result the hemoglobin has a lowered affinity for oxygen, and the molecules tend to clump together when the concentration of oxgyen is very low. During the clumping reaction, the red blood cells are distorted and assume "sickled" shapes.

If you did not study sickle cell anemia in Topic 12, do so now. Read the section entitled SICKLE CELL ANEMIA, and complete the exercise at the end of that section.

BLOOD PRESSURE

Blood pressure varies throughout the circulatory system, so it is usually measured in the upper arm for comparison of different individuals. The pressure is the hydrostatic pressure exerted on the walls of the blood vessels by the blood under the force of the contracting heart. It is greatest in the vessels adjacent to the heart, it drops off in the arterioles and capillaries due to the friction caused by the blood moving along the walls, and it is very low in the venules and veins leading back to the heart. Muscular contraction is required to squeeze the blood in the veins from the extremities to the heart because the pressure in these vessels is less than the force of gravity. **Valves** in the veins are extremely important in aiding the return of the blood to the heart, because its pressure is so low.

The blood pressure in the arteries follows a rhythmic cycle with each beat of the heart. When the ventricles contract, exerting the strongest pressure, the resulting high-pressure phase is termed **systole.** While the ventricles of the heart relax, the pressure in the arteries decreases in a low-pressure phase called **diastole.** Blood pressure is always measured at both systole and diastole, and the result is reported as the systolic value/diastolic value. The blood pressure reading is generally higher in older people and lower in people in good physical condition or in athletic training.

Working in pairs, use this procedure to take your partner's blood pressure while he or she is seated. Record your results in Table 25–3 at the end of this topic.

- Wrap the blood pressure cuff around your partner's upper arm so that it is neither tight nor loose, and secure it with the Velcro fasteners.
- Squeeze the bulb several times while watching the pressure gauge until the pressure reaches 180–200 mm Hg (760 mm is 1 atm).
- Place the stethoscope in the hollow at the inner side of the elbow, and be ready to listen for the sound of the pulse.
- Slowly release the pressure while watching the gauge, and make a mental note of the pressure at which you first hear a single sound with each heartbeat; this is the systolic pressure.
- Continue to release the pressure until you hear the double sound of a normal heartbeat (lub-dup . . . lub-dup . . .); this is the diastolic pressure.
- Record the pressure values for systole and diastole, and repeat the measurement twice more to be sure your value is reasonably accurate.
- Next, ask your partner to lie down and rest for a few minutes. Take the blood pressure three times, and record the results.

Did the value for the blood pressure depend on the position of the subject? ______________

- Repeat the measurement of blood pressure under other conditions.

The blood pressure will usually be higher if the person has recently had a cup of coffee, smoked a cigarette, or been subjected to stress. It is normally regulated by the activity of the autonomic nervous system. If the blood pressure is consistently too high (condition of **hypertension**), a stroke or small hemorrhage in the brain, or kidney failure may result. Hypertension has many causes, the most common of which is atherosclerosis. If your diastolic blood pressure is above 90 mm, have your blood pressure checked by a health professional. Hypertension can be controlled by drugs or in some cases by conscious effort through biofeedback training.

During exercise more blood is sent to the heart and muscles because their blood vessels dilate. At the same time, vessels in other areas (lungs, abdominal organs) constrict so that blood pressure does not fall and may even increase. If a person loses blood through injury or during blood donation, veins will automatically constrict to reduce the volume available for the circulating blood and thereby maintain normal blood pressure.

CAPILLARIES

The active functioning of the blood as a transport tissue goes on only in the capillaries. Here the actual exchange of gases, nutrients, and wastes takes place, and white blood cells can enter and leave the circulatory system. The smallest of the capillaries are so narrow that the cells within them move along in single file.

If you did not see living capillaries in Topic 24 (section entitled CAPILLARIES), carry out the observation of capillaries described there.

HEART RATE

The beating pattern of the heart is not due to nervous stimulation, but is rather due to spontaneous electrical events starting with the pacemaker of the **sinoatrial (SA) node** in the right atrium. The electrical impulse spreads quickly and smoothly through the cardiac muscle of the atria to trigger the **atrioventricular (AV) node,** which in turn stimulates the ventricles to contract (see Text Figure 31–3). The rate of the intrinsic heartbeat, however, is very responsive to changes in conditions within the body and to external stimuli. It speeds up when the demand for oxygen is high and also under the influence of the hormone **epinephrine (adrenalin).** Epinephrine and its close relative **norepinephrine** may be released into the blood by the adrenal gland, and norepinephrine is the final neurotransmitter of the sympathetic division of the autonomic nervous system (see Topic 27). The heart slows down, especially when we lie down, in response to **acetylcholine,** the neurotransmitter released by the **vagus nerve,** part of the parasympathetic division of the autonomic nervous system.

Human Heart Rate

- Place the fingertips of your left hand in the hollow of your wrist at the base of your right thumb. If you have trouble detecting your pulse, use the carotid artery alongside your larynx instead.
- Count your heartbeats for 3 min, calculate the rate

in beats per minute, and record the results in Table 25–4.

- Jump up and down until you feel out of breath or tired, and again record your heart rate.
- After your heart rate is back to normal, lie down for 5 min, and measure the rate while resting. Try to relax as fully as possible.

Response to exercise is diagrammed in Text Figure 25–13.

During exercise, epinephrine is released and acts to stimulate the heart rate. Exercise is supposed to benefit your circulatory system because the capillaries in the muscles, especially those of the heart, dilate in order to increase the supply of oxygen to the active tissues. After the exercise is over, the capillaries remain a little bit more open than before so that oxygenation of the tissues is improved all the time. An athlete in top condition will often have a very low heart rate of 50 beats/min or less because the pumping heart meets very little resistance in the peripheral capillaries. In order for the exercise to help significantly, however, the heart rate must reach 150 beats/min for at least a few minutes during each exercise session. Activities such as golf and gardening are just that: They do not increase the heart rate enough to be rated as exercise.

- During the coming week, test your more strenuous activities to see whether they raise your heart rate to 150 beats/min for a significant period of time.

The Frog Heart

The effects of different factors on the rate at which the frog heart beats will help you understand how the rate is normally adjusted to changing conditions.

Before starting this experiment, you and your partner should have access to all of the following:

Ice bath for the frog
Dissecting needle
Paper towels
Frog Ringer's solution
Scissors
Thread
Forceps
Pipets
5×10^{-5} M acetylcholine
5×10^{-5} M epinephrine
5×10^{-5} atropine
Electrical stimulator (on *low voltage*)

PITHING THE FROG

- Chill the frog in ice thoroughly. Remove and pat it dry, and carry out the following steps quickly.
- Hold the frog firmly in your left hand with its front legs between your middle and ring fingers and the hind legs dangling.
- Put your index finger over the head, and press down to bend the spinal cord.
- Run your right index finger along the head until you locate the depression at the base of the skull where the spinal column begins. This is where you will insert the dissecting needle.
- With a steady hand, insert the needle into the spinal cord, and sever it with a quick side-to-side movement. In this way you will be able to kill the frog as painlessly as possible.

If the frog shows signs of reviving, put it back into the ice bath and repeat the severing of the spinal cord.

- Push the needle forward parallel with the external surface into the brain. Move it back and forth to sever all connections with the spinal cord so that the frog is neurologically dead.
- Insert the needle into the spinal cord, and twist it to destroy the cord and the spinal reflexes. The legs will extend sharply as the needle contacts nerves within the cord.
- When you remove the needle, the legs should be completely limp, and the pithing operation is finished.

EXPOSING THE HEART

- Lay the frog on its back on a paper towel wet with Ringer's solution. The frog's skin must be kept moist at all times from this point on.
- Use the forceps to pinch the skin over the abdomen, slightly to side of center, and snip a hole with the scissors.
- Insert the point of the scissors carefully, being careful not to poke into the abdomen, and make an I-shaped incision by cutting in the order and direction shown in Figure 25–3.
- Once you can see the underlying muscle, repeat the incision, cutting through the muscle as well. Fold back the flaps to expose the viscera.
- Note that the beating heart is covered by the membrane of the **pericardium.**
- Slit the pericardium *carefully* to expose the heart.

Check that the frog has warmed up to room temperature before beginning your measurements.

- Record the rate of beating for a 3-min interval, and calculate the beats per minute. Record the results in Table 25–5 at the end of this topic.

You are now ready to study the effect of the drugs and electrical stimulation on the heart.

ACETYLCHOLINE

This neurotransmitter is released by the vagus nerve, which is part of the parasympathetic nervous system.

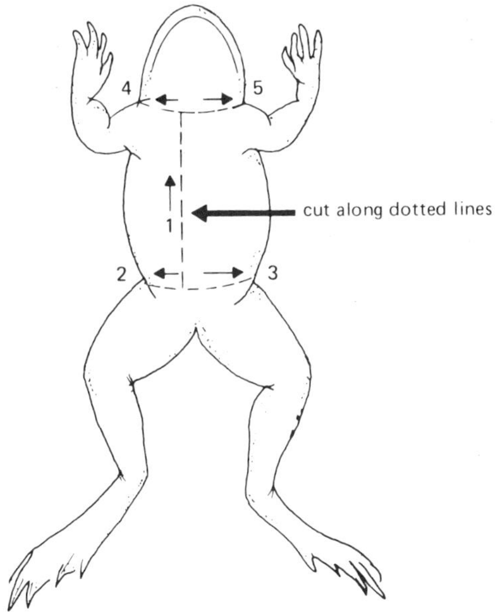

Figure 25–3. Frog dissection guide. Keep the frog moist inside and out at all times.

It should have the same effect as electrical stimulation of the vagus itself. The parasympathetic nervous system causes a "relax and digest" response in the internal organs.

- Add 3–4 drops of the acetylcholine solution to the fluid around the heart, not to the heart itself.
- Observe and record the rate of beating for each 15-sec interval until there is a perceptible change.
- Calculate the beats per minute by multiplying your final value by a factor of 4.
- Rinse the heart with Ringer's two or three times, removing the fluid from around the heart with a pipet.
- Give the heart a 5-min rest, and record a normal heart rate before going on.

EPINEPHRINE

This hormone (or its close relative norepinephrine) is released into the blood by the adrenal glands and by the sympathetic nervous system. It prepares the entire body for an emergency "fight or flight" response.

- Add 2 drops of the epinephrine solution to the fluid around the heart, and observe the effect as you did for acetylcholine.
- Record the results and allow the rinsed heart to rest.

ELECTRICAL STIMULATION OF THE VAGUS NERVE

Beneath the heart on its dorsal side, locate the bundle of stringlike nerves and connective tissue that joins the sinus venosus as shown in Figure 25–4. The vagus nerve passes through this bundle to ganglia (nerve cell clusters) in the sinoatrial node and atrial walls.

- Gently draw the bundle to one side with your forceps, and tie a thread to it loosely so that it can be pulled to one side.
- Take the frog to the stimulator or vice versa. Be sure to keep the frog moist at all times.
- Turn the stimulator on.
- Draw the nerve bundle to one side, and touch the points of the electrode to the nerve (*not* to the heart) until there is a change in the rate of the heartbeat. You may have to try several locations along the bundle.

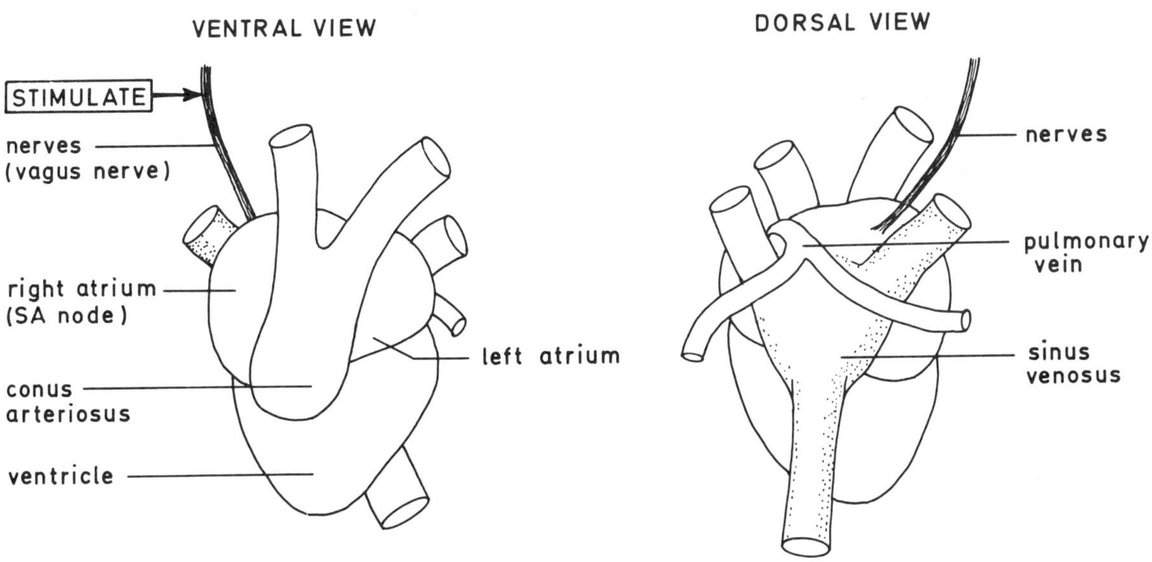

Figure 25–4. Frog heart. The vagus nerve is important in regulating the rate at which the heart beats.

The two points of the electrode must not touch each other.

- Observe the new rate of beating and record it.

Did electrical stimulation of the vagus have an effect similar to that of acetylcholine? ______________

- Allow the rinsed heart to rest until the beat is again close to the normal rate.

ATROPINE

This drug is an alkaloid poison that binds to and blocks the receptors for acetylcholine. Do not use this drug until you have completed your study of the effects of the neurotransmitters and electrical stimulation.

- Add 3–4 drops of atropine to the fluid around the heart, and record the new beating rate in Table 25–5.
- Apply electrical stimulation as you did before, and record the rate during electrical stimulation in the presence of atropine.

Did atropine change the effect of electrical stimulation? ______________

Explain your observation:

____

____

____

____

How would atropine affect the functioning of the rest of the parasympathetic nervous system?

____

____

____

____

When you have finished with the frog, dispose of it properly, and clean your dissecting tools for the next student.

REFERENCES AND FURTHER READING

ALSO SEE TOPICS 22 AND 24

Erslev, Allan J., and Thomas G. Gabuzda. *Basic Pathophysiology of the Blood,* 3rd ed. Philadelphia: W.B. Saunders, 1985.

Fox, E.G. *Sports Physiology,* 2nd ed. Philadelphia: Saunders College Publishing, 1984.

Golub, E.S. *The Cellular Basis of the Immune Response,* 2nd ed. Sunderland, MA: Sinauer Associates, 1981.

Guyton, A. *A Textbook of Medical Physiology,* 7th ed. Philadelphia, W.B. Saunders, 1986.

MacFarlane, R.G., and A. Robb-Smith. *Functions of the Blood.* New York: Academic Press, 1961.

Schmidt-Nielsen, Knut. *Animal Physiology: Adaptation and Environment,* 3rd ed. New York: Cambridge University Press, 1983.

Journal Articles

Adolph, E.F. "The heart's pacemaker." *Scientific American,* 32, March 1967 (#1067).

Axelrod, J. "Neurotransmitters." *Scientific American,* 58, June 1974 (#1297).

Cerami, A., and C.M. Peterson. "Cyanate and sickle cell disease." *Scientific American,* 44, April 1975 (#1319).

Changeaux, J.P., A. Devillers-Thiéry, and P. Chemouilli. "Acetylcholine receptor: An allosteric protein." *Science* 225:1335, 21 Sept., 1984.

Chapman, C.B., and J.H. Mitchell. "The physiology of exercise." *Scientific American,* 88, May 1965 (#1011).

Comroe, J.H. Jr. "Doctor, you have six minutes." *Science 84,* 62, January–February 1984.

Cooper, M.D., and A.R. Lawton III. "The development of the immune system." *Scientific American,* 58, November 1974 (#1306).

DiCara, L.V. "Learning in the autonomic nervous system." *Scientific American,* 30, January 1970 (#525).

Doolittle, Russell F. "Fibrinogen and fibrin." *Scientific American,* 126, December 1981.

Friedman, Milton J., and William Trager. "The biochemistry of resistance to malaria." *Scientific American,* 154, March 1981.

Hock, Raymond J. "The physiology of high altitude." *Scientific American,* 52, February 1970 (#1168).

Karpas, Abraham. "Viruses and leukemia." *American Scientist* 70(3): 277, May–June 1982.

Kolata, G. "Fetal hemoglobin genes turned on in adults." *Science* 218: 1295, 24 December 1982.

Leder, P. "The genetics of antibody diversity." *Scientific American,* 102, May 1982.

Mayerson, H.S. "The lymphatic system." *Scientific American,* 80, June 1963 (#158).

Morse, Edward E., and Barry Jacobs. "Blood donation and its aftereffects." *American Scientist* 73:68, January 1985.

Perutz, M.F. "Hemoglobin structure and respiratory transport." *Scientific American,* 92, December 1978 (#1413).

Rodgers, J.C. "Catching cancer strays." *Science 83,* 42, July–August 1983.

Rose, Noel R. "Autoimmune diseases." *Scientific American,* 80, February 1981.

Till, James E. "Cellular diversity in the blood-forming system." *American Scientist* 69(5): 522, September 1981.

Trost, C. "The blue people of Troublesome Creek." *Science 82,* 35, November 1982.

Wessel, Morris A., and A. Dominski. "Our children's daily lead." *American Scientist* 65: 294, May–June, 1977.

Zucker, Marjorie B. "The functioning of blood platelets." *Scientific American,* 86, June 1980 (#1472).

Student Name ______________________ **Date** ______________

TABLE 25–1. Plasma Buffering

Solution	Initial pH	Final pH	Drops HCl Added
Beaker #1 (control)			
Beaker #2 (plasma)			

TABLE 25–2. Determination of Hemoglobin Concentration

Normal Male Range	16 ± 2 g/100 mL
Normal Female Range	14 ± 2 g/100 mL
Absorbance at 545 nm	
Hemoglobin concentration (Absorbance × 28.8)	g/100 mL
Your Sex (Male or Female)	
Your Range (Low, Normal or High)	

TABLE 25–3. Blood Pressure

Condition	Reading	Blood pressure Systolic/diastolic
Initial pressure	1	/
	2	/
	3	/
Resting pressure	1	/
	2	/
	3	/
Other condition: ________________	1	/
	2	/
	3	/
________________	1	/
	2	/
	3	/

TABLE 25–4. Human Heart Rate

Condition	Beats/time interval		Beats/min
Initial normal			
Jumping			
Normal			
Lying down			
Other activities: ________________			
________________			
________________			
________________			

Student Name ______________________ **Date** ______________

TABLE 25–5. Frog Heart Rate

Condition	Beats/time interval		Beats/min
Initial rate			
Acetylcholine			
After rest			
Epinephrine			
After rest			
Vagus nerve			
After rest			
Atropine			
Atropine + Electrical Stimulation			

TOPIC 26

Excretion and Sexual Reproduction

OBJECTIVES

When you have completed this topic, you should be able to:

1. Define the process of excretion in mammals and list four substances that must be excreted.
2. Name the major nitrogenous waste product of mammals; name the type of food molecule that gives rise to it, and describe how and where it is produced.
3. Describe the nephron, and relate its structure to the internal structure of the kidney.
4. Trace the path of a waste molecule from the bloodstream to the opening of the urethra.
5. Name and contrast the position and functions of primary sex organs (gonads) in the male and female.
6. Trace the path an egg follows from its formation to the exterior of the body, assuming no fertilization.
7. Trace the path of sperm cells from their site of origin to the exterior, and give the function of each accessory organ along the way.
8. If copulation and fertilization occur in the pig, state where the semen is deposited, where fertilization occurs, and where development of the young piglets takes place.
9. Describe how fertilization and early development in the human differ from those processes in the pig.
10. Distinguish between: ureter/urethra; ejaculation/fertilization; ovulation/implantation; oviduct/uterine horn; vasectomy/castration.
11. For each of the major types of contraception, explain how the method works, give the exact step in conception that is blocked, and specify the anatomical site where the method has its effect.

TEXT REFERENCES Chapter 27; Chapter 28.

KEY TOPICS

nitrogenous wastes	nephron	gamete formation
male urinary system	male reproductive system	fertilization
female urinary system	female reproductive system	conception
kidney function	menstrual cycle	contraception

INTRODUCTION

Wastes from within cells are removed from the body in the process of **excretion.** The wastes are chemicals that are formed by the metabolic activity of cells and cannot be gotten rid of except as dissolved substances. They include carbon dioxide, water, nitrogenous wastes, salts, and hydrogen ions.

Substances foreign to the body, such as drugs and spices, must also be excreted.

Nitrogenous wastes are formed whenever proteins or amino acids in the diet or in cells are broken down to carbon skeletons (keto acids) and ammonia (NH_3) in the process of **deamination.** In mammals, the main nitrogenous waste is **urea,** a simple molecule formed in the liver from CO_2 and NH_3 by the enzymes of the **urea cycle.** Urea is soluble and relatively nontoxic, so it can be eliminated in urine in a fairly concentrated form. This conservation of water has a price, however, in that ATP energy is required to synthesize urea.

Although the kidney is the main excretory organ, some wastes are also excreted by the skin in the sweat, by the lungs, by the liver, and by the large intestine. The lungs and liver are important in detoxifying harmful substances so that they can be excreted by the kidney.

Food remains that are not and have never been inside of cells are eliminated from the body through the rectum as feces. This process is **egestion** and is basically different from excretion. Excretory wastes from cells of the liver and large intestine do enter the gut, but they make only a small contribution to the fecal wastes.

The reproductive system is anatomically associated with the excretory system and, in fact, complete separation of the two is rare among vertebrates. In female and male fetal pigs both systems have a common opening to the exterior, so you will study them both at the same time. You will be responsible for understanding both the male and female systems, so exchange your pig for one of the opposite sex when you are finished to complete your study of the reproductive system.

You will note that many of the parts of the male and female reproductive systems are counterparts of one another. This is not surprising because male and female embryos are anatomically alike early in development. Male or female organs then develop from the same rudimentary reproductive structures.

As you study the anatomy of reproduction, you will also be learning how it is possible to prevent conception, or use **contraception,** in several different ways. Permanent contraception, usually called sterilization, prevents the egg or sperm cells from ever reaching their intended destination. Other methods are temporary and can be started and stopped again whenever desired. The common methods are mechanical, chemical, or hormonal. In many vertebrates other than humans, contraception of some sort is an important part of reproduction. These animals have behavioral or physiological mechanisms for limiting reproduction, whereas human beings must spend millions of dollars to achieve the same end through contraception.

URINARY SYSTEM

- Push aside or remove the intestines of your pig and locate the large paired **kidneys** (see Figure 26–2 or 26–3). These are reddish organs located deep within the abdominal cavity, underneath the peritoneum on either side of the aorta. Unlike the other abdominal organs, they are not enclosed within the peritoneum.
- Before you clear away the membrane from one of the kidneys, locate the **adrenal gland** pressed against the inner side of the anterior part. The delicate gland is a narrow strip of light brown or yellow tissue and is difficult to find because it is easily destroyed. The **adrenal medulla** (inner part) is a neuroendocrine organ that secrets the hormones epinephrine and norepinephrine in the bloodstream, whereas the **adrenal cortex** (outer part) secretes a variety of steroid hormones.
- Now, expose the kidney and study its external structure. Identify the following parts (see Figure 26–1).

Renal Hilum. Where the blood vessels and **ureter** are attached.

Renal Artery. Carries blood from the aorta to the kidney.

Renal Vein. Carries blood from the kidney to the posterior vena cava.

Ureter. Hollow tube that carries urine from the kidney to the **urinary bladder.** The ureter lies underneath the peritoneum.

Kidney tissue is very rich in mitochondria and has an enormous metabolic rate. Much of its energy is spent in **resorption** of molecules from urine or in **secretion** of molecules into urine (see Color Plate 4). Within the kidney are about one million **nephrons,** which are richly supplied with blood vessels and are the functional units (see Text Figures 27–11 and 27–13).

- Slice through the kidney from right to left with a razor blade and study the internal structure (see Figure 26–1). Notice that there are two clearly demarcated zones.
- The outer zone is the **cortex.** It contains microscopic capillary tufts (glomeruli), each of which fits tightly against the first segment of a kidney tubule. The liquid portion of the blood, rich in ions and nutrients, filters out through the wall of a glomerulus, passes through the wall of a kidney tubule, and thus enters the tubule. Once inside, the filtrate is slowly changed into urine by a series of complex processes that occur along the length of

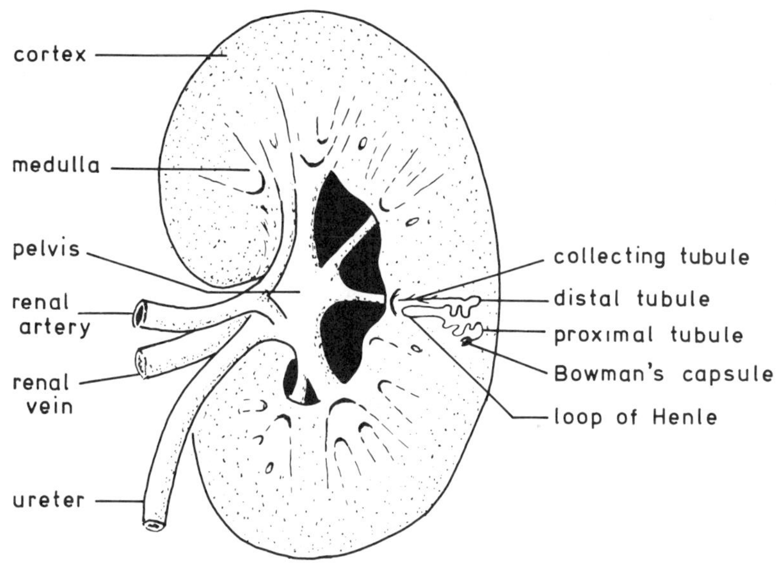

Figure 26–1. Internal structure of a kidney. The detail shows the position of the parts of a single nephron. Each kidney will have about one million nephrons. Urine flows out of the collecting tubule into the hollow pelvis of the kidney and then exits through the ureter to the bladder. Capillaries permeate the kidney and are closely associated with each individual nephron.

the nephron. The proximal and distal convoluted parts of the nephron are located here in the cortex.

- The inner zone is the **medulla.** It looks different from the cortex because the segments of the nephrons that are located in the medulla are straight instead of coiled as they are in the cortex. Henle's loop is found here. It is in the medulla that the urine takes on its final chemical composition as it passes through the collecting tubules. The collecting tubules can be seen as rays radiating from the central part of the kidney. Urine from thousands of tiny nephrons then empties into the ureter, which begins at the hilum.
- View the section of kidney cortex on demonstration in the microscope. The larger dark structures surrounded by membranes are the **glomeruli** surrounded by the capsules of the tubules. The round hollow tubes are the **proximal tubules,** and the irregularly shaped tubes are the **distal tubules.**

The composition of urine is controlled by hormones (see Text Figure 27–15). **Vasopressin** stimulates water resorption by making the collecting ducts more permeable, and **aldosterone** stimulates salt resorption by the distal convoluted tubule. **Angiotensin** constricts the blood vessels and increases aldosterone if the blood pressure or salt concentration falls too low.

Precipitates (solids) may form in urine and grow either into kidney stones (within the pelvis of the kidney) or into bladder stones. Very large kidney stones may fill the pelvis and take the shape of a "staghorn." When a stone becomes lodged within the ureter (or urethra), urine pressure builds up causing the victim of stones excruciating pain and creating a medical emergency. Damage to the kidney could result if the stone is not passed or removed surgically within a short time.

- Follow the ureter posteriorly to where it joins the **urinary bladder,** which lies between the umbilical arteries. The structure of the urinary tract is shown in Figures 26–2 and 26–3. After birth, urine leaves the bladder by the urethra.
- Locate the **urethra,** and clear away enough tissue so that you can see how it disappears deep within the girdle of pelvic bones. You will be able to follow it all the way to the outside when you have finished studying the reproductive system.

FEMALE REPRODUCTIVE SYSTEM

Refer to Figure 26–2 as you study the female reproductive system. Also see Text Figures 28–4 and 28–5.

- Locate the **ovaries,** which are the gonads or primary sex organs of the female. They are light-colored, small spheres found posterior to the kidneys and held in place by mesenteries.

Ovaries have a double function: They contain all the **egg cells** that the female will ever produce, and they are endocrine glands that release the female sex hormones **estrogen** and **progesterone** into the bloodstream. (See Color Plate 4.) Their development and functioning is controlled by the pituitary hormones **FSH** and **LH** (see Text Figure 28–9 and Text Table 28–1).

- Locate the **oviduct** (sometimes called the Fallopian tube). In the pig it is a *tiny,* twisted white tube closely attached to the top of the ovary.

The oviduct does not actually attach directly to the ovary. Rather each oviduct ends in an open funnel-shaped structure (difficult to see here) that partially encloses the ovary. When a ripe egg cell is released from the ovary, it generally moves into the ostium (opening) of the oviduct and passes down the oviduct

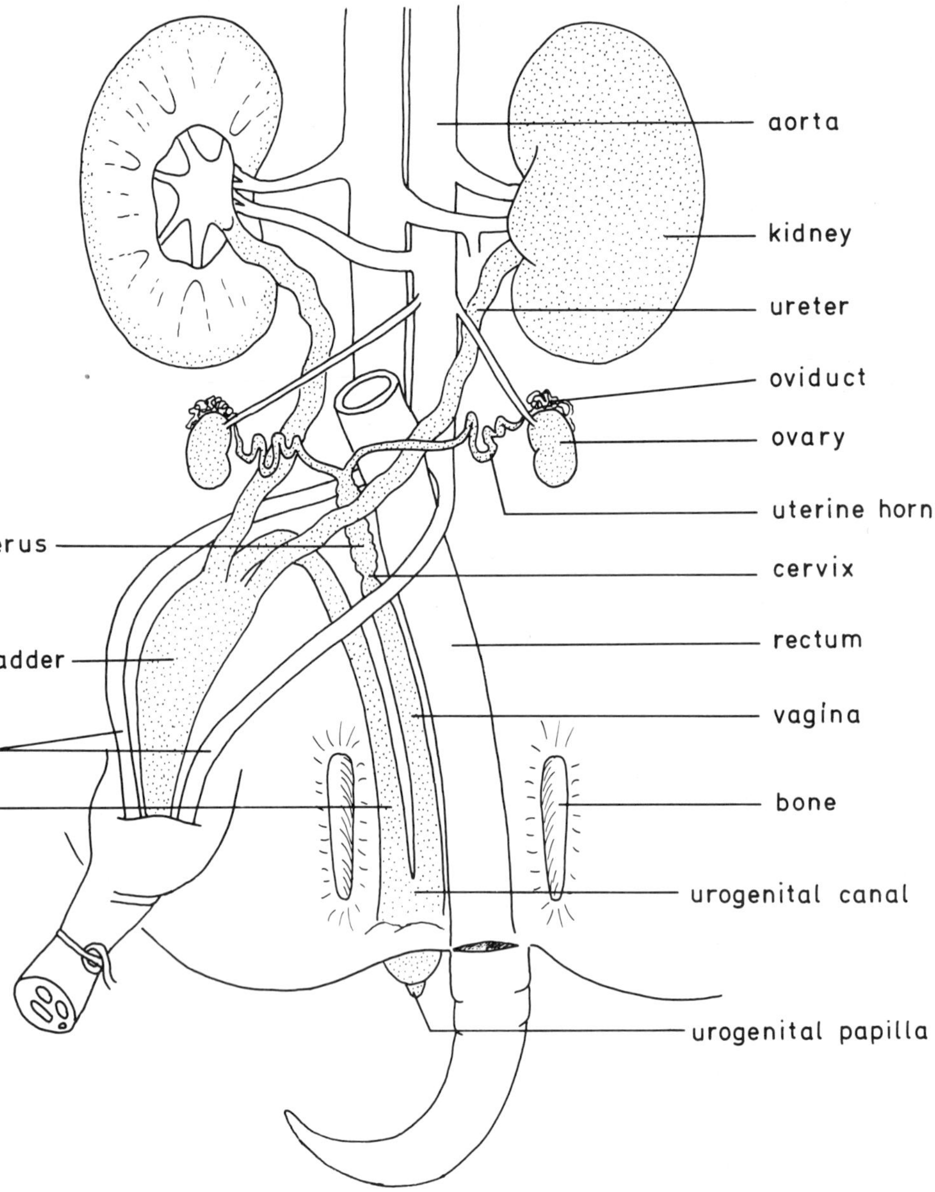

Figure 26–2. Female urogenital system. Dissection of a female fetal pig to show the structures of the urogenital system.

to the uterus. Fertilization normally occurs in the oviduct, when sperm are present. In **tubal ligation** or female sterilization, the oviducts are cut and blocked so that eggs and sperm do not meet. Production of the female hormones estrogen and progesterone and the menstrual cycle are unaffected by tubal ligation.

- The pig's **uterus** is quite different from that of the human female. It consists of two long **horns,** in which the developing eggs implant, and a small **body.** (In the human female, the uterus is a single median structure without horns, about the size and shape of a pear.)
- Locate the long horns, which are larger than the oviducts, and begin close to the ovaries. Locate the body of the uterus where the two horns join. The body ends at the **cervix,** a strong sphincter muscle.

The lining of the uterus, the **endometrium,** becomes soft and rich in blood vessels in preparation for implantation. Human pregnancy is maintained because the young embryo produces the hormone **human chorionic gonadotrophin (HCG),** which causes the ovaries to maintain the level of progesterone and suppresses the release of FSH and LH. Pregnancy tests rely on the detection of HCG and can be positive as soon as 7 days after fertilization. After implantation, the endometrium and part of the embryo form the **placenta,** which is connected to the embryo by the umbilical cord. Nutrients, wastes, gases, nicotine, and drugs are exchanged between mother and fetus across the capillaries of the placenta.

- The uterus leads into the **vagina,** or birth canal, and the beginning of the vagina can be seen as a constriction soon after the uterine horns join.
- Follow the vagina posteriorly until it disappears into the pelvic girdle.

At this point you should be able to show the rectum, urethra, and vagina all entering the pelvic girdle. It is now necessary to cut through the muscle and the bone of the pelvic girdle to see what happens next.

- Carefully make a cut with your razor blade along the midline of the pig's pelvic girdle.
- Cut almost all the way back to the urogenital papilla.
- Deepen the cut to the bone, and then cut through the bone, trying not to cut the structures located underneath it.
- Spread the legs apart, and force the bones of the girdle open to expose the structures within it. Follow the vagina and urethra posteriorly, and see where they join to form the **urogenital canal.**

This canal leads to the urogenital opening of the female pig. As the pig develops, the vagina and urethra become more and more separated so that finally there will be two openings: a ventral one for the urethra and a more dorsal one for the vagina. The rectal opening is most dorsal, just beneath the tail.

On each side of the urogenital opening are the rounded lips, or **labia.** They would have developed into the scrotum if the pig had been male.

- Slit the side of the urogenital canal, and look on the ventral surface of the canal for the **clitoris,** the female organ that is the counterpart of the male penis. In human females the clitoris is the most sexually sensitive area and produces a pleasurable sensation when stimulated by touch.

MALE REPRODUCTIVE SYSTEM

Before starting to dissect, study Figure 26–3 to familiarize yourself with the overall anatomy. Also see Text Figure 28–7. Notice that the pathways for urine and sperm join in a common urethra, which passes down through the pelvic girdle, then ventrally and anteriorly along the abdominal wall until it finally exits through the urogenital opening just below the umbilical cord. You will need to dissect quite a bit to see the entire pathway.

- Raise the flap of abdominal wall where the umbilical cord enters, and relocate the **bladder** between the umbilical arteries.
- Use your probe to separate the **ureters,** which drain from the kidney into the bladder.
- If it helps, cut across the middle of the bladder and umbilical arteries, and separate them from the abdominal wall so that you can locate the beginning of the **urethra,** which leads out of the bladder. This area is called the neck of the bladder.
- Notice the two small tubes that are looped over the ureters and which join the urethra at the bladder neck. They are the **vasa deferentia,** or "tubes," of the male, and carry sperm from the testes to the urethra. In **vasectomy** or male sterilization the vas is cut and blocked close to where it leaves the testis so that no sperm reach the urethra.
- To find the **testes,** or male gonads, investigate the scrotum of the pig to see whether they have already moved down from the abdomen, or descended. If not, as is usually the case, they can be found in the groin area, on the inner side of the upper part of the hind leg.
- Cut through the skin and muscle to locate each testis, which will be enclosed in a tough, white, opaque membrane. Be careful not to cut between the testis and the vas deferens that leads from it.
- Cut open the outer membrane so that you can see the reddish testis itself.
- Covering part of the testis is a whitish mass of tightly coiled tubules, the **epididymis.** Sperm cells produced in the testis pass into the coils of the epididymis where they are stored before entering the vas deferens.

Like the ovaries, the testes have two functions: They produce the **sperm cells** of the male that enter the vas deferens and they produce the male sex hormones, principally **testosterone.** (See Color Plate 4.) Testosterone is released directly into the bloodstream and leaves the testis in the spermatic vein, *not* in the vas deferens. It determines the sex of the pig and allows it to function sexually as a male.

Do you think vasectomy would interfere with the release of testosterone by the testis? _______________

The testes, again like the ovaries, develop and function in response to the pituitary hormones **FSH** and **LH.**

- Dissect as much as you need in this area to show the path of the vas deferens leading from the testis to the urethra.

Each vas deferens passes from outside the abdomen through the abdominal wall to the inside through the **inguinal canal.** In humans this canal is a weak spot and may rupture or split open, forming a hernia. Testes, like kidneys, are outside the peritoneum. When the testis is within the scrotum, it is supported by a ligament and has its own blood vessels leading to and from the abdomen. The vessels, ligament, and vas

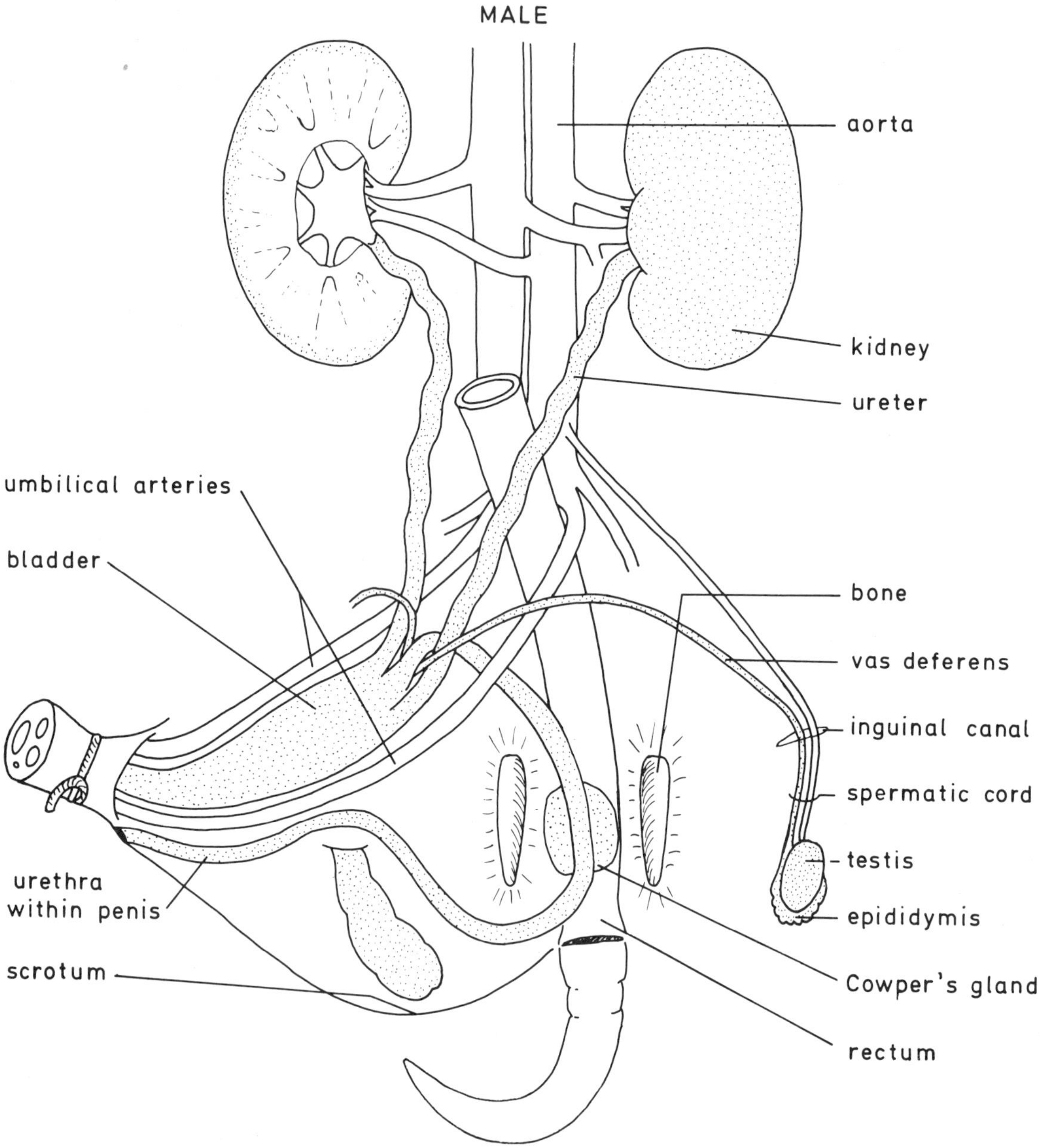

Figure 26–3. Male urogenital system. Dissection of a male fetal pig to show the structures of the urogenital system.

deferens form the **spermatic cord** from which the testis is suspended.

- Follow the vas deferens to the bladder, and observe where it enters the urethra.

The situation here is very complicated and is diagrammed in Figure 26–4.

- Pull the bladder toward you, and try to locate the following structures on the *dorsal* side:

  Seminal Vesicles. Small, paired, light-colored glands that branch off from the urethra just posterior to the vasa deferentia.

  Prostate. Very small, light-colored, smooth, round gland at the junction of the seminal vesicles. It is sometimes too small to be seen.

As sperm cells pass along the urethra, fluids from these glands are released to help make up the **semen.**

- Push the flap of abdominal wall containing the umbilical cord to one side: You will need it momentarily.
- To expose the urethra as it leaves the abdomen, cut through the skin, muscle, and bone of the pelvic girdle slightly to one side of the midline. Try not to cut beneath the bone.
- Locate the urethra just beneath the bone: It is now quite large and well developed, and is the combined pathway for both the excretory and reproductive systems.
- Pry apart the bones of the pelvic girdle, and locate the bean-shaped, white **Cowper's glands** (Figure

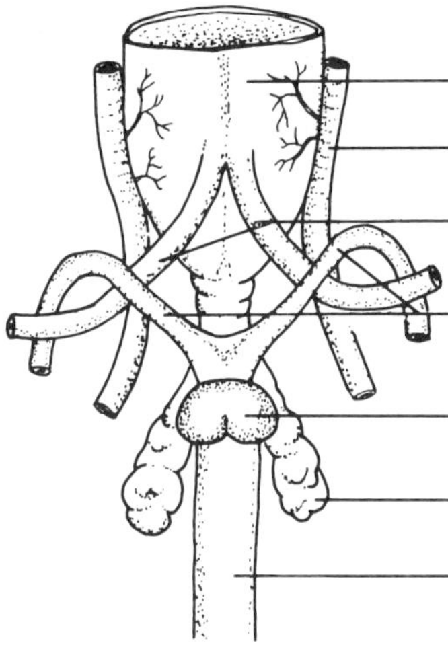

Figure 26–4. Male reproductive system. Dorsal view showing the structures in the region of the neck of the bladder. If you dissect carefully, you should be able to see all of these structures except the prostate, which may be too undeveloped to see.

26–3) on either side of the urethra. They also secrete fluids into the semen.

The urethra leaves the abdomen and enters the long muscular **penis** within the abdominal wall. (It separates from the wall later in development.) This organ, particularly the glans of the penis, is the most sexually sensitive area in the male and becomes stiff and erect when stimulated (see Text Figure 28–8). During **ejaculation,** muscular contractions in the vas deferens, urethra, and accessory glands expel semen out through the penis.

- Feel that free strip of abdominal wall to find the penis and dissect it free with your probe. You can now trace the excretory system from the kidney and the reproductive system from the testis through the common urethra to the external urogenital opening.
- Finally, follow the digestive tract as it exits through the pelvic girdle. The **rectum** lies just dorsal to the urethra and ends in the external opening, the **anus.**

What color is the rectum? ______________

Varicose veins within the human rectum are called hemorrhoids.

When you are finished, seal the pig and any extra tissue into the bag, and dispose of it properly. Clean your dissecting pan and tools, and oil your scissors for the next student.

HUMAN REPRODUCTION: CONCEPTION

Conception is the fertilization of an egg and the establishment of a pregnancy. Each human female has about 300,000 eggs that are set aside very early in development, about 3 weeks after conception. Some of these potential eggs mature, one at a time, once the woman is sexually mature. Thus any damage to the eggs or ovaries of a woman, such as from X-rays or toxic chemicals, is permanent. No new eggs can ever be formed. (In the testes, on the other hand, sperm are formed continuously and damaged sperm are quickly replaced.) The physiological conditions that cause eggs to be released and make the reproductive tract receptive to fertilization and implantation of the resulting embryo are under hormonal control. Regular changes in the levels of these hormones cause the events of the menstrual cycle: The egg is released halfway through the cycle, midway between two menstrual periods (see Figure 26–5). If you are unsure about the details of the menstrual cycle, you should turn to section 28-D in your text and review them so that you can understand conception.

The egg cell is rather short-lived and lives only about 36 hours after it has been released from the ovary. Fertilization normally takes place in the oviduct within half a day of egg release, or **ovulation.** Sperm cells are viable for several days, so regular sexual activity means that the egg has a good chance of being fertilized. In fact, about 80 out of 100 sexually active couples will conceive within a year when no contraceptive is used. This is defined as a conception rate of 80%. The majority of fertilized eggs fail to implant in the uterus and develop properly and so do not result in a recognized pregnancy. A large number of recognized pregnancies (up to 15%) fail to progress normally and end in spontaneous abortion or miscarriage. Usually the embryo is defective in such cases, with about half of the miscarriages due to abnormalities in the chromosomal karyotype.

If a couple had intercourse (coitus) 2 days before ovulation, could conception occur? ______________

If a couple had intercourse 2 days after ovulation, could conception occur? ______________

HUMAN REPRODUCTION: CONTRACEPTION

If natural events were allowed to take their course, most women would be pregnant most of the time. In our society, personal freedom, good medical care, improved general health and nutrition, and widespread preference for bottle rather than breast feeding all help contribute to this potentially high pregnancy rate. Of the many ways that pregnancy can be prevented, the only one that is 100% sure is sexual abstinence. Although this method is used to a degree by many societies, it has almost no role in our own. All the other methods of birth control or contraception, including sterilization, do have some finite risk of pregnancy. This risk is given as the chance of preg-

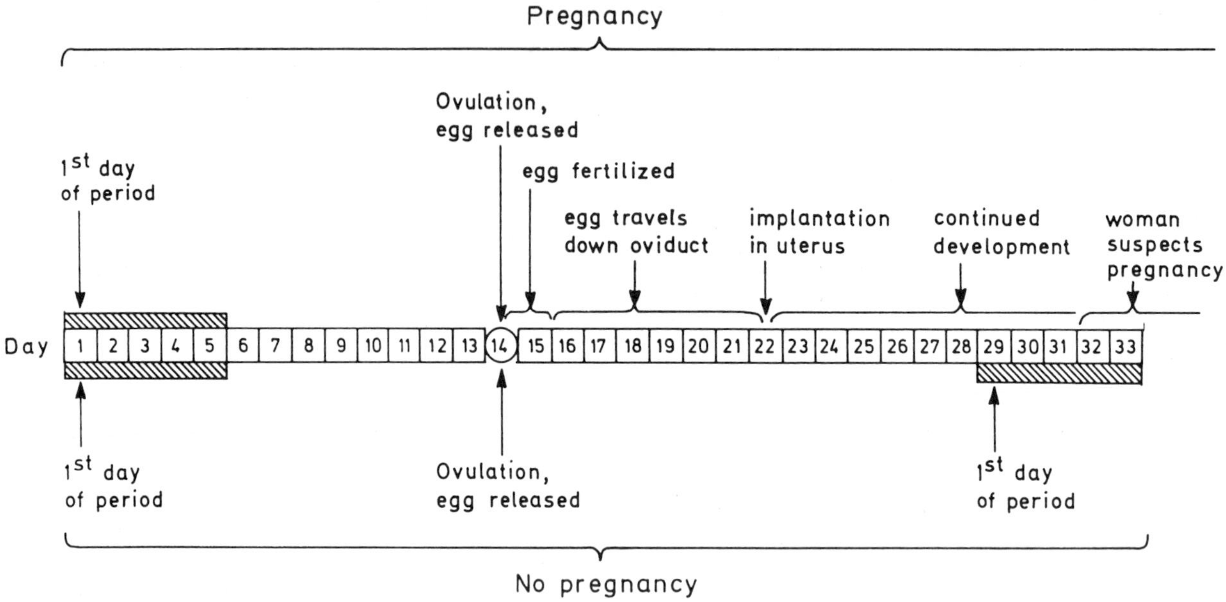

Figure 26–5. Menstrual cycle. If pregnancy occurs, the menstrual period is suspended because the hormone HCG (human chorionic gonadotrophin) from the embryo stimulates the ovary to keep secreting progesterone.

nancy in a group of couples using the method to prevent pregnancy during the first year of use.* On the average, the couples had a pregnancy rate of 3.7%; they would have expected a rate of about 80% if they had not used any contraception.

Some common methods of contraception are shown in Figure 26–6.

Mechanical Methods

CONDOM

■ Look at the various types of condoms on demonstration.

The rubber sheath is rolled over the penis before any genital contact during intercourse. It prevents sperm from entering the vagina. Thinner condoms are supposed to be more "natural," but are also more likely to break, releasing sperm into the vagina. Lubricants such as Vaseline cause the rubber of the condom to deteriorate and weaken. Condoms are available without prescription and are the only method that gives protection against sexually transmitted or venereal disease (STD). (Common diseases transmitted by sexual contact include syphilis, gonorrhea, *Chlamydia,* Type II herpes, and autoimmune deficiency syndrome or AIDS.) This method is coitus-related, and requires some practice, care, and self-control when the condom is put on and removed.

Conception rate is 7%. If the condom is used carelessly, the rate will be much higher than if the condom is used carefully with foam.

DIAPHRAGM

■ Look at the one size of diaphragm on demonstration.

This rubber shield is used with spermicidal jelly and inserted into the vagina before intercourse. To be effective, it must fit properly over the cervix (tip of the uterus), so each woman must be individually fitted for her diaphragm by a health professional. Diaphragms are available by prescription only and must be used with a spermicidal agent. They also have to be used at the time of intercourse and require practice in learning to insert them properly. Conception rate is 10%.

SPONGE

The soft sponge (Today®) is saturated with spermicide and is inserted into the vagina to cover the cervix before coitus. There is just one size, so no fitting or prescription is necessary. It is effective for 24 hours, so its use does not have to interfere with sexual foreplay or repeated intercourse. The main disadvantage is expense—each sponge costs about $1. Conception rate, if the sponge is used properly, is about 10–13%. Use of a condom as well is recommended.

IUD (INTRAUTERINE DEVICE)

■ Look at the various types on demonstration.

* Table 2 in B. Vaughan, J. Trussell, J. Menken, and E.F. Jones, "Contraceptive failure among married women in the United States, 1970–73," *Family Planning Perspectives* 9:6, 1977.

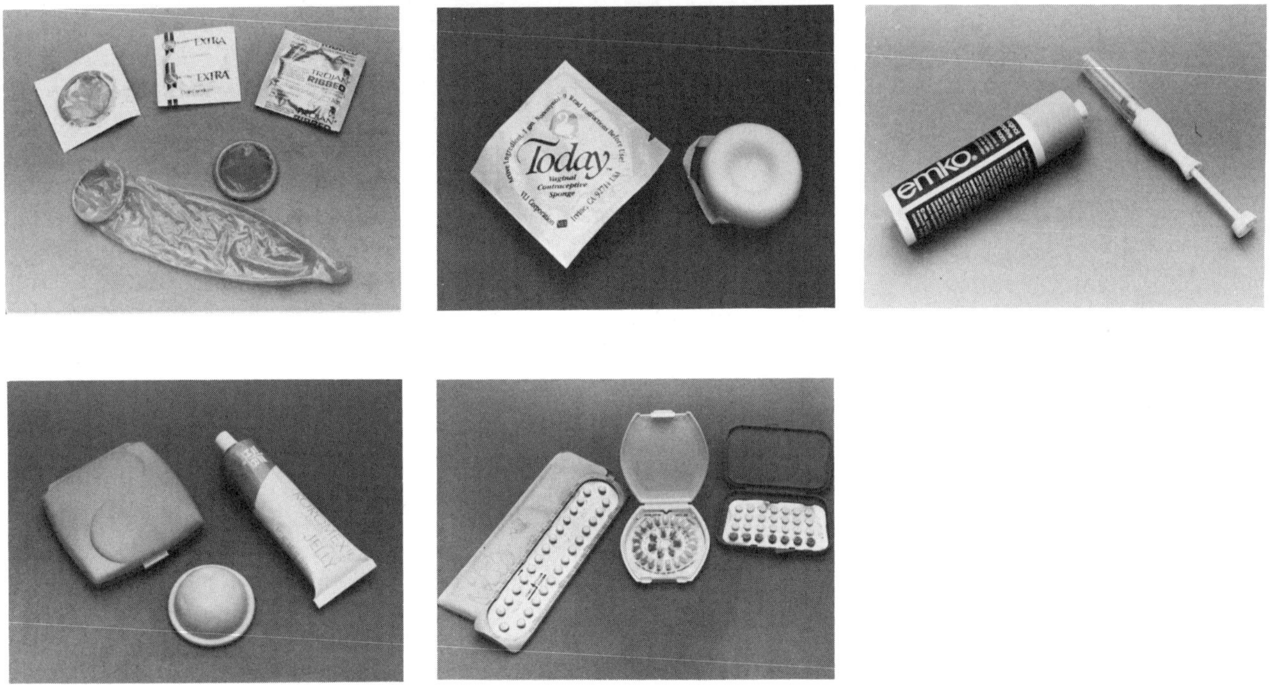

Figure 26–6. Contraceptive methods. These methods are widely used in the United States. (Photos: C. Eberhard)

IUDs come in all shapes and sizes. The device is inserted into the uterus through the cervix, and so long as it is in place, it will block pregnancy by preventing implantation of any eggs that happen to be fertilized. An IUD must be inserted by a physician or nurse-clinician, and it can be removed any time the woman desires. It has the great advantage that its insertion is not related to intercourse, but many women experience undesirable side effects such as cramps, excessive bleeding, and an increased susceptibility to infection. The woman must regularly check the IUD to make sure that it has not been expelled. IUDs are not recommended unless the woman has borne a child and they have been associated with problems of infertility. For these reasons, some IUDs are no longer marketed. Conception rate is 3% if the device is not expelled accidentally.

Chemical Methods

FOAMS, CREAMS, AND JELLIES

Foams from pressurized cans are put into the vagina with an applicator something like a tampon. Creams are applied in the same way. Foams, as well as creams and jellies, all contain chemicals that effectively kill sperm cells without irritating the vagina. They are available without prescription and must be used each time before intercourse. Some products can be used without a diaphragm, but the effectiveness of all such products can be increased if a diaphragm or condom is used in addition. Jellies are usually designed only for use with a diaphragm. Conception rate is 13%.

Feminine hygiene products (including douches) are not effective contraceptives.

Hormones

"THE PILL"

The most commonly used birth control pill is a combination of synthetic estrogen and progesterone. It works by blocking the release of anterior pituitary hormones (FSH and LH) that are needed for the egg to mature, be released, and implant properly. The pills are taken for 3 weeks to prevent ovulation, and then stopped for 1 week so that a menstrual period occurs. The pill is popular because it is cheap, effective, not used at the time of intercourse, and often makes the menstrual periods less bothersome. On the other hand, some women experience annoying side effects, it must be taken every day, and older women, especially smokers, will have an increased chance of blood clots (thromboembolism). Smokers and women over 35 should not take the pill. The pill may also increase the risk of minor vaginal infections, and it does not prevent venereal disease. Serious side effects, such as cancer, liver disease, and infertility, however, have not proved to be a significant risk. Conception rate is less than 2% and probably 0% if the woman never forgets to take the pills exactly as prescribed.

IMPLANTS

If there is a problem in taking the pill every day, the same effect can be achieved by implanting a capsule under the skin of the arm or thigh. The capsule will slowly release hormones into the body and thereby prevent pregnancy. This method is reversible, but the implant must be inserted and removed by a physician. Implants are less widely used than the other methods.

Breast feeding is not an effective birth control measure.

After birth it often takes the woman's hormones and menstrual cycle some time to get back to normal. Ovulation may be inhibited for a few months, particularly in nursing mothers, because prolactin suppresses the release of the pituitary hormones FSH and LH. This process is so variable, however, that it cannot be relied upon to prevent pregnancy. Some women ovulate very soon after giving birth, and many mothers have become pregnant while they are still nursing.

Surgical Methods

Sterilization is the most common method of contraception in the United States. It should not be confused with castration, which is the removal of the gonads, usually in cases of ovarian or testicular cancer.

FEMALE STERILIZATION

A surgical incision is made in the abdomen (or vagina) and the oviducts, or "tubes," are cut and tied off. As a result, any eggs released will just stay in the far end of the tube and will never be able to reach the uterus. This procedure is easy to do after childbirth but can also be done in nonpregnant women through a very small incision. It has the great advantage of being very effective and having no side effects. There is no change in the woman's menstrual cycle or secondary sex characteristics. On the other hand, it is relatively expensive and is limited to women who do not want (more) children. It can only rarely be reversed. Conception rate is almost 0%. Once in a great while the tubes manage to grow back together.

MALE STERILIZATION

The vasa deferentia, or "tubes," are cut and tied off through a small incision in the scrotum. See Text Figure 28–11. This procedure can usually be done in the doctor's office and is much cheaper and safer than female sterilization because the testes are outside the abdominal cavity. The other advantages and disadvantages of the two procedures are the same. After sterilization, the man's semen eventually will contain no sperm cells (the stored sperm must first be used up), but otherwise the semen will be the same as before the operation. Sexual functioning is just the same except that the man will not be able to make the woman pregnant. In some cases, the tubes have been surgically rejoined so that the man is again fertile, but this method is at present limited to men who do not plan to have children in the future. Conception rate is almost 0%. The male tubes have an even smaller chance of rejoining than do the female tubes.

Biological Methods

RHYTHM

The couple using this method tries to predict when the woman will ovulate, and avoids having intercourse around that time. Ovulation is supposed to occur just about 14 days before the next menstrual period begins, but the time between the end of the preceding menstrual period and the time of ovulation tends to be rather variable. If the couple relies strictly on the calendar, the method is quite poor, with a conception rate of roughly 35%. Attention to physical changes can give some improvement. Hints about the timing of ovulation can be obtained by noting an increase in body temperature and changes in the stickiness of cervical mucus. Even under the best of conditions, however, sickness might prematurely raise the body temperature, and all sorts of physical and psychological changes will affect the timing of ovulation, so that it is never possible to predict the correct time with absolute certainty. This method cannot be used by the many women with irregular or somewhat variable cycles and it always involves periods of sexual abstinence. Conception rate is 7% under the most favorable conditions. Usually the rate is higher than with any other method. Reported success depends strongly on the woman's physiology and the motivation of the couple involved.

Pregnancy Termination (Abortion)

Many pregnancies, perhaps 30–60%, end in **spontaneous abortions** (miscarriages) soon after conception. Usually this results from a defective embryo or imperfect implantation, and the woman may not even have realized that she had conceived. After pregnancy has been recognized, about 15% of the time there will be a miscarriage and about half of these miscarriages will involve abnormal embryonic chromosomes. In this way the body avoids spending energy in nourishing an embryo that has no chance of surviving.

Induced abortion, carried out by a physician, is not a satisfactory means of contraception by itself because of the strain being pregnant for even a short time puts on the woman's system, and because of the finite risk of infection or other consequence of the procedure. Abortion is, however, a necessary component of contraception because every method used has some risk of pregnancy. There are several types of abortions used, depending on the stage of the pregnancy:

1. A shot of estrogen within 24 hours of unprotected sex will prevent implantation. The side effects are unpleasant.
2. Insertion of an IUD after unprotected sex will prevent pregnancy because several days must elapse before the embryo is ready to implant.
3. Vacuum aspiration, an outpatient procedure in which the cervix is dilated and the embryo removed under suction, can be used for up to 12 weeks.
4. Dilation and curettage (D and C) is used during the first 12 weeks and sometimes later. The cervix is dilated, and the contents of the uterus are scraped out.
5. A saline solution can be injected into the amniotic sac, and as a result, the uterus expels the embryo in a few hours. This more complicated procedure can be used through up to 6 months of pregnancy.
6. Under unusual circumstances, a pregnancy can be ended even later by an operation (hysterotomy) in which the abdomen is surgically opened and the fetus removed. This is a major operation, but the uterus is not removed as it is in a hysterectomy.

Risk of Complication

Many people do not realize that for women under 40 the overall risk to health of any form of contraception is much smaller than the risk involved if the woman used no contraception and carried each pregnancy that occurred to term. Unprotected women under 25 have a risk of pregnancy-related death of 5 in 100,000 per year, whereas any combination of birth control methods, including IUD, pills, abortion, and so on, has a risk of 1–2 in 100,000 per year. Once a woman has become pregnant, the risk of serious illness or death is much smaller for having an abortion compared with carrying out the pregnancy.

- Review what you have learned about human reproduction by tracing the path of an egg and a sperm cell from formation to the site of fertilization.
- Locate each of the following sites:

  Where egg and sperm cells are formed.
  Where sperm are deposited.
  Where fertilization normally occurs.
  Where each of the following acts to interfere with conception:
  - Male sterilization
  - Female sterilization
  - Condom
  - Foam
  - Diaphragm
  - Sponge
  - IUD

REFERENCES AND FURTHER READING

ALSO SEE TOPIC 22

Boston Women's Health Collective. *The New Our Bodies, Ourselves.* New York: Simon and Schuster, 1985.

Cox, Klaudia, ed. *Contraceptive Efficacy Among Married Women in the United States.* National Center for Health Statistics (ser. 23, no. 5), 1979.

Demarest, Robert, and John Sciarra. *Conception, Birth and Contraception,* 2nd ed. New York: McGraw-Hill, 1976.

Ganong, W.F. "Renal Function." *Review of Medical Physiology,* 11th ed. Los Altos, Calif., Lange Medical Publishers, 1983.

Hatcher, R.A. *et al. Contraceptive Technology 1986–1987. Special Section on Sexually Transmitted Infections.* New York: Irvington Publishers, 1986.

Huff, Robert W., and Carl J. Pauerstein. *Human Reproduction: Physiology and Pathophysiology.* New York: John Wiley and Sons, 1979.

Nilsson, L. *A Child Is Born,* rev. ed. New York: Delacorte, 1977.

Silber, S.J. *How to Get Pregnant.* New York: Warner Books, 1980.

Turner, C.D., and J.T. Bagnara. *General Endocrinology,* 6th ed. Philadelphia: W.B. Saunders, 1976.

van Tienhoven, A. *Reproductive Physiology of Vertebrates.* Ithaca, NY: Comstock Publishing Associates, 1983.

Journal Articles

Beaconsfield, Peter, George Birdwood, and Rebecca Beaconsfield. "The placenta." *Scientific American,* 94, August 1980 (#1478).

Birnholz, J.C., and E.E. Farrell. "Ultrasound images of human fetal development." *American Scientist* 72: 608, November 1984.

Fuchs, Fritz. "Genetic amniocentesis." *Scientific American,* 47, June 1980 (#1471).

Gold, Michael. "The baby makers." *Science 85:* 26, April 1985.

Grobstein, Clifford. "External human fertilization." *Scientific American,* 33, June 1979 (#1429).

Langone, J. "The quest for the male pill." *Discover,* 26, October 1982.

Short, R.V. "Breast feeding." *Scientific American,* 35, April 1984.

Smith, Homer W. "The kidney." *Scientific American,* 40, January 1953 (#37).

Tietze, C. "New estimates of mortality associated with fertility control." *Family Planning Perspectives IX* 2, 1977.

Tietze, C., and S. Lewit. "Legal abortion." *Scientific American,* 21, January 1977 (#1348).

Vaughan, B., J. Trussell, J. Menken, and E. F. Jones. "Contraceptive failure among married women in the United States 1970–73." *Family Planning Perspectives* 9: 6, 1977.

TOPIC 27

Neurons, Nervous System, and Sense Organs

OBJECTIVES

When you have completed this topic, you should be able to:

1. Describe the basic structure of a neuron.
2. Show in which direction(s), if any, an impulse travels if the neuron is stimulated at any point.
3. Describe the function of a myelin sheath, and give the location of myelinated neurons in the nervous system.
4. Describe the structure of the mammalian brain, and give the functions controlled by the medulla, cerebellum, hypothalamus, pituitary, thalamus, reticular formation, colliculi, and cerebral cortex.
5. Describe the structure of the spinal cord, and give the main types of neurons entering, leaving, or functioning within it.
6. Distinguish between the central/peripheral nervous systems, the somatic/autonomic nervous systems and the sympathetic/parasympathetic nervous systems.
7. Name the neurotransmitters associated with synapses in each of the following: muscle, sympathetic nervous system, parasympathetic nervous system, brain, and spinal cord.
8. Describe the five main types of receptor cells, and name some sense organs in which each is found.
9. Describe the structure of the vertebrate eye and the types of receptors found in it.
10. Explain how the transducers in the inner ear work.
11. Describe where proprioceptors are found in the body, and explain their function.
12. Trace the path of the withdrawal reflex when you touch a hot stove.

TEXT REFERENCES Chapter 29 (B–H); Chapter 30; Chapter 32 (C).

KEY TOPICS

neurons	peripheral nervous system	taste
synapses	vision	smell
central nervous system	hearing	touch

INTRODUCTION

In vertebrates the nervous system is **bilateral** and shows **cephalization** in which much of the nervous system, namely the brain and sense organs, is concentrated in the head region. The function of the nervous system is to coordinate the activities of the different parts of the body. **Receptor cells** located in internal and external sense organs send it information, allowing it to respond by means of its **ef-**

fectors, which are usually muscles or glands. The nervous system maintains equilibrium, or **homeostasis,** within the body and allows the organism to respond to external stimuli with appropriate **behavior** patterns. In contrast to the endocrine system, which provides slow and long-lasting regulation, the nervous system provides for a rapid response to stimuli.

The **neurons,** or impulse-conducting nerve cells, of the vertebrate nervous system and their supporting cells are organized into the central nervous system and the peripheral nervous system. The brain and spinal cord make up the **central nervous system (CNS),** which receives input from the major sense organs and coordinates most of the body's activities. Many types of chemicals act as the **neurotransmitters** by which the individual neurons interact with each other within the CNS: acetylcholine, norepinephrine, dopamine, serotonin, glutamic acid, GABA, glycine, endorphins and enkephalins.

The **peripheral nervous system** consists of all nervous tissue outside of the CNS. Acetylcholine and norepinephrine, the neurotransmitters of the peripheral nervous system, are the best understood of the many known neurotransmitters. **Cholinergic neurons** are those that release acetylcholine and **adrenergic neurons** release norepinephrine. The **somatic** part of the peripheral nervous system involves the cranial and spinal nerves and their associated sense organs. This system controls voluntary activities such as muscular movement and uses acetylcholine as the neurotransmitter. The **autonomic** part of the peripheral nervous system controls those internal functions that are usually carried out at the subconscious level. This system operates with its own neurons and nerves, but also communicates with the CNS via the cranial and spinal nerves. It includes the **sympathetic** nervous system, which uses norepinephrine as the final neurotransmitter, and the **parasympathetic** nervous system, which uses acetylcholine.

NEURONS

We humans have about 10^{11} **neurons** or nerve cells. They conduct electrochemical impulses from one site to another through long cell processes. Those processes leading toward the cell body are called **dendrites,** and those leading away are called **axons** (see Figure 27–1). Excitation of the dendrite or of the

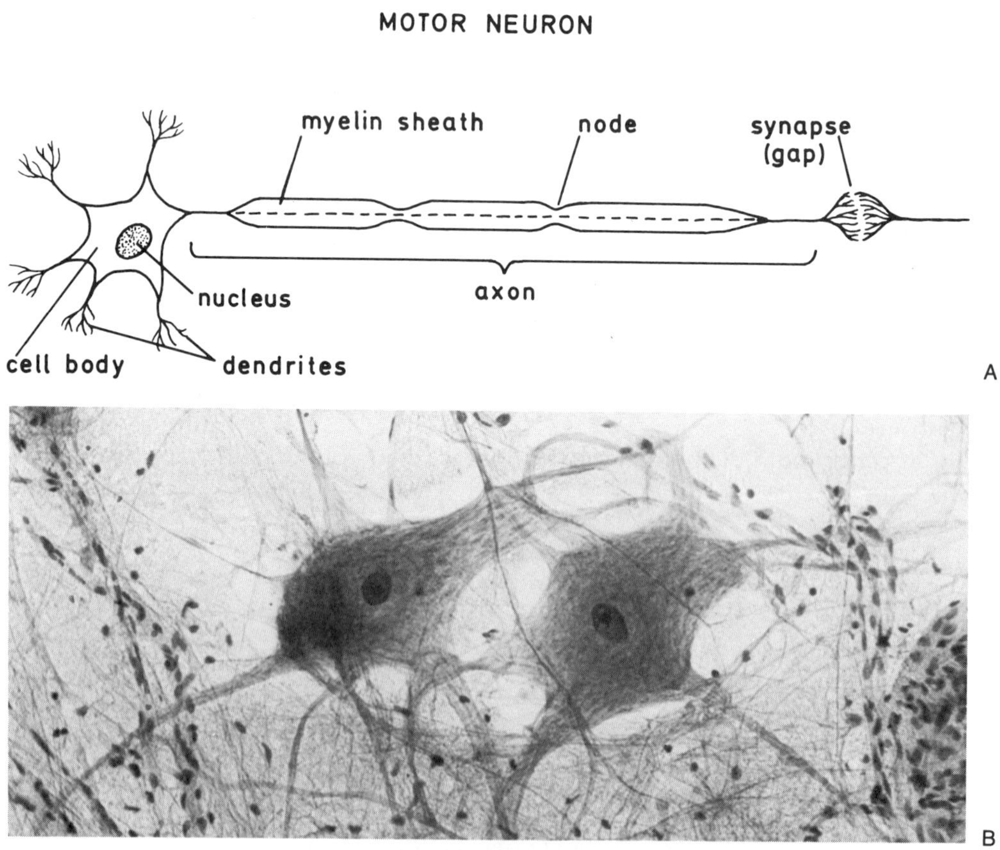

Figure 27–1. Neurons. The motor neuron shown in (A) is adapted for rapid transmission of impulses over its long axon. Its many dendrites receive impulses from neighboring neurons. In (B) you see the cell bodies of two motor neurons and the nuclei of many glial cells. (Photo by Carolina Biological Supply Company.)

membrane of the cell body results in a localized response proportional to the stimulus called a **generator potential** or **postsynaptic potential.** These potentials can add up or show summation and they fade as they spread over the membrane of the neuron. Excitation of the axon causes an **action potential.** This is an all-or-none response to a generator potential that is above a certain **threshold** when it reaches the base of the axon, and it maintains its size as it travels along the axon. An action potential can be repeated only after a **refractory period,** and it never shows summation with other action potentials.

Both types of processes usually interact with those of adjacent neurons, with receptor cells or effector cells across narrow gaps called **synapses,** typically 1,000–10,000 per neuron. **Neurotransmitters** are those chemicals mentioned above that conduct nervous signals from one neuron to the next by diffusing across the gap of the synapse (synaptic cleft). Synaptic transmission is described in Text Figure 29–11. Neurotransmitters are removed from the synaptic cleft by enzymes or by readsorption into synaptic vesicles within the neuron that released them. Dendrites generally receive neurotransmitter signals across their synapses from neighbor neurons, whereas axons release neurotransmitters to stimulate adjacent neurons.

Neurosecretion occurs when neurons secrete hormones rather than neurotransmitters. Like other hormones, these have a longer-lasting effect than neurotransmitters. Good examples of neurosecretory neurons are found in the hypothalamus of the brain and the posterior pituitary. When a hormone acts only locally, close to the neuron that secreted it, it is known as a **neuroregulator.** An example is the hormone that brings on sleep.

- Examine a prepared slide of neurons under low then high power in the microscope.
- Locate the **cell body, nucleus, axon,** and **dendrites.**

Where does the neuron receive neurotransmitters travelling *toward* it?

____

Where in a neuron does integration or summation of incoming impulses occur?

____

Where is an all-or-nothing action potential generated?

____

What part of the neuron releases neurotransmitters travelling *away* from it?

____

In vertebrates, some axons are surrounded by special Schwann cells, which are filled with a fatty insulating material called **myelin** (see Text Figures 4A and 29–9). The effect of the myelin sheath is to speed up the rate of impulse transmission along the axon. At intervals along the length of the axon, there are interruptions in the myelin sheath known as the nodes of Ranvier. Instead of travelling at the normal rate, the impulse skips from node to node at a much faster rate (see Text Figure 29–10). Myelinated nervous tissue appears white, whereas unmyelinated tissue is gray.

Each neuron is supported by many **glial cells,** which are very important in nervous tissue but do not conduct impulses (see Figure 27–1). Other types of cells are also important in the development and proper functioning of the nervous system.

THE CENTRAL NERVOUS SYSTEM

The Brain

The vertebrate brain is divided into the forebrain, the midbrain, and the hindbrain. In mammals the forebrain is the most prominent part, and it is considerably more developed than it is in lower vertebrates. The external features of a sheep brain are shown in Figure 27–2.

The brain is covered by a triple layer of membranes called the **meninges.** The skull, meninges, and a cushion of **cerebrospinal fluid** protect the brain. The hollow spaces within the brain are called **ventricles.** Both they and the hollow interior of the spinal cord are filled with cerebrospinal fluid.

The **forebrain** consists of the cerebral hemispheres and the underlying parts: the thalamus, hypothalamus, hippocampus, and pineal gland. The **midbrain** in mammals is a small region near the medulla (anterior and posterior colliculi) that is concerned with certain responses of the eye and ear. The **hindbrain** is the cerebellum and medulla.

- Examine a preserved sheep brain, and note the two **cerebral hemispheres,** which are the brain's largest structures. Although the hemispheres appear to be symmetrical here, in humans the two halves differ in size, and some functions are carried out only in one or the other hemisphere.
- Locate the longitudinal **cerebral fissure** between the two hemispheres.
- Spread the two hemispheres apart and find the thick, white **corpus callosum.** It consists of bundles of myelinated and unmyelinated axons, and connects the two hemispheres to each other. It is considered to be **white matter** because of its high myelin content.
- The **cortex** of the highly convoluted cerebrum is the thin outer layer of **gray matter** consisting of

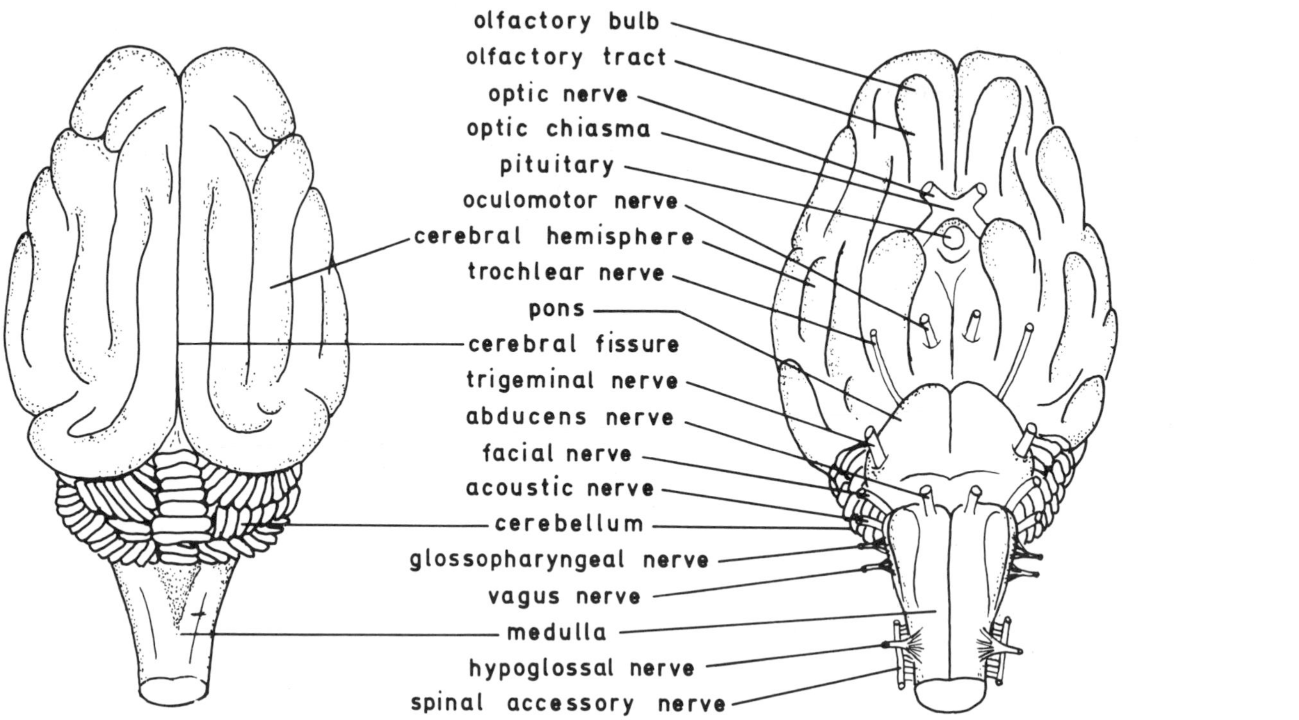

Figure 27–2. Sheep brain. The cerebral hemispheres are much smaller relative to the rest of the brain than in humans. The pituitary may be missing in your specimen.

cell bodies and unmyelinated axons. The cerebral cortex is important for associations, memory, and higher brain functions such as consciousness, language, and abstract thinking. Conscious sensations reach the **sensory cortex,** and the **motor cortex** controls voluntary movement (see Text Figure 29–16). The right side of the cortex controls the left side of the body and vice versa. In the part of the cerebrum underneath the cortex are the **hippocampus** (affected in Alzheimer's disease) and the **limbic structures,** which regulate emotional states and may be involved in short term memory.

- Examine the ventral surface of the brain, and find the **olfactory bulbs** and **olfactory tracts,** which lead from the chambers of the skull near the nasal passages. These are parts of the forebrain.
- Find the **optic nerves,** which cross in the **optic chiasma** before entering the brain.
- Posterior to the optic chiasma, look for the **pituitary,** an endocrine and neuroendocrine gland suspended from the hypothalamus of the brain. Secretions from the hypothalamus regulate the release of hormones from the pituitary gland (see Text Figure 32–6). (The pituitary may be lost or damaged in your sheep brain.)
- Find the **cerebellum,** which is a rounded structure with a convoluted surface just posterior to the cerebrum. It coordinates muscular movement and is important in maintaining equilibrium.
- Identify the **medulla,** which is connected to and continuous with the anterior part of the spinal cord. The medulla is the location of vital centers for breathing (in the pons at the anterior end), circulation, and reflexes such as swallowing and vomiting. It contains clusters of cell bodies called **nuclei.** The medulla and cerebellum both develop from the hindbrain of the sheep embryo.

The internal structures of the brain are shown in Figure 27–3.

- Examine a sagittal section of the sheep brain, and find the location of the pituitary.
- The dense **hypothalamus,** just dorsal to the pituitary can now be seen better. It forms the floor of the third ventricle (hollow space) of the brain and is the principle center for regulating the internal organs. In addition to controlling the pituitary, it regulates hunger, thirst, osmoregulation, sexual maturity and sex drive, and body temperature.
- Just dorsal to the hypothalamus, on the sides of the third ventricle, find the white matter of the **thalamus.** This is an important area for sorting sensory messages before they enter the cerebrum.
- Locate the **midbrain,** posterior to the thalamus and anterior to the cerebellum. Its two anterior protuberances are the **anterior** (or superior) **colliculi,** which control visual reflexes. The posterior protuberances are the **posterior** (or inferior) **colliculi,** which receive auditory input. The ventral part of the midbrain, the cerebral peduncles, contains tracts of fibers that conduct impulses anteriorly and posteriorly.

The parts of the brain are connected by various tracts of fibers. The medulla transmits impulses between the spinal cord and the rest of the brain. The

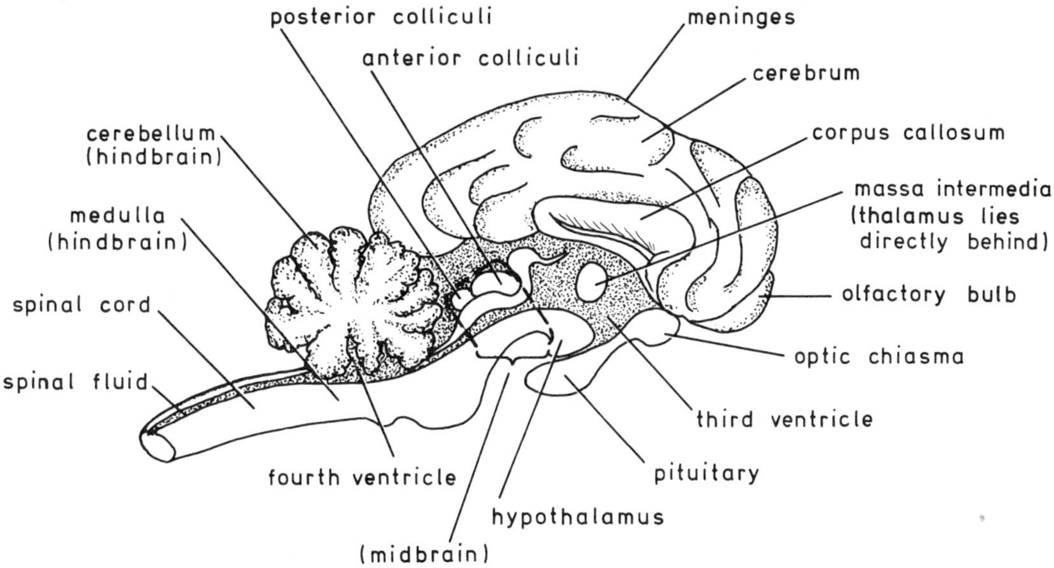

Figure 27–3. Sheep brain in sagittal section. The ventricles of the brain are continuous with the interior of the spinal cord and contain cerebrospinal fluid.

reticular formation is a complex neural pathway connecting the medulla and thalamus. It communicates with the cerebral cortex via the **reticular activating system (RAS),** which receives various kinds of sensory input and filters it before it can reach the cortex and be consciously perceived (see Text Figure 39–8). The RAS also controls the general arousal level of the animal and how strongly it responds to external stimuli.

Many drugs act by mimicking neurotransmitters in the brain. For example, amphetamine, which is very similar in structure to norepinephrine, interferes with the action of this neurotransmitter, so that the reticular activating system becomes abnormally stimulated. This causes euphoria and hyperactivity.

The Spinal Cord

A cross section of the spinal cord is illustrated in Figure 27–4 (also see Figure 29–19 in your text).

- Examine a slide of the spinal cord in cross section in your dissecting microscope, and orient it so that the **ventral fissure** is toward you.

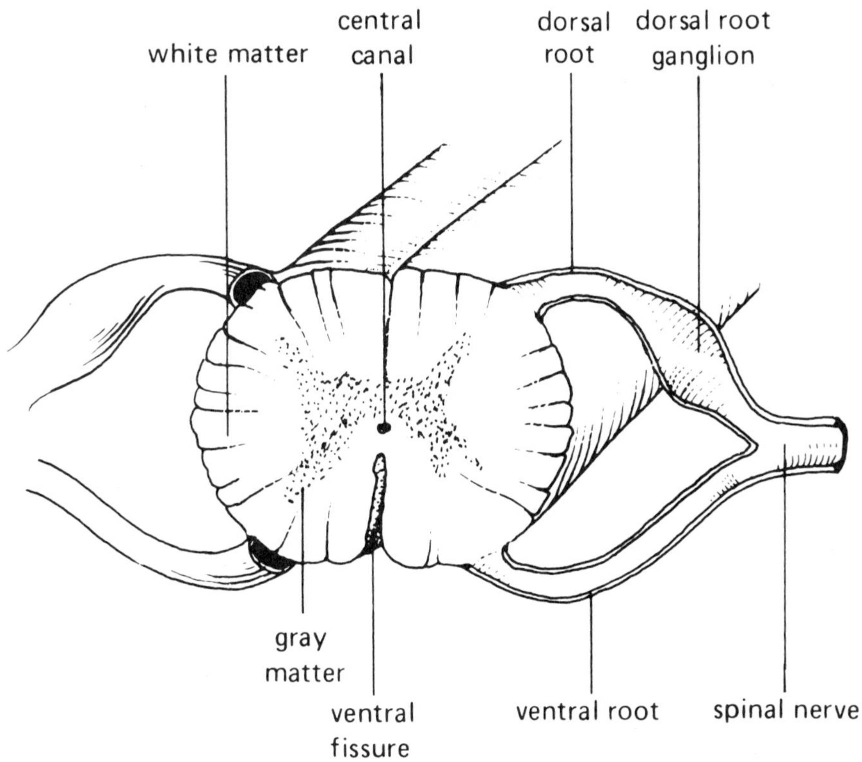

Figure 27–4. Spinal cord. This cross section is at the level of a spinal nerve and shows the dorsal and ventral roots. The white matter containing the axons surrounds the gray matter, which consists of cell bodies.

- Locate the H-shaped area of **gray matter** around the central canal; it consists mainly of cell bodies.
- Note the **central canal** that is filled with cerebrospinal fluid.
- Notice the **white matter** that makes up the rest of the spinal cord; it contains axons and dendrites.

How does the arrangement of gray and white matter here compare with the arrangement in the brain?

___

___

___

- Find the right and left **dorsal roots** through which sensory nerves enter the spinal cord. Each nerve is like a cable containing hundreds or thousands of axons or dendrites.
- Near the dorsal root, find the **dorsal root ganglion,** which contains the cell bodies of the **sensory neurons.** (A ganglion is a cluster of cell bodies found outside the CNS.)
- Locate the right and left **ventral roots** through which **motor neurons** carry impulses away from the spinal cord.
- Examine the ventral part of the gray matter in the microscope under low then high power.
- Locate the cell body of a motor neuron, a large, triangular cell that is darkly stained.

What is the function of the motor neurons?

___

___

- Locate an **interneuron** in the dorsal portion of the gray matter; it will have a more rounded cell body.

What is the function of the interneuron?

___

___

- Find the rounded body of a **sensory neuron** within the dorsal root ganglion.

In which direction does this cell conduct impulses?

___

- Examine the white matter to see cross sections of nerve fibers. The dark central part of each fiber is the axon of the neuron, and the space around it is where the myelin is located. (Myelin may be absent as a result of the process of making the slide.)
- Use the diagram in Figure 27–4 to sketch in a **polysynaptic reflex arc.** Include the sensory neuron, the interneuron, and the motor neuron in your sketch, and indicate the direction of impulse and synaptic transmission with arrows. Such a reflex action is automatic and depends only on the anatomy of the neuronal circuit involved.

Why is transmission in a polysynaptic reflex one-way?

___

___

Although reflexes are very important, our nervous system shows great **plasticity,** or the capacity to change its responses over time due to experience. In other words we **learn** and can modify our behavior and even some reflexes to suit changing situations. **Memory** is the capacity to retain over time what we have learned.

THE PERIPHERAL NERVOUS SYSTEM

The peripheral nervous system consists of nervous tissue outside of the CNS. It includes all of the **nerves** and **peripheral ganglia,** clusters of cells found alongside the spinal cord and elsewhere.

The **somatic nervous system** is the part of the peripheral nervous system concerned with perception of and response to the outside world. It involves the spinal and cranial nerves and their sensory receptors.

The dorsal and ventral roots join to form a **spinal nerve** as shown in Figure 27–4. There are 31 pairs of spinal nerves branching from the human spinal cord. The spinal nerves and spinal cord can control reflex behavior without any input from the brain. Normally most responses other than very basic reflexes are modified by impulses from the brain. Only some of the sensory input to the spinal cord will reach the brain.

The major sense organs and heart are served by 12 pairs of **cranial nerves,** which lead directly into the brain rather than first passing through the spinal cord. Their location is shown in Figure 27–1.

- Locate as many of these cranial nerves as you can on the sheep brain:

I	Olfactory	sensory	smell
II	Optic	sensory	sight
III	Oculomotor	motor*	eye muscles
IV	Trochlear	motor*	eye muscles
V	Trigeminal	mixed	face, mouth
VI	Abducens	motor*	eye muscles
VII	Facial	mixed	face, tongue
VIII	Auditory	sensory	hearing, balance
IX	Glossopharyngeal	mixed	mouth, throat
X	Vagus	mixed	neck, internal organs
XI	Spinal accessory	motor*	larynx, shoulder muscles
XII	Hypoglossal	motor*	tongue muscles

* Motor nerves always include sensory fibers that carry sensory messages from the muscle proprioceptors to the CNS.

The **autonomic nervous system** is that part of the peripheral nervous system concerned with maintaining **homeostasis** in the internal environment. It includes ganglia alongside the spinal cord (sympathetic cell bodies) or near target organs (parasympathetic cell bodies) and nerves connecting these ganglia to major organs, blood vessels, and skin. The **sympathetic** division triggers appropriate physiological responses to an emergency situation known as "**fight or flight**" responses (see Text Figure 32–5). The **parasympathetic** division generally has the opposite effect and stimulates "**relax and digest**" responses appropriate for routine activities. The autonomic nervous system communicates with the CNS via the cranial and spinal nerves. Although it usually works without conscious control, it can be influenced by the central nervous system. An emergency response, for instance, is often initiated by a terrifying visual or auditory stimulus. In conditioning by **biofeedback,** the responses of the autonomic nervous system are relayed to the conscious level of the CNS, which then allows the brain to exert a positive or negative influence on the response. In this way "autonomic functions" can be brought under conscious control to a certain extent.

SENSE ORGANS

The various forms of energy that affect an organism are called **stimuli.** Sense organs contain specialized cells or neurons called **receptors** and accessory structures that increase the efficiency of the receptors in perceiving stimuli. Receptors **transduce** physical stimuli into **generator potentials** and nerve impulses. For example, the eye transduces light into nerve impulses. The receptors synapse with sensory neurons that carry impulses through nerves to the CNS or to ganglia of the autonomic nervous system, or they carry the impulses themselves. We tend to think of sense organs as detecting stimuli external to the body, but many of our sense organs detect stimuli within the body. Sense organs that detect movement and position of the body parts are called **proprioceptors.**

The five types of receptors are:

1. **mechanoreceptors,** which respond to movement or physical stress.
2. **photoreceptors,** which respond to light.
3. **thermoreceptors,** which respond to heat or cold.
4. **chemoreceptors,** which respond to chemical substances.
5. **electroreceptors,** which respond to electrical fields.

In this exercise you will be studying the first four types. Electroreceptors are found in the lateral line organ of certain fish.

An important property of receptors is **adaptation** in which the response to a constant stimulus diminishes with time. Rapidly adapting receptors are suited for detecting only fast changing stimuli, while slowly adapting receptors accurately monitor stimuli that change only gradually.

The action potentials reaching the brain from the various sense organs are all alike. They are perceived as different sensations only because the impulses arrive in different specialized parts of the brain.

The Eye

In the eye, photons of a visual stimulus are transduced into nerve impulses by photoreceptors in the **retina.** In mammals there are two types of receptors: **rods** for perceiving the intensity of light and **cones** for detecting color or wavelength (see Color Plate 4). The cones are concentrated in the **fovea** of the retina, the area of most acute vision. The rods and cones synapse with neurons in the retina, the neurons interact in the initial processing of the visual input, and finally the information is transmitted to the brain via nerve impulses in the optic nerves. The actual transduction process occurs when the photon is absorbed by **retinaldehyde,** the vitamin A–derived part of the visual pigment **rhodopsin** and undergoes a structural change: 11-*cis* retinaldehyde changes to all *trans* retinaldehyde. This change causes a depolarization of the receptor cell membrane and sets up a generator potential. If enough photons strike the cell, an action potential will result.

The structure of the vertebrate eye is shown in Figure 27–5.

- Take a preserved sheep eye and note the outer covering, the **sclera,** to which muscles are attached to move the eye.
- Find the **optic nerve** on the posterior surface.
- The transparent part of the sclera over the pupil is the **cornea.** The cornea is thicker in the center and acts as a lens to help magnify the image.
- Beneath the cornea, find the **iris,** which surrounds the opening into the eye, or **pupil.** The amount of light entering the pupil is regulated by the contraction of muscles controlling the iris.
- Orient the eye so that the optic nerve is to your left and the pupil is horizontal.
- Cut through the eye with your scalpel from front to back leaving part of the sclera at the bottom intact to hold the halves together.
- Find the transparent **lens** just behind the pupil. It is hard in a preserved eye, but in life it is flexible and can be thickened or flattened by its associated muscles to focus the image on the retina. In most people over 40, the lens is no longer flexible enough to change shape, and corrective lenses must be worn for close vision.

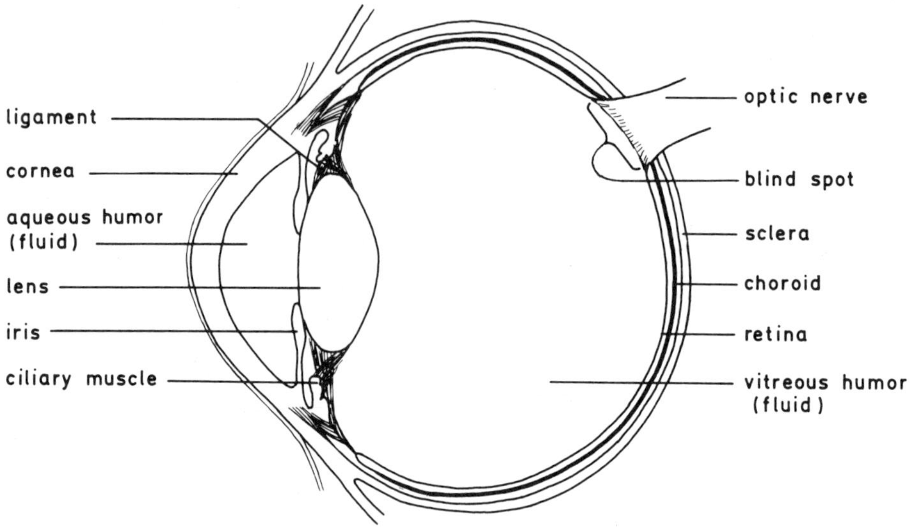

Figure 27–5. The vertebrate eye. Light passes through the transparent cornea and the pupil to the cells of the retina, where its energy is transduced into nerve impulses. The iris regulates the amount of light entering the eye by changing the pupil size.

- Note the **aqueous humor,** the fluid material that was contained between the cornea and lens.
- The main part of the eye is filled with the gel-like **vitreous humor.**
- Print something on a scrap of paper.
- Place a blob of vitreous humor on the printing.

Does the vitreous humor contribute to magnification of the image? ______________

- Note the black lining of the sides of the eye. The pigment is located in the **choroid** coat, which also contains blood vessels.
- Locate the **retina,** which lines the interior part of the eye opposite the pupil. The retina contains the rod and cone receptor cells, nerve cells, and blood vessels.
- Locate the white area in the retina where the optic nerve leaves the eye. This **blind spot** is filled with nerve fibers and has no receptor cells.

When would the iris be fully open?

____

When would the iris be tightly closed?

____

Why is the black lining important?

____

____

When light rays pass through the lens of the eye, they are bent so that the image of an object falls on the retina upside down. A thicker lens bends the rays more strongly. The lens of the eye thickens in order to bring nearby objects into focus. The shape of the lens is changed by the contraction of the **ciliary muscle** attached to it. **Accommodation** is the process of adjusting the shape of the lens so that the image falling on the retina is in focus. When the lens fails to focus properly, a person may be nearsighted, farsighted, or astigmatic.

- Sketch an object in front of the eye shown in Figure 27–5, and show how its image would fall on the retina.

If the object were moved farther from the eye, would the lens have to thicken or flatten to keep the image in focus? ______________

Why don't you see objects upside down?

____

____

The Ear

The structure of the human ear is shown in Figure 27–6. The outer ear is the part you are already familiar with: It consists of the **pinna** and the **external canal.** Water in the external canal gives you the sensation of "water in the ear," and too much ear wax in this canal will interfere with the transmission of sound to the **eardrum** (**tympanic membrane**) at its inner end.

What is the function of the pinna?

____

- Study the features of the middle ear and inner ear in a cutaway model.
- The **middle ear** consists of the space between the eardrum and the cochlea of the inner ear. It contains three small bones that conduct the sound, and it is connected to the pharynx by the **Eustachian tube.**

Why is a connection of the middle ear with the exterior necessary?

____

____

Respiratory infections can travel from the throat up

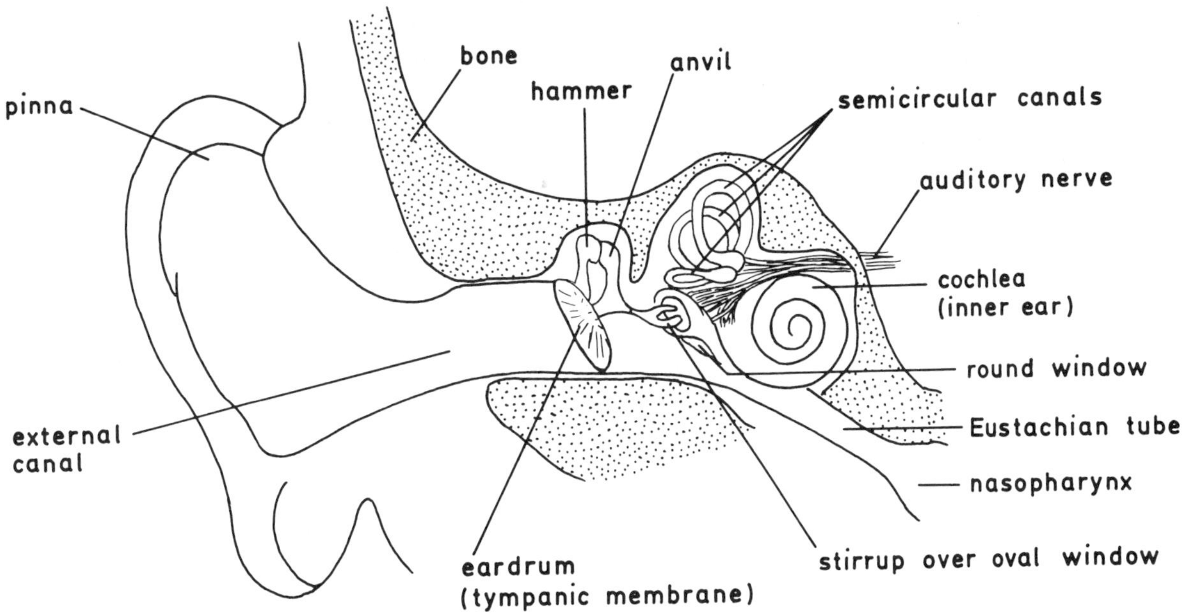

Figure 27–6. The human ear. The mechanical energy of sound waves is transduced into nerve impulses by cells of the organ of Corti within the cochlea. The labyrinth with its semicircular canals senses the movement and position of the head and is the organ of equilibrium.

the Eustachian tubes and cause painful inflammation and infection of the middle ear.

- Locate the **cochlea** in the inner ear. The cochlea contains the organ of Corti, which is the organ of hearing.

If the cochlea were uncoiled, a cross section of it would appear as shown in Figure 27–7.

Within the cochlea there are three fluid-filled canals: the **vestibular canal,** which connects to the oval window, the **cochlear canal,** and the **tympanic canal.** Sound waves reaching the oval window set up vibrations within the vestibular canal which then travel back through the tympanic canal to the round window. Indirect vibrations are set up in the cochlear canal, which contains the mechanoreceptor

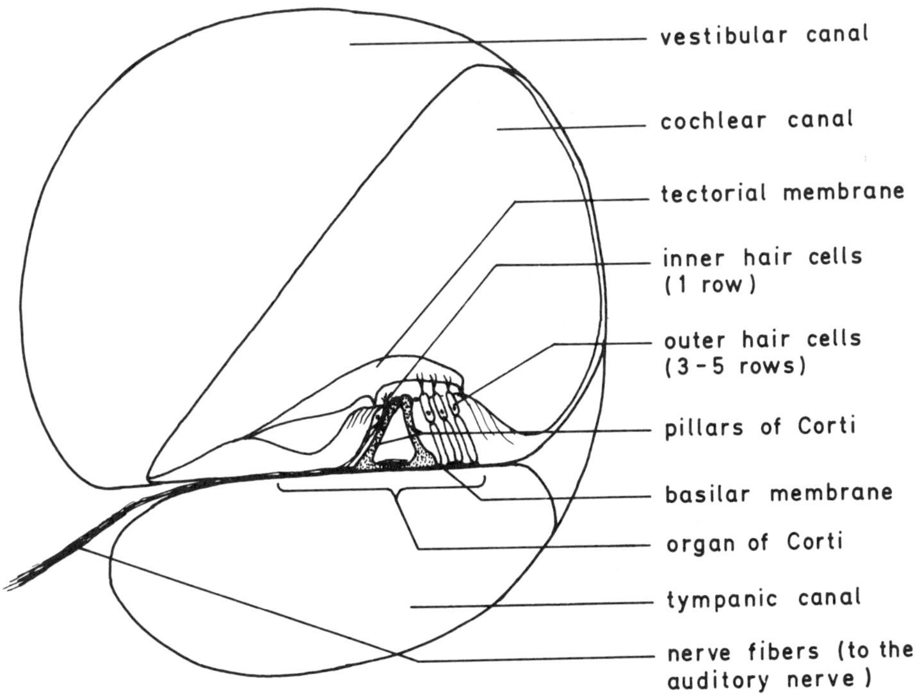

Figure 27–7. Cross section of one spiral of the cochlea. Vibrations from the bones of the middle ear travel through the vestibular canal and back through the tympanic canal. The organ of Corti, located between the basilar and tectorial membranes, consists of mechanoreceptors (hair cells) that detect the vibrations and transduce them into nerve impulses.

hair cells of the **organ of Corti** (see Text Figure 30–5). Vibrations of the tectorial membrane above these cells cause movements of the gel-like matrix around the hairs and excite the cells by moving the hairs. Impulses from the receptors are transmitted along the basilar membrane on which the cells are located and reach the auditory nerve, which carries them to the brain. Vibrations of different frequencies (pitches) will excite cells at different positions along the membrane, so that pitch as well as intensity can be sensed by the ear. Frequent exposure to loud noise will damage cells at the base of the organ of Corti and result in deafness to high-pitched tones. Most deafness in older people is caused by such environmental exposure.

- Examine a microscope slide of the **cochlea;** several sections through the three canals will be visible.
- Identify the **vestibular, cochlear,** and **tympanic canals** in one section.
- Locate the **organ of Corti** between the basilar membrane and the tectorial membrane.
- A second part of the inner ear, the **labyrinth,** has nothing to do with hearing, but also contains hair cells that are mechanoreceptors. This organ of equilibrium consists of three fluid-filled **semicircular canals** that are stimulated when the fluid flows through them in either direction. Each canal has an ampulla (sac) in which fluid pressure against a projecting cupula is detected by its hair cells (see Text Figure 40–12b). The semicircular canals are in three planes and allow the organ to sense **acceleration** (change in speed or direction) of the head as it is turned.
- At the base of the semicircular canals are two sacs, the **sacculus** and the **utriculus,** which are also fluid-filled and contain crystals of calcium carbonate (otoliths). The crystals rest on a bed of sensory hair cells and change position slightly as the head is moved (see Text Figure 40–11). The pattern of cells being stimulated while the head is at rest in any position allows the organ to sense **resting position** of the head relative to the force of gravity. It also detects acceleration in a straight line.

The sense of resting position in humans is very weak, as any pilot or scuba diver can testify, and is easily overridden by visual cues. The sense of movement is much stronger. When it conflicts with visual cues, as during a plane flight or boat trip, the result is the nausea of motion sickness.

HUMAN SENSORY PERCEPTION

The Blind Spot

- Obtain a sheet of white paper, and mark on it a heavy black cross and a heavy black dot about 8 cm apart.
- Cover your left eye, and hold the paper at arm's length with the dot at the left and the cross at the right.
- Stare hard at the dot while you slowly move the paper toward you.
- Note that the cross is visible at first, disappears completely, and then reappears.

What do you see instead when the cross disappears?

____

____

Map your blind spot as follows:

- Turn the paper sideways so that there is a clear space to the right of the black dot. Lay it flat on the desk.
- Cover your left eye, and place your right eye about 25 cm from the black dot. Stare at the dot intently while your partner moves a small, brightly colored object (such as a bead on a white pipecleaner) toward the dot from your right side.
- As soon as the object disappears, tell your partner to stop and mark the position of that edge of the blind spot.
- Next ask your partner to move the object from the black dot toward the blind spot, and then to stop and mark the spot as soon as the object disappears.
- Repeat several times, moving the object toward the blind spot from different directions, until your blind spot is well mapped, and then draw in its outline.

Does your blind spot have a regular or irregular shape?

____

Compare the size and shape of your blind spot with those of other members of the class. The distance from the dot to the blind spot will depend partly on the distance between your eye and the dot.

Explain why you have a blind spot in each eye.

____

____

Why don't you have an empty space in your field of vision all the time?

____

____

Afterimages

- Place a small square of solid blue paper or cardboard on a sheet of white paper, and put it under strong illumination.
- Working with your partner, stare intently at the blue paper while your partner times 60 sec. Then shift your gaze and stare intently at the plain white paper.

Do you see an afterimage? ______________

How long did the afterimage last? ______________

What color is it? ______________

- Place a small square of solid yellow paper on a white sheet, and repeat your observation.

What color is the afterimage when the object is a yellow square? ______________

While you are staring at the blue square, the cones that are sensitive to the blue light become **adapted** (less responsive), whereas those sensitive to red and green light are unstimulated. Then when you suddenly switch to white light, which contains all three colors, the blue receptors give a response much lower than normal, whereas the red and green receptors give a response stronger than normal. This is exactly the response that occurs with yellow light, and your brain interprets the plain white paper as being yellow. As the receptors return to their normal response levels, the afterimage slowly fades.

Chemoreception

Taste or **gustation** is due to **taste receptors** found in taste buds of the tongue papillae (see Text Figure 30–14). **Smell** or **olfaction** is due to the free endings of the **olfactory sensory neurons** of the olfactory epithelium that lines the roof of the nasal cavity. Often both taste and olfaction are involved in flavor discrimination. Other chemoreceptors monitor the levels of various substances in the body that must be regulated to maintain homeostasis.

Record your observations in Table 27–1 at the end of this topic.

TEST A

- Obtain slices of apple and potato cut to the same size.
- Close your eyes, pinch your nose shut, and have your partner place one slice on your tongue. Try to detect whether the slice is from an apple or from a potato without moving it around in your mouth and without breathing for 10 sec. Record your identification of the slice as correct (+) or incorrect (−) in Table 27–1.
- Repeat several times with slices of apple and potato in random order.

TEST B

- Suck on the slice, and move it around on your tongue. Record your observations.
- Repeat several times with slices of apple and potato in random order.

Can you distinguish apple and potato by taste alone?

____

TEST C

- Blot your tongue dry, and repeat the tests. Record your results in the table.

Is saliva necessary for tasting? ______________

TEST D

- Chew a slice of potato or apple with your nose pinched shut. Record your results.

Can you distinguish apple from potato without olfaction?

____

TEST E

- Again chew a slice of apple or potato, but this time breathe through your nose. Record your results.

Is it easier to distinguish them using olfaction?

____

TEST F

Proceed as follows to map the location of your taste receptors. Your observations will be a rough estimate of where the receptors are located, since a single taste papilla or even a single taste bud can respond to more than one of the four basic tastes. Spots to be tested are shown in Figure 27–8 at the end of this topic.

- Obtain dropping bottles of the following solutions:

5.0% sucrose	(sweet)
10% NaCl	(salty)
0.1% quinine sulfate	(bitter)
1.0% acetic acid	(sour)

- Open your mouth wide, without sticking out your tongue, and have your partner place a drop of one solution on spot #1 on the tip of your tongue. Repeat on the side and finally the back as shown in Figure 27–8. Rinse after each drop with plain water.
- Rate the taste for each drop as absent (0), faint (+), or strong (++), and record the results in the appropriate column in Table 27–2 at the end of this topic.
- Repeat for the other three solutions.

When you finish testing, map your areas sensitive to sweet, salty, bitter, and sour tastes on the outlines in Figure 27–8 at the end of this topic.

Proprioceptors

Proprioceptors in your muscles and joints continuously inform your nervous system about the location of your body and its appendages without visual input. These are mechanoreceptors such as muscle spindles (for muscle contraction), Golgi tendon organs (for tendon stretching), and joint receptors (for ligament movement). Stretching of the membrane of such a receptor sets up a generator potential. A certain posture can be maintained by reflexes involving antagonistic muscle pairs and their proprioceptors. Test the effectiveness of proprioception alone by performing these actions with your eyes closed. Record the results in Table 27–3 at the end of this topic.

TEST A

- Stand with your feet together and arms at your sides. Have your partner measure the extent of swaying and your success in attempting to compensate for it.

TEST B

- Stand on one leg with your arms extended. Have your partner time how long you can stand on one leg with your eyes closed, then repeat with your eyes open.

How important is visual input in supplementing your proprioceptors?

____

Other proprioceptors (kinesthetic receptors) allow detection of movement of body parts without the benefit of visual input. Test your kinesthetic sense as follows.

TEST C

- Close your eyes, and extend your arms fully to the sides at shoulder height.
- Move your arms together without bending your elbows, and attempt to touch your index fingers side to side.
- Have your partner measure how far apart your fingers are if your fingers fail to touch.

TEST D

- Close your eyes, and extend your arms sideways at shoulder height, but this time bend your elbows sharply so that your index fingers point at each other.
- Move the fingers together until they touch.
- Have your partner measure how far apart your fingers are if they fail to touch.

Thermoreceptors

Thermoreception in mammalian skin is due to free nerve endings and specialized **Krause's end bulbs** (for cold). Other sense organs in the skin detect touch, and free nerve endings detect pain.

MAPPING THERMORECEPTORS

- Stamp a grid on a hairless part of your forearm, or draw one 3 × 3 cm marked off with a grid 0.5 × 0.5 cm.
- Have your partner hold a probe in a boiling water bath, dry it, and quickly touch it *lightly* to one point on the grid while you look the other way.
- Indicate whether you felt a sensation of warmth, and if so, record the result on the grid in Table 27–4 at the end of this topic.
- Repeat for the other points on the grid, and calculate the number of heat-sensitive points per square centimeter in this area of skin.
- Next, place a probe in a bath of ice water; dry, and touch it to one point on the grid. Record whether there was a sensation of cold.
- Repeat for the other points on the grid, and calculate the number of cold-sensitive receptors per square centimeter.

Which type of receptor is more concentrated in this area of skin? ______________

ADAPTATION

- Obtain three beakers; fill one with cold water, one with warm water, and one with lukewarm water.
- Place the index finger of your left hand in the beaker of warm water and the index finger of your right hand in the beaker of cold water. Then count to 60.
- Now place both fingers in the beaker of lukewarm water.

What sensation do you feel with your left finger?

____

What sensation do you feel with your right finger?

____

Explain your sensations.

____

____

When you are finished with this lab, return all materials and store your microscope properly.

REFERENCES AND FURTHER READING

Bullock, T.H. *Introduction to Nervous Systems.* San Francisco: W.H. Freeman, 1977.

Eccles, J.C. *The Understanding of the Brain.* New York: McGraw-Hill, 1977.

Grollman, S. *A Laboratory Manual of Mammalian Anatomy and Physiology,* 4th ed. New York: Macmillan, 1978.

Jones, D.G. *Neurons and Synapses.* Baltimore: E. Arnold, 1981.

Julien, Robert M. *A Primer of Drug Action.* San Francisco: W.H. Freeman, 1985.

Katz, B. *Nerve, Muscle and Synapse.* New York: McGraw-Hill, 1966.

Kuffler, S.W. *et al. From Neuron to Brain,* 2nd ed. Sunderland, MA: Sinauer, 1984.

Schmidt-Nielsen, Knut. *Animal Physiology: Adaptation and Evolution,* 3rd ed. New York: Cambridge University Press, 1983.

Shepherd, Gordon M. *Neurobiology.* Oxford: Oxford University Press, 1983.

Springer, S.P., and G. Deutsch. *Left Brain, Right Brain.* San Francisco: W.H. Freeman, 1985.

Thompson, Richard E. *The Brain. An Introduction to Neuroscience.* San Francisco: W.H. Freeman, 1985

Journal Articles

The September 1979 issue of *Scientific American* is devoted to the brain.

Alkon, D.L. "Learning in a marine snail." *Scientific American,* 64, July 1983.

Arnold, A.P. "Sexual differences in the brain." *American Scientist* 68: 165, Mar.–Apr., 1980.

Axelrod, J. "Neurotransmitters." *Scientific American,* 58, June 1974 (#1297).

Bloom, Floyd E. "Neuropeptides." *Scientific American,* 148, October 1981.

Constantine-Paton, Martha, and Margaret I. Law. "The development of maps and stripes in the brain." *Scientific American,* 62, December 1982.

DiCara, L.Y. "Learning in the autonomic nervous system." *Scientific American,* 30, January 1970 (#525).

Dunant, Yves, and Maurice Israël. "The release of acetylcholine." *Scientific American,* 40, April 1985.

Favreau, Olga Eizner, and Michael C. Corballis. "Negative aftereffects in visual perception." *Scientific American,* 42, December 1976 (#574).

Gombrich, E.H. "The visual image." *Scientific American,* 82, September 1972 (#548).

Gordon, B. "The superior colliculus of the brain." *Scientific American,* 72, December 1974 (#553).

Guillemin, R., and R. Burgus. "The hormones of the hypothalamus." *Scientific American,* 24, November 1972 (#1260).

Held, R. "Plasticity in sensory-motor systems." *Scientific American,* 84, November 1965.

Hodgson, E.S. "Taste receptors." *Scientific American,* 135, May 1961 (no offprint).

Hoffman, Donald D. "The interpretation of visual illusions." *Scientific American,* 154, December 1983.

Huber, Franz, and John Thorson. "Cricket auditory communication." *Scientific American,* 46, December 1985.

Hudspeth, A.J. "The hair cells of the inner ear." *Scientific American,* 54, January 1983.

Jacobs, B.L., and M.E. Trulson. "Mechanisms of action of LSD." *American Scientist* 67: 396, July–Aug., 1979.

Jones, M.D., Jr., *et al.* "Oxygen delivery to the brain before and after birth." *Science* 216: 324, 16 April 1982.

Jouvet, Michel. "The states of sleep." *Scientific American,* 62, February 1967 (#504).

Land, Edwin H. "The retinex theory of color vision." *Scientific American,* December 1977 (#392).

Lassen, Niels A., David H. Ingvar, and Erik Skinhøj. "Brain function and blood flow." *Scientific American,* 62, October 1978 (#1410).

Llinas, Rodolfo R. "Calcium in synaptic transmission." *Scientific American,* 56, October 1982.

Loeb, Gerald E. "The functional replacement of the ear." *Scientific American,* 86, February 1985.

McEwen, B.S. "Interaction between hormones and nerve tissue." *Scientific American,* 48, July 1976.

Morell, Pierre, and William T. Norton, "Myelin." *Scientific American,* 88, May 1980 (#1469).

Nathanson, J.A., and P. Greengard. "Second messengers in the brain." *Scientific American,* August 1977 (#1368).

Nicholls, J.G., and D. Van Essen. "The nervous system of the leech." *Scientific American,* 38, January 1974.

Oscar-Berman, M. "The neuropsychological consequences of long-term chronic alcoholism." *American Scientist* 68: 410, July–Aug., 1980.

Owens, D. Alfred. "The resting state of the eyes." *American Scientist* 72:378, July 1984.

Parker, Donald E. "The vestibular apparatus." *Scientific American,* 118, November 1980.

Poggio, Tomaso. "Vision by man and machine." *Scientific American,* 106, April 1984.

Routtenberg, A. "The reward system of the brain." *Scientific American,* 122, November 1978.

Rushton, W.A.H. "Visual pigments and color blindness." *Scientific American,* 64, March 1975 (#1317).

Schwartz, J.H. "The transport of substances in nerve cells." *Scientific American,* 152, April 1980.

Shashoua, Victor E. "The role of extracellular proteins in learning and memory." *American Scientist* 73(4): 364, July–August 1985.

Shepherd, G.M. "Microcircuits in the nervous system." *Scientific American,* 92, February 1978.

Snyder, S.H. "Opiate receptors and internal opiates." *Scientific American,* 44, March 1977 (#1354).

Tosteson, Daniel C. "Lithium and mania." *Scientific American,* 164, April 1981.

von Bekesy, G. "The ear." *Scientific American,* 66, August 1957 (#44).

Wallach, Hans. "Perceiving a stable environment." *Scientific American,* 92, May 1985.

Winter, Peter M., and John N. Miller. "Anesthesiology." *Scientific American,* 124, April 1985.

Wurtman, Richard J. "Alzheimer's disease." *Scientific American,* 48, January 1985.

Wurtman, Richard J. "Nutrients that modify brain function." *Scientific American,* 50, April 1982.

Wu, Chau H. "Electric fish and the discovery of animal electricity." *American Scientist* 72(6): 598, November 1984.

Yellot, John I., Jr. "Binocular depth inversion." *Scientific American,* 148, July 1981.

Zwislocki, Jozef J. "Sound analysis in the ear: A history of discovery." *American Scientist* 69(2): 184, March 1981.

Student Name ______________________ **Date** ______________

TABLE 27–1. Data On Taste Chemoreception

Record (+) if the identification is correct, (−) if the identification is incorrect, and (0) if no identification can be made. Space has been provided in each test for up to six sets of observations.

Test		1	2	3	4	5	6
A. No sucking	Apple						
	Potato						
B. Sucking	Apple						
	Potato						
C. No saliva	Apple						
	Potato						
D. Chewing	Apple						
	Potato						
E. Breathing and chewing	Apple						
	Potato						

TABLE 27–2. Mapping Data for Taste Chemoreceptors

Rate each drop of solution as either no taste (0), mild taste (+), or strong taste (++). Map the results in Figure 27–8.

		Position on Tongue Being Tested						
		Front			Side		Back	
Receptor	**Solution**	*1*	*2*	*3*	*4*	*5*	*6*	*7*
Sweet	5.0% sucrose							
Salty	10% NaCl							
Bitter	0.1% quinine sulfate							
Sour	1.0% acetic acid							

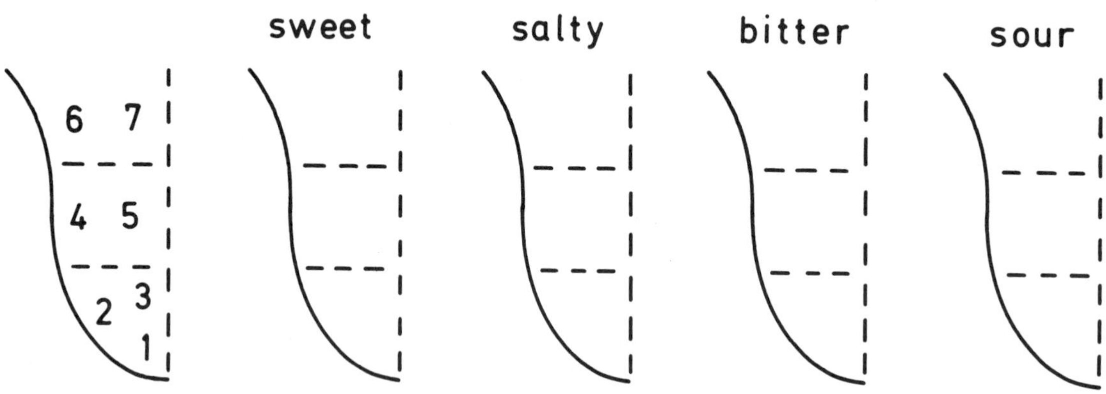

Figure 27–8. Results of taste test. Numbers label the spots to be tested. Mark the positive responses for each type of taste receptor where they occurred on the tongue.

Student Name ______________________________ **Date** ____________________

TABLE 27–3. Data for Proprioception of Static and Dynamic Equilibrium

Test	Observations
A. Feet together	
B. One leg, eyes closed One leg, eyes open	
C. Touch fingers	Fingers were ______ cm apart
D. Touch fingers	Fingers were ______ cm apart

TABLE 27–4. Data on Thermoreception in the Skin

Warmth Receptors ______________________________

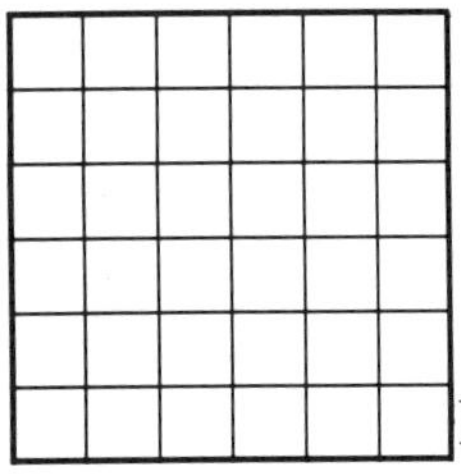

] .5 cm

concentration: ________/cm^2

Cold Receptors ______________________________

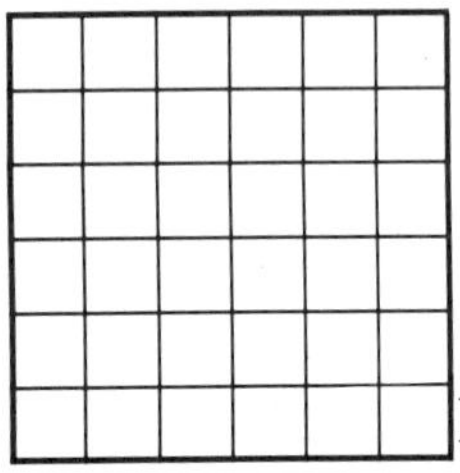

] .5 cm

concentration: ________/cm^2

TOPIC 28

Muscles and Skeletons

OBJECTIVES

When you have completed this topic, you should be able to:

1. Identify the major muscles of a vertebrate.
2. Identify the major bones of the human skeleton and distinguish whether they belong to the axial skeleton or the appendicular skeleton.
3. Give three functions of the skeleton and two nonskeletal functions of bone.
4. Describe the type of movement that is carried out by striated muscle, smooth muscle, and cardiac muscle; give the advantages of each muscle type.
5. Describe the major structural differences between bone and cartilage and give the advantageous properties of each tissue.
6. Explain what happens when a muscle is given an increasingly intense stimulus.
7. Explain what happens when a muscle is given an increasingly frequent stimulus.
8. Define: antagonist, motor unit, reciprocal inhibition, tetanus, fatigue, and twitch.
9. Distinguish between: tendon/ligament, cartilage/collagen, spongy bone/compact bone, tentanus/rigor mortis.

TEXT REFERENCES Chapter 31

KEY TOPICS

joints	striated muscle	bone
antagonistic muscles	smooth muscle	cartilage
skeletons	cardiac muscle	muscle physiology

INTRODUCTION

Muscles are **effectors.** They respond to stimuli detected by other cells. Glands, light organs, and electric organs are also effectors.

The major function of muscles is to move things by changing chemical energy to kinetic energy (motion). To be effective, muscles must exert force against a **skeleton.** This might be a shell, bones, or an internal space on which they can exert hydrostatic pressure. You have already studied the hydrostatic skeleton of the earthworm and the exoskeleton of the grasshopper. Vertebrates have a bony endoskeleton on which muscles act.

Locomotion is the most obvious form of movement in which the muscles in cooperation with the skeleton move the whole organism, but muscles also move blood, food, air, and eyeballs, regulate body temperature, and cause countless other physiological changes important in maintaining homeostasis. Many of our muscles work without being attached to bones. When they are in the form of a ring they can contract to close off a tube (pyloric sphincter), or when they are in the form of a tube they can contract in a ring or parallel to the length of the tube to change its shape and force the tube contents to move, as during peristalsis in the gut. Cardiac muscle contracts rhythmically for a lifetime

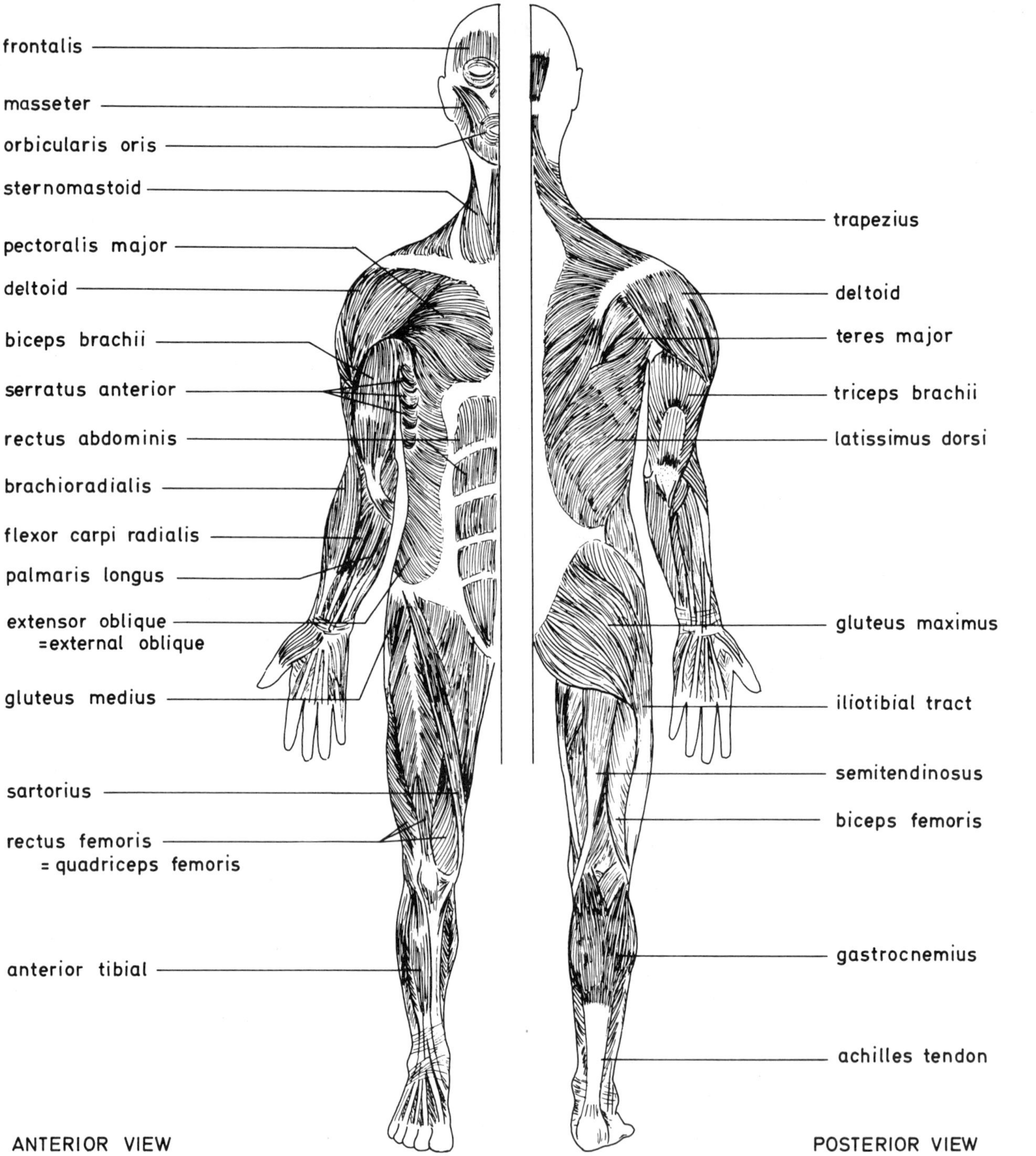

Figure 28–1. Human muscles. Skeletal muscles work in antagonistic pairs.

even without being stimulated to contract by individual nerve endings. Most of the activities of **smooth** and **cardiac** muscle are not under our conscious control, so we are not as aware of them as we are of **striated** muscle.

The endoskeleton also serves functions other than locomotion. It cradles and protects internal organs, especially the heart and brain, and it supports the body as a whole against the force of gravity. Antlers help some animals to protect themselves. The long bones are the main blood-generating organs and help maintain the immune system. Bones also serve as a depot for fat and calcium.

MUSCLE ANATOMY

Because muscles can react only by contracting and not by "expanding," they must work in antagonistic pairs and must be attached to the part to be moved, often a bone.

- Consider the bones that meet to form a **joint,** as shown in the sketch in Figure 28–9 at the end of this topic. They are held in place by connective tissue of the **ligaments.**
- Draw in a muscle attached to the bones of the joint that will draw them together so that the angle of the joint becomes smaller; this is the **flexor** of the joint. Label the upper attachment point the **origin** (nearest the body, moves the least) and the lower, the **insertion.** The tough connective tissue that attaches a muscle to a bone is called a **tendon.** Now draw in the antagonistic muscle, which widens the angle of the joint: This is the **extensor.** Label its origin and insertion as well.

Are the flexors or extensors important when you hit a baseball? ____________

Are flexors or extensors important when you pick up a book? ____________

Other pairs of antagonists work to draw in, raise, twist, or constrict body parts.

Because the forces developed by skeletal muscles are quite large, there must be reflex circuits that inhibit the contraction of a given flexor while the extensor is contracting, and vice versa. This set of reflexes is called **reciprocal inhibition.** Tension on a muscle triggers nerve impulses from its **muscle spindle receptors,** which inhibit the motor neurons of its antagonist. In this way the contraction of a muscle and relaxation of its antagonist are coordinated.

Study the human muscle pairs shown in Figure 28–1. If the **quadriceps femoris** muscle of the thigh contracted, the leg would be straight at the knee. If the **biceps femoris** contracted, the leg would be bent.

What pair of antagonists control the bending and straightening of the elbow?

____

____

What muscle is involved in each of the following activities?

Frowning: ______________________________

Puckering the mouth: ______________________________

Clenching the fist: ______________________________

Lifting a cup of coffee: ______________________________

Pushing a door shut: ______________________________

Rising from a deep knee bend: ______________________________

Shrugging your shoulders: ______________________________

Kicking a ball: ______________________________

The muscles of a frog are shown in Figure 28–2 for comparison. Also use this information for your dissection of the leg in the MUSCLE PHYSIOLOGY part of this laboratory.

SKELETONS

The vertebrate skeleton is divided into the **axial skeleton,** consisting of the skull, vertebral column, and sternum, and the **appendicular skeleton,** which consists of the pectoral and pelvic girdles and their appendages.

- Locate the axial and appendicular skeletons in Figure 28–3, which shows the bones of a human being.
- Study the major bones in Figure 28–3. Then complete Table 28–1 at the end of this topic by matching the technical names of the bones with their common names and deciding whether they are part of the axial or appendicular skeleton.
- If there are other vertebrate skeletons on demonstration in your laboratory, compare them to each other and to the human skeleton. Some good examples to use are a frog, cat, bat, pigeon, fish, snake, or turtle.
- Compare a bipedal animal with one that is quadrupedal.
- Notice the adaptations of the skeleton for the type of locomotion of the animal.
- Consider a particular bone such as the femur, ischium, or mandible, and see how it differs in different organisms. Pay particular attention to these features:

  Articulation of the spine and skull
  Structure of the vertebrae
  Ribs
  Scapula
  Clavicles
  Pelvic girdle
  Carpals and metacarpals
  Tarsals and metatarsals
  Phalanges of fore- and hindlimb

  Your instructor may ask you to make a table showing how these skeletal parts show different adaptations in different organisms.

- Study a long bone on demonstration or in one of the skeletons.

A typical long bone is shown in Figure 28–4. The ends are covered with smooth **cartilage** (hyaline cartilage) and rub against the cartilage of the adjacent bones. Muscles are attached to the bone by **tendons** made mostly of the tough structural protein **collagen.** During development, there are sections of cartilage within the end of the bone that grow as the bone lengthens. (The very ends of the bone, the **epiphyses,** have to be hard and bony or they would wear

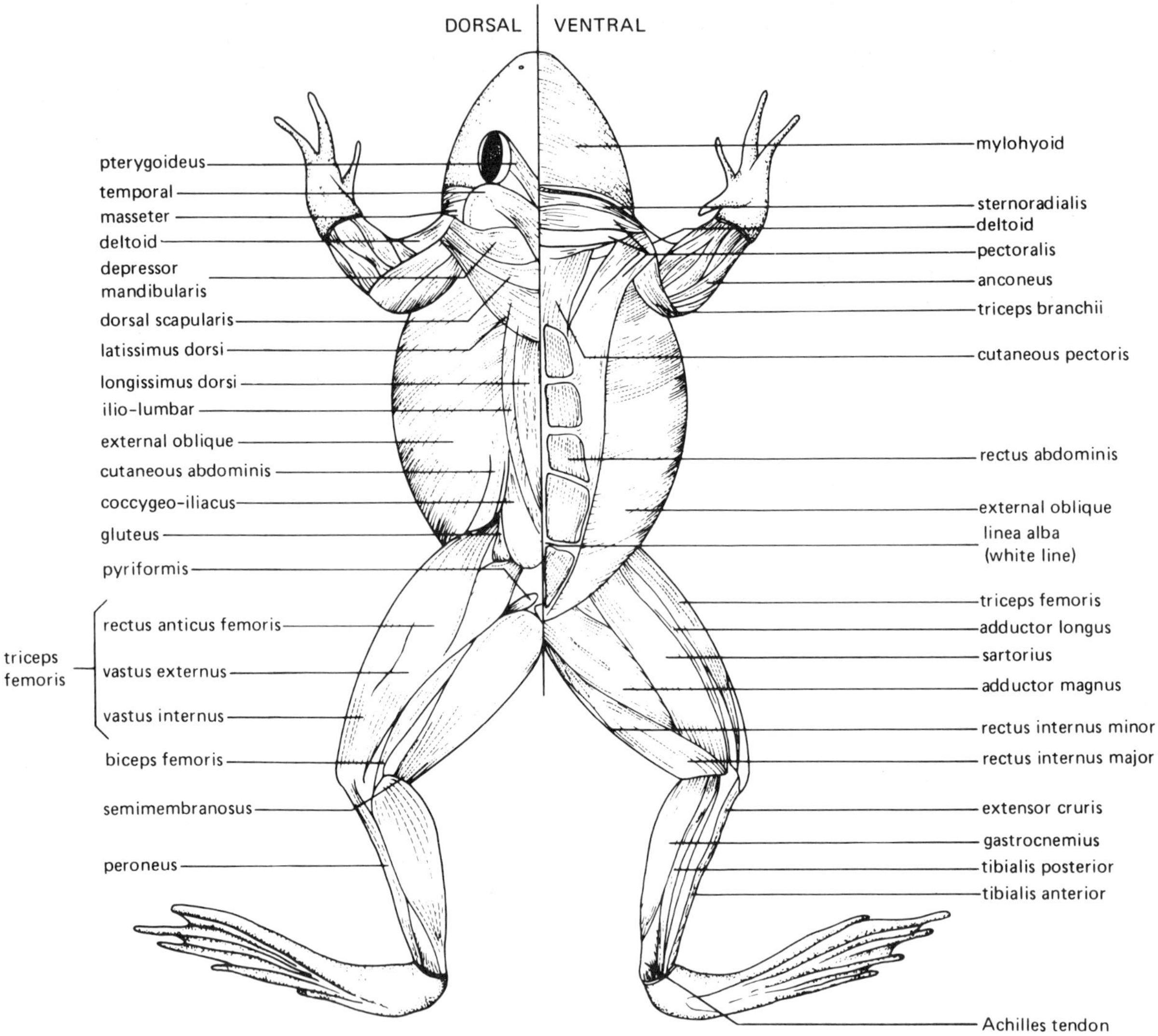

Figure 28–2. Frog Muscles. (Redrawn from Wodsedalek, J.E., and Charles F. Lytle. *General Zoology Laboratory Guide,* 8th ed. © 1981 Wm. C. Brown Company, Publishers, Dubuque, Iowa, Copyright © 1975 by Kendall/Hunt Publishing Co. Reprinted by permission of Kendall/Hunt Publishing Co.)

away.) When the growing cartilage becomes converted to bone so that the epiphyses and **shaft** (diaphysis) unite, growth in the length of the bone stops.

The hollow interior of the bone is the **marrow** and is filled with blood-forming cells and fat. In mammals, only the very outer shell of the bone is hard, or **compact bone.** The rest is **spongy bone,** which is rich in blood vessels.

The connection between any two bones is a joint. These are some different types of joints:

1. A **suture** is a joint where the bones don't move (example: skull).
2. A **hinge joint** is limited to movement in one plane (example: knee, finger joints).
3. A **ball and socket joint** allows rotation of the bones as well as flexion (example: hip, shoulder).
4. A **condyloid joint** allows movement in any direction but not rotation (example: joints between fingers and the hand).
5. A **plane joint** allows some slipping of two bones but very little flexon or rotation (example: joints between vertebrae).

The adjacent bones of a joint are held together by **ligaments** made of elastic connective tissue. Joints in which there is a lot of movement, such as the finger, may be surrounded by a sac filled with **synovial fluid** for lubrication (see Text Figure 31–16).

MICROSTRUCTURE OF MUSCLE

The three types of muscle cells are illustrated in Figure 28–5. Muscle cells always show **excitability,** like neurons, and most of them show **contractility** when they are excited.

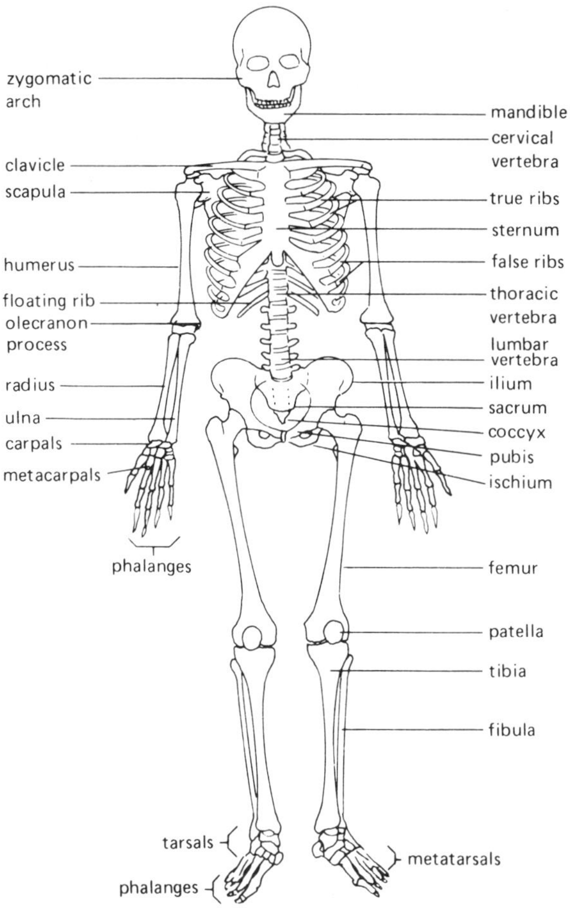

Figure 28–3. Human skeleton. Some of the 206 bones (29 in the skull alone) found in the endoskeleton. Bones are held together by elastic ligaments.

Striated Muscle

This type of muscle is also known as **skeletal** or **voluntary.** As the names imply, the muscles in which it is found are usually connected to the skeleton and always under voluntary or conscious control through the somatic nervous system. (See Color Plate 4.)

- Examine a slide of striated muscle in the microscope. Each strand is a muscle **fiber,** a large cell with many peripheral nuclei.
- Locate a muscle fiber in the slide, and identify some of its nuclei.

What is this kind of cell with many nuclei called?

____

The membrane of the muscle fiber is the **sarcolemma.** Within the fiber are the many longitudinal **myofibrils,** or bundles of thick and thin filaments. The interaction of thick and thin filaments provides the force for muscle contraction (see Text Figure 31–7). The relationship of the muscle fiber (a single cell) to the myofibrils and to the whole muscle is shown in Text Figure 31–5. The fine striations that give striated muscle its name are the result of the lining up of thick and thin filaments in exactly the same register in many adjacent myofibrils.

A striated muscle fiber gives an all-or-none twitch when it is stimulated by acetyl choline across its **neuromuscular junction** with a motor neuron. A motor neuron and the many muscle fibers with which it synapses are called a **motor unit.**

Smooth Muscle

- View a slide of smooth muscle

This type of muscle consists of bundles of spindle-shaped cells. Each cell has a central nucleus and is filled with myofibrils. There is no pattern of cross striations. Smooth muscle is designed for slow, sustained contractions and can produce a graded contraction with variable force. It usually exerts force on the space it surrounds and is under the control of the autonomic nervous system.

Where is smooth muscle found?

____

____

____

Cardiac Muscle

This type of muscle is found only in the walls of the vertebrate heart. It is very resistant to fatigue and is inherently contractile without stimulation from the nervous system. Some cardiac muscle cells have lost

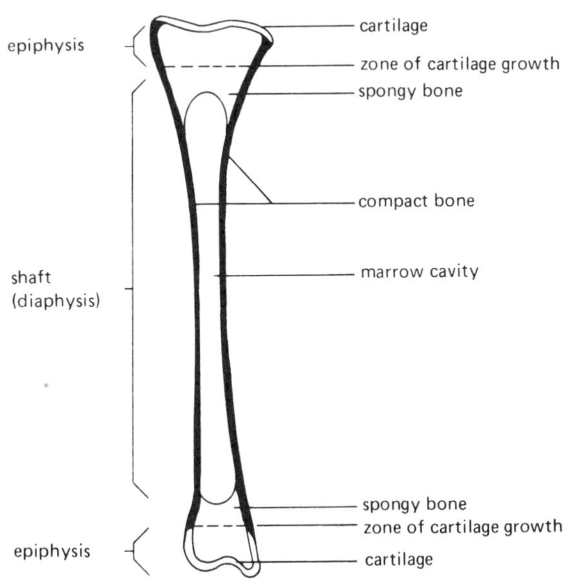

Figure 28–4. A long bone. During development the bone grows longer in the regions marked with dashes. When the epiphyses and shaft unite, growth stops.

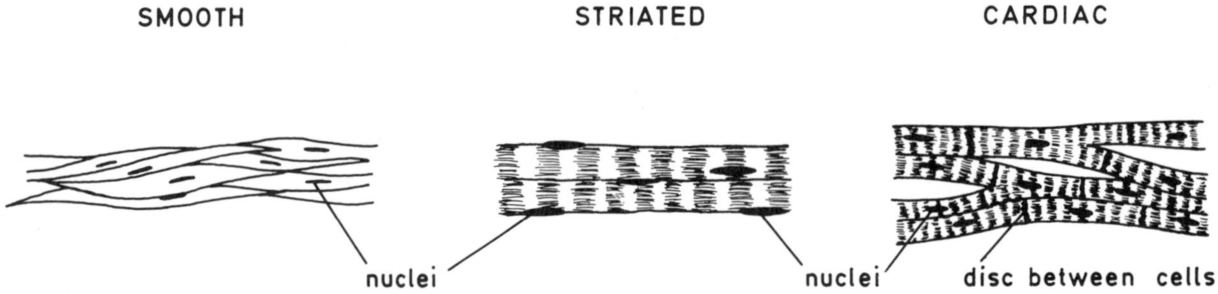

Figure 28–5. Muscle cells. The three types are smooth, striated (skeletal or voluntary), and cardiac. Only striated muscle is under voluntary control.

the ability to contract and are responsible for initiating and spreading the electrical impulses that regulate the heartbeat (see Text Figure 31–3).

- Examine a slide of cardiac muscle.

Is cardiac muscle striated? ______________

Does it have nuclei? ______________

What conspicuous features does it show that are not present in smooth or striated muscle?

____

____

- Locate the position of the **sarcolemma.** (This membrane is too thin to see.) The dark bands scattered throughout the tissue are the **intercalated discs,** which are gap junctions for nervous impulse transmission between individual muscle cells.

MICROSTRUCTURE OF BONE AND CARTILAGE

Compact Bone

Although compact bone is hard and impregnated with minerals, it contains living cells and blood vessels. The structure of compact bone is shown in Figure 28–6.

- Examine a slide of compact bone; this is bone ground thin enough so that light can penetrate. The concentric circles are the **Haversian systems.** Each contains a **Haversian canal** in which there are one or two blood vessels (a large capillary or a small arteriole and a venule) as well as a nerve. The rings are the lamellae or layers of the calcium phosphate **matrix** that makes up the substance of the bone. The matrix not only makes the bone hard, but is an important depot for the storage of calcium. The tiny open spaces around the edges of each circle are the **lacunae,** which contain the bone cells that produced all of the matrix. Tiny canals (canalicules) connect each lacuna with a Haversian canal, allowing extracellular fluid to reach and nourish the living cells.

Cartilage

Cartilage differs from bone in that it does not have mineralization in its matrix and it lacks nerves and a blood supply. Because this tissue is nourished only by diffusion from a distance, injuries to cartilage heal slowly or not at all. The fine structure of cartilage is shown in Figure 28–7.

- View a slide of cartilage in the microscope. The homogeneous area is the elastic **matrix,** which

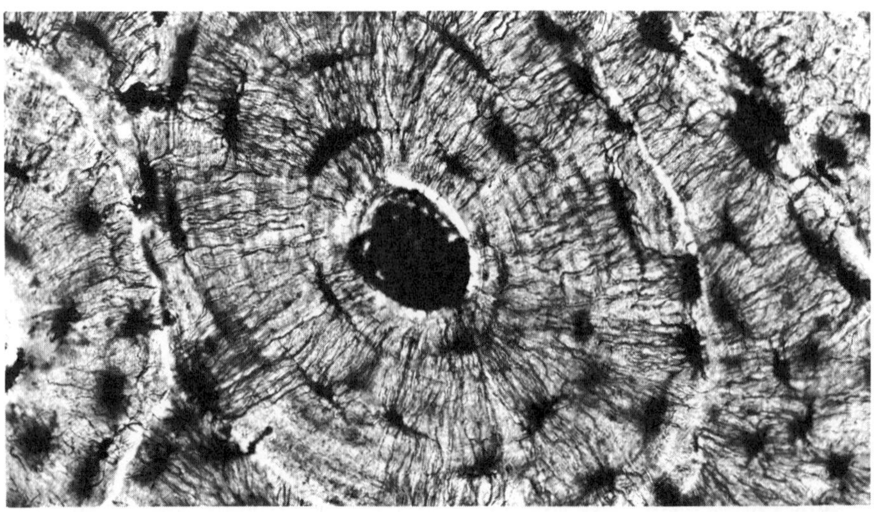

Figure 28–6. Microstructure of compact bone. The dark Haversian canal in the middle is a channel for blood vessels and nerves. The living bone cells are found in the small, dark lacunae in the layers of mineral matrix that they have secreted. (Photo by Carolina Biological Supply Company)

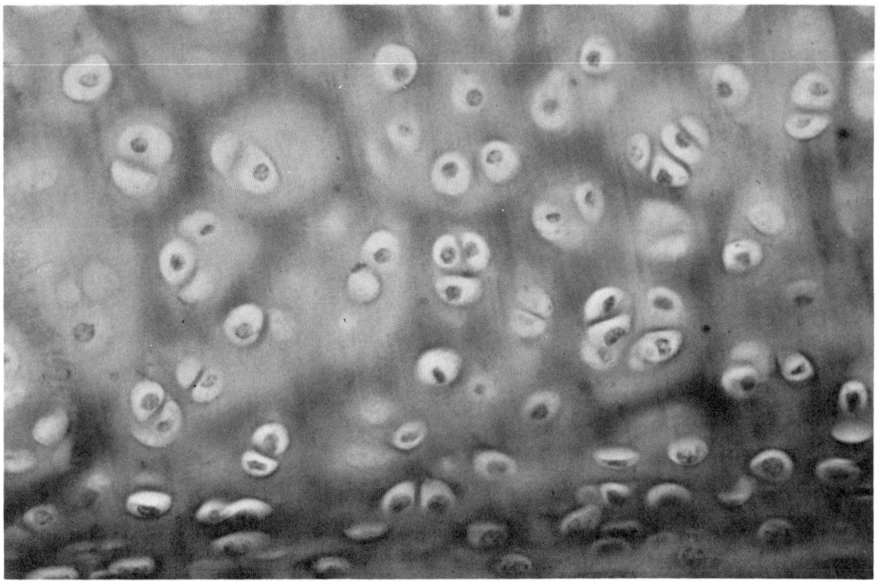

Figure 28–7. Microstructure of cartilage. The individual cells (white with darker nuclei) are found within lacunae scattered throughout the intercellular matrix. (Biophoto Associates)

contains the structural protein **collagen** and elastic fibers. The individual cells that secrete the matrix are scattered throughout the matrix in holes called **lacunae.** Each hole contains one or more cells.

- Examine your own head and the front of your neck, and list four sites where you find cartilage.

1. ______________________
2. ______________________
3. ______________________
4. ______________________

Where does the cartilage occur in the spinal column?

____

Where does it occur in the thoracic cage?

____

What parts of the pelvic girdle are made of cartilage?

____

MUSCLE PHYSIOLOGY

Preparation

Set up your experimental apparatus *before* beginning your frog dissection.

The setup for this experiment is shown in Figure 28–8. The **kymograph** is a drum that revolves at a constant rate. Your instructor will give you directions for operating it.

- Put a clean recording paper around the drum with the shiny side exposed. Lap the left edge of the

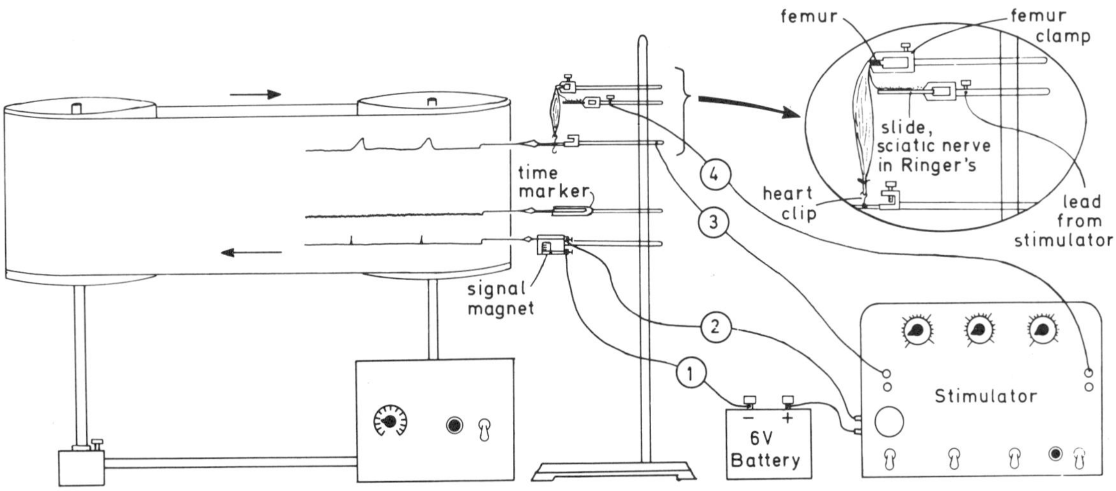

Figure 28–8. Muscle physiology experiment. Apparatus for stimulating a muscle and recording its response on a kymograph. The signal magnet marks the time of stimulation and the time marker shows the duration of the latent period and twitch. Set up the four leads according to your instructor's directions.

paper over the right so that the writing pen steps down when the drum is rotating, rather than catching on the edge of the paper.

- Set up the ring stand as shown in Figure 28–8. Attach a **femur clamp,** a utility clamp with a *clean* slide, and a muscle lever. The nerve leading to the frog's muscle will lie on the slide and *must be kept moist with frog Ringer's solution at all times,* or the sensitive nerve cells will quickly die.
- Set up the signal magnet and the time marker if they are provided with your experimental apparatus.
- Attach leads to the stimulator or inductorium as shown in Figure 28–8 or follow the directions given for your equipment.
- Lead 1 connects the signal marker with the 6-V battery and then goes to the MAG terminal of the stimulator.
- Lead 2 connects the signal marker with the MAG terminal directly.
- Lead 3 connects the heart clip clamp to the MOD IN terminal.
- Lead 4 connects the slide and utility clamp with the SYNCH IN terminal. It should be switched to the femur clamp if you wish to stimulate the muscle directly.

When your instructor has approved your setup, proceed with your frog dissection.

- Locate the **gastrocnemius** muscle and the Achilles tendon that attaches it to the bones of the foot as shown in Figure 28–2. This is the muscle that you will study. Locate the semimembranosus and the vastus externus muscles. Between them is the **sciatic nerve** (not visible in Figure 28–2), which you will stimulate to excite the gastrocnemius muscle.
- Obtain a pithed frog or pith your own according to your instructor's directions. The procedure for pithing a frog is given in Topic 25.
- Cut the skin completely around the "waist" of the frog, and pull it off, freeing it from the legs.
- Using a probe, *carefully* separate the large muscles (vastus externus and semimembranosus) on the *dorsal,* medial side of the upper leg. Locate the femur, along which you will find blood vessels and a white or light yellow fiber, which is the sciatic nerve.
- Use a glass hook to separate the connective tissue from the nerve, and trace it toward the frog's body to the point where it enters the body musculature. Cut the nerve close to that point. (Watch out for the contraction of the gastrocnemius muscle as you make the cut!)
- Avoid putting any tension on the delicate nerve. Lay the nerve over the lower leg so that it is out of the way.
- Pull the muscles of the upper leg away from the femur, and cut them off close to the knee.
- Cut the femur halfway between the hip and knee. Leave enough of a stump to grip with the femur clamp.
- Trace the broad Achilles tendon around the heel, and cut it free from the bottom of the foot. Free the gastrocnemius from the other muscles and bones of the lower leg.
- Cut the muscles and bones of the lower leg just below the knee, taking care not to damage the gastrocnemius or the sciatic nerve.
- Mount your preparation in the apparatus by attaching the femur to the femur clamp and attaching the Achilles tendon to the muscle lever with a heart clip and a piece of thread.
- Make these adjustments as necessary to get a good recording:

  Pen. The ink should be flowing freely to give a good tracing. If not, close the cap of the reservoir bottle, pinch it until the ink flows from the pen, and then loosen the cap slightly while pinching. Adjust the height of the bottle so that ink can flow out.

  Heart Clip String. Align the muscle and heart clip string so that they are vertical to the muscle lever.

  Muscle Tension. Raise or lower the muscle lever on the ring stand so that the Achilles tendon is under slight tension.

  Pen Arm. Move the position of the thread so that the arm responds freely to movement of the muscle.

  Kymograph Speed. Adjust the rotation rate so that a single twitch gives a record with a recognizable form.

Single Muscle Twitch

Your instructor will give you directions for stimulating the nerve.

Keep the nerve and muscle wet with frog Ringer's solution at all times.

Record the stimulus intensity used in Table 28–2, at the end of this topic.

- Stimulate the muscle through its nerve, wait 15 sec, and repeat.
- If the tracings are similar, copy the tracing of the

single twitch contraction in the space provided in Table 28–2.

- Compare the length of the twitch, beginning at the mark of the signal marker, with the number of cycles in the same interval. Calculate the duration of the twitch in msec according to Table 28–2.

For example, suppose your twitch lasted 12 cycles according to your time marker, and the time marker vibrated at 100 cycles/sec. Then the time that the twitch took would be:

$$\frac{12 \text{ cycles}}{100 \text{ cycles/sec}} = 0.12 \text{ sec} = 120 \text{ msec}$$

If you do not have a time marker, measure the circumference of the drum in cm. Then measure the time in sec for a complete rotation of the kymograph drum. Divide the length by the time in order to calculate the rate of rotation in cm/sec. The time that the twitch took would be:

$$\frac{\text{length of twitch in cm}}{\text{rate of rotation in cm/sec}}$$

The twitch consists of three parts: The **latent period** is the time from stimulus to the beginning of contraction; **contraction** is the period when the muscle is shortening; and **relaxation** is the time during which the muscle is lengthening.

- Calculate the duration of the latent period, contraction period, and relaxation period for your twitch, and enter the results in Table 28–2.

What chemical events go on during the latent period?

____

____

____

____

What molecules move to cause the contraction?

____

Is energy required for the relaxation process?

____

Did every muscle fiber in the muscle contract during the single twitch? __________

- Set the stimulator for a higher intensity, and repeat the single twitch.

Did a more intense stimulus cause more intense contraction? __________

Why or why not?

____

____

____

Fatigue

A muscle can contract indefinitely so long as it has an adequate rest period between twitches. If the stimuli are close together, however, the circulation won't be able to keep pace with the demands of contraction and the muscle will show **fatigue.** Fatigue could occur at the **neuromuscular junction** where the nerve axon synapses with the sarcolemma or within the muscle fibers.

What substances could run out and thereby cause muscle fatigue?

____

What substance builds up in the muscle when it is showing fatigue? __________

What could cause fatigue at the neuromuscular junction?

____

____

The muscle needs ATP energy to relax as well as to contract. In the extreme case of muscle fiber "fatigue," where all the ATP is used up permanently upon the death of the organism, the rigid muscle can neither contract nor relax and shows **rigor mortis.**

- Set the stimulator for multiple shocks at a frequency of 4/sec and an intensity of 3 V (0.3 V × 10 on the multiplier control). Set the kymograph for slow recording.
- Stimulate the nerve and record until the contractions are 2 mm high, and then turn off the stimulator.
- Record the result in Table 28–2.

Was the decrease in the strength of contraction gradual or sudden? __________

- Test whether the fatigue really occurred in the muscle, or whether it was due to inadequate stimulation of the muscle by the nerve by stimulating the muscle itself directly through the femur clamp using the same intensity of stimulus. If the muscle itself is fatigued, it will not show full strength contraction.

Was the muscle itself fatigued? __________

- Stimulate the muscle repeatedly with the same settings that you used for the nerve.

Does the muscle eventually show fatigue at this rate of stimulation? __________

- Rinse the muscle with Ringer's and give it a 5-min rest.

Temporal Summation

If the muscle is stimulated several times in a row, the intensity of contraction during the twitches may increase. This is called **temporal summation.**

How could the muscle show more contraction on a second or later twitch?

____

- Stimulate the muscle repeatedly at intervals shorter than the length of a single twitch.

Do you observe temporal summation?

____

- Record the results in Table 28–2.

Tetanus

Tetanus occurs when the stimuli are so close together that the muscle as a whole doesn't have time to relax between twitches and stays contracted (See Text Figure 31–11). Each individual motor unit twitches, goes through a **refractory period** during which it cannot respond to stimulation, and then immediately twitches again. The combined twitches of all the motor units keep the whole muscle fully contracted.

- Set the stimulator so that the frequency of stimuli is the maximum possible.
- First give the muscle a single stimulus so you have a basis for comparison.
- Then give the muscle the same strength stimulus at the high frequency, and record the muscle response for several seconds.
- Record the results in Table 28–2.

Does the muscle show temporal summation at first?

____

Does the strength of contraction during tetanus gradually decrease during the sustained contraction?

____

When might a tetanic contraction occur in your own body?

____

When you are finished with your experiment, dispose of the animal remains in the "animal wastes" container, then clean and store your equipment and microscope properly.

REFERENCES AND FURTHER READING

Grollman, Sigmund. *The Human Body: Its Structure and Physiology,* 4th ed. New York: Macmillan, 1978.

Kessel, Richard G., and Randy Kardon. *Tissues and Organs: A Text-Atlas of Scanning Electron Microscopy.* San Francisco: W.H. Freeman, 1979.

Leeson, C.R., T.S. Leeson, and A. A. Paparo. *Textbook of Histology,* 5th ed. Philadelphia: W.B. Saunders, 1985.

McMahon, T.A. *Muscles, Reflexes, and Locomotion.* Princeton, NJ: Princeton University Press, 1984.

Schmidt-Nielsen, Knut. *Animal Physiology: Adaptation and Environment,* 3rd ed. Cambridge, England: Cambridge University Press, 1983.

Wilkie, D.R. *Muscle.* New York: St. Martin's Press, 1968.

Journal Articles

Allman, W.F. "Steroids in sports: Do they work?" *Science 83,* 14, November 1983.

Allman, W.F. "Winning recipes." *Science 83,* 90, September, 1983.

Caplan, Arnold I. "Cartilage." *Scientific American,* 84, October 1984.

Cohen, Carolyn. "The protein switch of muscle contraction." *Scientific American,* 36, November 1975 (#1329).

Evarts, Edward V., "Brain mechanisms of movement." *Scientific American,* 164, September 1979 (#1443).

Feld, Michael S., Ronald E. McNair, and Stephen R. Wilk. "The physics of karate." *Scientific American,* 150, April 1979 (#3042).

Hildebrand, M. "How animals run." *Scientific American,* 148, May 1960 (no offprint).

Lester, Henry A. "The response to acetylcholine." *Scientific American,* 106, February 1977 (#1352).

McLean, F.C. "Bone." *Scientific American,* 84, February 1955 (no offprint).

McNeill, Alexander R. "Walking and running." *American Scientist* 72:348, July 1984.

Napier, John. "The antiquity of human walking." *Scientific American,* 56, April 1967 (#1070).

Oster, G. "Muscle sounds." *Scientific American,* 108, March 1984.

Ross, R., and P. Bornstein. "Elastic fibers in the body." *Scientific American,* 44, June 1971 (#1225).

Smith, D.S. "The flight muscles of insects." *Scientific American,* 76, June 1965 (no offprint).

Sonstegard, David A., Larry S. Matthews, and Herbert Kaufer. "The surgical replacement of the human knee joint." *Scientific American,* 44, January 1978 (#1378).

Student Name ______________________________ **Date** ______________

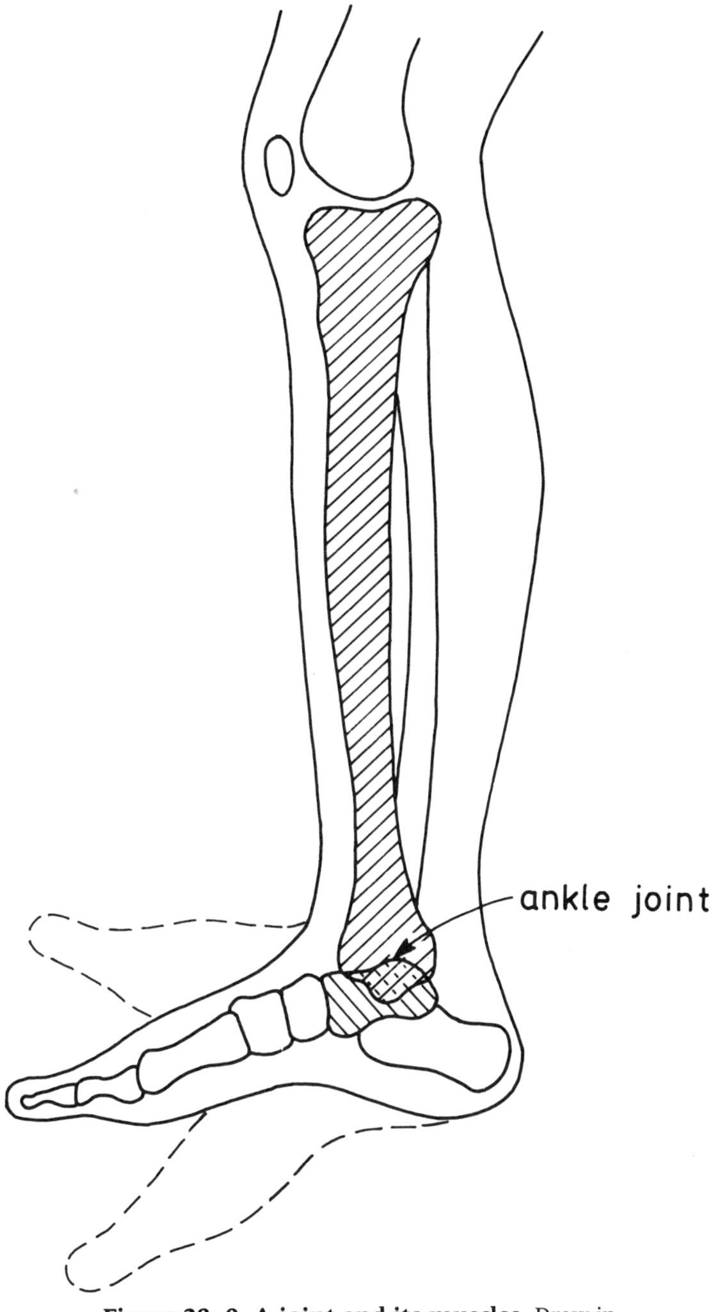

Figure 28–9. A joint and its muscles. Draw in a pair of extensor and flexor muscles that operate the ankle joint (ignore the knee and toe muscles). Label the extensor muscle, the flexor muscle, and the origin and insertion of each muscle.

TABLE 28–1. Common and Scientific Names of Some Human Bones

Common Names	Scientific Names	Axial	Appendicular
Toe bones			
Neck bones			
Hip bone			
Backbone			
Ankle bones			
Collar bone			
Shoulder blade			
Wrist bones			
Knee cap			
Jawbone			
Shin bone			
Thigh bone			
Foot bones			
Upper arm bone			
Forearm bones			
Bone of calf			
Hand bones			
Finger bones			
Funny bone			
Breast bone			
Tail bone			
Cheek bone			

Student Name ______________________ **Date** ______________

TABLE 28–2. Data from the Frog Muscle Physiology Experiment

A. DIAGRAM OF A SINGLE TWITCH CONTRACTION:

Stimulus intensity was: __________

B. DATA FOR A SINGLE MUSCLE TWITCH:

TIME MARKER CALCULATION

cycles/twitch: __________ cycles

Frequency of time marker: __________ cycles/sec

Duration of twitch = #/frequency: __________ sec = __________ msec

ALTERNATIVE CALCULATION

$$\text{Rate of drum rotation} = \frac{\text{drum circumference: \_\_\_\_\_\_\_\_\_\_ cm}}{\text{time for one rotation: \_\_\_\_\_\_\_\_\_\_ sec}} = \_\_\_\_\_\_\_\_\_\_ \text{ cm/sec}$$

Duration of twitch = length of twitch/rate = __________ sec = __________ msec

Phase	# Cycles or Length	Duration (msec)
Latent period		
Contraction		
Relaxation		

TABLE 28–2 (Cont.). Data From the Frog Physiology Experiment

C. DIAGRAM OF FATIGUE:

D. DIAGRAM OF TEMPORAL SUMMATION:

E. DIAGRAM OF TETANUS:

TOPIC 29

Behavior

OBJECTIVES

When you have completed this topic, you should be able to:

1. Define orientation, kinesis, sign stimulus, fixed action pattern, ritual, communication.
2. Define taxis and state the three major types of stimuli that result in the response of taxis.
3. Describe an experiment to test for the effectiveness of chemical stimuli in chemotaxis.
4. Explain the basis for the stimulus used in the optomotor response experiment.
5. Define conflict behavior and explain why it is important in survival.
6. Describe the fight and flight components of conflict displays in the Siamese fighting fish.

TEXT REFERENCE Chapter 33

KEY TOPICS

hypothesis	learned behavior	conflict displays
experiment	orientation behavior	courtship rituals
innate behavior	territoriality	communication

INTRODUCTION

Behavior is the response of an organism to any internal or external stimulus. In the broadest sense it includes just about anything that the animal or plant does. It is convenient to describe behavior as **innate** when it seems to be an automatic and consistent response to a certain stimulus and **learned** when the response tends to change with experience, so that it becomes more and more appropriate or elaborate when the stimulus has been given many times. Both types of behavior are important in animals with at least moderately complex nervous systems, including humans. In this exercise, you will be studying mainly innate behavior.

Orientation is a type of behavior pattern in which an organism turns its body or moves either toward or away from a certain stimulus. This type of behavior is basic to the success of the animal. Whatever it does, whether it is seeking food, water, sunlight, a mate; interacting with other members of a group; or avoiding predators, there must be some sort of orientation.

Kinesis is a simple type of orientation behavior in which the organism changes its *rate* of locomotion in response to the stimulus. If the response is **positive kinesis,** the animal will *slow down* its locomotion in response to the stimulus and speed up if the stimulus is absent. In a way, this is a primitive form of seeking behavior: If a necessary environmental condition or factor is missing, the animal moves about quickly, thereby increasing its chances of finding better conditions. If the conditions are favorable, it stays close to that spot by slowing down. As a result of kinesis, animals will accumulate near the stimulus as if they were "attracted" by it. Actually, however, the animals are not "attracted" by the stimulus and do *not* change the *direction* of their locomotion.

If individuals in a culture of *Paramecium* tend to accumulate near a bubble of CO_2 as a result of kinesis, does lower pH make them swim slower or faster? ______________

In organisms that can detect stimuli at a distance, the orientation behavior may be a form of **taxis,** in which there is a sustained, *directed* movement either toward or away from the stimulus. This kind of response may be shown for light, heat, cold, moisture, gravity, sound, or chemicals. The response will be termed **phototaxis** if the stimulus is light, **geotaxis** if the stimulus is gravity, and **chemotaxis** if the response involves a chemical stimulus.

One way in which an organism can show taxis is to try to balance the stimuli reaching a pair of sense organs, such as its photoreceptors or chemoreceptors. If less stimulus reaches the right side, the animal turns to the left until the stimulus on its right increases to the level of the left side (or vice versa). Even if the receptors are very crude, this strategy (called tropotaxis) will allow movement toward or away from the stimulus. If one of the paired receptors is destroyed, the animal will receive no stimulus from the damaged receptor and will continuously move in futile circles *away* from that side. It is trying to increase the stimulus to the damaged side. These futile circular movements are called "circus movements" and are diagnostic for organisms that use tropotaxis.

Many stimuli cause very predictable **stereotyped** responses. These stimuli are called **sign stimuli,** or releasers, and the response is called a **fixed action pattern.** The advantage of an innate fixed action pattern is that an appropriate response can be made quickly and accurately the first time without the need for learning. The disadvantage is that it will occur even if the animal has been tricked by nature or a biologist into making the response under inappropriate conditions. For example, a robin, whose red breast is a sign stimulus warning other males away from its territory, will peck at red feathers that bear no resemblance to another bird.

Not all such stereotyped behavior is innate. Animals can program learned responses into the nervous system so that they can be efficiently controlled by a few neurons and will predictably follow sign stimuli. Furthermore, a stereotyped response doesn't necessarily *always* occur. The nervous system has some leeway to adjust to environmental and physiological conditions.

In addition to responding to environmental conditions, animals must usually interact with other animals of the same or different species. Interactions between members of the same species, usually by visual, auditory, or chemical cues, are called **communication.** Many of the signals used in communication are sign stimuli, which cause very predictable responses in the other individuals.

Sign stimuli are important in courtship, in mating, and in **conflict displays** over territory, food, or mates. A display is behavior used for the purpose of communication. Conflict displays allow members of the same species to settle disputes over territory, food, or mates with minimal physical contact and usually without damage to either party in the dispute. The presence of a competitor will release responses that show the readiness of the opponent to fight. Signs of the intention to fight will cause the opponent either to show its own fight intention or to withdraw. Signs of timidity or **appeasement** will tend to reduce the aggressiveness of the opponent. Sometimes the animal will respond even more strongly to an artificial stimulus than to another organism.

The conflict display will be a mixture of "fight" and "flee" signals, and the proportion of each will be influenced by the **motivation** of the individual. Males will be more motivated to fight when they are in their own territory and when it is the breeding season. They will be less motivated to fight when the female approaches during the breeding season, and will therefore be able to begin courtship. In many species courtship is a means of overcoming the conflict behavior of the male to the extent that mating is possible.

The displays of sign stimuli are often built up into chain reactions called **rituals,** in which the response to a sign stimulus acts as a second sign stimulus, that response as a third stimulus, and so forth. Courtship rituals, for example, ensure coordination of physiological readiness to mate, mating within the same species, and sometimes care of the young.

In carrying out experiments on animal behavior, it is important to avoid **anthropomorphism,** or attributing human emotions and conscious feelings to animals. We cannot know whether an animal is "thinking" or "deciding" but we can obtain objective evidence about how it acts under various conditions. The most reliable evidence about animal behavior is obtained when the animal is observed under a **control situation** and again under an **experimental situation,** which differs by a single factor. Quantitative results are obtained, if possible, and enough data are gathered to be statistically meaningful.

In the following experiments, you should clearly understand the hypothesis being tested.

Hypothesis: A trial, unproved idea put forward as an explanation of data that have been collected; an explanation of cause and effect or of a certain chain of events. It must make predictions that can be tested experimentally.

Example of a hypothesis: Sowbugs are found in damp conditions because they show kinesis in response to moisture.

This hypothesis is an explanation which could be tested by observing whether sowbugs show a difference in rate of locomotion under dry or moist conditions. The results you obtain will either **support** or **disprove** the hypothesis; a hypothesis can never be proved.

Example of a nonhypothesis: Sowbugs are found in damp conditions because they enjoy the company of other sowbugs.

This explanation is anthropomorphic and cannot be tested.

Example of a nonhypothesis: Sowbugs will show faster locomotion under dry versus moist conditions.

This statement is a prediction that can be tested but it does not attempt to explain anything. The prediction follows from the above hypothesis which could also be stated in this way:

IF: Sowbugs show kinesis with respect to moisture . . .
THEN: Sowbugs will show faster locomotion under dry versus moist conditions.

ORIENTATION BEHAVIOR: KINESIS IN SOWBUGS

In **kinesis,** the response to a stimulus is movement at a different rate with no directionality. If you correctly analyzed the previous question about *Paramecium,* you predicted that lower pH makes them swim more *slowly,* so that they congregate around a bubble of CO_2. How does that happen? Suppose that the pH is low (acidic) near the bubble due to carbonic acid and high away from the bubble and initially the *Paramecium* are evenly distributed over both areas. The "basic" individuals will move faster and, on the average, tend to leave the basic area and enter the "acidic" area at a high rate. The "acidic" individuals will move slowly and leave the acidic area at a slow rate. The *net movement of Paramecium* will be from the basic area to the acidic area, where they will accumulate. They are *not attracted* to this area, but rather end up there as a result of different rates of locomotion under different conditions.

The sowbug is a small terrestrial crustacean often found under rotting wood (see Figure 29–1).

There are many possible reasons to explain why sowbugs are found there: They are born there and never leave, they find their way by positive chemotaxis toward wood molecules, they show negative phototaxis or they show kinesis with respect to moisture. We will hypothesize that they accumulate in damp places by kinesis in response to moisture. Our experiment will investigate the locomotion of sowbugs in response to dry compared with moist conditions. Keep in mind that more than one factor or type of behavior may result in the location of the sowbugs, but only one factor at a time can be studied.

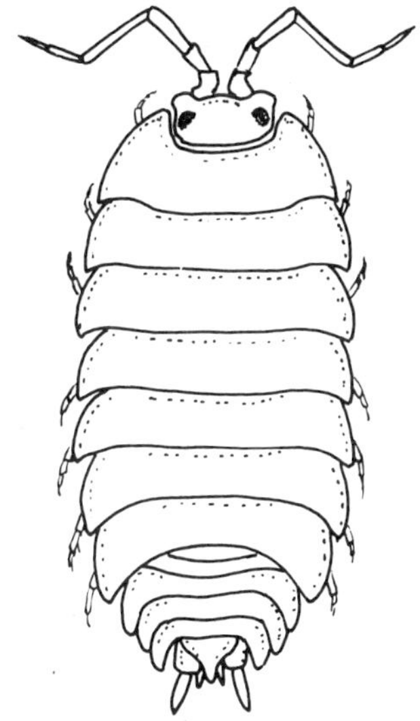

Figure 29–1. A sowbug. This terrestrial crustacean must live under moist conditions and is often found in rotting wood.

Your instructor has set up an experiment in which one set of 10 sowbugs is in the dry environment of **DISH A** and another set is in the moist environment of **DISH B.** Other conditions that might affect the sowbugs, such as temperature, light, and the physical environment, are the same for both dishes. As a class you will observe the sowbugs every 15 minutes during the experimental period. Record your results in Table 29–1 at the end of this topic.

Light disturbs sowbugs, so make your observations as quickly as possible.

- When it is your turn, count and record the number of moving sowbugs and the total number of sowbugs in dish A. Any animals that are moving their legs but not going anywhere should be counted as stationary because they do not show locomotion.
- Next count and record the number of moving sowbugs and the total number of sowbugs in dish B.
- When everyone has had a chance to observe the sowbugs, record the combined class results and calculate the percent moving in each dish.

What results would you expect if the sowbugs accumulate in the damp conditions by kinesis?

What results would you expect if the sowbugs do not

show kinesis but rather positive chemotaxis toward moisture?

____

Did the class results support either possibility?

____

ORIENTATION BEHAVIOR: TAXIS IN *DROSOPHILA*

Taxis, in contrast to kinesis, involves *directed* motion toward or away from a stimulus and so requires that the organism be able to detect the stimulus from a distance. You may have noticed during the genetics experiment that the flies had the annoying habit of sitting on the cotton plug of the food vial. This behavior may have been due either to positive phototaxis, negative geotaxis, or both types of taxis.

Phototaxis

- Place five flies in a clean food vial (without the food), cover it with another clean vial, and tape the vials together as shown in Figure 29–2.
- Cover the vials with a black paper shield so that only one end is open to the light, and lay it on a horizontal surface.
- Place a light source near the open end of the tube, and wait 3 min.
- Quickly remove the shield, and count the flies in each half of the tube; record the results in Table 29–2 at the end of this topic.
- Tap the flies back into the center of the tube and reverse the tube so that the former right end is now the left end. Replace the shield and light source, which should be exactly as before.
- Wait 3 min, remove the shield, and count the flies once again.
- Repeat once more unless your results agreed very well in the first two trials.

Did the flies show phototaxis? _______________

If present, was it positive or negative?

____

Why did you have to reverse the vials between trials?

____

____

____

____

Why did you have to keep the vials horizontal?

____

Geotaxis

- Use the same vials or take five new flies, put them in a clean food vial, and tape it end to end to another clean food vial.
- Let the flies attach to the glass, and then cover the vials completely with a black shield.
- Hold the vials in a vertical position for 3 min.
- Quickly remove the shield, and count the flies in each vial. Record the results in Table 29–2.
- Tap the flies to the middle of the tube, cover it with the shield, and hold it in a vertical position, reversing the vials so the one that was held up at first is now down.
- Wait 3 min, remove the shield, and count the flies in each vial. Record the result.
- Repeat once more unless your results in the first two trials matched very well.

Did the flies show positive geotaxis or negative geotaxis?

____

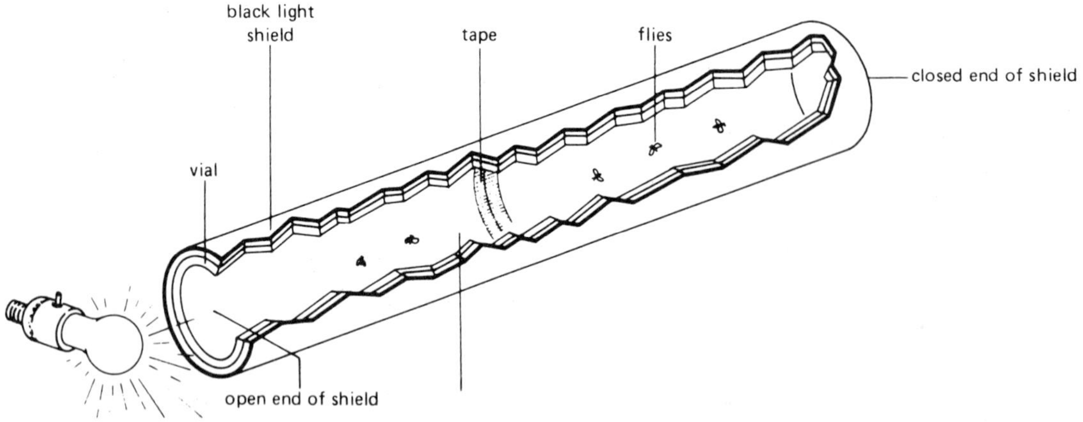

Figure 29–2. Phototaxis in *Drosophila*. In this experimental setup, the open ends of the vials are taped together, and the vials are covered with a black shield open to the light at one end only.

Why did you have to reverse the position of the vials between each trial?

____

____

Phototaxis and Geotaxis

Test the relative strengths of the response to light and gravity by opposing these stimuli.

- Enclose the vials with five flies in the black shield that is open at one end.
- If you observed positive phototaxis, hold the vials in a vertical position so that the covered end is the end where you would expect the flies to be under the influence of gravity alone. Reverse if you observed negative phototaxis.
- Hold a light source near the open end of the tube, and wait 3 min.
- Remove the shield, and count the flies in each vial. Record the results.
- Repeat two or three more times until you see a definite trend.

Was phototaxis strong enough to overcome geotaxis?

____

Chemotaxis

Orientation toward volatile substances helps *Drosophila* locate food and a mate and avoid unpleasant or harmful substances. A simple T-maze for testing chemotaxis is shown in Figure 29–3. The substance to be tested is placed in one vial, the second vial acts as the control, and the flies are placed in the third vial. The progress of the flies can be viewed through the glass.

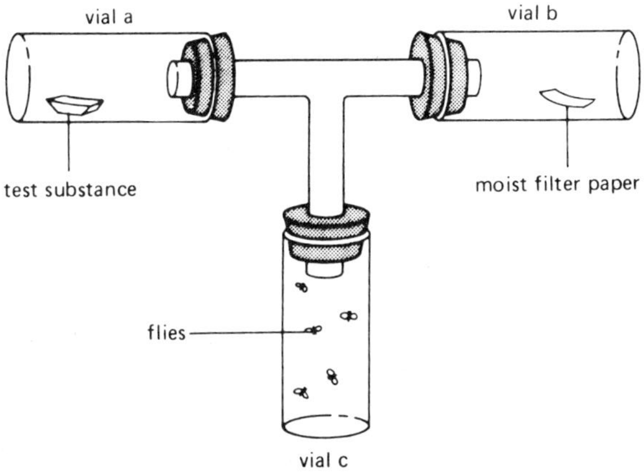

Figure 29–3. Chemotaxis in *Drosophila*. In the T-maze, vial (a) contains the substance to be tested, and the control, vial (b), contains moist filter paper. The flies are introduced into vial (c).

- Place some fermenting *Drosophila* food or fermenting fruit on a small piece of aluminum foil, and put it into vial a. Attach it to the T-tube.
- Place a piece of moist filter paper in vial b, and attach it to the T-tube.
- Place 12–15 flies in vial c, attach it to the T-maze, and place the maze in a dark place.
- Observe the flies after 5 min, and record the number of flies in each vial.
- Record the numbers of flies again after 10 min.

Do you see any trend in the movements of the flies?

____

- Put the maze back in a dark place for an hour, and record the number of flies in each vial after 60 min.

Did the flies show chemotaxis toward the food?

____

THE OPTOMOTOR RESPONSE

Some fish show social behavior in that they orient toward each other and form a motion group called a **school.** This is orientation behavior in which the stimulus is living as opposed to nonliving (as in phototaxis, etc.). The fish are mutually attracted to each other and are kept together by several factors, including vision, swimming movements in the water, and olfaction. Vision seems to be the most important factor and the appearance of other members of the same species seems to have the strongest attraction. Schooling is thus a form of communication between members of the same species, but fish of different species do sometimes school together.

Schooling behavior is partly innate and partly learned. The response of fish to the visual stimulus of a pattern of moving stripes can be tested in the laboratory. In this type of experiment, the fish is presented with a pattern of moving stripes in the apparatus shown in Figure 29–4. While swimming in a school, the fish must maintain its position with respect to a pattern of the surrounding bodies, a stimulus related to the experimental stimulus. An **optomotor response** to the moving stripes may be present before the young fish has learned to school and it has been found in both schooling and nonschooling fish. This seems to be an innate response found in fish in general. Schooling in response to other fish is presumably learned later by those fish that school.

Fish are sensitive, so handle and disturb them as little as possible.
Be patient.
Avoid unnecessary noise or shock.
Always handle the fish with nets.

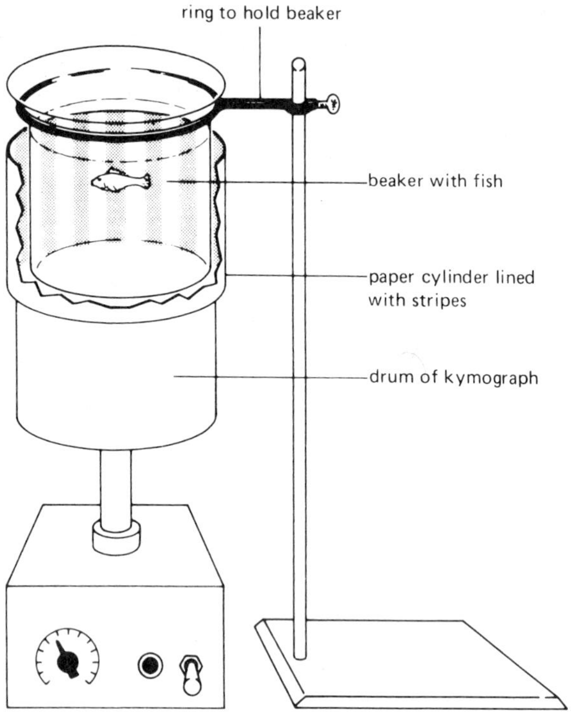

Figure 29–4. Optomotor response. A striped paper cylinder is the visual stimulus. It is attached to the drum of a kymograph and rotated around the outside of the beaker containing the test fish.

- Test for the presence of the optomotor response in a schooling fish with visual markings, such as a striped *Zebra danio* (A), a schooling fish without stripes (B), and a nonschooling fish (C).
- Place fish A in the beaker and observe its behavior with no special stimulus as the control.
- Record the results in Table 29–3 at the end of this topic.
- Attach a paper cylinder with vertical stripes to the drum of the kymograph, and observe the reaction of the fish to stationary vertical stripes.
- Start the drum rotating at about 0.5 revolutions per second, and observe the behavior of the fish for 3 min.

Did the rotating stripes change the behavior of the fish? ______________

- Change the paper cylinder to one with interrupted horizontal stripes, and observe the response of the fish to stationary horizontal stripes.
- Start the drum rotating at about 0.5 revolutions per second, and observe the behavior of the fish for 3 min.

Did the rotating stripes change the behavior of the fish? ______________

Be sure to return the fish to the aquarium as soon as you are finished.

- Place fish B in the beaker, and repeat your observations on the fish's response to no stimulus, stationary and moving vertical stripes, and stationary and moving horizontal stripes.
- Record the results, and return the fish to its tank.

Did the fish without stripes show the optomotor response? ______________

- Repeat for fish C, record the results, and return the fish to its tank.

Did the nonschooling fish show the optomotor response? ______________

CONFLICT BEHAVIOR IN SIAMESE FIGHTING FISH

Siamese fighting fish (*Betta splendens*) live in the freshwater tropical streams of Indochina and have been bred for exaggerated **conflict behavior.** The male fish shows by visual displays its conflicting and simultaneous urges to attack another male or to flee. You have often seen this flee/attack behavior in the dog, where his intentions are signaled by voice, posture, and the position of his ears and tail. One of the male fish eventually will get the upper hand, and bluff its competitor into leaving. When the fish are confined within the aquarium, however, the winner will chase and bite the losing fish.

All of the fighting fish in the lab are male. The presence of another male acts as a **sign stimulus** to release a **fixed action pattern** response. Although the sign stimulus in nature is another male, you can substitute the fish's own mirror image. Various artificial models can be used to investigate what features of the invading male's appearance are the strongest stimulus. The purpose of the display is to drive invaders out of the male's territory, and the defending male will have the strongest **motivation** to display if he is in his own territory. Thus it is best to experiment with the fish in their home tanks, if possible. The fish's motivation will be higher if he has just displayed to or defeated another male. The best displays occur between closely matched fish, where neither one is forced to flee.

The types of conflict displays are shown in Figure 29–5. The normal resting fish shown in (a) is drab in color but if another male appears, the intruder will be challenged with a **broadside display** (b). The intruder may respond in kind and the mutual broadside display may go on for several minutes. One fish will gradually give up the broadside display and the other may attempt an aggressive frontal attack, tearing and biting to drive him away. The defeated fish will swim away and assume the **appeasement** appearance of the normal resting fish.

The broadside display of the male is also used dur-

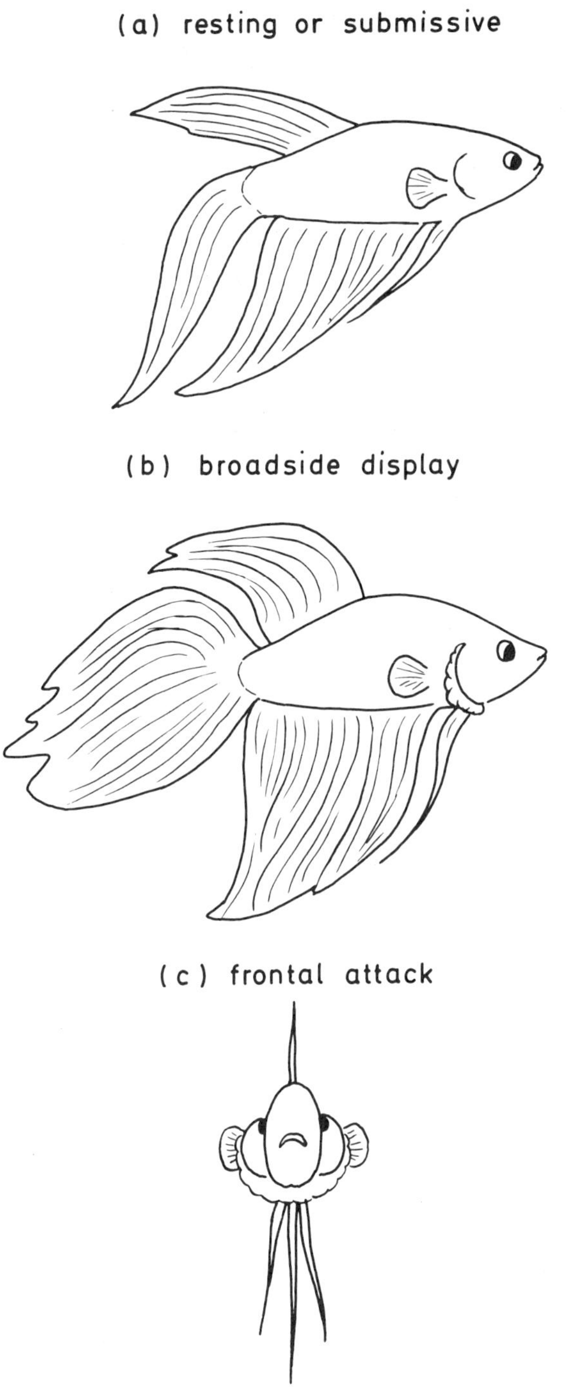

Figure 29–5. Conflict displays in a Siamese fighting fish. The display in (a) is the least aggressive and is shown by the resting or submissive fish. Its coloring is dull or moderately bright. The broadside display in (b) is commonly seen between closely matched equals, brilliant in color. The fish in (c) is dominant and about to drive the submissive fish away with its brilliant coloring and fierce frontal attack.

ing the courtship ritual to attract the female to the male's bubble nest. The male maintains the display during mating, and the female must make herself as inconspicuous as possible to avoid being attacked.

- Observe the normal appearance of the fish (*Betta splendens*) in its tank. These are some of the components of the displays to look for:

  swimming
  swimming away
  pectoral fin movement
  raising of dorsal fin
  lowering of ventral fin
  expansion of tail
  increased coloring
  extension of gill covers
  facing the intruder
  biting the intruder

- Record your control observations with no stimulus in Table 29–4 at the end of this topic.
- Use a mirror to stimulate the fish by his own appearance. Pay particular attention to the color, fins, gill covers, and the orientation of the fish toward the stimulus. A good rule of thumb is to allow the fish 5 minutes to respond to a stimulus. His resting image is not very threatening, but if he starts to display, the "intruder" will respond likewise and the display will rapidly escalate.
- Note your observations in Table 29–4.
- If your fish responds to the mirror, select a model fish and try the effect of moving it alongside the tank. If it is plastic, you can also try moving it in the tank water.
- Describe what features were present in the model (A) and note how your fish responds.
- Note the response to other models (B, C), or use data from other class members, and if possible draw conclusions about the effect of model shape and color.
- Next, put two fish, **in separate tanks** next to each other. If the two are evenly matched, there will be displays for a considerable time.

Do not put two fish in the same tank yourself. Wait for the demonstration.

Demonstration of Conflict Behavior

If two fish are removed from their tanks and put together in a third tank, they will both be "invaders," neither one will have the advantage of being on his home territory, and the displays are likely to be evenly matched. If a second fish is introduced into the home tank of a fish, the "resident" will have the advantage of home territory and will be more likely to attack the second fish. The fish are more likely to display if they have recently displayed to and exerted dominance over another fish.

To demonstrate conflict behavior, your laboratory instructor will place two evenly matched fish in the same tank where they can interact. During the demonstration, be ready to assist in removing the fish as soon as the possibility of physical contact becomes likely.

- Observe the demonstration in which the fish will probably begin to display within a short time and record what happens in Table 29–4.
- After one fish becomes dominant, record its appearance while attacking.
- Record the appearance of the submissive fish while it is trying to escape.

When you are finished with these experiments, it is very important to return all fish to their proper containers as directed by your instructor. Return your materials and clean your workplace.

REFERENCES AND FURTHER READING

Alcock, J. *Animal Behavior: An Evolutionary Approach,* 3rd ed. Stamford, CT: Sinauer, 1983.

Barash, D.P. *Sociobiology and Behavior,* 2nd ed. New York: Elsevier, 1982.

Bond, Carl E. *Biology of Fishes.* Philadelphia: W.B. Saunders, 1979.

Gould, J.L. *Ethology.* New York: W.W. Norton, 1982.

Kalmus, H. *101 Simple Experiments with Insects.* Garden City, NY: Doubleday, 1960.

Lorenz, K.Z. *King Solomon's Ring.* London: Methuen. 1942.

Manning, A. *An Introduction to Animal Behavior,* New York: Addison-Wesley, 1979.

Shaw, E. "Schooling in fishes: Critique and review," in *The Development and Evolution of Behavior.* Eds. L.R. Aronson, E. Tobach, D.S. Lehrman, and J.S. Rosenblatt. San Francisco: W.H. Freeman, 1970.

Journal Articles

Benzer, S. "Genetic dissection of behavior." *Scientific American,* December 1973 (#1285).

Crews, David. "The hormonal control of behavior in a lizard." *Scientific American,* 180, August 1979 (#1435).

Eibl-Eibesfeldt, Irenäus. "The fighting behavior of animals." *Scientific American,* 112, December 1961 (#470).

Erickson, Robert P. "On the neural bases of behavior." *American Scientist* 72(3): 233, May–June 1984.

Fernald, Russell D. "Vision and behavior in an African cichlid fish." *American Scientist* 72(1): 58, January–February 1984.

Gould, S.J. "The guano ring." *Natural History,* 12, January 1982.

Hailman, Jack P. "How an instinct is learned." *Scientific American,* 98, December 1969 (#1165).

MacNichol, Edward F., and Joseph S. Levine. "Color vision in fishes." *Scientific American,* 108, February 1982.

Moore, Janice. "Parasites that change the behavior of their host." *Scientific American,* 108, May 1984.

Mykytowycz, Roman. "Territorial marking by rabbits." *Scientific American,* 116, May 1968 (#1108).

Partridge, Brian L. "The structure and function of fish schools." *Scientific American,* 90, June 1982.

Ryker, Lee C. "Acoustic and chemical signals in the life cycle of a beetle." *Scientific American,* 112, June 1984.

Shaw, E., and B.D. Sachs. "Development of the optomotor response in the schooling fish *Menidia menidia.*" *J. Comp. Physiol. Psychol.* 63(3): 385, 1967.

Sheller, Richard H., and Richard Axel. "How genes control an innate behavior." *Scientific American,* 54, March 1984.

Tinbergen, N. "The curious behavior of the stickleback." *Scientific American,* 22, December 1952 (#414).

Tinbergen, N. "On war and peace in animals and man." *Science* 160: 1411, 28 June, 1968.

Todd, John H. "The chemical language of fishes." *Scientific American,* 98, May 1971 (#1222).

Webb, Paul F. "Form and function in fish swimming." *Scientific American,* 72, July 1984.

Whitehead, Hal. "Why whales leap." *Scientific American,* 84, March 1985.

Wilson, E.O. "Pheromones." *Scientific American,* 100, May 1963 (#157).

Student Name ______________________ **Date** ______________

TABLE 29–1. Results from Kinesis Experiment

Record the total number of sowbugs and the number moving about (not just moving their legs) in each dish.

	Dish A (Dry)	Dish B (Moist)
# Sowbugs in Dish		
# Sowbugs Moving		
CLASS TOTAL # Sowbugs in Dish		
CLASS TOTAL # Sowbugs Moving		
% Sowbugs Moving		

Results expected if orientation behavior is due to kinesis:

____

Results expected if behavior is due to chemotaxis:

____

Did the class total results support either possible explanation? ______________

TABLE 29–2. Results from Taxis Experiments

A. PHOTOTAXIS

	# Flies in Light End	# Flies in Dark End
Observation 1		
Observation 2		
Observation 3		

Did the flies show phototaxis? ______________

If present, was it positive or negative? ______________

B. GEOTAXIS

	# Flies in Upper End	# Flies in Lower End
Observation 1		
Observation 2		
Observation 3		

Did the flies show positive or negative geotaxis? ______________

C. PHOTOTAXIS AND GEOTAXIS

	# Flies in Upper End	# Flies in Lower End
Observation 1		
Observation 2		
Observation 3		

Was phototaxis strong enough to overcome geotaxis? ______________

D. CHEMOTAXIS

	# Flies in (a)	# Flies in (b)	# Flies in (c)
After 5 min			
After 10 min			
After 60 min			

Do the flies show chemotaxis toward the food? ______________

Student Name ______________________ **Date** ______________

TABLE 29–3. Results from Optomotor Experiment

Record the appearance of each fish under the various experimental conditions.

A. FIRST FISH

Type of Fish Used	
Control—No Stimulus	
Stationary Vertical Stripes	
Rotating Vertical Stripes	
Stationary Horizontal Stripes	
Rotating Horizontal Stripes	

Was the optomotor response present in this fish? __________

B. SECOND FISH

Type of Fish Used	
Control—No Stimulus	
Stationary Vertical Stripes	
Rotating Vertical Stripes	
Stationary Horizontal Stripes	
Rotating Horizontal Stripes	

Was the optomotor response present in this fish? __________

C. THIRD FISH

Type of Fish Used	
Control—No Stimulus	
Stationary Vertical Stripes	
Rotating Vertical Stripes	
Stationary Horizontal Stripes	
Rotating Horizontal Stripes	

Was the optomotor response present in this fish? __________

TABLE 29–4. Results from Conflict Behavior Experiments

A. ARTIFICIAL STIMULI

Record the displays shown by the Siamese fighting fish in each situation and the features of the models used as stimuli by yourself or by others. The fish used as a stimulus is in a separate tank, **not in the same tank with the experimental fish.** Allow the fish up to 5 minutes to respond to a given stimulus. Finally, record the most aggressive OVERALL RESPONSE shown by the fish to each stimulus.

Model A: ______

Model B: ______

Model C: ______

Reaction of Fish	Stimulus					
	None/Control	Mirror	Model A	Model B	Model C	Fish
No Response						
Swimming						
Swimming Away						
Pectoral Fin Movement						
Raising of Dorsal Fin						
Lowering of Ventral Fin						
Expansion of Tail						
Increased Coloring						
Extension of Gill Covers						
Facing Intruder						
Biting Intruder						
OVERALL RESPONSE:						
No Response						
Fled						
Broadside Display						
Frontal Attack						

Did model shape strongly influence the response of the fish to the models? ______

Did model color strongly influence the response? ______

Student Name ______________________________ **Date** ______________

TABLE 29–4 (Cont.). Results from Conflict Behavior Experiments

B. DEMONSTRATION USING NATURAL FISH STIMULUS

Describe the behavior of two fish placed in the same tank:

____

____

Appearance of dominant fish: ______________________________

____

Appearance of submissive fish: ______________________________

____

Drawing an analogy between the fish and a bullfighter with a red cape, would you expect the dominant fish to strike at the expanded fins or the body of the less dominant fish:

____

TOPIC 30

Structure and Growth of Vascular Plants

OBJECTIVES When you have completed this topic, you should be able to:

1. Explain why the division of labor between roots, stem, and leaves contributes to the success of a flowering plant.
2. Define meristem, and list the locations in a plant where it is found.
3. List four functions of a root and compare the advantages and disadvantages of a fibrous root system and a tap root system in carrying out those functions.
4. List five main functions of stems.
5. Given a twig, tell its age, and identify the external features, including the bud scales, axillary buds, leaf scars, bud scale scars, and lenticels.
6. Define girdling, list the layers that are removed, and describe the consequences to the tree.
7. Describe the differences between the external coverings and the internal structures of a stem and a root.
8. Compare the way that branches form in a stem and in a root.
9. Explain the difference between primary and secondary growth in vascular plants; describe the process of primary growth in a root and primary and secondary growth in a stem.
10. Distinguish between sapwood/heartwood; spring wood/summer wood; vascular cambium/cork cambium; outer bark/inner bark.
11. Give the functions of epidermis, mesophyll, veins, and guard cells in a leaf.
12. Distinguish between simple and compound leaves, and between alternate, opposite, or whorled leaf arrangements.
13. Explain the term apical dominance, and describe its role in the growth of a plant.
14. Define node, internode, axillary bud, intercalary meristem, pericycle, and cambium.
15. Trace the movement of a molecule of water from the soil to a guard cell, listing all the structures through which it passes.
16. Explain how the rate of photosynthesis regulates the rate of transpiration.
17. Given a cross section of a leaf, stem, or root, identify the following structures if they are present: epidermis, root hairs, cuticle, stomata, guard cells, endodermis, pericycle, xylem, phloem, pith, cortex, collenchyma, palisade mesophyll, spongy mesophyll, and air spaces.
18. List the important features that distinguish monocots and dicots.

TEXT REFERENCES Chapter 34; Chapter 35.

KEY TOPICS

roots	xylem	plant nutrition
stems	phloem	transpiration
woody stems	leaves	

INTRODUCTION

Angiosperms are characterized by their **flowers** and their highly efficient **vascular tissues.** The rapid uptake of water and minerals by the **root system** (below ground) and their distribution to the **shoot system** (above ground) allow an angiosperm to maintain large leaves with broad surface areas. The higher rate of photosynthesis carried on by these leaves allows the plant to grow much more quickly than a gymnosperm, and it may reach reproductive maturity in a few weeks. The details of reproduction will be considered later. Here you will study the structure of angiosperms and gymnosperms and the important role played by their vascular tissues.

Most plant growth is **indeterminate** and stops only when the plant dies. Not all parts of the plant undergo cell division, however. Only certain tissue, called **meristem,** contains actively dividing cells. Meristematic tissue is found at the tip of the **shoot** (the part above the ground, growing upward) and the tip of the **root** (the part underground, growing downward). Growth occurs there due to the actively dividing cells of the **apical meristem.**

Plants may survive for one, two, or many years. The **annuals** die after a season's growth and allow their seeds to continue the species the following spring. **Biennials** reproduce after two seasons and then die. **Perennials** live on for many years. The body of an annual grows mainly in length, from the tips of its shoot system and from the tips of its roots where actively dividing **apical meristem** is found. This type of growth is called **primary growth.** Some annuals and most perennials also increase in size sideways. This growth in girth due to the addition of new cells is called **secondary growth** and is due to **lateral meristem.** The stem of some perennials—the trunk of a tree, for example—continues to increase in diameter year after year. Other perennials, such as the chrysanthemum and bulb-forming plants, die back each year and produce little or no secondary growth.

When secondary growth occurs, large amounts of **secondary xylem** called **wood** may be formed. This tissue gives great strength to the stems and roots of **woody plants.** Plants that do not form wood are called **herbs. Secondary phloem** is also produced.

Angiosperm plants show great diversity, but they fall into two main groups: the **monocots** and the **dicots.** These names are derived from the structure of the seed. A monocot has one seed leaf (cotyledon), whereas a dicot has two. The characteristics of the roots, stems, and leaves are also different. Monocots are usually herbs, whereas dicots include both herbs and woody plants. When you finish with the laboratory work in this exercise, state the main characteristics of monocots and dicots in Table 30–1 at the end of this topic.

TRANSPIRATION

When plants are actively photosynthesizing, they need to use water and carbon dioxide. Air must be allowed to enter the leaf interior by opening the **stomata,** and water reaches the leaf by rising through the **xylem** from the roots. The stomata open when their guard cells, responding to low CO_2 during active photosynthesis, become turgid. (See Color Plate 5.) When the stomata are open, however, a great amount of water is lost from the leaves, especially on a hot, sunny day with the breeze blowing. This lost water must be replaced by rapid uptake from the soil around the roots, or the plant will wilt. The process of losing water by evaporation from the leaves is called **transpiration.** The "transpiration stream" of water from roots to leaves depends on a continuous column of water molecules from the soil, through xylem vessels, to the leaf. Evaporation in the leaves is the driving force behind **transpiration pull,** pulling up on the water column. Thus the plant does not need to expend energy during transpiration. The continuous column of water is under **tension** or negative pressure but it can be maintained intact because water molecules have very high **cohesion** or tendency to stick to one another. (See Text Figure 35–18.)

Demonstrate transpiration by carrying out this simple experiment:

- Add a drop of dye to a small amount of water in a beaker or flask.
- Make a clean cut with a sharp razor in the stem of a white flower, such as a daisy or carnation.
- Place the flower in the beaker, set the beaker in a well-illuminated location, and check the petals from time to time during the laboratory for the presence of the dye.
- Wait 1–2 hr.
- After the dye has reached the flower, note which part of the petals stained more darkly.
- Make a sketch in the space provided in Figure 30–11 at the end of this topic to show the pattern of staining of the petal. Label the part of the petal where transpiration was most rapid.
- Make a thin cross section of the stem with a sharp razor blade, and examine it under the dissecting microscope.
- Note the xylem vessels through which the dye rose.
- Make a sketch of your cross section in Figure 30–11 and label the location of the xylem.

You can also measure the amount of water lost by a plant quantitatively by measuring the rate at which it is taken up by the stem. Record your results in Table 30–3, at the end of this topic.

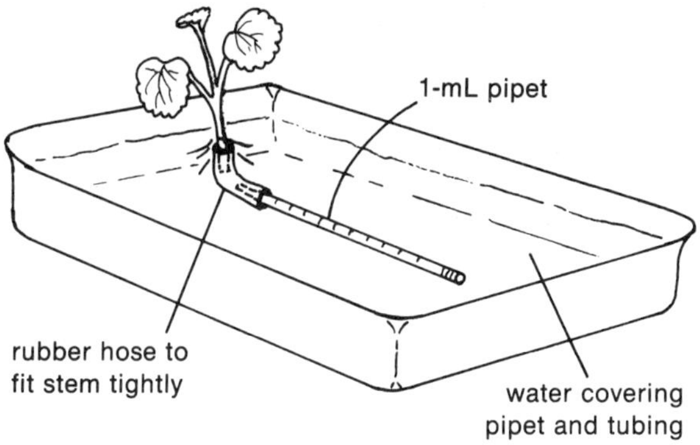

Figure 30–1. Transpiration experiment. Once all air has been excluded from the system by immersing it in water, the plant cutting is lifted out and placed in strong light so that the uptake of water by the stem can be measured.

- Assemble the materials shown in Figure 30–1 and a sharp razor blade.
- Hold the stem of a plant such as part of a geranium plant beneath the water, and cut the stem off without crushing it.

Do not wet the leaves.

- Hold the piece of rubber tubing under the water, and insert the cut stem into it. (The tubing must be just a bit narrower than the stem.)
- Immerse the pipet in the pan of water, and attach it to the other end of the tube.
- Gently squeeze the tubing to expel any air bubbles that might have been trapped below the stem.

If the experiment is to work properly, all air bubbles must be removed.

- Holding the pipet in a horizontal position, lift the plant and the pipet out of the water, and set the pipet on a flat surface in a good source of light.
- Record the position of the water column in the pipet every 15 min until transpiration is well established.
- Calculate the rate of water uptake by the stem in mL/hr.
- Place the plant in darkness, and continue to follow the uptake of water.
- Calculate the rate of water uptake by the stem in mL/hr in the dark.

How did the absence of light affect the rate of transpiration?

____

ROOTS

Types of Root Systems

When a seed germinates, it sends a single **primary root,** or **radicle,** down into the soil. This root sends out side branches called **lateral roots,** and these branch out in turn until a **root system** is formed.

Why must the root develop before the rest of the plant?

____

____

If the primary root continues to be the largest and most important part of the system, the plant is said to have a **tap root** system. If many main roots are formed, the plant has a **fibrous** root system (see Color Plate 5). Most monocots, such as grass, and some dicots have a fibrous root system. Some trees, dandelions, and carrots have a tap root.

- Examine the demonstrations of a fibrous root system and a tap root system.
- In Table 30–2, at the end of this topic, describe how well each type of root system is adapted for carrying out the major functions of a root.

Root Tips

Root growth is **indeterminate** and continues throughout the life of the plant. Apical meristem tissue is found only at the tips of the roots.

- Examine a longitudinal section of a root tip from a plant such as an onion under low power in your microscope.
- Referring to Figure 30–2, identify the **root cap,** the zone of **cell division,** the **quiescent center,** the zone of **elongation,** and the zone of **maturation.**
- Note the **root hairs** in the zone of maturation.

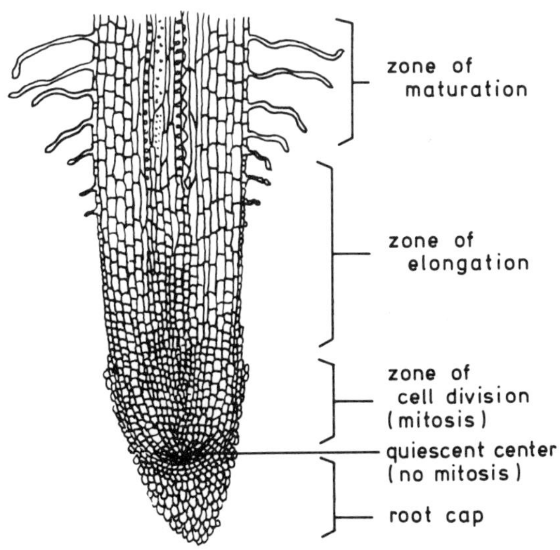

Figure 30–2. A growing root tip. Mitosis occurs mainly in the outer columns of the zone of cell division. Cells of the root cap are sloughed off as the root grows through the soil.

They are outgrowths of the epidermal cells and increase the surface area of the root for absorption.

- Switch to high power, and examine the zone of cell division more carefully.
- Note the many mitotic figures in the columns of **meristem** cells along the sides of the root and the absence of mitosis in the region of the quiescent center.

The new cells formed by mitosis elongate and push the root tip away from the older section of the root. The mature cells remain fixed in position relative to the soil while the tip of the root moves down continuously.

Why are the delicate root hairs found only in the mature part of the root tip?

____

____

Why should the foliage of a shrub be reduced after it has lost many root hairs during bare root transplantation?

____

____

The inner layer of root cap cells constantly divides to produce more cells. These replace the cells sloughed off from the outer layers of the root cap as the growing root forces its way through the soil.

What is the purpose of having a root cap?

____

Certain cells of the root cap sense gravity and aid the **geotropic** response so that the root grows downward.

- Examine a 2-day-old radish seedling that has sprouted.
- Note the fuzzy appearance of the radicle, which is covered with hundreds of root hairs.

Are the root hairs longer at the top or bottom of the root? ______________

How does the growth of the root cause this pattern?

____

____

In many plants, especially trees, the roots are closely associated with fungi known as **mycorrhizae.** They greatly increase the area for absorption and assist the plant by preferentially absorbing needed minerals.

Root Structure

Use the diagram in Figure 30–3 to study the internal structure of a dicot root.

- Examine a cross section of a root under low power. Locate the inner **stele,** which is separated from the outer **cortex** by the layer of endodermis.
- If you are looking at a dicot root, note the X-shaped core of **xylem** vessels within the stele. Xylem in monocots forms a wavy ring.
- Locate the **phloem** tissue, which is also vascular tissue.
- Find the **endodermis,** the layer of cells surrounding the stele and regulating the fluid entering it.

Because a major function of the root is absorption, the surface of young roots is not covered with a waterproof layer but is somewhat permeable to the water and dissolved nutrients with which the root is in contact. Roots also need oxygen to respire and must usually be in contact with air pockets in the soil so that they can carry on gas exchange. Soil solution and oxygen can readily enter the cell walls and spaces of the endodermis and cortex known as the **symplast.** The living cells, known as the **apoplast,** thus have a greatly increased surface area for absorption and they also absorb soil solution. Solution from both the sym-

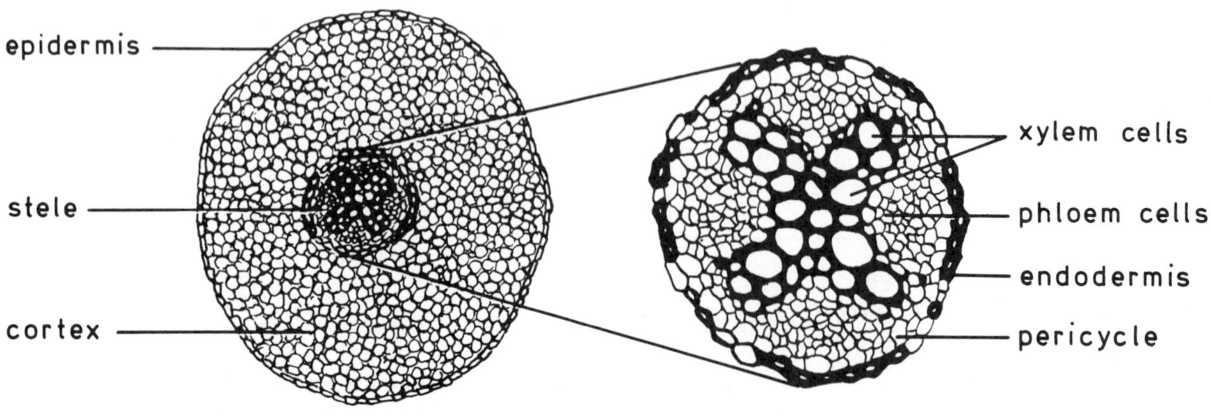

Figure 30–3. A dicot root. The vascular tissue is located within the stele.

plast and apoplast then enters the central stele of the root (see Text Figure 35–4).

The mineral content of the water that is sent to the rest of the plant is controlled to some extent by the layer of endodermal cells that surrounds the central stele and through which the solution must pass. This layer is made waterproof by the **Casparian strip,** which surrounds the endodermal cells. The vascular tissue is located within the central region, so that it is isolated and does not passively take up whatever is in the water around the plant. Thus **sap,** the fluid leaving the root in the vascular tissue, has passed through living endodermal cells that concentrate certain things needed by the plant.

- Note that the cells making up the cortex all look similar. They are **parenchyma cells** and may contain grains of stored starch within organelles called **amyloplasts.** Food storage is a major function of some roots.
- Study the outermost layer of cells, the **epidermis,** and note that **root hairs** are actually extensions of the epidermal cells.
- Just inside the endodermis locate the cells of the **pericycle,** which retain the ability to divide.

During primary growth, the root may send out side branches, or **lateral roots.** These begin growing from the meristematic tissue of the pericycle and push their way out through the cells of the cortex into the soil (see Text Figure 34–5).

- Examine a cross section of a root, showing a developing lateral.

Does the new root have a root cap? ______________

Does it have vascular tissue yet? ______________

Sometimes the plant will form additional roots called **adventitious roots** from the stem or leaves. They might help prop up the plant's stem (corn) or help a climbing plant stick to a surface.

If the root is from a woody plant, it will form a layer of lateral meristem between the xylem and phloem called the **vascular cambium.** Division of cells within the cambium will allow the formation of secondary xylem and secondary phloem, so that the root can grow in diameter as well as in length.

All roots contain food for their own needs, and the roots of overwintering plants must contain the nutrients to support the growth of new shoots the following spring. Some roots are especially modified to store large amounts of food and, as a result, are popular human food as well. The carrot, for example, is like a nonstorage root, except that the parenchyma in the secondary xylem and phloem has been greatly expanded.

- Obtain a piece of a carrot and make a very thin cross section with a razor blade.
- Place the slice on a slide and add a few drops of iodine stain.
- After the stain has soaked in for a few minutes, examine the slice under the microscope.

Where are the grains of starch located?

____

____

- While you are waiting for the results of the starch test, make a longitudinal cut through the section of carrot.
- Note the light-colored ring that divides the innermost layer of phloem from the outermost layer of xylem.
- Locate the light-colored streaks from the inner part of the carrot to the edge. These are the connections to the lateral roots that were present along the tap root before it was pulled up out of the ground.

Although nutrients are required for a plant's healthy growth, too much fertilizer can be harmful.

- Place a piece of carrot root in a solution of 2% ammonium phosphate or ammonium nitrate for 1 hr, and place another piece in distilled water.
- Describe the condition of the two pieces after 1 hr.

Carrot in 2% fertilizer: ______________________________

Carrot in distilled water: ______________________________

If fertilization of plants is overdone, they may be "burned" by the high concentrations of salts that dehydrate them by osmosis and die as a result. Only certain plants such as mangrove are adapted to tolerate high salt concentrations.

STEMS

The Pattern of Stem Growth

As in the root, growth in the stem is **indeterminate.** The growing part is the **apical meristem** tissue found at the tip of the shoot. As the shoot grows, it produces not only the lengthening stem, but also **leaves** that branch from the stem, and side or **lateral branches** (see Text Figure 34–10). The point where a leaf or branch joins the stem is called a **node,** and the section of stem between the nodes is called an **internode.**

- Examine a young plant, such as a bean plant, and locate the nodes and internodes.

- Examine closely the junction of a leaf or the lateral branch with the stem, and note the **lateral** or **axillary bud** that is found in the axil (axilla = armpit) or angle between the leaf or lateral and stem.
- Locate the **apical bud** at the tip of the growing shoot.

The Shoot Apex

The apical bud produces a hormone (auxin) that keeps the adjacent lateral buds dormant for a time. Preferential uptake of nutrients and other hormones (e.g. cytokinin) by the rapidly growing apical meristem may also contribute to lateral bud dormancy. This inhibition of the lateral bud development is called **apical dominance.** Eventually the apical bud may grow far enough away so that its influence lessens, and the lateral buds will develop into lateral branches. If the apical bud is removed by a grazing animal or for experimental purposes, it will no longer inhibit the lateral buds, and they will immediately begin to grow.

- Examine a prepared slide showing the apical bud of a plant such as a lilac or *Coleus* in longitudinal section.
- Identify the **apical meristem,** where the most actively dividing cells are found. In contrast to the root, there is no protective cap over this mound of tissue and no quiescent zone.
- Note the **leaf primordia** adjacent to the meristem; each one (or pair) is attached to the stem at a node, and the small spaces between them are the internodes.
- Look for **vascular tissue** in some of the older leaf primordia.
- Lumps of tissue above the leaf primordia are **bud primordia,** which will form the buds that will eventually give rise to new laterals or to flowers.

As the apical meristem grows, the stem between the leaf primordia lengthens and the new leaves become fully developed. The process is something like pulling out the sections of a collapsible telescope, except that the number of sections keeps growing and can be extended indefinitely.

Internal Structure of a Stem

In stems, the vascular tissues are arranged in bundles that may be located around the edge of the stem (in dicots), or scattered throughout the interior of the stem (in monocots). The vascular bundles are surrounded by the tissues of the cortex, and the epidermis surrounds the whole stem. The structure of a dicot stem is shown in Figure 30–4. A monocot stem for comparison is shown in Color Plate 5.

- Examine a cross section of the stem of an herb under low power, and locate the **vascular bundles,** the cells of the vascular tissue.

Is the stem from a monocot or dicot?

____

- Note the central **pith,** which may have been pulled apart during growth to make the interior of the stem hollow. It is composed of **parenchyma** cells.
- Note the layer of **epidermis** around the outside of the stem. The epidermis is only a single cell layer thick.
- Switch to high power, and examine the edge of the stem closely.
- Note the transparent layer of the **cuticle.** The stem and leaves but not the root are covered with a cuticle formed from the substance **cutin,** which is secreted by the cells of the epidermis.

What is the function of the cuticle?

____

____

____

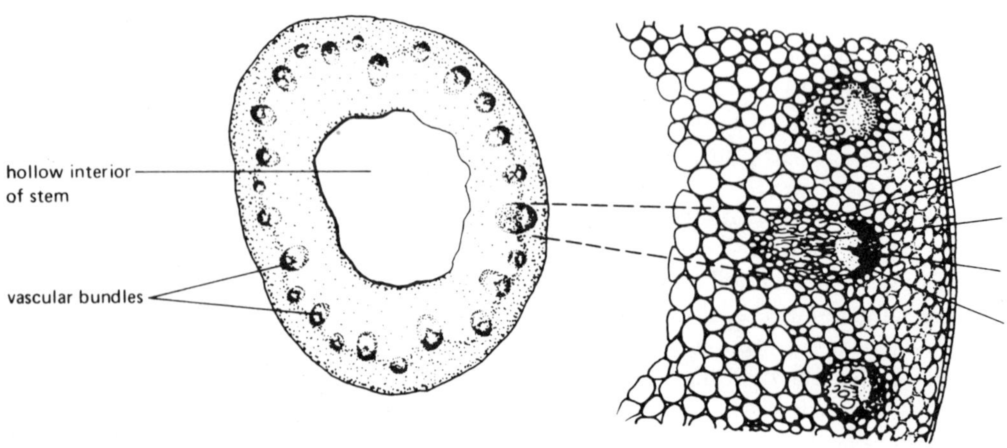

Figure 30–4. A dicot stem. The many vascular bundles form a ring.

- Look for **epidermal hairs** that may be present on the surface of the stem. Hairs on stems and leaves usually have a protective function and may produce irritating secretions to discourage predators (see Color Plate 6).
- Locate the **cortex,** the nonvascular tissue within the epidermis. **Turgor** in the cells of the cortex helps support the shoot.
- Study the layer of cells just inside the epidermis. It consists of **collenchyma** cells, which have characteristically thickened corners and act as supporting columns for the stem.
- Note the inner part of the cortex, which is composed of **parenchyma** cells. Some of the stem cells are photosynthetic, and some may contain amyloplasts for storing starch.
- Examine a single vascular bundle. Locate the larger thick-walled cells of the **xylem** vessels in the inner part and the smaller thin-walled **phloem** tubes in the outer part of the bundle.
- Locate the cells of the **cambium** between the xylem and the phloem.
- Locate the thickly packed, darkly staining **fibers** along the outer edge of the vascular bundle. These are heavily walled sclerenchyma cells, which give added strength to the stem.
- Label the parts of a vascular bundle in Figure 30–4.

The conducting cells of the phloem tissue are the **sieve tube members.** These cells contain living cytoplasm but lack nuclei. Individual sieve tube members are connected end-to-end across perforated plates known as sieve plates. Strands of cytoplasm (plasmodesmata) pass through the holes in the sieves and connect the cytoplasm of all cells in a sieve tube. Sieve tube members are closely associated with **companion cells,** which do have nuclei (see Text Figures 35–20 and 35–21).

- Locate the small **companion** cells adjacent to the sieve tube members in your slide.

The major function of the phloem is to transport nutrients, often in the form of sucrose, from their **source** (leaves, roots) to **sinks** in the rest of the plant for nourishment (apical buds) or for storage (roots). It can conduct dissolved substances in both directions: Some sieve tubes may be conducting downward while others are conducting different substances upward. In this way, minerals and nutrients can be continuously redistributed among the organs of the plant to meet their changing needs. Conduction in phloem is called **translocation** and requires energy for active transport of sucrose into the veins at the source. The phloem contents are under positive pressure and move rapidly by a **mass flow** mechanism from source to sink (see Text Figure 35–23).

Storage Stems

In some plants, the stem is important for storing food or water. In the white potato, for example, part of the stem is a storage organ and also functions in vegetative reproduction.

- Examine a cross section of a potato through one of the "eyes."
- Note that the eyes are really buds that will send up shoots if the potato is planted in moist soil or exposed to the light. The presence of buds is a distinguishing feature of stems.

You should now be able to list five important functions of stems:

1. ____________________
2. ____________________
3. ____________________
4. ____________________
5. ____________________

The Woody Twig

In woody plants, unlike in herbs, the apical meristem has to survive the winter, and the green shoot must be protected from drying out or freezing. The young branches of the plant are called **twigs,** and they become covered with a protective layer of waterproof **cork.** At the end of the growing season, changing environmental conditions cause the terminal bud to cease vegetative leaf production and instead to form protective **bud scales.** Growth of the plant for that season comes to an end. In the spring, the bud scales will fall off, the terminal bud will grow into a new shoot of primary growth, and the "old" twig will continue secondary growth. It will continue cycle after cycle of secondary growth as long as the plant is alive.

- Examine the twig of a woody plant such as a hickory tree, and use Figure 30–5 as your guide to its structure.
- Note the **bud scales** covering the apical bud.
- Identify the **lateral buds** along the sides of the twig. Those in the older portions of the twig have become latent and will not develop unless the terminal bud is damaged.
- Just below each bud is a **leaf scar** where the old leaf was once attached. The bud was thus formed in the leaf axil.
- Look at the leaf scar under the dissecting microscope. The dots within it plugged vascular bundles that once led into the leaf before the leaf was dropped from the tree.
- Locate the **bud scale scars,** which are a ring of wrinkled tissue. This was the position of the terminal bud one year before its present position.

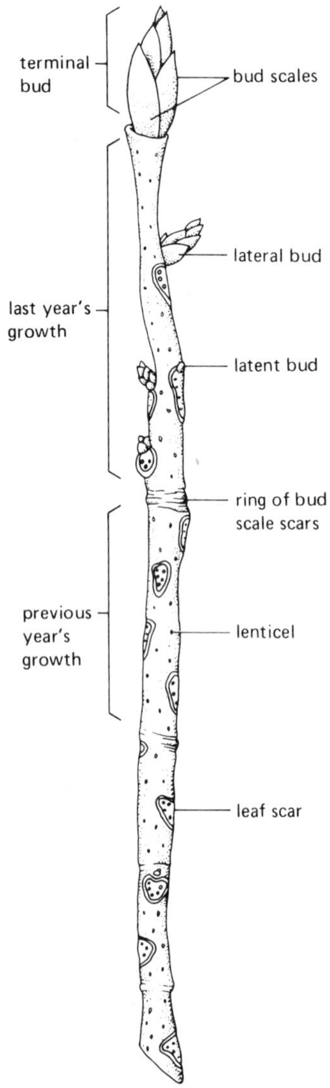

Figure 30–5. A woody twig. This hickory tree grew the length of twig between the ring of bud scale scars and the terminal bud in one season.

- Look further down the stem to find the next ring of bud scale scars. This was the position of the terminal bud two years before its present position.

The point where the twig joins the next larger branch marks the place where it was a lateral bud before it started developing into a new branch.

How many seasons has a twig been growing if it has four rings of bud scale scars between the terminal bud and the branch that it grows from? ______________

- Peel off a bit of the **bark** from the twig, and note that it is much too thick to allow gas exchange. Bark consists of all layers external to the vascular cambium. (See Figure 30–7.)
- Look for small rough spots on the surface of the bark called **lenticels,** and examine one with the dissecting microscope.
- Note that the surface is spongy and not smooth like that of the bark. Its porous structure allows gas exchange between the cells of the stem and the exterior. The internal structure is shown in Color Plate 5.

The internal structure of a woody twig that has grown for only part of one year is not much different from that of a herbaceous dicot. As the twig continues to grow, however, important changes must take place. The meristematic cells of the vascular cambium, located between the primary xylem and phloem in the vascular bundles, continue to divide and produce more vascular cells. As they do so, the vascular bundles become fused into a ring with phloem on the outside, xylem on the inside, and a layer of dividing cambium cells in between. A simplified diagram of this stage is shown in the left part of Figure 30–6.

Secondary growth begins when the vascular cambium becomes active and produces secondary phloem and secondary xylem. The layer of primary phloem is split apart as it is forced outward, and eventually it degenerates. The secondary phloem now functions to transport food. The primary xylem is left behind as the cambium moves outward in a slowly expanding cylinder, leaving the newly formed cells behind. The oldest xylem is in the center of the stem nearest the pith, and the newest is right next to the cambium. A diagram showing the structure of a twig that has grown for its first season, with both primary and secondary growth, is shown on the right in Figure 30–6. The layer of secondary xylem is called **wood,** and the twig now has an annual ring. After another season, it will have another **annual ring** of secondary xylem. The older stems, which you will examine next, may have dozens of annual rings.

Which is "older," outermost xylem next to the vascular cambium or innermost xylem next to the pith?

____

The Woody Stem

Just as the layer of phloem split apart as the vascular cambium grew outward, so the epidermis split and a mechanism was necessary to keep the outer covering of the stem intact. Some of the cells outside the phloem became meristematic tissue, and formed the cylinder of **cork cambium.** Cells produced by the cork cambium move outward and secrete the substance **suberin,** which acts to waterproof the surface of the stem. As the layers accumulate, the cells die and form much of the bark of the branch or tree trunk. The cork cambium and its derivatives are known as the periderm or outer bark of the stem and provide an actively growing protective covering.

- Examine a woody stem that is several years old under low power in your microscope.

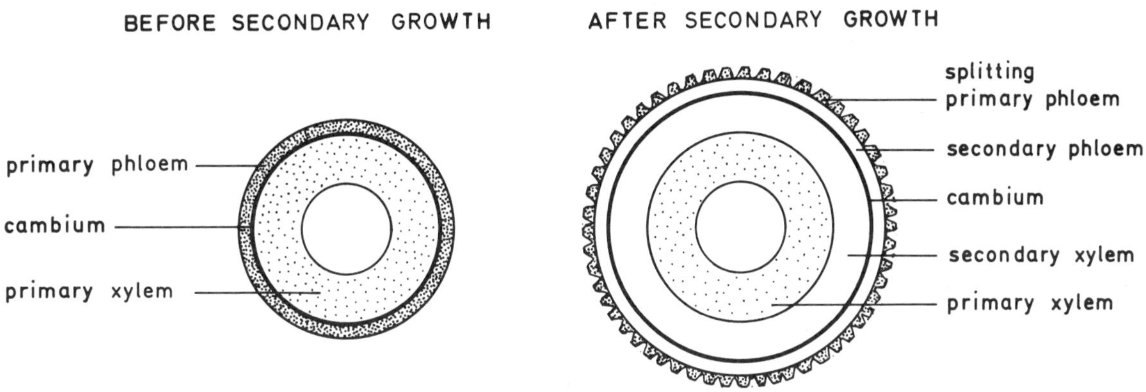

Figure 30–6. Secondary growth in a woody stem. Dividing cells are found only in the cambium, which produces the secondary tissue (shown in white). New xylem is pushed inward and new phloem outward with respect to the cambium. The cambium and phloem must enlarge in circumference as secondary xylem is laid down in successive years.

- Referring to Figure 30–7, identify the **vascular cambium,** which consists of the flattened cells along the outer edge of the largest annual ring.
- Note the annual rings of **secondary xylem** inside the vascular cambium.
- Counting inward from the vascular cambium, note the total number of layers of secondary xylem. (The inner layer nearest the pith is primary xylem.) The number of layers is equal to the age of this part of the stem in growing seasons.

How old is the stem you are studying?

____

- Note the scalloped layer of **secondary phloem** growing from the outer layer of the vascular cambium. This layer is always being split apart as the

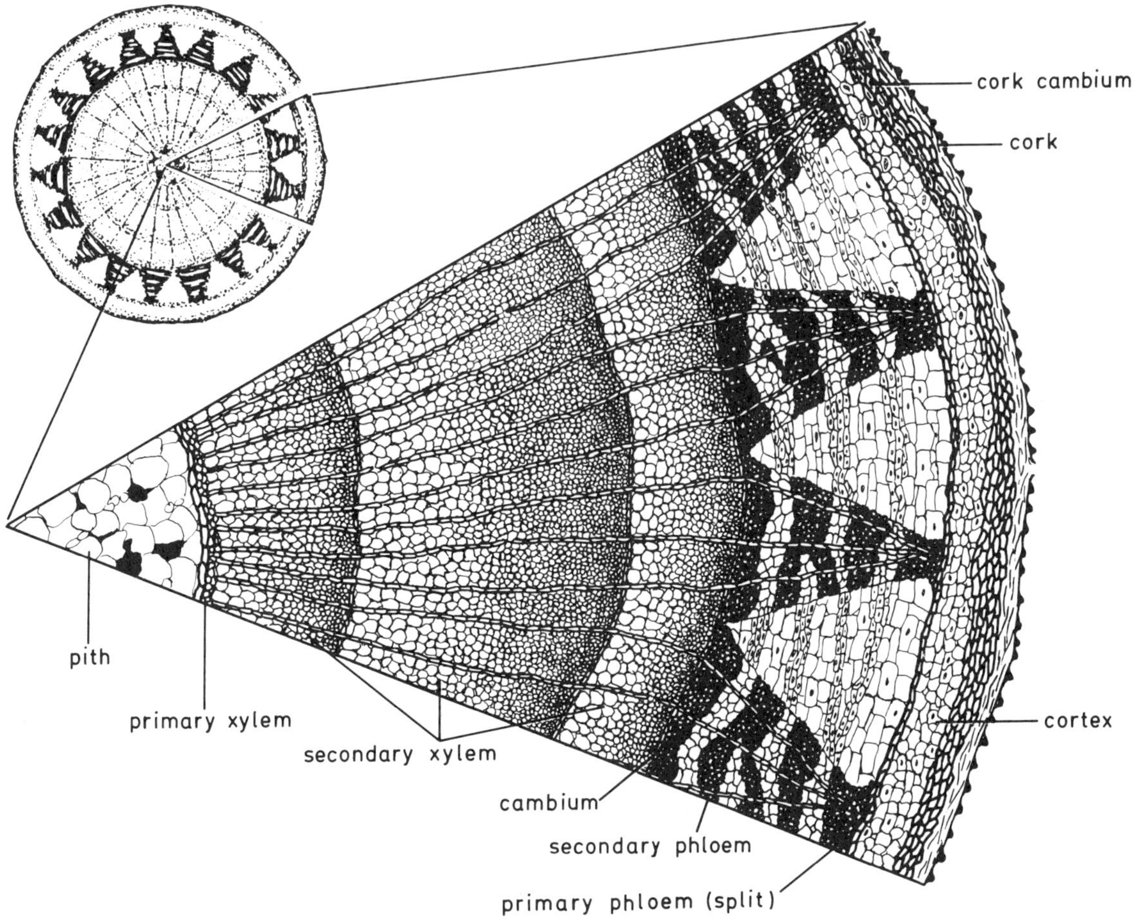

Figure 30–7. A woody twig in cross section. This twig has grown for three seasons. It has one ring of primary xylem and three rings of secondary xylem, one for each season.

cambium enlarges. The secondary phloem and cortex make up the inner bark.

- Note the pyramids of tissue within the splits of the phloem. These are called phloem rays.
- Outside the phloem, locate the **cortex** and the thin layer of **cork cambium** that surrounds it.
- Note the dead and dying cells on the outside of the cork cambium that will make up the **cork.**
- The remains of the original epidermis might still be seen on the outermost surface of the bark.

If you were to peel the bark from this twig, it would separate from the stem at the level of the vascular cambium. The bark includes not only cork and the cork cambium (outer bark), but also cortex and secondary phloem (inner bark). Removing the bark in this way cuts the supply lines through which the roots receive nourishment from the leaves. If the tree is **girdled** by removing a ring of bark that completely encircles the tree, the roots will starve, and the tree will eventually die (see Text Figure 35–6). If even a thin strip of bark remains to connect the top and bottom of the stem, however, the roots may receive enough food and the tree may continue to grow.

The trunk of the tree within the vascular cambium consists of secondary xylem, primary xylem, and pith in the center. As the stem grows older, the secondary xylem in the interior no longer conducts sap. It becomes the **heartwood** of the tree. The cells of the heartwood may become filled with metabolic wastes and resins and may turn a darker color than the still-active xylem, the **sapwood.** Heartwood is often the most beautiful and aromatic wood in a tree. Sometimes it starts to rot, or it may be eaten away by insects to make the tree hollow. The physiology is not upset as a result, but the tree is weakened and may split or break under stress.

- Examine a cross section of an older tree, such as you might come across while splitting firewood.
- Identify the **bark, sapwood,** and **heartwood,** and indicate the position of the **vascular cambium.**
- Label these structures in Figure 30–8a.

How old was your tree when it was cut down?

____

Wood

What is the botanical term for the tissue that we call wood? ______________________

Although the cells of wood are mostly dead, the structure of wood is rather complicated and involves interesting cell adaptations. Gymnosperm wood is shown in cross section in Figure 30–8b.

GYMNOSPERM WOOD

Tracheids are the only type of conducting cells found in gymnosperms. These cells are primitive because they have a small diameter, and because there are only small pores or pits connecting the cells end-to-end and sideways (see Text Figure 35–8).

- Examine a cross section of pine wood or a prepared slide showing tracheids in the microscope.

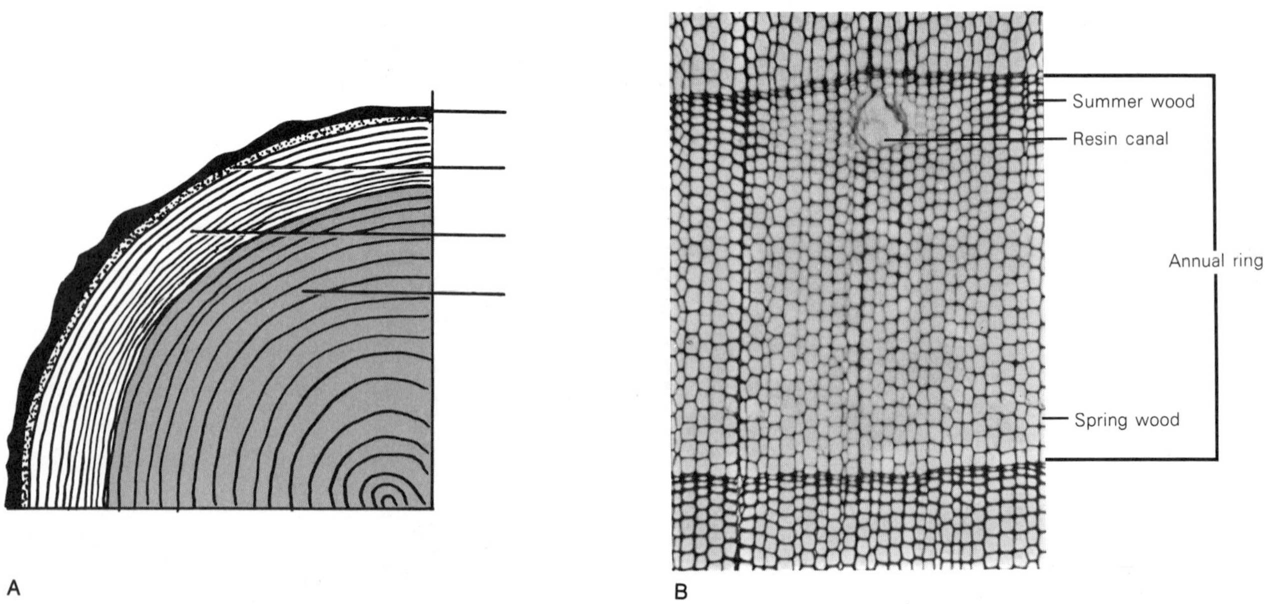

Figure 30–8. Woody stems. (A) A tree trunk in cross section. Label the parts with their correct names. (B) Pine wood in cross section showing the difference in tracheid size between spring wood and summer wood. (B, Biophoto Associates)

- Note the uniform appearance of the tracheids and their small diameter.
- Note also that the tracheids in each annual ring are somewhat larger toward the center of the tree and smaller and more closely packed toward the outside of each annual ring.
- Note the very large **resin canals** scattered through the wood. Resin is a sticky, pungent substance formed in response to trauma that protects the tree against invading fungi and bark beetles. It is secreted by columns of living parenchymal cells.

The inner layer of each ring, the **springwood,** is formed during the spring when the tree is growing faster. The outer layer of **summerwood** is more compact and was formed during the hot, dry summer months.

Is the outside of the tree toward the top or bottom of the photo in Figure 30–8b? ____________

ANGIOSPERM WOOD

Angiosperms have developed efficient conducting cells called **vessel members,** which are stacked vertically to form **vessels** (see Text Figure 35–8). They have a larger internal diameter and have sometimes completely lost the end walls between the ends of adjacent cells. The wood of these trees is a mixture of fibers, tracheids, and vessel cells.

- Examine a cross section of angiosperm wood such as maple, elm, or basswood under low power in your microscope.
- Note the very large vessel members characteristic of angiosperms.

Are the vessels scattered uniformly through the annual ring, or are they concentrated mainly in the springwood?

____

____

- Identify the **tracheids,** which have a much smaller diameter but do conduct some water.
- Locate the **fibers,** which are very long, thin cells with heavy walls. The presence of fibers makes the wood strong or "hard." Technically (according to the U.S. Department of Agriculture) gymnosperms are softwood and angiosperms are hardwood. Actually, the wood of many gymnosperms is harder than that of many angiosperms: Balsawood is from an angiosperm but lacks fibers and is extremely soft. The concentration of fibers determines the strength of the wood.

Within the sapwood are two structures containing living cells. Vertical sheets, or **rays,** of living parenchyma cells run out from the center of the tree to connect the living cells of the cambium with the conduits of the xylem in the sapwood.

- Find the **rays** in the cross section of wood.

There are also columns of living parenchymal cells that secrete **resin** into resin canals.

LEAVES

Types of Leaves

A leaf consists of a flat **blade** (lamina) and a stalk, the **petiole,** which attaches the leaf to the stem. This form provides a large surface area for photosynthesis and gas exchange that is lightweight enough to be supported by the stem. There will be a lateral bud in the leaf **axil** where the petiole joins the stem. Vascular tissue passes through the petiole to enter the **midrib** of the leaf and the leaf veins. Some leaves are **simple** and consist of just one blade and its petiole, but others are **compound** and consist of several **leaflets** that share a single petiole. The types of leaves are shown in Figure 30–9.

If it is unclear whether a leaf is a compound leaf or several simple leaves, look for buds. Because the buds are always found on stems, leaves that join without buds in their axils are part of a single compound leaf.

- Examine the leaves on demonstration, and decide whether each is simple or compound.

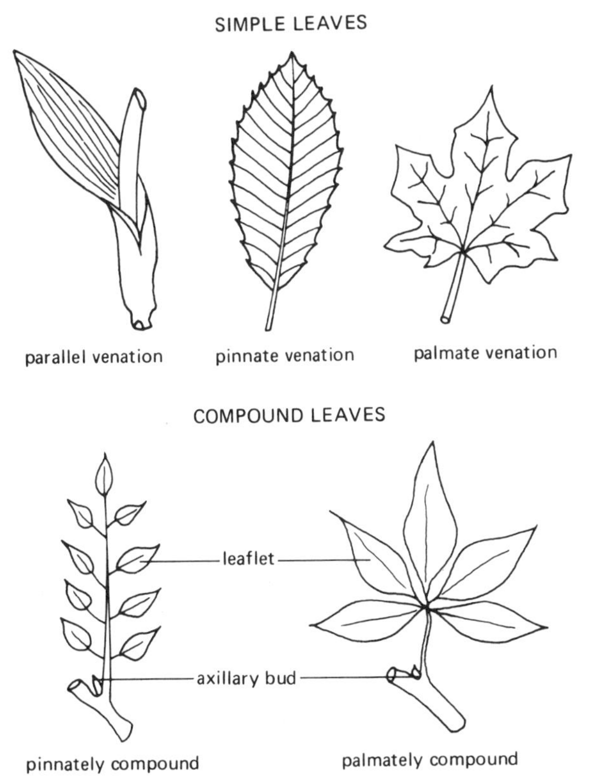

Figure 30–9. Leaves. A compound leaf has a bud only at its base, not next to its individual leaflets.

The vascular tissue within the leaf is in large bundles called **veins.** If the veins run from the petiole more or less parallel out to the tip of the leaf, the leaf has **parallel venation.** If there is one main vein with smaller veins branching off it and running parallel to each other, the venation is **pinnate.** If several main veins radiate out from the petiole and have many smaller branches, the leaf has **palmate venation.** Most monocots have parallel venation, and dicots tend to have pinnate or palmate venation. These types of venation are shown in Figure 30–9.

- Identify the three types of venation among the leaves on demonstration.

The arrangement of the leaves on the stem may be **opposite,** with two leaves on the opposite sides of the stem; **alternate,** with leaves appearing first on one side and then on the other; or **whorled,** with three or more leaves clustered together. If the leaves are opposite, the twigs and branches will also be opposite because they arise from buds in the leaf axils.

- Find an example of opposite, alternate, and whorled leaf arrangements among the leaves on demonstration.

Structure of the Leaf

(Refer back to Topic 5 and to Text Figure 34–14 for details.) Because leaves are covered by an impermeable cuticle, access to the leaf interior is through the pores called **stomata.** When the plant has sufficient water but needs more CO_2, the adjacent **guard cells** will open the stomata to permit gas exchange. (See Color Plate 5.)

- Obtain a dicot leaf, and study its upper and lower surface under high power in the microscope to locate the stomata.

On which surface are the stomata located?

____

Are they open or closed?

____

What ion pump is activated to open the stomata?

____

Stomata are generally found on both surfaces of monocot leaves.

- Review the internal structure of the dicot leaf by examining a cross section of a leaf such as a privet leaf.
- Identify the **cuticle** that covers the leaf and the cells of the **epidermis** that secrete it.
- Locate the **stomata** (openings in the epidermis) and the pair of **guard cells** that surrounds each stoma.
- Note the **mesophyll** (= middle leaf) in the interior of the leaf. It is fundamental tissue containing mainly parenchyma cells.
- Locate the **palisade mesophyll** (palisade parenchyma) and the **spongy mesophyll** (spongy parenchyma), which has many **air spaces.** The air spaces and stomata allow the mesophyll to carry on gas exchange with the air outside the leaf.
- Locate a bundle of vascular tissue, and identify the **xylem** and **phloem.**

The leaves of monocots tend to be long and pointed and oriented vertically. They may not have a palisade layer because the sunlight strikes them at an oblique angle. Often the leaf will be adapted to curl up when the amount of water lost through transpiration becomes too high. The curled leaf keeps the concentration of water vapor inside high, but also cuts down the efficiency of gas exchange. The cross section of a blade of grass that has curled up to prevent water loss is shown in Figure 30–10. Notice that the stomata are all on the inside where there is more water vapor, and that they are located in sunken pits.

- Examine a cross section of a leaf from a monocot such as a lily or oleander.
- Note where the stomata are located and whether air spaces are present or absent.

Sometimes monocots have a layer of **intercalary meristem** at the base of each leaf. The meristem continues to divide and can replace the leaf tissue as it becomes old, damaged, or cut off by a lawnmower. This is the process that allows grass to be mowed frequently without killing it.

- Turn to Table 30–1 and fill in the characteristics of monocots and dicots that you have observed. If the

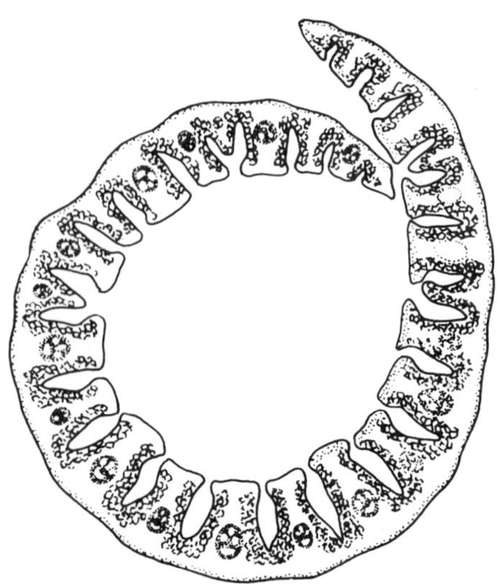

Figure 30–10. A blade of grass in cross section. The leaf is curled and the stomata are located in deep pits to minimize the loss of water.

potted plants that you studied were not in bloom, consult Figure 44–27 in your text for the difference between monocot and dicot flowers.

- Note the results of the transpiration experiments, and return all materials and equipment before leaving the laboratory.

REFERENCES AND FURTHER READING

Balbach, M.K., and L.C. Bliss. *A Laboratory Manual for General Botany,* 6th ed. Philadelphia: Saunders College Publishing, 1982.

Bold, H.C., C.J. Alexopoulos, and T. Delevoryas. *Morphology of Plants and Fungi,* 4th ed. New York: Harper and Row, 1980.

Cronshaw, J. "Support and protection in plants." in *Topics in the Study of Life.* New York: Harper and Row, 1971.

Esau, K. *Plant Anatomy,* 2nd ed. New York: John Wiley and Sons, 1965.

Galston, A.W., P.J. Davies, and R.L. Satter. *The Life of the Green Plant,* 3rd ed. Englewood Cliffs, NJ: Prentice-Hall, 1980.

Jensen, W.A., and F.B. Salisbury. *Botany: An Ecological Approach,* 2nd ed. Belmont, CA: Wadsworth, 1984.

Lee, A.E., and C. Heimsch. *Development and Structure of Plants: A Photographic Study.* New York: Holt, Rinehart and Winston, 1962.

Prance, G., and K. Sandved. *Leaves.* New York: Crown, 1985.

Raven, P.H., R.F. Evert, and S.E. Eichhorn. *Biology of Plants,* 4th ed. New York: Worth, 1986.

Ray, P., T.A. Steeves, and S.A. Fultz. *Botany.* Philadelphia: Saunders College Publishing, 1983.

Salisbury, F.B., and C.W. Ross. *Plant Physiology,* 2nd ed. Belmont, CA: Wadsworth, 1978.

Weier, T.E., C.R. Stocking, and M.G. Barbour. *Botany: An Introduction to Plant Biology,* 6th ed. New York: John Wiley and Sons, 1982.

Wilson, B.F. *The Growing Tree.* Amherst: University of Massachusetts Press, 1970.

Wooding, F.B.P. "Phloem." *Carolina Biology Reader.* Burlington, NC: Carolina Biological Supply Company, 1978.

Journal Articles

Albersheim, Peter. "The walls of growing plant cells." *Scientific American,* 80, April 1975 (#1320).

Baes, C.F. III, and S.B. McLaughlin. "Trace elements in tree rings: Evidence of recent and historical air pollution." *Science* 224:494, 4 May, 1984.

Biddulph, Susann, and Orlin Biddulph. "The circulatory system of plants." *Scientific American,* 44, February 1959 (#53).

Elias, T.S., and H.S. Irwin. "Urban trees." *Scientific American,* 110, November 1976 (no offprint).

Epstein, Emanuel. "Roots." *Scientific American,* 48, May 1973 (#1271).

Erickson, Ralph O., and Wendy Kuhn Silk. "The kinematics of plant growth." *Scientific American,* May 1980 (#1470).

Folkerts, George W. "The gulf coast pitcher plant bogs." *American Scientist* 70(3): 260, May–June 1982.

Fritts, Harold C. "Tree rings and climate." *Scientific American,* 92, May 1972 (#1250).

Galston, A.W., and C.L. Slayman. "The not-so-secret life of plants." *American Scientist* 67:337, May–June 1979.

Heslop-Harrison, Yolande. "Carnivorous plants." *Scientific American,* 104, February 1978 (#1382).

Hitch, Charles J. "Dendrochronology and serendipity." *American Scientist* 70(3): 300, 1982.

Kaplan, D.R. "The development of palm leaves." *Scientific American,* 98, July 1983.

Olsen, Ralph A., Ralph B. Clark, and Jesse H. Bennett. "The enhancement of soil fertility by plant roots." *American Scientist* 69(7): 378, July 1981.

Overbeek, Johannes. "The control of plant growth." *Scientific American,* 75, July 1968 (#1111).

Pillemer, E., and W. Tingey. "Hooked trichomes: A physical plant barrier to a major agricultural pest." *Science* 193:482, 6 August, 1976.

Shigo, Alex L. "Compartmentalization of decay in trees." *Scientific American,* 76, April 1985.

Tomlinson, P.B. "Tree architecture." *American Scientist* 71: 141, March–April, 1983.

Torrey, J.G., and L.J. Feldman. "The organization and function of the root apex." *American Scientist* 65:334, May–June 1977.

Uribe, Ernest G., and Ulrich Lüttge. "Solute transport and the life functions of plants." *American Scientist* 72:567, November 1984.

Zimmerman, Martin H. "How sap moves in trees." *Scientific American,* 132, March 1963 (#154).

Student Name ______________________ Date ______________

TABLE 30–1. Characteristics of Monocot and Dicot Angiosperms

Characteristic	Monocots	Dicots
Type of seed leaf (cotyledon)		
Type of leaf venation (vein pattern)		
Arrangement of stem vascular tissue		
Symmetry of flower parts (multiples of 3, 4, and so on)		
Form of plant (herb, shrub, tree)		
Formation of wood		
Type of root system		
Typical examples		

TABLE 30–2. Roots

Function	Tap Root	Fibrous Roots
1. Anchoring the plant		
2. Absorbing water and minerals (surface area)		
3. Transport to the stem		
4. Storage of nutrients		

Drawing of Petal After Transpiration	**Drawing of Xylem Cross Section After Transpiration**

Figure 30–11. Transpiration. Sketch the results from the dye experiment.

TABLE 30–3. Results from the Transpiration Experiment

Transpiration in Light			Transpiration in Dark		
Time (min)	Reading	Change (mL)	Time (min)	Reading	Change (mL)
0					
15					
30					
45					
60					
75					
90					
105					
120					
Average rate = ______ mL/hr			Average rate = ______ mL/hr		

TOPIC 31

Reproduction in Flowering Plants

OBJECTIVES

When you have completed this topic, you should be able to:

1. List the reproductive characteristics that distinguish angiosperms from other seed plants.
2. Given a diagram of a flower or an actual flower, identify it as a monocot or dicot and point out the following parts: sepal, petal, stamen, anther, pistil, ovary, style, stigma, ovule, pollen.
3. Define pollination and fertilization, and state where they occur in a flower.
4. Explain the role or origin of the following structures and identify them in a microscope slide: microspore, male gametophyte, pollen grain, pollen tube, megaspore, female gametophyte, embryo sac, micropyle.
5. Give the botanical definition of a fruit, and name some "vegetables" that are botanically fruits.
6. Given an actual fruit or a diagram, identify the following parts: ovule, locule, endocarp, mesocarp, exocarp, placenta, ovary, attachment sites of old sepals and petals.
7. List or identify the parts of a bean seed and of a corn kernel; state the function or fate of each part.
8. Describe the role of animals in pollination and seed dispersal.
9. State four factors that are important in seed germination.
10. Explain the role of plant hormones in dormancy, germination, tropisms, and apical dominance.
11. Explain the advantages and disadvantages of vegetative reproduction; give an example of vegetative reproduction from a root, from a stem, from a leaf, and from an infertile fruit.

TEXT REFERENCES Chapter 34 (A, F); Chapter 36 (C–E); Chapter 37.

KEY TOPICS

flowers	seeds	plant hormones
pollen	germination	vegetative reproduction
fruit	development	

INTRODUCTION

As we have already seen in surveying the plant kingdom, most seed plants have dispensed with a free flagellated sperm in their reproductive cycle and have instead developed the pollen grain as a carrier of the male gamete. The process of fertilization leads to the formation of a seed that contains a small embryo and also a source of nutrition within a tough seed coat. The seed is usually resistant to harsh environmental conditions and begins growing only when conditions become favorable.

Angiosperms differ from the other seed plants in that the seeds are formed within **flowers.** In many plants the flowers are showy, or have an odor so that animals are attracted to them and become involved in pollination. The seeds are enclosed within a fruit that may be a source of food for animals. The seeds are then scattered about when the animal feeds, or are carried away in its digestive tract and are later expelled far from the parent plant. Thus animals are an important factor in seed dispersal for many angiosperms.

The mutual adaptation of plants and animals relating to plant reproduction are the product of **coevolution.** (See Color Plate 7.) Flowers advertise themselves through color and odor and reward their animal visitors with pollen, nectar, or fruit. The more adapted a flower is to attracting specific pollinators, the more efficient pollination will be. Some flowers, the fig for example, are so highly evolved that they can be pollinated only by a particular species of animal.

The flower fits into the generalized plant life cycle in a rather complex way. If you refer back to this life cycle in Figure 16–1, you will be reminded that in plants meiosis gives rise to **spores.** The plants that we see are **sporophytes.** They are diploid (2N) and so are their flowers, where meiosis occurs. Separate male and female parts produce male spores (microspores) and female spores (megaspores). The spores remain in the flowers where they develop into haploid (N) **male gametophytes** (pollen) and **female gametophytes** (immature seeds). After fertilization, the diploid (2N) sporophyte embryo of the next generation develops within the seed. As a flower develops into a fruit, it includes three generations packed in together:

SPOROPHYTE →	GAMETOPHYTE →	SPOROPHYTE
outer fruit	pollen, immature seed	embryo in seed

Many angiosperms propagate by **vegetative** or **asexual reproduction** under appropriate circumstances. A **clone** of offspring is produced that is genetically identical to the parent. If environmental conditions are stable, it is advantageous for the offspring to resemble their successful parent and vegetative reproduction is favored. This method, however, foregoes the variation associated with sexual reproduction that is important if conditions change.

Before beginning the next section, turn to the sections entitled POLLEN and GERMINATION, and set up the demonstrations of pollen tubes and seedling greening, unless your instructor gives other directions.

FLOWERS

Angiosperms that are pollinated by animals have several strategies for attracting attention. Their flowers may have large, showy petals, which are brightly colored to attract the notice of potential pollinators. They may even have color patterns that are not visible to us but that attract insects such as bees because some insects can see into the ultraviolet range further than we can. Within the flower there may be nectar, which is an important food source for the pollinators. Some flowers produce volatile chemicals that give the flower a very attractive odor. (Putrid odors that are disgusting to humans may be very attractive to pollinators such as flies.) Many angiosperms have a sort of symbiotic relationship with humans, in which they are protected and propagated in exchange for their production of flowers.

Flowers are modified shoots with very short internodes in which the leaves have become modified to form the flower parts.

- Obtain a large, simple flower, such as a gladiolus. Cut it from the stem with a razor blade far enough down so that you take all of the green part at the base in addition to the petals.
- Identify the green **receptacle** at the base of the flower.
- Carefully dissect the flower from its outer whorls inward. The flower may be enclosed in a green bract, or leaflike part below the flower itself.
- Note the **sepals,** which make up the outer whorl. In the gladiolus, the sepals are the same color as the petals, but they are often green, as in a rose bud.
- Note the colored **petals,** which make up the next whorl called the corolla. (In the gladiolus, the sepals and petals are so similar they are collectively called the tepals.)
- Locate the **stamens,** the male reproductive structures that make up the next whorl.
- Locate the **pistil.** A pistil is made up of one or more modified leaves called **carpels,** which are the female reproductive organs. In this case the carpels are unfused at the top of the pistil so you can count them. You will be able to verify the number of carpels later, when you dissect the ovary at the base.

How many petals, sepals, stamens, and carpels does the flower have?

Petals: ______

Sepals: ______

Stamens: ______

Carpels: ______

Monocots tend to have the flower parts in threes, whereas dicots tend to have four or five of each part.

Considering the other characteristics that distinguish monocots and dicots, as well as the structure of their flowers, do you think this flower is from a monocot or from a dicot? ______________

Pollen is formed in the **anther,** which is the top part of the stamen supported by the stalklike **filament.** It is then deposited on the sticky surface of the **stigma** at the top of the pistil during **pollination.** The germinating pollen grain must grow a tube through the **style** of the pistil to the **ovary** at its base. The ovules (prospective seeds) within the ovary are fertilized when the pollen tubes reach them and sperm nuclei enter to unite with nuclei within each ovule.

When you have finished this part of the dissection, label the structures indicated in Figure 31–5 at the end of this topic.

- Make a cut across the base of the carpels where the green **ovary** is the widest, as shown in Figure 31–6 at the end of this topic.
- Make a second cut and remove the slice from the flower.
- Note the small **ovules** that will mature into seeds; each ovule will be fertilized by one pollen tube.
- Note the spaces containing the seeds, called **locules.** The number of locules is equal to the number of carpels that make up the pistil.

How many locules (carpels) does the ovary contain?

____

- Identify the wall of the ovary, which surrounds the locules. It will develop into part of the fruit.
- Make a sketch through the ovary in the space provided in Figure 31–6, and label its internal structures.

There are many variations in flower structure. The flower that you have just studied is **simple** because it contains a single ovary, and it is **complete** because it contains both stamens and carpels. A flower such as a daisy is a **composite** flower and consists of many tiny individual flowers. Each white petal is formed by a separate ray flower, and the central disc consists of many other tiny flowers. Sometimes the carpels and stamens are in separate flowers, which are termed **incomplete.** Such separate male and female flowers may occur on the same plant, as in corn and squash, or on separate plants, as in holly and willow trees.

Some flowers are self-fertilizing and have petals that close over the reproductive organs. The anther fertilizes the carpel right next to it, and the closed petals prevent the entry of pollen from another flower carried by wind or insects. Flowers that are normally pollinated by the wind rather than by animals are often inconspicuous and may lack petals altogether. The flowers of many trees and grasses lack petals.

- Examine other types of flowers that are on demonstration.

Such specialized flowers are considered to be **highly evolved.** Ancestral flowers tend to have many flower parts, which are arranged in spirals around the stem. They also tend to have radial symmetry. In more evolved flowers, the sepals, petals, anthers and carpels are fewer and are found in whorls, as in your gladiolus. The most highly evolved flowers are bilaterally symmetrical, as in an orchid or sweetpea.

The sepals usually form a whorl of their own outside the petals, although they may closely resemble the petals, as in the gladiolus. Sometimes they are green and inconspicuous, and sometimes they are large and showy compared to the rest of the flower. In other flowers, such as the poinsettia, the red "petals" are modified leaves called bracts that arise below and surround the tiny flowers in the center.

POLLEN

> *Note:* **At the beginning of the laboratory session, set up your demonstration in the following manner.**
>
> - **Obtain a Petri plate of agar or a slide with a drop of 10% sucrose.**
> - **Shake some pollen from the anther of a flower such as gladiolus or *Tradescantia* onto the agar or the drop.**
> - **Cover the pollen, and set it aside in a warm, moist place to allow pollen tubes to develop from the grains of pollen.**

The male reproductive organ, the **stamen,** consists of the **anther** (pollen sac), which is supported by the **filament.**

- Locate the filament and anther of your flower.
- Shake a few grains of pollen onto a slide, make a wet mount, and observe the pollen under high power in your microscope.
- Note the shape of the pollen grains.
- If the pollen from other sources is available, study it also, and compare the shapes of pollen grains from different sources. Otherwise consult Figure 37–5 in your text.

Is all pollen alike? ______________

The cells within the anther that undergo meiosis are the **microspore mother cells,** and the haploid cells that are produced are the microspores. This tetrad of microspores will produce four male gametophytes that are parasitic within the anther as they

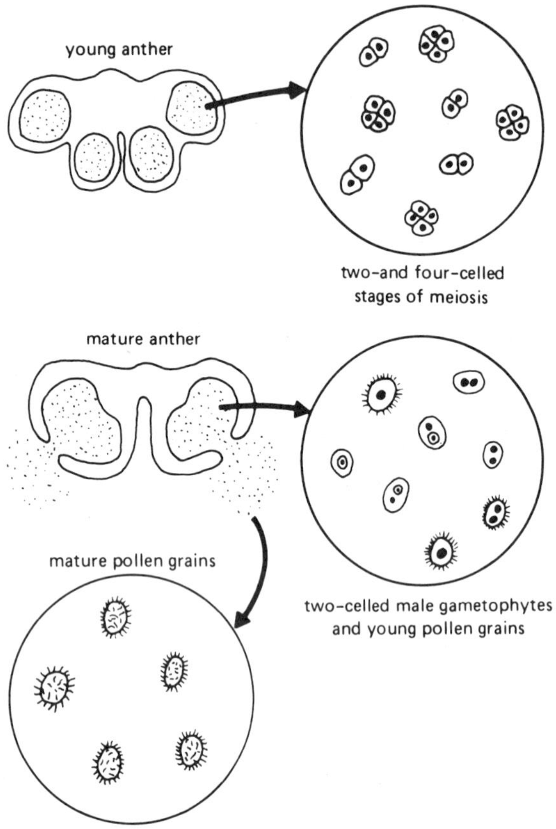

Figure 31–1. Production of pollen grains. A microspore mother cell produces four microspore daughters in a tetrad. Each of the haploid cells gives rise to a pollen grain with a generative nucleus (divides to form sperm nuclei) and a tube nucleus (degenerates). The mature pollen grain, the immature male gametophyte, has a characteristic spiny covering.

mature into pollen grains. The formation of pollen is shown in Figure 31–1.

- Examine a prepared slide of an anther of the lily containing pollen tetrads.
- Look for groups of four cells, which are the tetrads.
- Examine a slide showing a mature anther that is almost ready to shed pollen.
- Note that many grains have two functional cells, and there are some mature pollen grains that are dark and opaque.

A single grain of mature pollen contains a generative nucleus and a tube nucleus. The generative nucleus will participate in fertilization. The pollen is deposited on the stigma during pollination and the **tube nucleus** then directs the growth of the **pollen tube,** into which it moves. The **generative nucleus** divides to form two **sperm nuclei,** which also move out into the tube. Only the sperm nuclei will eventually enter the ovule and contribute genetic material to the new embryo through fertilization. In angiosperms, pollination and fertilization may be separated by a few hours or whatever time it takes for the pollen tube to reach its destination.

- Examine your demonstration of germinating pollen to see if any pollen tubes have grown.

Are the tubes straight or irregular? ______________

If there are no tubes yet, examine the grains again later. (See Text Figure 37–7.)

Pollen doesn't always germinate when it lands on a stigma. Mechanisms have evolved to prevent germination: for example, if the stigma is the wrong species, at the wrong stage of maturity, or from the same flower as the pollen. These adaptations tend to promote variation through cross-pollination.

OVARY

The ovary at the base of the pistil contains one or more ovules or **megasporangia.** The reproductive cell within the ovule, called the **megaspore mother cell,** undergoes meiosis to form four haploid cells; usually only one survives to form the female gametophyte or **embryo sac.** (The exact number of cells forming the female gametophyte varies, depending on the species of flower.) In the end the mature embryo sac will usually have eight haploid nuclei, produced from the megaspore mother cell by mitosis, and seven spore cells. There will be one **egg nucleus** near the micropyle and two **polar nuclei** that have migrated from the ends of the embryo sac to the large central cell. The development of the embryo sac is shown in Figure 31–2.

- Examine a prepared slide showing the cross section of the ovary of a lily under low power.
- Referring to Figure 31–2, scan the six ovules of one cross section to locate an **embryo sac,** a large structure with granular cytoplasm. (Many of the ovules will not show an embryo sac because the section was at the wrong level.)
- When you have located an embryo sac, switch to high power, and note the nuclei within it. Developing sacs will have two or four nuclei, and mature ones will have eight (rarely seen).

Are any of the cells in the embryo sac in the process of dividing? ______________

Fertilization occurs when the sperm nuclei of the pollen tube enter a pore at the bottom of the embryo sac called the **micropyle,** and fuse with the nuclei within. One sperm nucleus fuses with the egg nucleus to form the diploid zygote and thus the **embryo.** These are the only nuclei that contribute genetic information to the next generation. The second sperm nucleus fuses with the polar nuclei (which may have meanwhile fused) to form the triploid **endosperm.** Mitosis then produces the embryo and the tissues of the endosperm that will nourish it. This **double fertilization** occurs only in angiosperms.

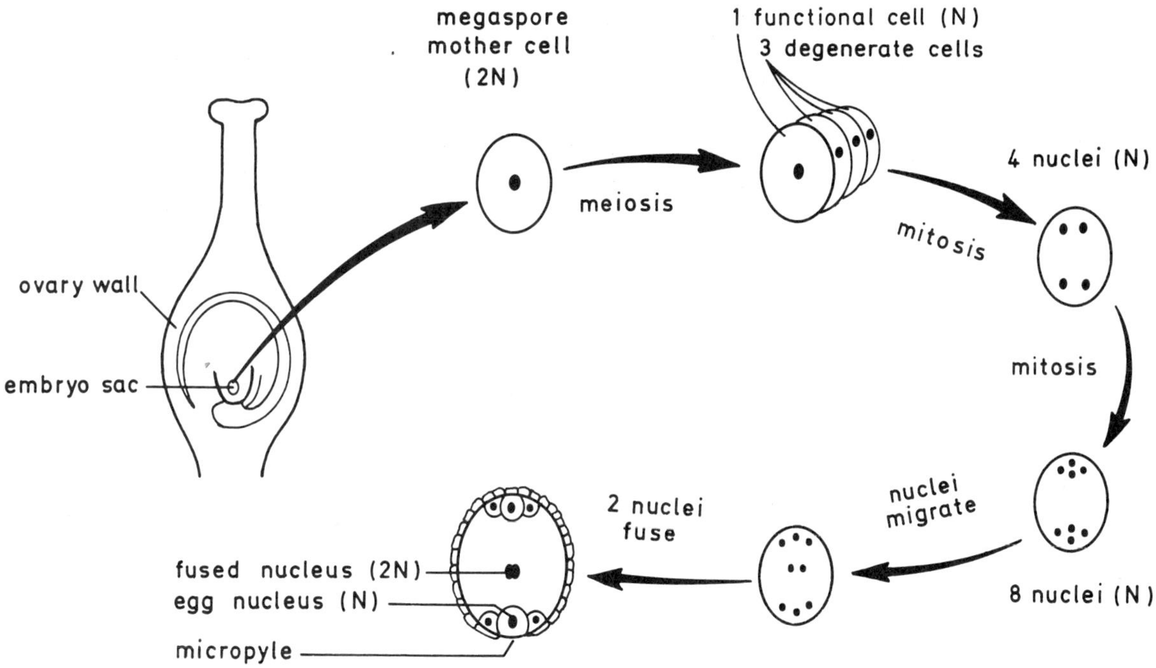

Figure 31–2. Development of the embryo sac. The developmental pattern shown is typical of most flowers. The embryo sac represents the female gametophyte and develops from the haploid nucleus of a megaspore. The egg nucleus will be fertilized to form the embryo. The two polar nuclei in the central cell fuse at some point and are fertilized to produce the endosperm.

Label the important structures in the diagram of fertilization shown in Figure 31–7 at the end of this topic, and give the function or fate of each part listed there in Table 31–1.

FRUIT

After fertilization, the ovule develops into a seed, and the entire ovary matures into the **fruit.** Botanically speaking, a fruit is a ripened ovary or an ovary fused with nearby tissue. Any plant organ that contains seeds is a fruit. The maturing process begins when pollen stimulates the production of the growth hormones auxin and gibberellin. Cytokinin also promotes growth of the fruit and ethylene causes it to ripen.

- Obtain a fruit such as a bean pod. If the fruit was recently picked, you may see old petals and sepals clinging to the end of the pod.
- Split the pod open and find the seeds. Each one represents an **ovule.**
- Note that each seed is attached to the pod at a single point. The part of the ovary where the ovule originates and remains attached is called the **placenta.** In ancestral plants, the ovules were attached to the edges of a leaf. If the edges were then rolled together so that the ovules were enclosed, the ovary would closely resemble the contemporary bean pod.
- Identify the space surrounding the seeds, known as the **locule.**

How many locules does the bean have?

____

- Note the pod, which represents the wall of the ovary and is called the **pericarp.**

While a bean pod is not very "fruitlike," it is a perfectly good example of a fruit. The pod itself is a compressed version of the kind of fruits you are used to: it has an inner layer, the **endocarp;** a middle, juicy layer, the **mesocarp;** and an outer layer, the **exocarp.**

- Make a cut through the pod of the bean, and examine a slice of pod material in the dissecting microscope.
- Note the lining of the pod, the endocarp, the juicy mesocarp, and the tough outer exocarp.
- Label the parts of the bean pod shown in Figure 31–8 at the end of this topic. (One ovule failed to get fertilized.)

What was the fate of the unfertilized ovule?

____

____

A more familiar fruit such as a peach has the same layered structure. The seed is enclosed within a space surrounded by the stony endocarp that we call the pit. The juicy flesh is the mesocarp, and the skin is the exocarp. Label the parts of a peach fruit in Figure 31–8 at the end of this topic.

Sometimes the whole pericarp is reduced to a dry skin that surrounds the seed. This is seen in a wheat grain or in the fruit of a dandelion. A strawberry also has dry fruits that we call the "seeds" and they are embedded in the fleshy receptacle of the flower that we call the "fruit."

Edible fruits are an excellent means of dispersing large seeds far and wide through animal carriers. Small, inedible fruits may have hooks so they can hitchhike long distances on animal fur. Fruits that depend on the wind for dispersal, such as the milkweed, are necessarily light and have a tuft of hairs for sailing. (See Color Plate 5.)

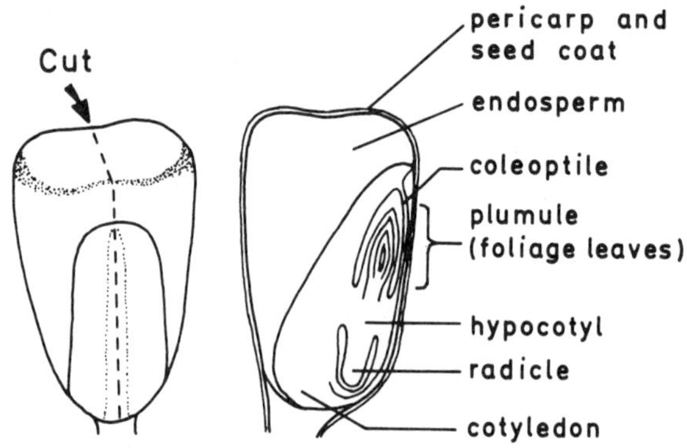

Figure 31–4. A corn kernel. The shoot is covered by the sheath of the coleoptile, and the pericarp has fused with the seed coat.

SEEDS

Anatomy

Take a closer look at the structure of a seed by dissecting a bean seed that has been soaked to loosen the skin and soften it. Its structure is shown in Figure 31–3.

- The brown skin of the bean seed is the **seed coat.** It developed from the ovule of the flower.
- Note the rough whitish area called the **raphe.** This is the remnant of the stalk that once attached the hilum of the seed to the pod's placenta.
- Examine the seed with your dissecting microscope, and find the tiny opening of the **micropyle.** Pollen entered the ovule through the micropyle.
- Carefully split the seed into two halves, and refer to Figure 31–3 to identify the parts: **cotyledon, plumule** or **epicotyl, hypocotyl, radicle,** and **foliage leaves.** (Epicotyl is that part above the attachment point of the cotyledons; hypocotyl is that part below.) Apical meristem is found at the tip of the radicle and within the foliage leaves. The process by which the zygote grew into the young embryo is shown in Text Figure 47–12. In contrast to animal embryos, the cells in plant embryos have cell walls and cannot migrate, so the geometry of cell division is strictly programmed.

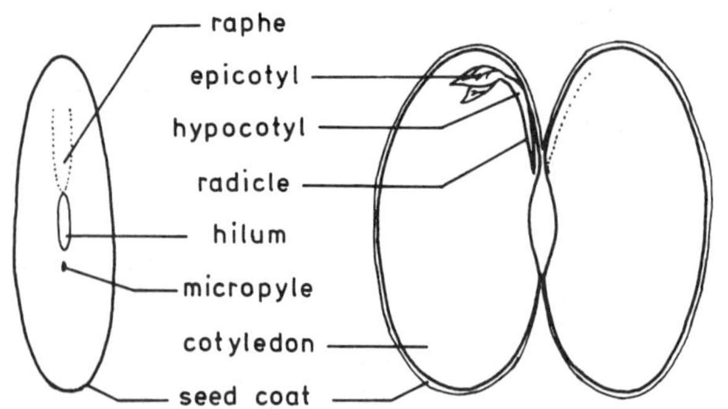

Figure 31–3. A bean seed. The bean on the right has been split open to reveal the embryo (epicotyl and hypocotyl). The apical meristem of the shoot is hidden within the first foliage leaves.

The bean is a dicot, and its seed has two cotyledons. The seeds of monocots have only one. The endosperm tissue of the bean has been completely absorbed by the cotyledons.

- Obtain a soaked corn kernel, and cut it in half as shown in Figure 31–4.
- Identify the single **cotyledon,** the **endosperm,** the **hypocotyl, epicotyl,** and the **foliage leaves** (plumule) of the embryo. The shoot is covered by a sheath called the **coleoptile.** It protects the shoot during germination.

In the corn kernel, the cotyledon has absorbed only part of the endosperm tissue, and there is still a substantial amount of endosperm left. The ovary wall has become the thin, dry pericarp fused with the seed coat. A corn "seed" is thus really a fruit!

- Add a few drops of iodine stain to the cut surface of the corn kernel to stain the stored starch.

Does the endosperm or cotyledon contain more stored starch?

____

Metabolism

The seed has lost most of its water, and carries on metabolism at a very slow rate. This **dormancy** is maintained by the hormone abscisic acid. Some seeds will be able to germinate under the right conditions, and are **viable** whereas others are **inviable.** Seeds must be kept dry to retain viability. It is easy to test for metabolic activity in seeds by soaking them and observing the reduction of a dye by the living tissue. The dye is tetrazolium, which is normally colorless, but turns pink when it is reduced by living tissue.

- Cut several soaked yellow corn kernels in half, and put them into a test tube.
- Add enough 0.5% tetrazolium solution to cover the kernels, and set them aside.
- Observe the cut surfaces from time to time for the production of a pink color.
- After the pink color is well developed, remove the kernels, and examine them to see which part of the seed has the darkest color and therefore the greatest metabolic activity.

Where was there the greatest metabolic activity?

____

Were any of the seeds totally inactive with the dye (dead)? ____________

GERMINATION AND EARLY DEVELOPMENT

> *Note:* **At the beginning of the laboratory session, set up your demonstration in the following manner.**
>
> - **Obtain a few alfalfa sprouts, a Petri dish, and some filter paper or paper toweling.**
> - **Cut a circle of toweling to fit the Petri dish, and moisten it with tap water.**
> - **Place the sprouts on the toweling, and cover the dish.**
> - **Note the color of the cotyledons and hypocotyls of the sprouts.**
> - **Place the dish in strong sunlight or artificial light.**

Several factors may be important in the process of germination: a period of cold (vernalization), absorption of water, abrasion or digestion of the seed coat, and an increase in the content of the hormone gibberellin. Seeds differ in their requirements, but they all require a supply of water in order to germinate. The absorption of water that starts germination is called **imbibition.**

- Examine some bean seeds that have been soaked in water overnight, and compare them with dry seeds.

In what two ways does the seed change during imbibition?

____

____

During the first days of development, the seed must rely on its own stored food present in its cotyledons. It must reach sunlight and a source of water within a limited time span or it will not survive. In order to increase its chance for success, its growth is closely controlled by hormones. These cause appropriate growth responses, called **tropisms,** to environmental gradients encountered by the plant.

The level of abscisic acid is high in the seed and maintains dormancy. After imbibition has taken place, gibberellin production increases and stimulates elongation of the stem. Inhibitory concentrations of auxin in the root tip cause the root to show **positive geotropism** (gravitropism) and to grow downward toward the center of the earth, i.e., toward water (see Text Figure 36–11). Auxin also causes the shoot to show **negative geotropism** and to grow toward the soil surface. Once the shoot breaks the soil surface, auxin causes it to show **positive phototropism** and to grow toward the light. Gravity can be sensed by root cap cells that contain heavy organelles called **amyloplasts** (shown in Text Figure 36–11); light is detected by the colored protein **phytochrome.** The mechanism by which these detectors change the hormone balance to cause the tropic response, however, is not understood.

- Examine some young bean seedlings that have been allowed to sprout on agar plates or paper towels.
- Note that the hypocotyl has grown downward, whereas the plumule has turned up toward the light.
- Look at the oldest plants, and note how the cotyledons have shriveled up as their food was removed by the embryo. They will soon fall off the plant. The normal development of bean seedlings is shown in Text Figure 34–7.

An interesting example of the vital importance of hormone balance can be seen in seedlings that have been kept in complete darkness so that they have become **etiolated.**

- Examine some etiolated seedlings, and compare them with normal seedlings of the same age.

What color are the seedlings? ____________

What is their appearance compared with that of normal seedlings?

____

Gibberellin secreted by a fungus has long been known to cause "foolish seedling" disease of rice, in which the seedlings grow tall and spindly and then fall over and die. Excessive gibberellin is also responsible for the abnormal growth of etiolated seedlings. During the first days of growth, a seedling must reach the light or it will die. The concentration of gibberellin is high to stimulate rapid growth of the stem while the seedling is in the darkness underground. As soon as it reaches the light, the red wavelengths of light transform phytochrome P_r to P_{fr}, and the production of gibberellin falls off as a result. The levels of gibberellin and auxin reach the proper ratio needed for balanced growth. The stem's growth slows down, and normal development follows. If, however, the shoot is unlucky and never reaches the sunlight, the phytochrome stays in the form of P_r and continues to stimulate the production of large amounts of gibberellin. The seedling grows and grows until it becomes severely etiolated and dies. A seedling that is being kept in complete darkness can be saved by irradiating it with red light to reduce the amount of P_r and decrease gibberellin production.

What other process in angiosperms is under the control of P_r and P_{fr}? ______________________

As the seedling emerges into the light, it has the potential for carrying on photosynthesis, and will turn green within a short time. One of the steps in the production of chlorophyll is a photoreaction, which requires light. As long as the seedling is in the dark, the reaction is blocked, and there is no chlorophyll. There is, however, a large amount of the compound preceding the light-sensitive step. When light becomes available, the production of chlorophyll proceeds rapidly, and the seedling turns green, as the chlorophyll builds up.

- Examine the alfalfa sprouts that you set up at the beginning of the period for signs of greening.
- If they are not yet green, examine them once more before you leave the laboratory.

How long did it take the sprouts to turn visibly green?

____

Why is it advantageous for the seedling to postpone formation of chlorophyll until it reaches the light?

____

____

____

____

VEGETATIVE REPRODUCTION

Although there is a great deal of concern about the possibility of cloning human beings, growing **clones** of plants in the laboratory and in agricultural practice has already become widespread: One or a few meristematic cells can be used to grow plant tissue and even whole plants that are genetically identical to the parent cell and to each other. Vegetative, or asexual, reproduction is basically the same process: Cells of the parent plant produce progeny that are genetically identical to the parent. Vegetative reproduction is a way of life among plants, and often it is the only means by which a plant can reproduce at all. When a successful plant reproduces asexually, the progeny will most likely also be successful because they will be well adapted to local conditions.

The parent plant makes a larger investment in each offspring produced vegetatively, compared to a single seed, but that offspring has a much better chance of surviving. Commonly a plant will send out stems called **stolons** (underground) or **runners** (above ground) that take root and send up shoots to form new plants (see Color Plate 2 and Text Figure 7–22A). These offspring will be assured of a good growing site and all the nourishment they need to get well established. On the other hand, the offspring must grow close to and therefore compete with the parent plant. Also there will be no variability to assure that some offspring survive if environmental conditions change.

In some cases the new plant is physically separate from the parent. If an African violet leaf falls to the ground, it takes root and forms a new plant. *Kalanchoe* ("Mother of Thousands" plant) is unusual in that it produces a great many offspring that grow as tiny plantlets around the edges of the leaves. They eventually drop to the ground and grow into new plants (see Text Figure 37–18). The white potato parent plant dies, but leaves the potato, an enlarged underground stem, to overwinter. Its buds will grow into a new plant the next spring.

Sometimes plants have it both ways: They produce reproductive organs and seeds so that they have good dispersal and a lot of offspring, but the seeds are genetically identical to the parent and their formation does not involve meiosis (see Color Plate 2). This process of **agamospermy** happens in dandelions, some grasses, and well-adapted plants that have given up sexual reproduction but have retained their reproductive organs. Other plants produce fruits, but have no fertile seeds, and the plants must be propagated vegetatively. Some examples are the pineapple, banana, and tangelo.

- Examine the interior of a banana.
- Note the tiny black dots. These are degenerative ovules that failed to form viable seeds.
- Examine the various examples of vegetative reproduction on demonstration in the laboratory.
- Match the examples that you see with the corresponding type of vegetative reproduction shown in Figure 31–9 at the end of this topic.

Check your demonstrations of pollen tubes and alfalfa sprouts, if you have not already done so, and return all materials. Clean up your work area, and store your microscope properly before leaving the laboratory.

REFERENCES AND FURTHER READING

ALSO SEE TOPIC 30

Emboden, William A. *Bizarre Plants: Magical, Monstrous and Mythical.* New York: Macmillan, 1974.

Evans, L.T. *Daylength and the Flowering of Plants.* Menlo Park, CA: W.A. Benjamin, 1975.

Faegri, K., and L. vander Pijl. *The Principles of Pollination Ecology,* 3rd. ed. New York: Pergamon Press, 1979.

Hillman, W.S. "Photoperiodism in plants and animals." Carolina Biology Readers. Burlington, NC: Carolina Biological Supply Co., 1979.

Hutchinson, J. *The Genera of Flowering Plants,* Vol. I and II. Monticello, NY: Lubrecht and Kramer, 1979.

Johri, B.M., ed. *Embryology of Angiosperms.* New York: Springer-Verlag, 1984.

Meeuse, B., and S. Morris. *The Sex Life of Flowers.* New York: Facts on File, 1984.

Moore, T.C. *Biochemistry and Physiology of Plant Hormones.* New York: Springer-Verlag, 1979.

Peterson, R.T., and M. McKenny. *A Field Guide to Wildflowers of Northeastern and North Central North America.* Boston: Houghton-Mifflin, 1974.

Proctor, M., and P. Yeo. *The Pollination of Flowers.* New York: Taplinger, 1973.

Thimann, K.V. *Hormone Action in the Whole Life of Plants.* Amherst: University of Massachusetts Press, 1977.

Vince-Prue, D. *Photoperiodism in Plants.* New York: McGraw-Hill, 1975.

Journal Articles

Albersheim, P., and A.G. Darvill. "Oligosaccharins." *Scientific American,* 44, September 1985.

Altshul, Siri von Reis. "Exploring the herbarium." *Scientific American,* 96, May 1977 (#1359).

Beard, Benjamin H. "The sunflower crop." *Scientific American,* 150, May 1981.

Chilton, Mary-Dell. "A vector for introducing new genes into plants." *Scientific American,* 50, June 1983.

Cook, Robert E. "Clonal plant populations." *American Scientist* 71(3): 244, May 1983.

Echlin, Patrick. "Pollen." *Scientific American,* 80, April 1968 (#1105).

Grant, Verne. "The fertilization of flowers." *Scientific American,* 52, June 1951 (#12).

Heinrich, B. "The energetics of the bumblebee." *Scientific American,* 96, April 1973.

Juniper, B.E. "Geotropism." *Ann. Rev. Plant Physiol.* 27:385, 1976.

Kaplan, Donald R. "The development of palm leaves." *Scientific American,* 98, July 1983.

Koller, D. "Germination." *Scientific American,* 75, April 1959 (no offprint).

Miller, J.A. "Somaclonal variation." *Science News* 128:120, 1985.

Morse, Douglass H. "Milkweeds and their visitors." *Scientific American,* 90, July 1985.

Niklas, Karl J. "Wind pollination—A study in controlled chaos." *American Scientist* 73(5): 462, September–October 1985.

Pettitt, John, Sophie Ducker, and Bruce Knox. "Submarine pollination." *Scientific American,* 134, March 1981.

Rick, C.M. "The tomato." *Scientific American,* 76, August 1978 (#1397).

Shepard, J.F. "The regeneration of potato plants from leaf-cell protoplasts." *Scientific American,* 154, May 1982.

Swaminathan, M.S. "Rice." *Scientific American,* 80, January 1984.

Torrey, John G. "The development of plant biotechnology." *American Scientist* 73(4): 354, July–August 1985.

Student Name ______________________________ **Date** ________________

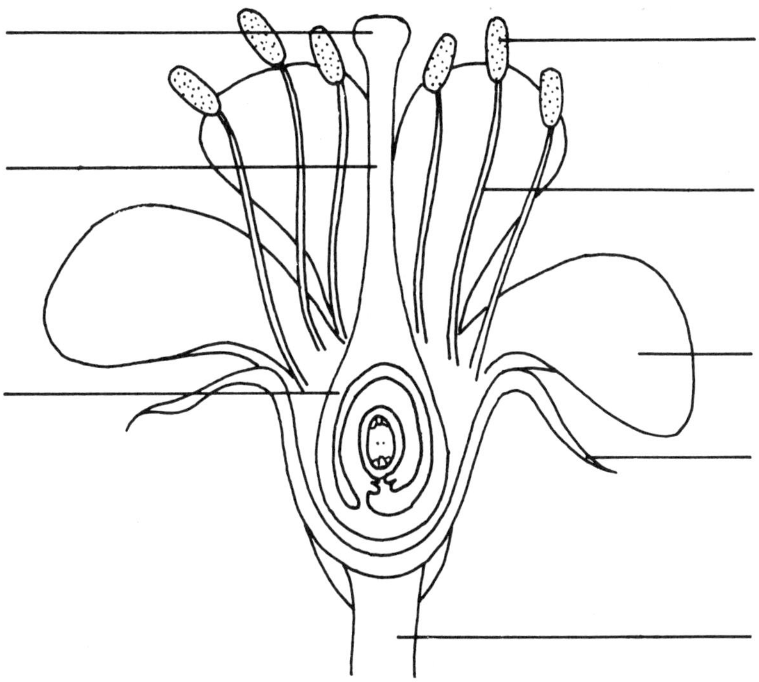

Figure 31–5. A simple, single flower. Label its parts.

CUTTING THE OVARY

SKETCH OF OVARY IN CROSS SECTION

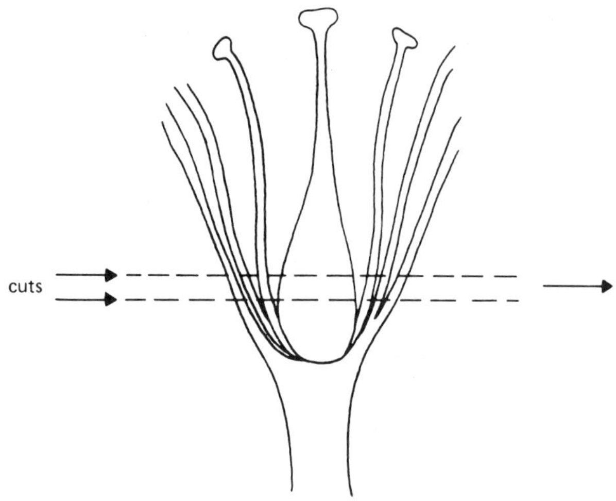

Figure 31–6. Cutting the ovary. Make a cross section by cutting at the positions shown.

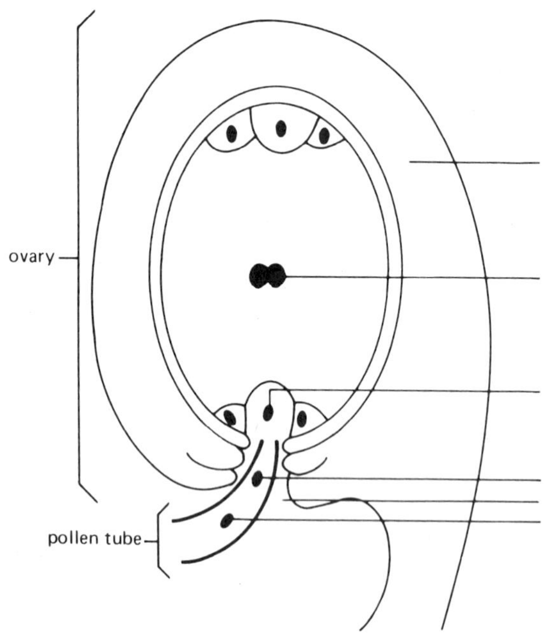

Figure 31–7. Fertilization. Label the parts indicated.

Student Name ______________________________ **Date** ______________

TABLE 31–1. Fertilization

Structure	Function or Fate
Egg nucleus	
Polar nuclei	
Ovule wall	
1st sperm nucleus	
2nd sperm nucleus	
Micropyle	

BEAN

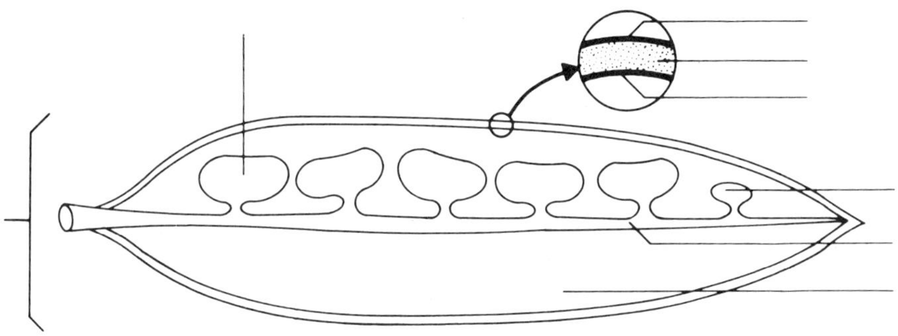

PEACH

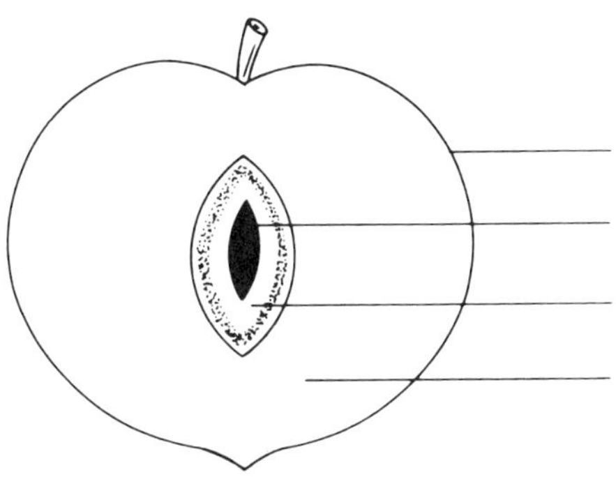

Figure 31–8. Fruits. Label the parts indicated with their correct botanical names.

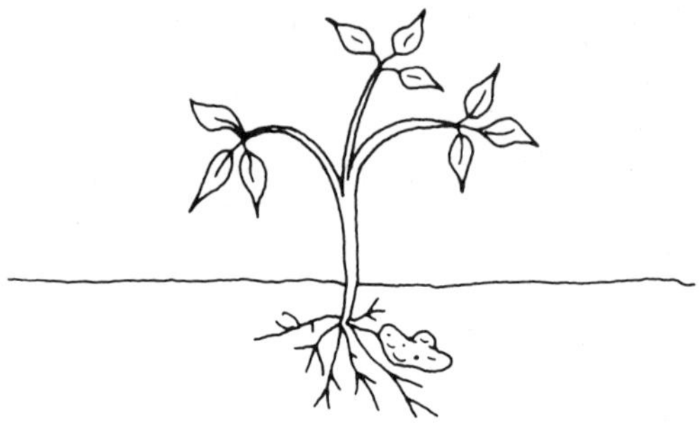

type: ____________________
example: ________________

type: ____________________
example: ________________

type: ____________________
example: ________________

type: ____________________
example: ________________

type: ____________________
example: ________________

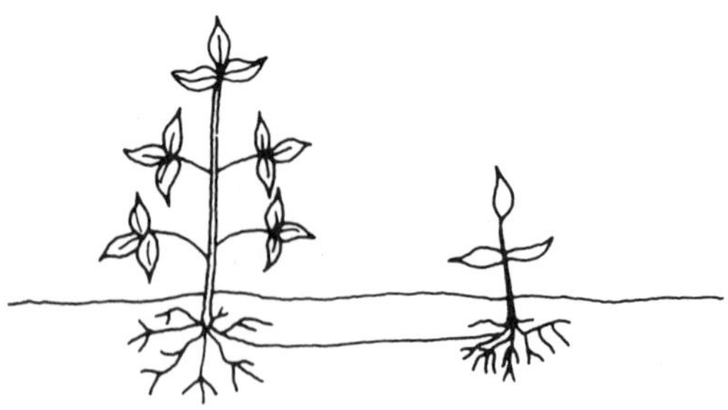

type: ____________________
example: ________________

Figure 31–9. Vegetative reproduction. Label each type with its name or description, and note an example.

Introduction to the Projects

The following projects require more than a single laboratory period of experimentation. Project A requires you to prepare a proposed experiment before laboratory, and the other projects require extended observation periods. Your instructor will give you specific directions as to how the project will be carried out. Because you will be working more independently than usual, it is important that you understand the rationale behind the experiments and exactly how they are to be done before you begin work.

You and your partner or group are about to become researchers in animal behavior, plant physiology, or ecology and will carry out some experiments in order to answer certain questions. Because you may be asked to design the experiments yourself, you should read over the discussion of the scientific method given below. Then, because the first step in any research is a review of the current knowledge, you should read over the section in your textbook relevant to your project and refer to any other sources that you can find. You will then be ready to prepare a written description of the planned work. You should bring the written plan with you to the laboratory, or hand it in ahead of time. Your instructor will check it over before you actually start your experimental work.

During the project, be careful to **record all of the conditions, manipulations, and observations at the time.** Do not rely on your memory alone, or you may end up in serious difficulty when you attempt to write up the results in the laboratory report. Label all records with your name and the date they were made, since your instructor may ask you to include the raw data in your report.

When the project is finished, write up the results in a written report according to the directions given in Appendix A, or prepare to discuss them orally, as your instructor directs.

THE SCIENTIFIC METHOD

In your study of the scientific method you learned that the scientific method is a formalized way of explaining cause and effect in the natural world. In principle, there are three steps, although a working scientist may not be involved in all three steps as discussed here:

1. **observations**
2. **hypotheses**
3. **experiments**

Scientists are expected to use reasoning or "thinking" in their work. This is how the rules of logic, a formal way of reasoning, apply to the steps of the scientific method. Scientists are using **inductive logic** when they gather **observations** on a phenomenon and then formulate a generalization or scientific law from these observations. The generalization represents new knowledge and will be used to make predictions until it is shown to be invalid by a new observation. The more observations that have been made, the more likely the generalization is to be valid. For example, the discovery of chlorophyll in a number of plants might lead to this generalization:

Generalization: All green plants contain chlorophyll.

The generalization can be used to generate a **hypothesis** that can be tested by applying **deductive logic.** A hypothesis is a trial, unproven proposition put forward as a possible explanation of observed events to provide a basis for further study. It must make a testable prediction. This is where a really good scientist is distinguished from a plodder, because it takes creativity to use the information available to come up with a hypothesis that makes powerful predictions. The hypothesis itself is stated as the **major premise,** its application to a particular case is the **minor premise,** and the results obtained by applying it to a particular case lead to the **conclusion.** To continue our example:

Major premise: Chlorophyll is the pigment responsible for the color of all green plants.

Minor premise: This mint plant is green.

Conclusion: This mint plant contains chlorophyll.

Deduction does not generate new knowledge, but

it proposes **experiments** that will support or disprove the major premise. In this example, if a single plant were found to be green but not to contain chlorophyll, the major premise would have to be discarded because it had been proved to be false. If, on the other hand, many mint plants were tested and were all found to contain chlorophyll, the major premise would be supported. The major premise *would not be proved* because it is not deductively valid to argue it is proved by one or more supporting observations. (There might be exceptions yet to be discovered.) When many supporting experiments have been done, the hypothesis is regarded as very probably true, and it will gain the status of a **theory** and may eventually become a **scientific law,** or "fact."

In summary, one observation that is inconsistent with the conclusion is sufficient to disprove the major premise. Many observations consistent with the conclusion will support but not prove the major premise.

When hypotheses are formulated, they often take the form:

If: A is true

Then: B is true

A is the generalization or hypothesis to be tested and B is the experimental observation to be made. According to the preceding discussion, if B is false, A *must* be false. If B is true, however, A may or may not be true. The example just given can be restated:

If: chlorophyll is responsible for the color of all green plants

Then: this green mint plant contains chlorophyll

An experiment would be done to test for the presence of chlorophyll. If it were present, the hypothesis would be supported but not proved, and if it were absent, the hypothesis would be disproved according to deductive logic. In the latter case, the scientist would have to turn to a new hypothesis or develop a more sensitive method of testing for chlorophyll.

The experiment will often be set up so that there is an **experimental situation** and a **control situation,** which differ by only a single variable. In this case we would test a tube with chlorophyll as a **positive control,** a tube without chlorophyll as a **negative control,** and an **experimental** tube with material from the mint plant. Without controls we can't be sure that the results that we see are really due to the factor that we are testing, namely the presence of chlorophyll.

PROJECT A

Fish Schooling Behavior

OBJECTIVES

When you have completed this project you should be able to:

1. Distinguish a hypothesis from a non-hypothesis.
2. Draw conclusions from experimental results as to whether a hypothesis has been supported or disproved.
3. List the factors that may be involved in fish schooling behavior.
4. Design an experiment to test each of the factors in Objective 3.
5. Define orientation, sign stimulus, fixed action pattern and communication.

TEXT REFERENCES Chapter 1 (A–D); Chapter 33 (D, E, J).

KEY TOPICS

scientific method	experiment	communication
hypothesis	orientation behavior	fish schools

INTRODUCTION

The laboratory work for this project can be done in one laboratory period. After reading over INTRODUCTION TO THE PROJECTS and this project, and before coming to laboratory, propose a hypothesis relating to fish schooling behavior that can be tested with the experimental equipment provided. Design and describe the experiment you could do to test the hypothesis, two possible outcomes, and the conclusions you would draw from each. Record your proposed experiment in Table P–1 at the end of this project.

Orientation is a type of behavior in which the organism turns its body or moves either toward or away from a certain stimulus. This type of behavior is basic to the success of the animal. Whatever it does, whether it is seeking food, water, sunlight, a mate, interacting with other members of a group, or avoiding predators, there must be some sort of orientation.

Fish schooling is a type of orientation behavior in which the stimulus is other fish. The fish swim as a **school** because they are mutually attracted to one another. This same type of mutual interaction exists in a **herd** of mammals or a **flock** of birds. It is different from an **aggregate,** which is a group of animals attracted to one spot by food, light, or some other environmental stimulus. The interaction of the fish in the school is a form of **communication** in which signals from one or more members of the school cause a predictable schooling response. Schooling fish show a primitive type of social behavior that is egalitarian because there is no leader and practically no division of labor. Most fish school as juveniles and about half of the known species continue to school as adults.

The signals between members of a school may be visual, chemical, or the hydrostatic disturbance of swimming motions in the water detected by the lateral line system. Vision seems to be especially important in attracting the fish toward each other, but olfaction and vibration help maintain the cohesion and spacing of an established school. Apparently the strongest attraction is for members of the same species and size. Most schools consist of fish of the same species, but fish of different species do sometimes school together. The presence of the other fish in the school acts as a **sign stimulus** because it predictably causes the schooling response. The response is a **fixed action pattern** in that it is a stereotyped response that is programmed into the fish's nervous system and occurs automatically. No input from the forebrain or learning is required for a fish to maintain its position parallel to the other fish in a school (the optomotor response). As in other forms of stereotyped behavior, however, the response is not absolutely rigid and there is some room for adjustment to various environmental and physiological conditions. All fish seem to require a learning period of interaction with other fish to show fully coordinated schooling behavior.

Schooling is very prevalent among all sorts of fish, from very primitive ones to the more highly evolved species. One advantage is that members of the school may be better protected from predators than single fish. A tightly balled school may appear to be a large organism and its density may make it physically difficult for a predator to attack. Some schools swim in a confusing defensive pattern when a predator is about to strike. Also, the mere presence of a large number of similar fish makes it difficult for the predator to aim for a certain fish. The presence of a predator (or food) is quickly communicated to all of the fish in the school. For fish that swim much of the time, another advantage is that they are able to swim more efficiently due to favorable hydrodynamics when they are spaced out at certain intervals from one another in the school. Even predators such as hammerhead sharks school, probably because a school has many eyes on the look out for prey and can cooperate to concentrate and surround prey. Another reason fish might school is because they can learn adaptive behavior patterns more quickly as a school. Finally, fish in a school never have to make a big investment in finding a suitable mate.

Schooling also has its disadvantages. For predators, there is *less* food for each fish, when they do find prey. Plankton-feeders would be expected to get *less* food as a school, since their food is available all around them but would be reduced for the fish further back in the school. During reproduction there would be more competition for mates or gametes. The biggest disadvantage of schooling behavior, however, is that it has made the fishing industry very successful and efficient and may ultimately lead to the decline or extinction of many fish species.

In this project you will be investigating the interaction of fish by visual cues. Try to maintain constant physical conditions, such as water temperature, light, and noise levels. Chemical signals will be minimized because the fish will be separated by the glass wall of a beaker. Sound communication can't be eliminated, because sound travels through glass. For this reason, it is important to keep noise and vibrations in the vicinity of the tank to a minimum.

Behavior experiments will not work or be valid unless the fish are treated with care and patience.

Do:

1. Always use a net to transfer the fish, and treat them with care. The skin covering their scales is very delicate and easily gets infected.
2. Allow time for the fish to adjust to new conditions before beginning your observations.
3. Wash and rinse your hands before reaching into the experimental tanks to place beakers and so forth. (Grime and hand creams are not good for the fish, and in any case are unwanted chemical signals.)
4. Report any sick-looking or dead fish to the instructor.
5. Return all fish to the proper tanks when you are finished.

Don't:

1. Disturb the fish any more than is necessary for the experiments.
2. Expect the fish to respond instantaneously to a new stimulus.
3. Reach into the stock tanks where the fish are kept; use a net.
4. Leave fish stranded in the experimental tank or a beaker.

Depending on your equipment, you will work with a partner or in a group of four. Your experimental setup will be similar to that shown in Figure P–1. A test fish is placed in the tank and given two choices: A or B. One can be the experimental stimulus and the other a control, or you could compare

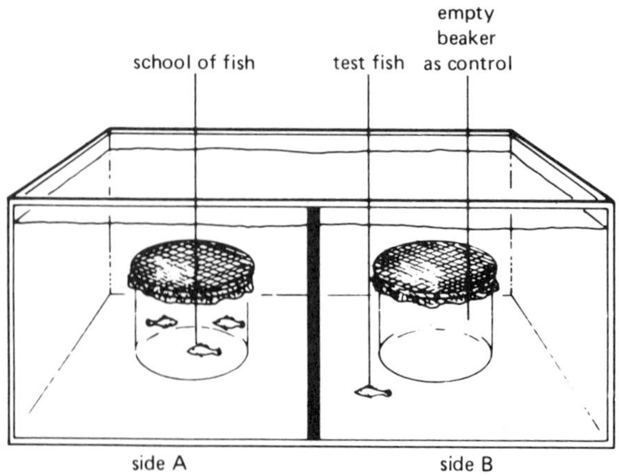

Figure P–1. Experimental setup for schooling behavior. The aquarium is divided into side A and side B. A school of fish may be contained in a beaker with a screen cover as shown on side A. An identical empty beaker may serve as a control on side B. The test fish is free to swim throughout the aquarium. The stimuli should be exchanged between runs to control for a possible tank bias of the test fish.

two different stimuli. The attraction of the stimuli is measured by comparing how much time the fish spends on side A of the tank versus side B.

Everyone will run a tank test in which one test fish is placed in the tank alone with no stimulus on either side. It is observed to see whether it shows an unexpected bias toward one end of the tank. This important control helps establish that your experimental stimulus is really causing the fish to be attracted to that side of the tank.

These are some of the variables that you can introduce into your experiments:

Test fish could be species 1.
Test fish could be species 2, etc.
Empty beaker could be a control.
Aquatic plants could be placed on one side of the tank as a refuge.
Beaker could contain a "school" of 2–5 fish of species 1.
Beaker could contain a "school" of 2–5 fish of species 2, etc.

You may place various numbers of fish in the tanks or in the beakers.

Before coming to laboratory, complete the PROPOSED EXPERIMENT form in Table P–1 at the end of this project. After discussing your hypothesis and experimental setup with your group and with your instructor, carry out that experiment or another approved experiment. Be prepared to present and discuss your results orally or in a formal written report (See Appendix A: WRITING A LAB REPORT).

TANK TEST

- Obtain a stopwatch and a container with a test fish. Wind the watch and practice starting and stopping it a few times.
- When you gain access to an experimental tank divided into sides A and B (see Figure P–1), place the fish in the water by submerging the container and drawing it away from the fish. Allow the fish a few minutes to adjust.
- Then run a 5-min (300 sec) observation period to see whether the fish spends more than half of its time on one side of the tank.

The first partner should start the experiment by starting the stopwatch; the second partner should watch the fish and order the watch stopped during the time when the fish has crossed to the other side of the tank. The first partner should stop the experiment when 5 min have elapsed. If your stopwatch must be reset after each interval, be sure to quickly record the length of each interval, reset, and be ready to start timing the next interval.

How many seconds did the fish spend on side A?

____

On side B? ____________

What percentage of the time did the fish spend on side A? ____________

On side B? ____________

If your fish spent more than about 75% of the time on one side, it will be hard to draw valid conclusions from any experiment that you might try. You should probably proceed with one of the following alternatives:

a) If the fish did not actively swim about or stayed on one side, place it on the opposite side of the tank and repeat the test.
b) Move the fish to another tank and repeat the test.
c) Obtain another fish and repeat the test in your tank.

Consult with the instructor if you can't demonstrate that your fish visits both sides of the tank. When your fish does visit both sides, go on to the experiments.

EXPERIMENTS ON FISH SCHOOLING

- Working with your partner or in a group of four, decide which hypothesis you will test.

  Example: In species 1, a school of two fish is sufficient to cause schooling behavior in a test fish of the same species.

- Decide how you will set up the tank.

  Example: Two fish of species 1 are placed in a beaker on one side of the tank and an empty control beaker is placed on the other side.

- Decide what results you expect if your hypothesis is correct.

  Example: *If* two fish cause schooling behavior in the test fish, *then* the test fish will spend more time on the side of the tank with the fish than on the side with the control beaker. NOTE: The statement "The test fish will spend more time on the side of the tank with the fish than on the side with the control beaker" is a prediction, not a hypothesis. It does not explain anything.

- Show your written plan to your lab instructor.

Depending on the equipment available and the other experiments being run, your instructor will direct you to carry out this or another experiment.

Obtain the approval of your instructor before starting your experiment.

- Run your experiment three times, allowing the test fish to rest in between. It might be worthwhile to reverse the conditions on sides A and B for one of the experiments, in case the fish begins to develop a preference not related to the test conditions.
- Calculate and record the mean ("average") percent of the time the test fish spent on the side corresponding to each stimulus (see Appendix B).

Example: Test Fish Species 1

Side With 2 Fish (Species 1)	Empty Beaker Control
200 sec	100 sec
150 sec	150 sec
90 sec	210 sec
147 mean sec	153 mean sec
49%	51%

- State your results in quantitative and qualitative terms understandable to the rest of the class.

  Example: Test fish species 1 spent 49% of the time on the side with two fish of the same species and 51% of the time with the empty beaker. Test fish species 1 showed the same attraction toward a school of two fish as toward the control.

- State your conclusion.

  Example: The hypothesis that in species 1, a school of two fish is a sufficient stimulus to cause schooling behavior in the test fish was not supported.

When you discuss your results and conclusion, describe what experiment you would carry out next, whether or not you actually do so. Describe how you would continue or extend your work, whether or not your hypothesis was supported.

  Example: If we know that species 1 is a schooling fish, we can also conclude that schooling behavior depends on school size. We propose that a larger number of fish in a school is required to cause schooling behavior.

- If time permits run another experiment or try your experiment with a different species of fish.

When you are finished with your laboratory work, return all fish to the proper storage tanks, return all equipment, and wipe up any spills around the tanks.

If you are asked to prepare a written report, follow the instructions given in Appendix A.

REFERENCES AND FURTHER READING

Alcock, J. *Animal Behavior: An Evolutionary Approach,* 3rd ed. Stamford, CT: Sinauer, 1983.

Bond, Carl E. *Biology of Fishes.* Philadelphia: W.B. Saunders, 1979. Chapter 16.

Bone, Q., and N.B. Marshall. *Biology of Fishes.* New York: Methuen, 1983.

Gould, J.L. *Ethology.* New York: W.W. Norton, 1982.

Huntingford, F. *The Study of Animal Behavior.* London: Chapman and Hall, 1984.

Manning, A. *An Introduction to Animal Behavior.* New York: Addison-Wesley, 1979.

Shaw, E. "Schooling in fishes: Critique and review." in *The Development and Evolution of Behavior.* Eds. L.R. Aronson, E. Tobach, D.S. Lehrman, and J.S. Rosenblatt. San Francisco: Freeman, 1970, pp. 452–480.

Journal Articles

Fernald, Russell D. "Vision and behavior in an African chiclid fish." *American Scientist* 72(1): 58, January–February 1984.

MacNichol, Edward F., and Joseph S. Levine. "Color vision in fishes." *Scientific American,* 108, February 1982.

Partridge, Brian L. "The structure and function of fish schools." *Scientific American,* 90, June 1982.

Shaw, Evelyn. "Schooling fishes." *American Scientist* 66: 166, March 1978.

Todd, John H. "The chemical language of fishes." *Scientific American,* 98, May 1971 (#1222).

Webb, Paul F. "Form and function in fish swimming." *Scientific American,* 72, July 1984.

Student Name ______________________________ **Date** ______________

TABLE P–1. Proposed Experiment.

Propose a hypothesis relating to fish schooling behavior that can be tested with the available experimental equipment.

Hypothesis: ______________________________

Experimental Setup: ______________________________

First Possible Outcome: ______________________________

Second Possible Outcome: ______________________________

Conclusion for First Possible Outcome: ______________________________

Conclusion for Second Possible Outcome: ______________________________

PROJECT B

Plant Hormones

OBJECTIVES

When you have completed this project, you should be able to:

1. Define hormone, and describe how plant hormones tend to differ from animal hormones.
2. Explain the role of the plant hormone auxin in cell elongation, lateral bud development, ethylene production, root formation, and fruit formation.
3. Given a living plant or a diagram of a plant, show where each of the effects of auxin occurs.
4. Describe five criteria for establishing that a certain process is under the control of a plant hormone.
5. Given the laboratory facilities and equipment at your disposal, design and carry out an experiment to test the hypothesis that apical dominance is due to production of auxin by the apical meristem.
6. Write up your results in a formal laboratory report.
7. Describe several other experimental approaches that might be used to test this hypothesis.

TEXT REFERENCES Chapter 1 (A–D); Chapter 34 (C); Chapter 36.

KEY TOPICS

scientific method
hypothesis
auxin
apical dominance
statistics

INTRODUCTION

The observations for this experiment must be made over a period of at least 2 weeks. The experimental procedures and measurements themselves do not take much time.

Hormones are chemical messengers synthesized in one part of an organism that are transported to a different part of the organisms's body. There they have an effect on its development or physiology. Because animals spend a large portion of their lives in a stable adult form without much growth or differentiation, many animal hormones are directed toward maintaining the status quo, or **homeostasis.** In most plants, on the other hand, growth and differentiation tend to continue throughout the life of the organism, so plant hormones tend to be directed toward controlling growth and differentiation in the developing tissues. In some plants, aging and senescence are also closely regulated by hormones. As in animals, certain hormones regulate the events leading up to the reproduction of a mature plant. In animals, hormones are produced in specific organs, the endocrine organs, and they usually act on specific organs called the target organs. Plants, on the other hand, do not have organs with the specific function of producing hormones. Their hormones are produced by parts of the plant that have other functions as well. The hormones may then act on several different organs of the plant. Since several hormones may act together, their proper ratio is critical in promoting normal growth and development. (See Table 36–1 in your text.)

IDENTIFICATION OF PLANT HORMONES

If a certain substance is a plant hormone, you must be able to show that:

1. The substance is produced naturally by the plant.
2. It exerts a physiological effect on another part of the plant in low concentrations.
3. It is normally transported between the two parts of the plant.

The generally accepted experimental criteria for establishing that a process is under hormonal control were first articulated by Jacobs in 1959 (see *Journal Articles,* at the end of this topic). They can be summarized as follows:

1. **Parallel Variation.** The substance must be present in the intact plant that shows the response or process, and the degree of response must vary with the concentration of the substance.
2. **Excision.** Removal of the substance from the plant should result in a loss of the regulatory effect. This may be done by any of these procedures:
   a. Surgically removing the part of the plant producing the hormone
   b. Blocking the production of the hormone by poisoning the metabolic pathway
   c. Blocking the effect of the hormone by using competitive antagonists
   d. Developing a mutant strain that lacks the suspected hormone
3. **Substitution.** Replacement of the suspected hormone should reinstate the regulatory effect. For a new hormone whose structure may not be known, this can be done by using an extract from or homogenate of the tissue thought to produce the hormone. If such a substitution experiment is successful, the next step is to isolate and identify the chemicals in the preparation that are responsible for its biological activity.
4. **Specificity.** It should be shown that other naturally occurring substances do not have the effect shown by the substance isolated in step 3. Testing of other compounds might identify factors that contribute to the observed effect and will help clarify the exact role that the isolated substance plays in the regulatory process.
5. **Generality.** Once hormonal control of a process has been demonstrated in one species, it is desirable to test whether the substance tentatively identified as the controlling hormone will exert similar control on the same process in other species.

AUXIN

Auxins are a group of structurally related small molecules derived from the amino acid tryptophan. The most commonly studied auxin is indoleacetic acid. Auxin is a stimulatory hormone that causes plant cells to elongate if it is present in the right concentration. It works in cooperation with the other growth hormones, gibberellins and cytokinins, and its effect is counteracted by an oligosaccharide hormone. It acts partly by stimulating the activity of **cellulases,** enzymes that soften the material in the cell walls and allow the cells to expand. If there is a greater concentration of auxin on one side of a plant stem, that side will elongate faster, causing the plant to bend toward the *other* side. Sunlight causes auxin to accumulate on the shady side of a plant stem so that the plant bends *toward* the source of the sunlight and shows **positive phototropism.**

Auxin is stimulatory only when it is present in a certain concentration range and the concentrations of other hormones are appropriate. This range is different for different tissues and, if the concentration is above the stimulatory range, the auxin will act to *inhibit* elongation and growth. In the plant, auxin is produced by the tissue of the **apical meristem** at the tips of the branches of the shoot. It is transported from the meristematic tissues to the rest of the plant. Buds are stimulated by a lower range of auxin concentrations than are stems, so the concentrations of auxin that stimulate growth of the stem will tend to inhibit the development of adjacent buds. This is the basis of **apical dominance.** Auxin also causes the buds to produce the hormone ethylene, which further inhibits bud development. As long as the apical meristem is growing, it will produce auxin in sufficient concentration to inhibit the development of the lateral buds next to it. If the meristem is removed, the auxin concentration reaching the buds will decrease, and the buds will start to develop and will now be much less sensitive to inhibition. Once a bud has broken dormancy and has started to grow into a new stem, it will be stimulated by the same concentration of auxin that inhibited the bud.

Auxin is also involved to some extent in the response of roots to gravity. If a root happens to be in a horizontal position, low concentrations of auxin on the lower side will promote ethylene production, which in turn slows growth on the underneath side. As a result, the root will bend downward and grow toward the center of the earth. The root of a plant thus shows **positive geotropism.** Abscisic acid also accumulates on the lower side and contributes to geotropism by inhibition.

In addition to stimulating elongation, inhibiting lateral bud development, and controlling the response of plant stems to light and gravity, auxin is important in inducing the formation of lateral and adventitious roots, stimulating cell division in the cambium to form reaction wood in response to stress, stimulating cytoplasmic streaming, inducing development of the fleshy parts of some fruits, and promoting the development of female flower parts. Auxin does this partly by stimulating production of another hormone, ethylene.

Study a diagram of a plant, or better yet, a living plant to review the sites of production and sites of action of auxin and the other plant hormones.

EXPERIMENTS ON APICAL DOMINANCE

Apply your understanding of the scientific method to the problem of apical dominance. Read over **INTRODUCTION TO THE PROJECTS** and the following instructions before you begin to plan your project. You must then prepare a detailed written description of your experimental plan before coming to laboratory to start the experiment.

Obtain the instructor's approval of your plan before starting the experiment.

Hypothesis

The hypothesis of this project is that the growth of lateral buds in bean plants is normally inhibited by auxin produced in the apical meristem.

Implications

Use your knowledge of the effects of hormones on plant development to make a list of testable or measurable phenomena you could study to help support or disprove the hypothesis:

1. ______________________________

____

2. ______________________________

____

3. ______________________________

____

4. ______________________________

____

5. ______________________________

____

Which of these implications could be most easily tested experimentally? ______________

If you wished to demonstrate **parallel variation,** what kinds of experimental problems would you have to overcome?

____

____

Specificity might be difficult to demonstrate; in terms of what you know about the complexities of plant physiology, why might this criterion be hard to meet?

____

____

How would you demonstrate **generality?**

____

____

Why is it important to do so?

____

____

Materials and Methods

In this exercise you will do only **excision** and **substitution** experiments using the following materials and any others that may be available to you:

Young bean plants
Auxin in lanolin paste
Plain lanolin paste
Razor blades
Plant tags

For each of the experiments to be done, describe in your written plan the manipulations that you must do to the experimental plants, the proper controls that you must include for comparison, and the measurements that you will make.

Because you are studying the effect of the manipulation that you will perform on the development of the plant, all other factors should be kept constant.

List in your written plan all the factors that might affect the growth of your plants.

In addition, because you are going to compare experimental and control plants during the experiment, you will need to start out with plants that are as nearly identical as possible and certainly the same species, age, and size. The lateral buds on the plants must be dormant.

Why?

____

____

____

Do not choose plants in which the dormancy of these buds has been broken. Carry out your treatments quickly, so that their dormancy will not be broken by the experimental procedure. Your plants should be about 10–14 days old and should look like the young plant shown in Figure P–2.

The anatomy of the apical meristem and lateral buds of a young bean plant is shown in Figure P–3. The apical shoot of the plant should be about 2–3 cm

Figure P–2. A bean seedling. The lateral buds on either side of the shoot must be dormant at the beginning of the experiment. The shoot must be long enough so that lanolin paste can be applied to it after excision of the apical bud, without having the paste slide down directly onto the lateral buds as well. (Photo by C. Eberhard)

long, and lateral buds must be dormant in order for the experimental procedure to be valid.

Once the experimental manipulations have been performed, follow the growth of the lateral buds for at least a 2-week period, taking measurements as often as necessary.

How often will you be able to measure your plants?

____

Design a format for recording your data in the same way each time you take measurements. Include all measurements in your laboratory report.

Be sure each entry includes your name and the date.

When you are finished with your experimental work, dispose of the plants according to the directions given by your instructor.

Results

The number of plants in your individual experiment is too small to yield statistically valid results. When a large number of such results are combined, however, the results may become statistically meaningful. Read the information on statistics given in Appendix B so that you will be able to understand the analysis of the class data. Your instructor may ask you to calculate the standard deviation for the class results. After the analysis, answer these questions:

Are the class's results statistically significant?

____

What does the statistical analysis tell us about the experiment?

____

____

____

____

Discussion and Conclusions

Is the hypothesis supported? ______________

Is the hypothesis proved? ______________

Is there an alternative hypothesis that could explain the results?

____

____

What experiments could you do to test the alternative hypothesis?

____

____

If an externally applied hormone can be shown to affect a given phenomenon, does this necessarily mean that the hormone controls the process in the intact plant?

____

What else would you want to know in support of the conclusion?

____

____

____

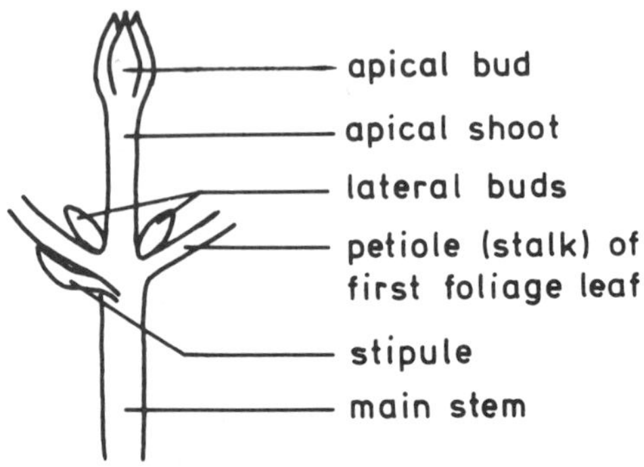

Figure P–3. Apical shoot. The shoot of a 12-day-old bean plant will show apical dominance of the lateral buds by the apical meristem located within the apical bud.

Consider these points in formulating your discussion:

1. The apical meristem may draw nutrients away from the lateral buds.
2. Cytokinin from the roots may be preferentially transported to the apical meristem.
3. The apical meristem may receive more light than do the lateral buds.

Write up this experiment according to the instructions given in Appendix A, including the class analysis and an explanation of what the analysis means.

REFERENCES AND FURTHER READING

Bidwell, R.G.S. *Plant Physiology,* 2nd ed. New York: Macmillan, 1979.

Galston, Arthur W., and Peter J. Davies. *Control Mechanisms in Plant Development.* Englewood Cliffs, NJ: Prentice-Hall, 1970.

Galston, Arthur W., Peter J. Davies, and Ruth L. Satter. *The Life of the Green Plant,* 3rd ed. Englewood Cliffs, NJ: Prentice-Hall, 1980.

Moore, T.C. *Biochemistry and Physiology of Plant Hormones.* New York: Springer-Verlag, 1979.

Newcomb, E.H., G.C. Gerloff, and W.F. Whittingham. *Plants in Perspective.* San Francisco: W.H. Freeman, 1965.

Raven, P.H., R.F. Evert, and S.E. Eichhorn. *Biology of Plants,* 4th ed. New York: Worth, 1986.

Ray, P., T.A. Steeves, and S.A. Fultz. *Botany.* Philadelphia: Saunders College Publishing, 1983.

Salisbury, Frank B., and Cleon W. Ross. *Plant Physiology,* 2nd ed. New York: Harper and Row, 1978.

Thimann, K.V. *Hormone Action in the Whole Life of Plants.* Amherst: University of Massachusetts Press, 1977.

Wareing, P.F., and I.D.J. Phillips. *Control of Growth and Differentiation in Plants,* 3rd ed. Elmsford, NY: Pergamon Press, 1981.

Weier, T.E., C.R. Stocking, and M.G. Barbour. *Botany: An Introduction to Plant Biology,* 6th ed. New York: John Wiley and Sons, 1982.

Journal Articles

Albersheim, P., and A.G. Darvill. "Oligosaccharins." *Scientific American,* 44, September 1985.

Flagg, R.O. "Plant growth regulators." *Carolina Tips* 26(3, 9), 1963, and 27(9, 10). Burlington, NC: Carolina Biological Laboratory, 1964 (reprint 1965).

Juniper, B.E. "Geotropism." *Ann. Rev. of Plant Physiol.* 27:385, 1976.

Jacobs, W.P. "What substance normally controls a given biological process?" *Dev. Biol.* 1: 6, 1959.

PROJECT C

Plant Nutrition

OBJECTIVES

When you have completed this project, you should be able to:

1. List the six elements that are macronutrients needed by all plants.
2. Give at least one example of how each macronutrient is used.
3. Give some examples of micronutrients, and describe why each is needed.
4. Describe how a plant could show a deficiency of a nutrient known to be present in its soil.
5. Describe two ways in which a plant might obtain a nutrient other than from its soil or from its water supply.
6. Give two ways in which a plant can compensate for a limited supply of a nutrient; explain how such compensation would affect nutritional experiments.
7. Identify the deficient nutrient for a seedling by comparing its appearance with that of seedlings grown under conditions of known nutrient deficiencies.

TEXT REFERENCES Chapter 1 (A–D); Chapter 35 (I); Chapter 36.

KEY TOPICS

scientific method	macronutrients	soil solution
hypothesis	micronutrients	absorption

INTRODUCTION

The laboratory work for this project continues for 3 to 4 weeks. Once the experiment has been set up, brief measurements (15 min) must be made every few days.

Plants obtain energy reserves and the bulk of their structural material from the carbon dioxide that they take in from the air. In order to carry out the metabolic processes necessary to convert CO_2 to carbohydrates and cellulose and to simply maintain the components of their cytoplasm, however, they need many other substances in lesser amounts. The specific functions of macro- and micronutrients in the plant are described in Text Table 46–1.

Macronutrients: Nitrogen (N) and phosphorus (P) are elements needed for synthesis of amino acids and nucleotides. Potassium (K) is the main cation (ion with a + charge) in plant tissue and has many functions. These elements are needed in the greatest amount. Sulfur (S) is needed in lesser amounts for amino acids. Calcium (Ca) is required for many enzyme reactions (as are iron, manganese, and magnesium), while magnesium (Mg) forms part of the structure of the chlorophyll molecule.

Micronutrients: Many elements are needed in tiny amounts by all plants, and others are especially needed by certain plants for specialized functions. For example, when plants are associated with nitrogen-fixing bacteria, molybdenum is required because it is necessary for that reaction. This element is also needed by plants for their enzyme nitrate reductase when nitrate is the source of nitrogen. Cobalt, a component of vitamin B_{12}, is also needed for nitrogen fixation. Iron (Fe) is required by many enzymes, including those of the electron transport chain and those that synthesize chlorophyll. Zinc is necessary for the synthesis of auxins. Silicon, boron, copper, sodium, chlorine, and selenium are all important to at least some plants, but are needed in only tiny amounts.

Plants generally take up the minerals they need from the soil solution through their roots, although certain plants are also able to take up nutrients through their leaves. Since the concentration of the minerals in soil solution is low, most of them (except Ca^{++} and Mg^{++}) are actively transported into the root cells. In many cases fungal mycorrhizae associated with the roots assist in absorbing minerals and water. Poor soils are those in which minerals are insufficient or are unavailable to the plant. Acid soils tend to be especially deficient in nutrients, because some minerals dissolve readily in acid and are leached (washed out) from the soil. Carnivorous plants can survive in acidic conditions because of their unusual adaptations for supplementing their intake of minerals, especially nitrogen: Their leaves are highly specialized to trap, digest, and absorb insect prey.

Plants that are deficient in a mineral will grow poorly and will also show signs of deficiency characteristic of that mineral (see Text Figure 35–1). Chlorosis, an unhealthy pale green or yellow color, is a common deficiency symptom. In fact, if a plant is greatly deficient in one particular mineral, it is possible to identify which mineral from the appearance of the plant. An experienced person can describe the deficiencies found in a certain sample of soil by studying the appearance of the plants growing in it.

The characteristic effects that certain mineral deficiencies have on growing plants are shown in Table P–2. In nature, it is possible for the plant to show deficiency symptoms even though the nutrient is actually present in the soil around the plant. The nutrient may be in a form held tightly to the soil

TABLE P–2. Symptoms of Mineral Deficiencies in Plants

Mineral that is deficient:	Symptoms that result:
Nitrogen	Leaves change from a green-yellow at the top to yellow to brown, dead leaves at the bottom. In many species a red or purpling occurs along the veins. Plant shows stunting.
Potassium	Limited growth, weak stems, and a yellow mottling of the older leaves; ultimately necrotic areas on leaf tips and edges; a general overall yellowing appearance of the leaves.
Phosphorus	Plant stunted; leaves small and dark green; occasionally production of anthocyanins causes a red or purple color. Dead areas develop on leaves, petioles, and (in older plants) fruits, causing some dropping. Roots are poorly developed.
Sulfur	Yellowing of the younger leaves in the early stages; ultimately an overall pale green may dominate (chlorosis). Roots are poorly developed.
Calcium	First symptom usually deformation of younger leaves, and then a disintegration of terminal growing areas, due to the death of the meristem tissue.
Magnesium	Lower leaves show symptoms first, mottled yellowing (chlorosis) from the tip, and eventually falling. Veins remain green longer than the rest of the leaf.
Micronutrients	*Iron:* Young leaves light green or almost white (chlorosis), while older leaves are green. Unlike most other elements, iron cannot be withdrawn from the older leaves, so they retain a normal appearance. The yellowing of the younger leaves is most obvious between the veins. *Boron:* Abnormally dark green leaves; abnormal growth with malformations; root tip growth slows. *Manganese:* Leaves become spotted with dead areas and fall. *Copper:* Tips of young leaves wither; plant may wilt even when it is watered; impaired protein synthesis. *Zinc:* Yellowing of lower leaves at tips and margins; leaves small and deformed; stems stunted. *Chlorine:* Retarded photosynthesis causes very small leaves that become wilted, chlorotic, or necrotic; growth slows. *Molybdenum:* Required for nitrogen metabolism, so symptoms resemble those of nitrogen deficiency.

particles or insoluble at the soil pH. Even if it enters the plant, it may not be in a form that can be transported to the site where it is needed. Finally, it may be available in the needed location, but the combination of other nutrients may not be adequate for the nutrient to serve its necessary function. In this project, you will provide the plants with the nutrients in a readily available form and do not need to consider problems of availability, uptake, or transport. Iron, for example, will be provided bound to the chelator EDTA to ensure its solubility.

In some cases a plant will actually be deficient for a certain nutrient but will compensate so that the effect of the deficiency is minimal. If, for example, calcium is required for a certain reaction but is present only in small amounts, magnesium or manganese, also divalent cations, will be able to substitute. Sometimes minerals can be removed from an older part of a plant and relocated to the new, growing part where they are more urgently needed. This type of recycling can prolong the active life of a plant growing under poor conditions. Compensation will complicate nutritional studies because the differences between a healthy plant and one with a deficiency will tend to be masked. Since most nutrients are needed for many processes, however, a deficiency will usually at least result in a slower growth rate, even if no overt deficiency symptoms appear in the experiment.

EFFECT OF NUTRITIONAL DEFICIENCIES

Hypothesis

We will hypothesize that a particular mineral is needed for at least one process critical for the normal growth and development of the seedling.

Experiment

In this project you will work with a partner and test the effect of two different deficiencies on young seedlings. You will measure the growth rate and observe the appearance of the experimental seedlings and compare your observations with similar ones on seedlings deficient for an unknown nutrient. The control seedlings will be given all the necessary nutrients. All of the seedlings will be grown under identical conditions of light, temperature, and moisture.

PROCEDURE

Obtain 12 seedlings that have been grown in washed sand. Obtain 4 growth jars and label them properly with the labels provided:

A = experimental solution deficient in ____________

B = experimental solution deficient in ____________

C = experimental solution with unknown deficiency ____________

D = control solution with no deficiency

The components of the experimental solutions are given in Table P–3 at the end of this project. Follow these instructions exactly in filling your bottles:

1. Add 200 mL of distilled water to each bottle.
2. Locate the column in the table for the first mineral deficiency that you have been assigned and mark it A.
3. Add each of the ingredients called for in turn, *MIXING THOROUGHLY AFTER EACH ADDITION.*
4. Fill bottle A with distilled water.
5. Locate the column in the table for the second mineral deficiency that you have been assigned and mark it B.
6. Add the ingredients called for as before and fill bottle B with distilled water.
7. Add 15 mL of an unknown solution to bottle C and fill it with distilled water. Record the unknown number in your procedure.
8. Add the components called for under "complete" in the table to bottle D and fill it with distilled water.

Measure the pH of one of your solutions by adding a drop of it to a piece of pH paper:

pH = ____________

Iron is insoluble at neutral or basic pH values. Why is the iron provided in the form of FeNaEDTA?

____

What kind of compound is EDTA? ____________

Obtain four aluminum lids for your bottles, punch four holes in each, and cover your bottles as shown in Figure P–4. One hole is for adding distilled water to your plants and the other three are for the plants themselves. Label three holes in each lid 1, 2, and 3, so you will be able to identify the individual plants.

When your bottles are ready, obtain twelve 10-day-old corn or sunflower seedlings that have been grown in washed sand, and a few pieces of cotton. The plants should be about 8 cm long. Rinse the roots of the plants carefully in distilled water to remove the sand. Wrap a piece of cotton around the stem of each plant and insert three plants into the lid of each bottle. Arrange the bottles neatly in the greenhouse space assigned to your class. All bottles must be labeled with your name and the date.

Results

When the plants are in place in the greenhouse, measure the height of each plant and record the data in

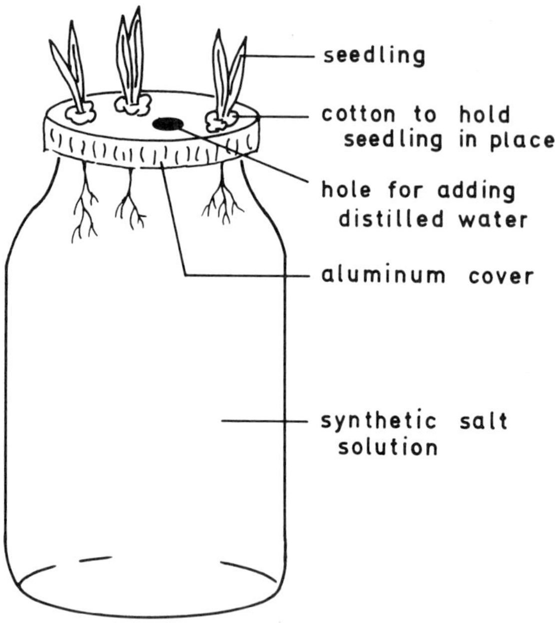

Figure P–4. Nutritional testing. Seedlings are grown in a bottle to test for nutritional requirements. The bottle must be kept completely filled by adding distilled water to replace water lost from the nutrient solution.

Table P–4 at the end of this topic. Repeat the measurements once or twice each week for four weeks and note the appearance of the plants in the table. Notice the appearance of the experimental seedlings of the other students as well.

At the end of the experiment summarize your results and obtain the results for the other deficiencies that you did not test yourself. Summarize the data in Table P–5 at the end of this topic.

Conclusions

a. Which single mineral deficiency had the greatest effect on the seedlings? ______

b. Did the seedlings in plain distilled water show any growth at all? ______

c. If there was growth in these seedlings, how could you explain it? ______

______

______

______

d. If your class studied both corn seedlings and sunflower seedlings, which type was more sensitive to iron deficiency? ______

e. Were there other differences in corn and sunflower seedling responses to different deficiencies?

______

______

______

______

Compare the appearance and growth rate of the seedlings in the solution with an unknown deficiency (bottle C) to those of the other seedlings studied by the class.

f. Was the growth of these seedlings similar to that of the control (no deficiency) or did the unknown solution appear to be deficient in one or more minerals?

______

______

______

______

When the experiment is over, dispose of the plant remains properly and wash the bottles very carefully in hot soapy water. Rinse them six times in tap water and three times in distilled water so that all traces of the mineral solutions have been removed.

If you are asked to write a formal laboratory report, follow the instructions in Appendix A and include all of your data tables. If you are asked to include a statistical analysis in your report, refer to Appendix B and calculate the mean final heights of a group of control and a group of deficient plants. Include data for as many plants as you can. Determine whether there is a statistically significant difference between the two groups using Student's *t* Test.

REFERENCES AND FURTHER READING

Bowling, D.J.F. *Uptake of Ions by Plant Roots.* London: Chapman and Hall, 1976.

Epstein, E. *Mineral Nutrition of Plants: Principles and Perspectives.* New York: John Wiley and Sons, 1972.

Galston, Arthur W., and Peter J. Davies. *Control Mechanisms in Plant Development.* Englewood Cliffs, NJ: Prentice-Hall, 1970.

Galston, Arthur W., Peter J. Davies, and Ruth L. Satter. *The Life of the Green Plant,* 3rd ed. Englewood Cliffs, NJ: Prentice-Hall, 1980.

Kaufman, Peter B., John Labovitch, and Anne Anderson. *Laboratory Experiments in Plant Physiology.* New York: Macmillan, 1975.

Raven, P.H., R.F. Evert, and S.E. Eichhorn. *Biology of Plants,* 4th ed. New York: Worth, 1986.

Ray, P., T.A. Steeves, and S.A. Fultz. *Botany.* Philadelphia: Saunders College Publishing, 1983.

Salisbury, F.B., and C.W. Ross. *Plant Physiology,* 2nd ed. Belmont, CA: Wadsworth, 1978.

Stefferud, A., ed. *Soil: The 1957 Yearbook of Agriculture.* U.S. Department of Agriculture. Washington, DC: U.S. Government Printing Office, 1957.

Journal Articles

Emery, Thomas. "Iron metabolism in humans and plants." *American Scientist* 70(6): 626, November 1982.

Epstein, E. "Roots." *Scientific American,* 48, May 1973 (#1271).

Jordan, Carl F. "Amazon rain forests." *American Scientist* 70(4): 394, July–August 1982.

Olsen, Ralph A., Ralph B. Clark, and Jesse H. Bennett. "The enhancement of soil fertility by plant roots." *American Scientist* 69:378, July–August 1981.

Student Name ______________________ Date ______________

TABLE P–3. Preparation of Experimental Solutions*

Stock Solution	Solutions Deficient In:									
	No minerals = COMPLETE	Ca	S	Mg	K	N	P	Fe	Micronutrients	All minerals = DISTILLED WATER
		Add (mL)	*Add (mL)*	*Add (mL)*	*Add (mL)*	*Add (mL)*	*Add (mL)*	*Add (mL)*	*Add (mL)*	*No additions*
0.5 M $CaNO_3$	5	0	5	5	5	0	5	5	5	0
0.5 M KNO_3	5	5	5	5	0	0	5	5	5	0
0.5 M $MgSO_4$	2	2	0	0	2	2	2	2	2	0
0.5 M KH_2PO_4	1	1	1	1	0	1	0	1	1	0
FeNa EDTA	1	1	1	1	1	1	1	0	1	0
Microelements	1	1	1	1	1	1	1	1	0	0
0.5 M $NaNO_3$	0	5	0	0	5	0	0	0	0	0
0.5 M $MgCl_2$	0	0	2	0	0	0	0	0	0	0
0.5 M Na_2SO_4	0	0	0	2	0	0	0	0	0	0
0.5 M NaH_2PO_4	0	0	0	0	1	0	0	0	0	0
0.5 M $CaCl_2$	0	0	0	0	0	5	0	0	0	0
0.5 M KCl	0	0	0	0	0	5	1	0	0	0
	15	15	15	15	15	15	15	14	14	0

* Reprinted with permission of Macmillan Publishing Co. from *Laboratory Experiments in Plant Physiology* by Peter B. Kaufman, John Labovitch, and Anne Anderson.

TABLE P–4. Experimental Data on Mineral-Deficient Plants.
(Experiment started on (date): ______________)

		Date: __________		Date: __________		Date: __________	
Plant	**Deficiency**	Height	Appearance	Height	Appearance	Height	Appearance
A-1							
A-2							
A-3							
B-1							
B-2							
B-3							
C-1	Unknown #: __________						
C-2	Unknown #: __________						
C-3	Unknown #: __________						
D-1	Complete						
D-2	Complete						
D-3	Complete						

Student Name ______________________ **Date** ______________

TABLE P–4. Experimental Data on Mineral-Deficient Plants. (contd.)

	Date: ________		Date: ________		Date: ________		Date: ________	
Plant	Height	Appearance	Height	Appearance	Height	Appearance	Height	Appearance
A-1								
A-2								
A-3								
B-1								
B-2								
B-3								
C-1								
C-2								
C-3								
D-1								
D-2								
D-3								

TABLE P–5. Summary of Effects of Nutritional Deficiencies

Plants	Deficiency	Corn Seedlings		Sunflower Seedlings	
		Total Growth (cm)	Final Appearance	Total Growth (cm)	Final Appearance
Controls	none				
Expt.	Ca				
Expt.	S				
Expt.	Mg				
Expt.	K				
Expt.	N				
Expt.	P				
Expt.	Fe				
Expt.	micronutrients				
Expt.	all minerals				
Unknown #					

Unknown is probably deficient: ______________/probably not deficient: ______________

Unknown might be deficient in: ______________________________

Unknown is probably not deficient in: ______________________________

PROJECT D

The Pond Ecosystem

OBJECTIVES

When you have completed this project, you should be able to:

1. Define ecosystem, community, and succession.
2. Explain the advantages of a microecosystem (microcosm) over an outdoor ecosystem.
3. Give examples of the following that might be found in a pond: abiotic substance, producer organism, macroconsumer organism, saprobe, decomposer, detritus, scavenger, primary carnivore, secondary (top) carnivore.
4. Define species, species diversity, and climax community.
5. Describe a laboratory experiment designed to compare succession in two pond microcosms.
6. Analyze the results of such an experiment and present them in a formal laboratory report.

TEXT REFERENCES Chapter 19 (A, B); Chapter 38 (A–H); Chapter 39.

KEY TOPICS

binomial nomenclature	populations	competition
ecosystems	communities	succession

INTRODUCTION

The laboratory work for this project extends over a period of at least 3 weeks. One laboratory period is required to set up the experiment, and the time spent in following it is open-ended (10–20 hours).

The Microecosystem

An **ecosystem** includes the organisms in a certain area and their interactions with the physical environment with which they exchange energy and materials. The group of organisms living in an ecosystem is called a **community.** The energy flowing through the ecosystem will result in a diversity of species feeding at different trophic levels and in a cycling of materials through the system. The **microcosms,** or microecosystems, that we will establish in the laboratory fit the definition of ecosystems. However, the microcosms also have properties unlike those of most natural ecosystems, which must be taken into account.

Today many ecologists are turning to laboratory and field microecosystems that have discrete boundaries and can be manipulated and replicated at will, because natural ecosystems are complex, hard to delineate, and often difficult to study by the traditional means of "experiment and control." Whereas you cannot remove the sediments from a pond to see what effects the action would have, you can set up a microcosm without sediment and observe the results. Another advantage of a microcosm is that the time scale of events is shortened. Normally, **succession,** which is the pattern of changing communities over time, is very slow. This is particularly true of succession in which organisms first colonize a nonliving habitat. In the microcosm, however, succession to a stable system takes place in a few weeks. Microcosms also generally have fewer species as well as simpler communities than their parent ecosystems.

In what other ways are microcosm environments unlike the ecosystems that they are supposed to mimic? Size and isolation are two obvious differences. Organisms in the microcosms tend to diverge from those in their parent communities and from each other, but this problem can be eliminated to an extent by cross-seeding the jars: We mix the water in each jar with water from other jars after they have stabilized for a few days.

The Pond

Let us first consider the natural pond ecosystem (see Text Figure 39–15). Not only is the pond a place where organisms live, but organisms make the pond what it is. Despite the complexities of this system, we can reduce it to several basic factors.

PHYSICAL FACTORS

Light from the sun is the main source of energy. It is captured by photosynthetic plants, flows through the ecosystem, and is lost as infra-red radiation (heat). Energy cannot be recycled, and a percentage is lost at every conversion. The amount of sunlight is basically determined by the latitude of the pond. Sunlight and elevation help determine the pond's temperature. Light and temperature are the main factors controlling the **climate** experienced by the pond community. A pond of sufficient size will have a boundary called a **thermocline** separating it into a warmer upper layer (epilimnion) and a cooler lower layer (hypolimnion). Higher temperatures in the pond will be associated with greater production of organic matter and may lead to anaerobic conditions.

ABIOTIC SUBSTANCES

These are simply **organic** and **inorganic compounds** such as H_2O, CO_2, O_2, calcium, nitrogen, phosphorus, and other salts. A small portion of these vital nutrients is in solution and immediately available to organisms, but a much larger portion is held in reserve in particulate matter (especially in the bottom sediments) as well as in the organisms themselves. Nutrients continuously pass through geochemical cycles and can be reused many times. The rate of nutrient release, and the solar input, temperature, and other climatic factors are the most important in regulating the rate of function of the entire ecosystem on a day-to-day basis.

PRODUCERS

Producers are the autotrophs of the community. They obtain all of the energy entering the ecosystem by trapping sunlight during photosynthesis. Generally speaking, there are two main types of producers in a pond: rooted or large floating **plants** and minute floating plants called **phytoplankton.** Rooted producers live in the **littoral zone** along the shore, and phytoplankton float in the open surface water called the **limnetic zone.** Phytoplankton are usually much more important than rooted vegetation in the production of food for the ecosystem. Pollution of the pond, in the form of excess nutrients or higher temperature, will increase the numbers of producers and lead to **eutrophication.** The **trophic level** of the other organisms in the pond is determined by the number of feeding steps in the food web between an organism and a producer.

CONSUMERS

Unicellular protozoans and zooplankton are the **microconsumers.** They feed on bacteria and planktonic organisms smaller than themselves. The **primary macroconsumers** are herbivores that feed directly on living plants or plant remains and are of two main types. **Zooplankton** are the small animal plankton, and **benthic** organisms are bottom-dwellers. The **secondary consumers** are carnivores such as predacious insects and game fish, which feed on the primary consumers. **Tertiary consumers** feed in turn on the secondary consumers. Any given animal can feed at more than one of these trophic levels.

DECOMPOSERS

Decomposers are heterotrophs that feed on nonliving organic matter. They perform the vital function of recycling nutrients in the ecosystem. All of these organisms are called saprotrophic (= "dead-eating") because they feed on **detritus** or dead remains. The **saprobes** are the aquatic bacteria, flagellates, and fungi that cannot make or ingest food and so absorb it through their cell membranes. They are distributed throughout the pond but are especially abundant at the mud-water interface along the bottom, where the bodies of plants and animals accumulate. Larger detritus feeders that ingest dead or decaying matter are called **scavengers.** If the producers outstrip the decomposers, organic remains accumulate, causing anaerobic conditions on the pond bottom. Eventually this will adversely affect aerobic benthic organisms and fish.

It is important to remember that although the pond may seem to be a self-contained unit, its productivity and stability over a period of years is very much determined by the input of solar energy and especially by the rate of inflow of water and materials from the watershed. The **watershed** is the whole drainage basin, that it, all the land that contributes runoff to the pond.

Succession

Succession, the more or less orderly process of community change, involves replacement, in the course of time, of the dominant species within a given area by other species. It often represents the process of returning to equilibrium after a disturbance. The ecological succession that you will be observing in your microcosms is a temporal succession. The microcosms will develop a series of different communities as the physical environment becomes modified by the community. Therefore, succession is community-controlled even though the physical environment determines the pattern and the rate of change, and often sets the limits as to how far development will go. This is very true of your microcosms because they are limited by the size of the container, the amount of incoming solar radiation, and so on. As succession proceeds, you should notice a change in species composition and an increase in species diversity. The diversity of species will increase because rare species that were unnoticed at first may increase in abundance so that they are eventually included. Typically there will be a few dominant species and many relatively rare species. The successional pattern will culminate in a **climax community,** whatever that may be for a particular ecosystem. This is the community that forms and maintains itself if left undisturbed. It tends to have fewer species than the communities where succession is in progress.

Aquatic microecosystems derived from ponds will maintain themselves in a climax state indefinitely with only light input. They will usually reach this level in less than 100 days.

- Be sure to examine the microcosms that have been growing for 1 or 2 years if they are on display in the laboratory.

PLANNING YOUR EXPERIMENT

Lab Groups

Work in groups of four. Each group will have two microcosm jars—one to be used for an experimental treatment, the other for a control. Two students will be in charge of each jar. There is a lot of work involved in this project, but a smoothly functioning team can handle it with a reasonable expenditure of time and effort. Because you will need to exchange data among yourselves, you should choose your group with two things in mind:

1. You should all be able to come to lab at the same time to set up your experiment and to take your subsequent observations. This way, you will all be referring to one kind of organism by the same name!
2. You should be people who see each other frequently (have classes together, live in the same dormitories, and so on) if possible. Then you will be able to get hold of each other easily to exchange data.

Materials

The water you will use comes from a nearby pond. Three types of microcosms have been set up, all containing about 3 liters of water:

1. Microcosm—no sediment. There will be little exchange of nutrients, no sediment biota, and no species refuge.
2. Microcosm—sand. No exchange of nutrients, no sediment biota, but a refuge for other species.
3. Microcosm—sediment. The "natural look." There will be nutrient exchange, a refuge for species, and a good healthy sediment biota.

You may wish to use additional chemicals, lights, aerators, heaters, and the like. If you plan in advance what you will need and notify the staff, they will let you know whether it can be obtained. You may wish to bring additional materials of your own on the day that you set up the experiment.

Experimental Design

Prepare a detailed written description of your experimental plan before coming to laboratory to start the experiment, and obtain the instructor's approval of your plan before starting the experiment. This plan should include:

1. A hypothesis giving the rationale for the experimental treatment.
2. The nature of the experimental and control microcosms.
3. The amounts of components and concentrations of added materials or chemicals.
4. A schedule for gathering data.
5. Criteria for deciding whether or not the experimental treatment had an effect.

As a group, you should discuss and decide on a semioriginal experimental design. Here are some suggestions to mull over:

Have a microcosm with sediment as your control and one with no sediment as the experimental model. Then compare and contrast trends.

Expose one microcosm to constant light and one to the natural diurnal pattern. Again, compare and contrast trends.

Enrich one microcosm with glucose as an immediately available carbon source. Use 3 g of glucose for 3 L of water. Please be aware that this treatment creates quite a stench in the experimental jar!

Be sure that both your microcosms are receiving the *same light exposure* (both in the south window, etc.) unless you are deliberately using different exposures. Also, cross-seed your jars before you begin so that they have the same organisms at the beginning of the experiment.

STARTING YOUR EXPERIMENT

- Choose two microcosm jars.
- Label each microcosm with your names, the treatment, the date, and a number.
- Record the qualitative aspects of your microcosms. What organisms can you see with the naked eye? Is there anything living in the sediment? Are there snails present? Whenever you record data, label it with your name and the date on which the observations were made. Do not rely on your memory alone, or you may end up in a state of great confusion when you attempt to write up the results. If possible, use a notebook for your data.
- Remove a 5-mL sample from the center of your microcosm using the pipets and bulbs provided. If you are not going to examine the sample immediately, transfer it to a small sample bottle and add one drop of acid Lugol's solution to preserve the sample. Cork the sample tightly, and label it completely (name, date, microcosm, number).
- Examine your sample in the microscope by making a wet mount of one drop or by placing a drop in a depression slide and adding a coverslip. The depression slide will allow wiggly animals more freedom of movement but it will also be more difficult to keep them in focus as they move vertically in the field.
- While viewing the sample under low power, move the slide around to locate as many different kinds of organisms as you can. Switch to high power to search for smaller organisms. If the motile organisms are swimming too fast, add a drop of Protoslo, which will slow them down.

The search you just finished was designed to familiarize you with the different organisms. Obviously classification of the organisms is the most difficult part of this lab. Using the posters and books in the lab, try to identify the organisms that you examine. But *do not become overly concerned* with scientific identification of the organisms. It is *more* important to be able to distinguish one kind from another and to be able to communicate with other members of your group so that all of you are using the same name for the same thing.

- When you find a new organism:

1. Draw it in your data chart, which should also have space to record the date, the kind of organism, the number in each of five fields, and the total number.
2. Place it into a taxonomic group, for example, bacterium, cyanobacterium, diatom, protozoan, multicellular animal, or multicellular plant.
3. Try to find its picture in one of the posters or books in the lab, and call it by its accepted name. For bacteria, classification as rod-, sphere-, or spiral-shaped will suffice.
4. If you cannot locate a picture that looks like the organism in one of the sources provided, give it your own name, and make sure the others in your group have a good look at it in your microscope so that all of you will be calling it the same thing. At the same time, try to distinguish and name as many different kinds of organisms as possible.

- Now that you have identified the organisms in your sample, you will try to count them.

The following procedure is to provide a random count that gives an accurate representation of the relative abundances of organisms in the sample. You should avoid the "cheating" that would result from deliberately seeking out the most interesting or abundant organisms. Always record exactly what you see.

- Looking at the microscope from one side, center the coverslip of your slide on the microscope stage. View the slide under low power. Count the number of each kind of organism in the field, and record it in the data table. If there are too many of some kinds of organisms to count, estimate the order of magnitude (10^2, 10^3, and so on) of individuals in each field.
- Switch to high power to count bacteria, and be sure to make adjustments for the change in field size at the higher magnification when you record the number present.
- Move the slide in turn to the four corners of the coverslip (again watching from the side, *not* through the microscope), and record the numbers of each type of organism present in each field.
- In the right-hand column of the data sheet, record the total number of each type of organism observed.
- Examine a larger sample of water in a watch glass under the dissecting microscope to identify and observe zooplankton present. You may be able to see *Daphnia* (water flea) with young in their brood pouch as well as many other interesting organisms. Carefully return this sample to your microcosm.
- Do a chlorophyll analysis to give you some idea of the overall size of the producer trophic level. (See Figure P–5.)

1. **Sampling.** Dip a 100-mL beaker into the middle of your microcosm to get a 100-mL sample of phy-

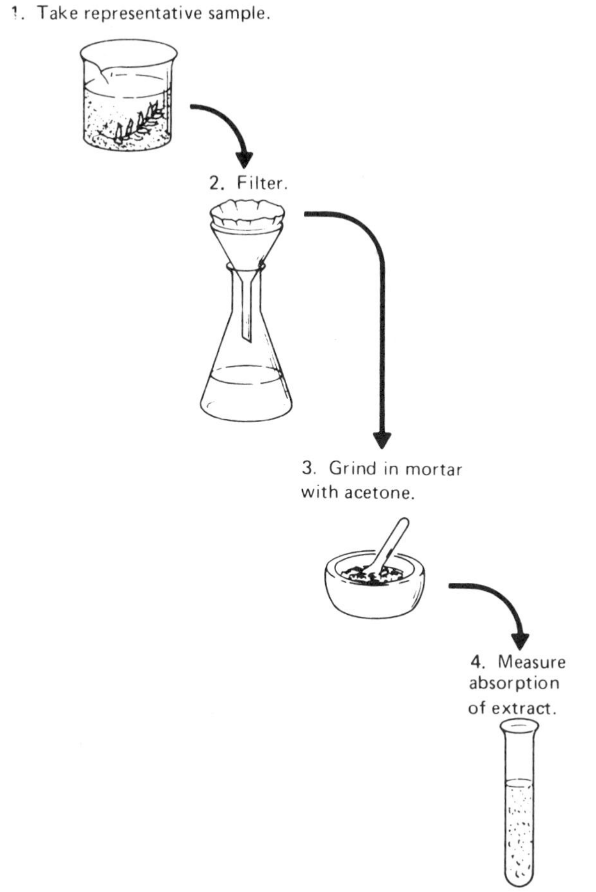

Figure P–5. Chlorophyll analysis. The concentration of chlorophyll will be proportional to the difference between the absorbance at 663 nm and that at 750 nm.

toplankton. Estimate what proportion of your water is occupied by green masses of algae floating or attached to the side of the jar, and then add to your 100-mL sample of water a mass of algae to take up the same proportion of space in the beaker. (For example, if your microcosm has a clump of algae that occupies 10% of the jar, you will take out a piece of the clump that will occupy 10% of the beaker.)

2. **Filtration.** Flute a piece of Whatman filter paper, place it in a funnel, and place the stem of the funnel into an Erlenmeyer flask. Pour the contents of your beaker into the filter paper; then wait till all the liquid has gone through. Return the liquid in the flask to the microcosm.

3. **Pigment extraction.** The algae are now clinging to the filter paper. Place the paper in the mortar and add 5 mL of acetone. Gently grind with the pestle, hard enough to break the algal cells but not hard enough to shred the paper. Remove the filter paper and let any paper fibers settle. Pour off the acetone, containing the algal pigments, into a spectrophotometer cuvette.

4. **Chlorophyll measurement.** Fill a second cuvette with acetone for your blank.

Be careful not to spill acetone onto the spectrophotometer.

Following the directions in Topic 2 for operating the spectrophotometer, take readings of both tubes at 663 nm (the absorption maximum of chlorophyll *a* in acetone solution) and at 750 nm (where chlorophyll *a* absorption is negligible). The second reading will give you some idea of the contamination of your sample by pigments other than chlorophyll and by light-scattering debris.

FOLLOWING YOUR EXPERIMENT

Sample your microcosms every 2–3 days for 3–5 weeks, making sure to keep accurate records of the distribution of large and small organisms.

Repeat the chlorophyll analysis a total of three to four times during the course of your experiment. If you expect the producers to be experiencing major population changes because of your particular experimental treatment, you should do it more often.

At the end of the experiment, follow your instructor's directions for disposing of the microcosms (some may be kept for a longer period). Clean up any equipment that you have used.

ANALYZING YOUR EXPERIMENT

Now that you have all these data, what are you going to do with them?

1. Summarize your data in tables and graphs, when possible. Graph the abundance of different kinds versus time following the example shown in Figure P–6. To simplify your data summary, you may want to lump organisms of one type (ciliated protozoan, diatom, green alga, rod-shaped bacteria, and so on) together.

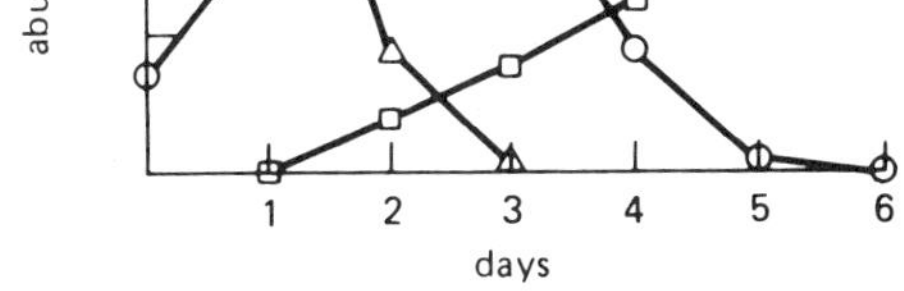

Figure P–6. Changing abundances.

2. Calculate a simple diversity index:

$$D = \frac{\text{total number of kinds}}{\log_{10} \text{ of total number of individuals}}$$

3. Classify the organisms present in your habitat as rare (10 or fewer individuals), common (11–50), or abundant (more than 50).
4. Write a laboratory report on this experiment according to the standard form given in Appendix A of this manual. Try to answer the following questions in the course of your discussion:
   a. Compare and contrast your microcosm with the pond this water came from, or ponds in general (food chains, size, diversity, and so on). What are the properties peculiar to the microcosm?
   b. Can you draw any conclusions from your data about ponds or ecosystems in general? Did you learn anything about aquatic ecosystems or is the microcosm simply an anomaly?
   c. How could you conduct the same experiment in the field? (The answer "it is not possible" will not be accepted.)
   d. Can you explain your results in terms of a food web?

REFERENCES AND FURTHER READING

Abramoff, P., and R.G. Thomson. "The microscope." in *Laboratory Outlines in Biology III.* San Francisco: W.H. Freeman, 1982. (Freeman laboratory separate #846.)

Bold, Harold C., and Michael J. Wynne. *Introduction to the Algae: Structure and Reproduction,* 2nd ed. Englewood Cliffs, NJ: Prentice-Hall, 1985.

Coker, R.E. *Streams, Lakes and Ponds.* New York: Harper and Row, 1968.

Jahn, T., and F. Jahn. *How to Know the Protozoa,* 2nd ed. Dubuque, IA: W.C. Brown, 1979.

Klots, E.B. *The New Field Book of Freshwater Life.* New York: Putnam, 1966.

Krebs, C.J. *Ecology: The Experimental Analysis of Distribution and Abundance,* 2nd ed. New York: Harper and Row, 1978.

Odum, E.P. *Basic Ecology,* 3rd ed. Philadelphia: Saunders College Publishing, 1983.

Prescott, G.W. *How to Know the Freshwater Algae,* 3rd ed. Dubuque, IA: W.C. Brown, 1978.

Reid, George K. *Pond Life: A Guide to Common Plants and Animals of North American Ponds and Lakes.* New York: Golden Press, 1967.

Southwood, T.R.E. *Ecological Methods,* 2nd ed. London: Chapman and Hall, 1978.

Turk, J. *Introduction to Environmental Studies,* 2nd ed. Philadelphia: Saunders College Publishing, 1985.

Whittaker, R.H. *Communities and Ecosystems,* 2nd ed. New York: Macmillan, 1975.

Journal Articles

Bormann, F.H., and G.E. Likens. "The nutrient cycles of an ecosystem." *Scientific American,* 92, October 1970 (#1202).

Clark, John R. "Thermal pollution and aquatic life." *Scientific American,* 18, March 1969 (#1135).

Forbes, S.A. "The lake as microcosm." *Bulletin of the Illinois Natural History Survey* 15: 537, 1887.

Gosz, J.R., R.T. Holmes, G.E. Likens, and F.H. Bormann. "The flow of energy in a forest ecosystem." *Scientific American,* 92, March 1978 (#1384).

Janick, J., C.H. Nöller, and C.L. Rhykerd. "The cycles of plant and animal nutrition." *Scientific American,* 74, September 1976 (no offprint).

Likens, Gene E., Richard F. Wright, James N. Galloway, and Thomas J. Butler. "Acid rain." *Scientific American,* 43, October 1979 (#941).

Schindler, D.W. "Eutrophication and recovery in experimental lakes: Implications for lake management." *Science* 184: 897, 24 May 1974.

Woodwell, George M. "Toxic substances and ecological cycles." *Scientific American,* 24, March 1967 (#1066).

APPENDIX A

Writing a Lab Report

FORMAT

A formal lab report should follow the general format used for a research report published in a scientific journal. Although different journals require different formats, all papers have a roughly similar outline. They reflect the basic scientific method of asking a question, formulating hypotheses, conducting experiments to test the hypotheses, and interpreting the results. You may want to go to the library to look at articles in journals such as *Bioscience, Proceedings of the National Academy of Sciences,* or *Limnology* to get an idea of how to write a scientific paper. (*Scientific American* is *not* considered a scientific journal because its purpose is not communication between research scientists engaged in related work.)

Although not all journals require authors to divide their papers into labeled sections, doing so will help make sure that your work is reported in an organized and clear fashion. Therefore, you are asked to divide your report into these six clearly labeled sections (do not label the title): **Abstract, Introduction, Material and Methods, Results, Discussion,** and **References.**

Title

The title should consist of a few well-chosen words indicating the subject of the report. "Well-chosen" means that the title reflects the scope of your report accurately: Do not call your paper "Effects of Chemicals on Animals" if you really studied how feeding dietary supplements of vitamins A and D affected the appearance of the fur of three white rats.

Abstract

The abstract is a short paragraph of 150 words or less that summarizes your experiment, including pertinent information about your experimental subjects, materials and methods, results, and conclusions. This is the part of the paper scientists read to decide whether they are interested in looking at the rest of the paper. Thus the information the abstract contains should help people decide whether your research sheds any useful light on what they are studying.

Introduction

The introduction gives the background of the experiment. It should include an explanation of the background of the general problem or area being investigated, telling why this problem is of interest and outlining what information is already known about the problem.

The problem does not necessarily have to be relevant to the health and welfare of human beings or of immediate economic importance to be of interest. You do not need to force a human perspective on a problem that is of interest because it is related to broad underlying principles of the structure or functioning of living things. In building up this part of your report you may want to consult outside references or, at the very least, reread the relevant parts of your text. But be sure to keep track of where you get this information and list all references used, including your textbook and laboratory guide, in a reference list at the end of the paper.

The introduction should also present the question you are trying to answer or the hypothesis you are testing. You should include what outcome you expect, and how it would help support or refute your hypothesis or answer your question. Be sure to distinguish between the hypothesis itself and the experiment you will do to test the hypothesis.

Materials and Methods

This section should include a concise description of the materials, procedures, and equipment used. It should clearly describe the experimental situation, the control situation(s), and the type of observations you made. There should be enough detail so that someone else could repeat your work. Therefore, brand names of equipment, concentrations and amounts of solutions, species, size or age, sex, and other information about the experimental subjects should be included. If you follow directions from a book or paper, just say so. You need not repeat them in your paper. If, however, you change the procedure, you should describe exactly what you did differently.

Do not include the rationale for your work in this section. Also, be sure to report your procedure as a past event rather than writing this section as a set of directions to your reader. You do not need to report attempts at the experiment with techniques that failed unless these techniques are very likely to be tried by other people in the future.

Results

If you worked with others, give their names here. Include your raw data sheets or refer to the report where they may be found.

Present your findings in a logical, not chronological, order. Give the results that you found, not what you think you should have found. The organism is always right! You may have to do some thinking, however, to find out why the results came out as they did. Do not explain your results in this section. Many papers are valuable for their correct results, although the interpretation of the results may be completely wrong and may later be modified in light of further information.

Results can often be reported more effectively in the form of one or more tables or graphs or drawings. These should have clear labels and captions. Be sure to indicate whether the data reported are single readings or averages. Statistical analyses of your data should also be included here if appropriate. In addition, a written description should summarize the results illustrated in the graph or figures. This should point out trends or inconsistencies but should not include explanations or opinions.

Discussion

Here you should give your interpretations of the data and relate them to the questions you posed in the introduction. Be careful to avoid making this section just a repetition of the introduction. If you have any data to explain away, do it here or make a new hypothesis as to why the results came out in a way you did not expect. Did the results answer your question? Did they support or disprove your hypothesis? Draw some conclusions, supporting them with your data. What is the significance of your results in the general area you studied? What are the main principles demonstrated by your results?

What further experiments should be performed to clear up discrepancies or ambiguities in your results? How might your work best be continued or extended?

Conclusion

This should be a separate section only if you have a lot of conclusions. Otherwise you should simply round off your discussion section with the conclusions that you can draw from your experiments.

References

You must refer to all the sources of information that you used in your paper. Failure to do so is a serious offense, **plagiarism,** and will result in rejection of your laboratory report or worse. All of the sources in the reference list must be cited in the paper and vice versa. Usually your sources will be books or journals, but you must also cite any persons from whom you obtained information directly as "personal communication."

Citation of a reference does *not* entitle you to "lift" passages of text for use in your report. Any such direct quotation must be set off by quotation marks and the source and page number indicated after the quote. It is very much preferable to digest the information and to present it in your own words.

There are several possible systems for organizing your references. You may choose to list the references in the order in which you cite them in the text. In this case you place a number in your paper after a comment from the reference: for example, "The Loch Ness monster is a big worm (1)," and then list the reference from which you obtained this information as number 1 in your reference list. Alternatively, you may alphabetize your list of references and then number the list when you have finished preparing your report. In that case you can use either the number or the author's last name(s) to refer to the source.

Whichever system you use, you may sometimes want to cite the author and date of the source in your text, and you must cite the source of any direct quote. This has the advantage of letting your reader know something about the authority and how recent your information is without flipping to the end of the report all the time. For example, you could say, "The Loch Ness monster is a giant worm (Smith, 1970)," or "Smith (1970) has found that the Loch Ness monster is a huge worm," or "In 1970, Smith found that the Loch Ness monster is a large worm." With three or more authors, name them all the first time that you cite them and then use "Smith et al." subsequently (et al. means "and others"). You do not need to repeat the date each time unless you have more than one reference from the same author.

SUGGESTED FORMAT FOR REFERENCE LIST ENTRIES

Here are examples of three types of entries that you might make in your reference list. You may also use the format used by a prominent scientific journal or one suggested by your instructor. Once you have decided on the format, use it consistently for all of your references. Consult your reference librarian for the standard abbreviations used for the titles of the journals in your reference list.

A book reference might read as follows:
Vince-Prue, D. *Photoperiodism in Plants.* New York: McGraw-Hill, 1975. pp. 104 and 194.

A reference to a journal article:
Raymond, J.A., and A.L. DeVries. "Adsorption inhibition as a mechanism of freezing resistance in polar fishes." *Proc. Nat. Acad. Sci., U.S.A.* 74: 2589–2593, 1977.

A reference to an article that appeared in a book:
Rehn, S. "Chemical defenses of ant acacias." In J. Pearson, ed., *Phytochemistry.* New York: Academic Press, 1973.

A reference to a discussion with or letter from an individual:
Arms, K. Personal communication.

WRITING TIPS

Your report will be easier to read and understand if you follow these tips and try to conform to the accepted style of scientific writing that is required for scientific papers. Conformity can be a good thing if it increases communication!

1. Write in the past tense, not present or future: "Two rats were anesthetized and . . ."
2. Use metric units (gram, meter, liter, second) and the Celsius (Centrigrade)* temperature scale.
3. Use correct abbreviations *without* periods:

10 milliters	=	10 mL
12 liters	=	12 L
3 micrograms	=	3 μg
7 grams	=	7 g
24 kilograms	=	24 kg
663 nanometers	=	663 nm
8 micrometers	=	8 μm
21 centimeters	=	21 cm
9 meters	=	9 m
52 kilometers	=	52 km
37 degrees Celsius*	=	37° (37°C)

4. If you start a sentence with a number, write out the number: "Twelve grams of minced toad brain was . . ."
5. For solutions, use molarity rather than normality:

   5 molar = 5 M
   .005 molar = 5×10^{-3} M or 5 mM
   .000005 molar = 5×10^{-6} M or 5 μM

   If the solution is given as a percent solution, indicate whether it was made up on the basis of weight or volume:

   5 g glycerol + 95 g H_2O = 5% glycerol (w/w)
   5 mL glycerol + 95 mL H_2O = 5% glycerol (v/v)
   5 g glycerol brought up to 100 mL with H_2O = 5% glycerol (w/v)

   A w/v solution is very useful because each 1-mL aliquot (portion) of this particular solution, for example, will contain exactly 50 mg of the solute, glycerol; small weights can thus be accurately measured out with a pipet.
6. Refer to animals and plants by their scientific names. (This is not necessary when referring to humans and common laboratory animals.) Always underline or italicize the genus and species names of organisms, for example, *Stentor coeruleus.* Be as specific as you can in naming the organism, but when the name is not known completely (as in your Pond Ecosystem report), write *sp.* (meaning "some species") after the generic name, for example, *Oscillatoria sp.*
7. Your report should be written in grammatically correct English. It should be concise and logically sound. Avoid boring the reader with copious verbiage and excessively formal writing. It is perfectly all right to sound excited about your work if you have worked carefully and have interesting results to report.

When in doubt, ask your instructor or try to find an example in a recent journal.

* The Centigrade and Celsius scales are the same: Water freezes at 0°C and boils at 100°C.

APPENDIX B

Biological Experiments and Statistics

INTRODUCTION

Many scientific experiments lead to observations that take the form of numerical data. An experiment is designed to make a limited number of observations from which the scientist would like to generalize to the whole of the population that is under study. (A statistical population is a collection of organisms or phenomena that have something in common.)

The most direct way of showing how to deal with a large body of data is to use an example to illustrate the steps in analyzing and interpreting the results. We will use a familiar case: the results of a biology exam given to 100 biology students. Suppose the biology exam consisted of 40 questions. The scores of all the students constitute the **raw data.**

If the course is graded on a "curve," the instructor and students are eager to see the plot of the **frequency distribution.** This is a graph in which the number of students receiving each score (the frequency) is plotted against the individual scores (40, 39, 38, and so on). The basis for making the graph is a table of score frequencies as is shown in Table B–1.

The table shows that 13 students received a score of 38, 4 received a score of 30, and so forth. The high score was 40 and the low score was 24, so the **range** of the scores was 24–40.

TABLE B–1. Frequency Distribution and $f \cdot x$

Scores (x)	Frequency (f)	$f \cdot x$
40	6	240
39	13	507
38	13	494
37	13	481
36	9	324
35	15	525
34	8	272
33	8	264
32	3	96
31	4	124
30	4	120
29	1	29
28	1	28
27	1	27
26	0	0
25	0	0
24	1	24
		Total: 3555

ARITHMETIC MEAN

The **arithmetic mean** is the average of all scores: All scores are added together, and the total is divided by the number of scores. If the data are in the form of Table B–1, you know that 6 students received a grade of 40. You should thus multiply 40 × 6, 13 × 39 and so on to give $f \cdot x$ and then add up all the products. The final total is divided by the number of students, which was 100. The sum of $f \cdot x$ is given in Table B–1.

$$\text{Mean} = \bar{x} = \Sigma x/N, \text{ or } = \Sigma f \cdot x/N,$$

where
$\bar{x}$ is the average or mean;
$\Sigma f \cdot x$ is the sum of all $f \cdot x$;
N is the number of scores.

What is the arithmetic mean for this test? ________

STANDARD DEVIATION AND VARIANCE

Variance is a measure of how far the scores tended to stray away from the average score of 35.55. One class might all score in the 30s with a mean of 35.55, and another class might include many 40s and some scores in the 20s and still have a mean of 35.55. The first class would have a lower variance than the second.

The formula for variance is:

$$\text{variance} = s^2 = \frac{\Sigma f(x - \bar{x})^2}{N - 1}$$

where

s is the standard deviation of the data points;
f is the frequency of a score x;
$\bar{x}$ is the mean of all scores;
N is the number of scores.

The difference between each score and the mean was determined, squared, and multiplied by the frequency of that score. All the products were added up, and the total was divided by the number of scores minus 1. In our example,

$$s^2 = 1068.5/99 = 10.8$$

The **standard deviation** is the square root of the variance. Thus in our example:

$$s = \sqrt{10.8} = 3.3$$

Any scores within 1 standard deviation of the mean (3.3 points, in this example) are said to be within 1 **standard deviation unit** of the mean. When there are many data points, as in our example, the mean should be expressed as mean ± standard deviation:

35.6 ± 3.3*
*standard deviation

This form of presenting experimental data in a report tells the reader how much variance there was in the data gathered, and, therefore, how accurate the mean value really is. The smaller the standard deviation, the smaller the variance in the sample, the more confident we are that the reported result is close to the true value of the mean and standard deviation for the population from which our sample was drawn.

For example, when the class data for the plant hormone experiment (Project B) are analyzed, the results will be expressed by calculating the mean and standard deviation for each group of data. There will be a mean ± standard deviation for the lengths of the axillary buds on the control plants, another for the buds on plants treated with lanolin, and so forth.

If the frequency distribution is a normal distribution, it will form a graph that is symmetrical about the mean and that approximates a bell-shaped curve. The standard deviation then takes on additional meaning. In that case, about 68% of the values, that is, the bulk of the data, will fall within 1 standard deviation unit on either side of the mean. Furthermore, 95% of the data will fall within 2 standard deviation units, and 99% will fall within 3 units of the mean. In the case of the test scores, we would expect 99% of all students taking the test to score within 9.9 points (3s) of the mean, or above 25.7. The standard deviation can be used in this way only if the sample was drawn from a "bell-shaped" population. (Persons who didn't study for the test or who had taken an advanced course in the subject would tend to invalidate the test results.)

SIGNIFICANT FIGURES

The **significant figures** in a number are those that convey information. It is important that your reported data reflect the accuracy of your measurements and calculations, no more and no less. The number of significant figures is the number of digits except for those zeroes used to indicate the decimal point position. By convention, all digits to the left of an indicated decimal point (including zeroes) are significant.

Here are some examples:

Number	Sig Figs	
27	2	
270	2	no decimal indicated
270.	3	
0.27	2	
0.2700	4	2 zeroes added solely to indicate sig figs
0.027	2	

In measurements, you should report your results to the smallest unit of which you are absolutely sure, plus one significant figure that is an estimate. For example, if you are measuring length with a ruler marked in millimeter increments, measure the millimeters, estimate the tenths of millimeters, and report your data in tenths as 27.1 mm or 2.7 mm.

In reporting the mean of data, the number of figures to be presented is determined by the **standard deviation of the mean.** By convention, the standard deviation of the mean is reported to two significant figures (or one for small samples). The overall mean is reported to the **same number of decimal places** as the standard deviation of the mean:

standard deviation by the mean = standard deviation of the data points/$\sqrt{N}$
where N is the number of data points.

For the data points in Table B–1:

mean = 35.55
standard deviation = 3.3
standard deviation of the mean to two significant figures will be $3.3/\sqrt{100.} = 3.3/10. = 0.33$

In this case you are entitled to report the mean to two decimal places, i.e. hundredths:

mean = 35.55 ± 0.33*

*standard deviation of the mean

The type of deviation must always be specified, as by the footnote (*) above. (If using standard deviation of the data points, specify that this is the case and follow the same rule as for the standard deviation of the mean: Report a mean of 35.6 ± 3.3, using a standard deviation with two significant figures.)

If, however, these exam data were instead a few experimental measurements, the mean would be reported very differently. Suppose nine measurements gave the same mean and standard deviation:

mean = 35.55
standard deviation = 3.3
$N = 9$
standard deviation of the mean = $3.3/\sqrt{N} = 3.3/3$
= 1.1 *reported as 1* (one significant figure)

mean = 36 ± 1*

*standard deviation of the mean

Note that when experimental data are presented according to the convention, the last digit of the mean is in fact not significant, since it is completely uncertain.

SIGNIFICANT DIFFERENCE

When we examine the results of an experiment, we would like to find out whether the results are "significant" or not. There are two ways of looking at significance.

1. We predict the results that we would obtain according to a certain hypothesis. Then we see how far away from the expected our results actually were. This approach is the one you will use in testing the results from your genetics experiment to see whether they could have varied from expected due to random chance or are so far from expected that your hypothesis is doubtful (chi-square test).
2. We compare an experimental group with a control group using the **null hypothesis:** The two groups will be the same unless the experimental treatment had an effect. The more the results from the two groups differ, the less likely it is that the null hypothesis is correct.

In Project B, we would like to decide whether the axillary buds in the plant group treated by removing the apical buds grew longer than the buds in the untreated group. This can be done by using the student's *t* test or the Wilcoxon test to determine whether there is a significant difference between the results for the two groups.

All the plants originally came from the same population. If samples had been drawn from this population before the experiment, they might have had different means, but the difference would be due to chance alone. Such a difference would not be significant. After one group of plants has been treated in the experiment, the means may differ significantly as a result of the experimental treatment. We would like to know whether the difference is due to just random chance or to the experimental treatment. The statistical test will show whether the difference in the means could be due to random chance, and is therefore not significant, or is not due to random chance but rather to the experimental treatment, and is significant.

Statistical significance is usually expressed as a decimal. A significance level of 0.05 or 5% means that we could expect to find such a difference by random chance only about 1 out of 20 times the experiment is conducted *if there is really no difference between the populations.* The exact significance level at which populations are assumed to be really different is fuzzy and depends on the size and type of study being conducted. As a rough guideline, however, a significance level of 0.05 is usually accepted as sufficient to assume that the two groups are likely to be from different populations. A level of 0.01 or 1% is considered highly significant.

THE CHI-SQUARE TEST

The chi-square statistic is a mathematical comparison of the data obtained with the data expected according to the rules of probability. A chi-square (χ^2) value represents the degree to which the data fit the hypothesis on which the experiment was based. If chi-square is small, the data fit the hypothesis well; a large chi-square value represents a large deviation and throws doubt on the hypothesis. The actual formula for chi-square need not concern us here. In all problems where we wish to compare the frequencies in an observed sample with an expected frequency distribution we calculate a value known as x^2, which for our purposes may be regarded as a good approximation to χ^2 for large sample sizes.

The formula for x^2 is:

$$x^2 = \sum_{i=1}^{c} \frac{(o_i - e_i)^2}{e_i}$$

where

e_i is the number expected in each class;
o_i is the number observed in each class;
c is the number of classes; and

$\sum_{i=1}^{c}$ means that the expression's value is summed for each class from 1 to c.

To calculate x^2 for a genetics experiment:

1. For each class of offspring, subtract the number expected in the class (e) from the number observed in the class (o) to find ($o - e$).
2. Square this difference $(o - e)^2$.
3. Divide the result by the number expected in the class (e).
4. Repeat for all classes of offspring.
5. Add up (Σ) all the results; the sum is x^2.

Let us apply this to a concrete example.

Problem: A genetics student crosses two fruit flies and obtains 150 offspring. He or she sorts the offspring according to sex and finds 85 females and 65 males. Is this close enough to the expected 75 of each sex so that the observed sex distribution can be regarded as due to chance? (An alternative hypothesis might be that males die as embryos more often than females.)

Solution: The experimenter obtained 150 offspring of which 85 were female and 65 male where you would have expected 75 of each sex. To calculate the x^2 value for these results we can conveniently make a table like Table B–2. The observed and expected numbers of each sex of offspring are entered in the two left columns of the table. Now we subtract the expected from the observed offspring in each class to get the deviation for each class ($o - e$). In the next column we square each deviation. Finally we divide the squared deviation for each class by the expected number in each class and add up all the results to obtain x^2, which in this case is 2.6 ($\Sigma = x^2 = 2.6$).

To determine whether this x^2 value, which we treat as chi-square, is significant, we look at a table of chi-square values. Table B–3 gives an abridged version of such a table. You will see that there is one item in the table that we have not yet mentioned and that we need before we can read off the probability: **degrees of freedom.** The degrees of freedom in a system are usually defined as the number of classes minus one. In this case, with two classes of offspring (male and female), there is only 1 degree of freedom. One way to look at degrees of freedom is to see that it represents the number of choices you can make in assigning an individual to a class. In this case, you have only one decision to make. If you decide that a particular fruit fly is not male, it must be female. If degrees of freedom are thought of as a number of decisions to be made, you can make one less decision than the number of classes, because after that number of decisions, the last class has been determined automatically. In our fruit fly problem we have an x^2 value of 2.6 and 1 degree of freedom. From the table we see that this value falls between 1.6 in the 0.2 (20%) column and 2.7 in the 0.10 (10%) column. This means that we would obtain a chi-square value of 2.6, slightly more than 10% of the time if we performed numerous crosses and counted the sex of the offspring. We say that the chi-square value is **not significant.** The hypothesis that half the offspring of a fruit fly are male and half are female need not be questioned. If we obtained a chi-square value that fell in the 5% or 1% columns, we would suspect that another hypothesis would explain our results better than the one we used. A value in the 5% column would be **significant,** and a value in the 1% column would be **highly significant.**

Here is another example of the application of the chi-square statistic to a genetics problem:

Problem: A polled (hornless) bull with a straight coat (both of which are traits determined by dominant characteristics) is mated with 48 cows that show the recessive alleles of horns and curly coat. The following offspring are produced:

17 polled straight
15 horned curly
8 polled curly } recombinants?
8 horned straight }

TABLE B–2. Genetics Experiment

Class	Observed (o)	Expected (e)	Deviation ($o - e$)	$(o - e)^2$	$\frac{(o-e)^2}{e}$
Females	85	75	10	100	100/75
Males	65	75	−10	100	100/75

TABLE B–3. Chi-Square Values

Degrees of Freedom (n)	Probability Values (P) Deviation from Hypothesis Not Significant								Deviation Significant	Deviation Highly Significant
	0.95	0.90	0.80	0.70	0.50	0.30	0.20	0.10	0.05	0.01
1	0.004	0.016	0.06	0.15	0.46	1.1	1.6	2.7	3.8	6.6
2	0.1	0.2	0.4	0.7	1.4	2.4	3.2	4.6	6.0	9.2
3	0.4	0.6	1.0	1.4	2.4	3.7	4.6	6.3	7.8	11.3
	Chi-square value consistent with hypothesis								Not consistent	

(Modified from Table IV in R.A. Fisher and F. Yates, *Statistical Tables for Biological, Agricultural, and Medical Research* (6th ed., 1974). Published by Longman Group Ltd., London. Previously published by Oliver and Boyd, Ltd., Edinburgh. Reprinted by permission of the authors and publishers.)

The third and fourth phenotypes might represent recombinants.

Do these results provide convincing evidence of linkage between the two loci involved?

Solution: This problem presents us with the following two alternative hypotheses, of which only the second can be tested statistically.

1. The loci are linked and $\frac{16}{48}$ = 33% of the offspring are recombinants. This hypothesis can be tested directly only by carrying out additional breeding experiments.
2. The loci are unlinked, in which case we would expect equal numbers of the four classes of offspring. We can test this hypothesis statistically.

If we can eliminate one of these alternate hypotheses, we shall have provided strong evidence for the other. Since the only hypothesis we can test at the moment is hypothesis 2, we test it. If it must be rejected, we are left with only one alternative, that the first hypothesis is correct.

To test hypothesis 2, we assume that 12 offspring were expected in each of the four classes and calculate a value of x^2 equal to 5.49. In this example there are 3 degrees of freedom because there are 4 possible classes of offspring. From the chi-square table, we find that we would obtain a x^2 value of 5.49 in a system with 3 degrees of freedom between 10% and 20% of the time. By convention this deviation is *not significant* because it would occur by chance more than 5% of the time. We cannot reject hypothesis 2. In this particular case, this means that even if the loci are unlinked, the random fusion of the gametes at fertilization has resulted in offspring whose numbers differ from the 1 : 1 : 1 : 1 ratio of offspring expected for unlinked loci. But the chi-square test tells us that this deviation is not statistically significant. Because in science we can disprove, but never prove, a hypothesis, the answer to the problem is that these results do not provide convincing evidence of linkage between the loci. Only a larger sample size (that is, more calves) would provide the data that would permit us to make the determination whether the loci are linked or not.

Sample size has an important effect on the observed and expected results of an experiment. Intuitively, the larger the sample size taken, the closer you would come to the expected ratio. On the other hand, when sample sizes are small, you would not be surprised that results appear to deviate considerably (but *not significantly!*) from the expected.

There are two limitations on using the chi-square statistic to test closeness of fit between observed and expected results:

1. The chi-square method can be applied only to actual frequencies, not to percentages.
2. Chi-square is not reliable if applied to distributions in which the expected value is less than 5 for any class.

STUDENT'S *t* TEST

This test assumes that the populations under study each fit a normal distribution (bell-shaped curve), that the two populations have common variance, and that the samples within the group and between groups are independent of each other. It may be used to analyze the results from Project B or Project C, as long as data from many plants are included.

The values of t for a significance level of 0.05 are given in Table B–4.

The degrees of freedom, n, is the sum of the degrees of freedom for each class. The degrees of freedom for a class is the number of plants in the class minus 1. If an experiment included 50 experimental and 49 control plants, n would equal $(50 - 1) + (49 - 1) = 97$. The value closest to this in the table would be 120. If your instructor provides you with the value of t for each experiment, you can interpret the results by following these steps:

TABLE B–4. Values for *t* at a Significance Level of 0.05 (one-tailed test)*

Degrees of Freedom (*n*)	Value of *t* (0.05 Significance Level)
1	6.314
2	2.920
3	2.353
4	2.132
5	2.015
6	1.943
7	1.895
8	1.860
9	1.833
10	1.812
15	1.753
20	1.725
25	1.708
30	1.697
40	1.684
60	1.671
120	1.658
∞	1.645

* In student's *t* test, the calculated value of *t* is compared with that in the table. If it is higher than the value in the table, the groups being compared are probably different.

(Modified from Table III in R.A. Fisher and F. Yates, *Statistical Tables for Biological, Agricultural, and Medical Research* (6th ed., 1974). Published by Longman Group Ltd., London. Previously published by Oliver and Boyd, Ltd., Edinburgh. Reprinted by permission of the authors and publishers.)

1. Calculate the degrees of freedom for the experiment from the number of plants involved.
2. Note the value of t that you find in the table for that value of n at the 0.05 significance level.
3. Compare the value of t calculated from the experimental results with the value of t in the table.

If the magnitude of the observed t is *less* than the value in the table, there is *no significant difference* between the groups. We conclude that the experimental treatment of the plants had no effect. If the magnitude of the observed t is *greater* than the value in the table, there is a *significant difference* between the groups. We conclude that the experimental treatment did have an effect.

In our original example, suppose that there were 50 men with a mean of 35.10 ± 3.28*, and 50 women with a mean of 36.00 ± 3.06*. Is this difference in the means significant? The degrees of freedom would be 49 (men) + 49 (women) = 98. Reading from the closest value (120) in Table B–4, we find that the value of t must be at least 1.658 if there is a significant difference at the 0.05 level. In this case, $t = 1.418$, so the difference is not significant.

If you, rather than the instructor, are going to calculate the value of t, use the following formula, which applies to cases where the two groups being compared are equal in size so that $N_1 = N_2 = N$:

$$t = \frac{(\bar{x}_1 - \bar{x}_2)\sqrt{N}}{\sqrt{(s_1)^2 + (s_2)^2}}$$

where

N = number of plants in *EACH* group;
$\bar{x}_1$ = mean of one group of plants;
$\bar{x}_2$ = mean of the second group;
s_1 = standard deviation of the first group;
s_2 = standard deviation of the second group.

Once you have the value for the magnitude of t, follow the steps given above to analyze the results of your experiment.

THE WILCOXON TEST

If your class data for Project B consisted of only a few values, or if the results did not fit a bell-shaped curve, you should use another test, the Wilcoxon (or Mann-Whitney) test, which does not depend on the frequency distribution. This test uses a ranking system to compare data from two samples.

Suppose that an excision experiment were carried out with 12 plants, 6 excised and 6 control, with the results shown in Table B–5. The first step in applying the test is to arrange the data from each group in increasing order of magnitude.

Next, all the data are ranked, keeping track of the data from one of the groups (the control plants) (Table B–6). The rank numbers of the control and excised groups are then added:

control ranks = 1 + 2 + 4 + 5 + 6 + 7 = 25
excised ranks = 3 + 8 + 9 + 10 + 11 + 12 = 53

The lower value, in this case 25 (sum of the control groups ranks), is designated the Wilcoxon statistic, W. The value calculated for W is compared with the critical values given in Table B–7. If the calculated value is equal to or less than the critical value, there is probably a significant difference between the two groups with a significance level of 0.05.

In this example, read down the column of N_1 val-

* Standard deviation of the data points.

TABLE B–5. Excision Experiment

Bud Length in Excised Plants (mm)	Bud Length in Control Plants (mm)
1.0	0.8
2.5	0.9
3.6	1.1
4.2	1.4
7.1	1.5
9.2	1.9

ues (for the group with the smaller mean) to $N_1 = 6$. Then read across the rows of N_2, the second group, to $N_2 = 6$. The critical value of W is 28. Our value of $W = 25$ is less than 28, so the two groups of plants are probably different within our significance level of 0.05.

If W had been 29 or greater, we would have concluded that there was no significant difference between them.

TABLE B–6. Ranked Data

Rank	Length (mm)	C = Control Group
1	0.8	C
2	0.9	C
3	1.0	
4	1.1	C
5	1.4	C
6	1.5	C
7	1.9	C
8	2.5	
9	3.6	
10	4.2	
11	7.1	
12	9.2	

TABLE B–7. Critical Values of *W**

N_1 \ N_2	1	2	3	4	5	6	7	8	9	10	11	12	13	14	15	16	17	18	19	20	21	22	23	24	25
1	—	—	—	—	—	—	—	—	—	—	—	—	—	—	—	—	—	—	1	1	1	1	1	1	1
2	—	—	—	—	3	3	3	4	4	4	4	5	5	6	6	6	6	7	7	7	8	8	8	9	9
3	—	—	6	6	7	8	8	9	10	10	11	11	12	13	13	14	15	15	16	17	17	18	19	19	20
4	—	—	—	11	12	13	14	15	16	17	18	19	20	21	22	24	25	26	27	28	29	30	31	32	33
5	—	—	—	—	19	20	21	23	24	26	27	28	30	31	33	34	35	37	38	40	41	43	44	45	47
6	—	—	—	—	—	28	29	31	33	35	37	38	40	42	44	46	47	49	51	53	55	57	58	60	62
7	—	—	—	—	—	—	39	41	43	45	47	49	52	54	56	58	61	63	65	67	69	72	74	76	78
8	—	—	—	—	—	—	—	51	54	56	59	62	64	67	69	72	75	77	80	83	85	88	90	93	96
9	—	—	—	—	—	—	—	—	66	69	72	75	78	81	84	87	90	93	96	99	102	105	108	111	114
10	—	—	—	—	—	—	—	—	—	82	86	89	92	96	99	103	106	110	113	117	120	123	127	130	134
11	—	—	—	—	—	—	—	—	—	—	100	104	108	112	116	120	123	127	131	135	139	143	147	151	155
12	—	—	—	—	—	—	—	—	—	—	—	120	125	129	133	138	142	146	150	155	159	163	168	172	176
13	—	—	—	—	—	—	—	—	—	—	—	—	142	147	152	156	161	166	171	175	180	185	189	194	199
14	—	—	—	—	—	—	—	—	—	—	—	—	—	166	171	176	182	187	192	197	202	207	212	218	223
15	—	—	—	—	—	—	—	—	—	—	—	—	—	—	192	197	203	208	214	220	225	231	236	242	248
16	—	—	—	—	—	—	—	—	—	—	—	—	—	—	—	219	225	231	237	243	249	255	261	267	273
17	—	—	—	—	—	—	—	—	—	—	—	—	—	—	—	—	249	255	262	268	274	281	287	294	300
18	—	—	—	—	—	—	—	—	—	—	—	—	—	—	—	—	—	280	287	294	301	307	314	321	328
19	—	—	—	—	—	—	—	—	—	—	—	—	—	—	—	—	—	—	313	320	328	335	342	350	357
20	—	—	—	—	—	—	—	—	—	—	—	—	—	—	—	—	—	—	—	348	356	364	371	379	387
21	—	—	—	—	—	—	—	—	—	—	—	—	—	—	—	—	—	—	—	—	385	393	401	410	418
22	—	—	—	—	—	—	—	—	—	—	—	—	—	—	—	—	—	—	—	—	—	424	432	441	450
23	—	—	—	—	—	—	—	—	—	—	—	—	—	—	—	—	—	—	—	—	—	—	465	474	483
24	—	—	—	—	—	—	—	—	—	—	—	—	—	—	—	—	—	—	—	—	—	—	—	507	517
25	—	—	—	—	—	—	—	—	—	—	—	—	—	—	—	—	—	—	—	—	—	—	—	—	552

* If the calculated value of W is equal to or less than the value shown in the table, the two groups are statistically different at a significance level of 0.05. The values of N_1 are the numbers of entries in the first group and the values of N_2 are the numbers in the second group.

(Adapted from Table 1 in L.R. Verdooren, "Extended tables of critical values for Wilcoxon's test statistic," *Biometrika* 50: 177, 1963. Used with permission of the Biometrika Trust.)

APPENDIX C

Laboratory Equipment

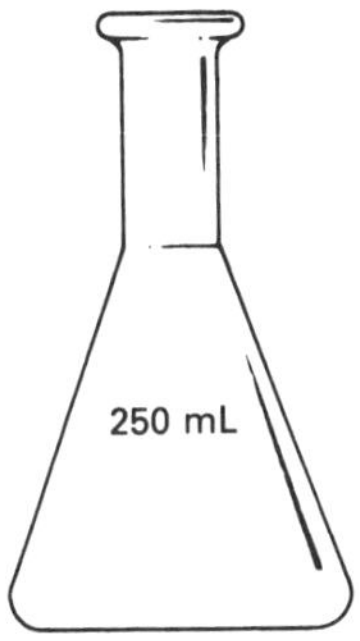

Erlenmeyer flask

test tube

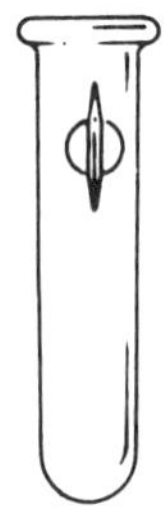
spectrophotometer tube or cuvette

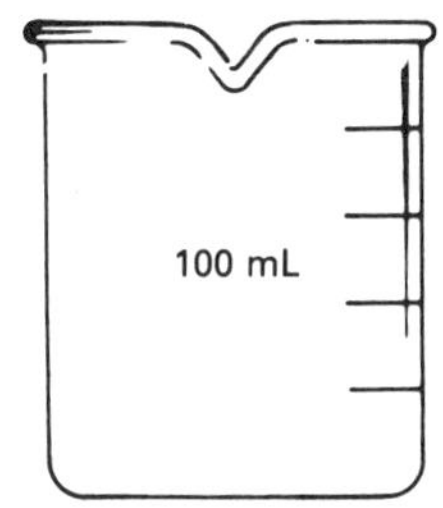

beaker

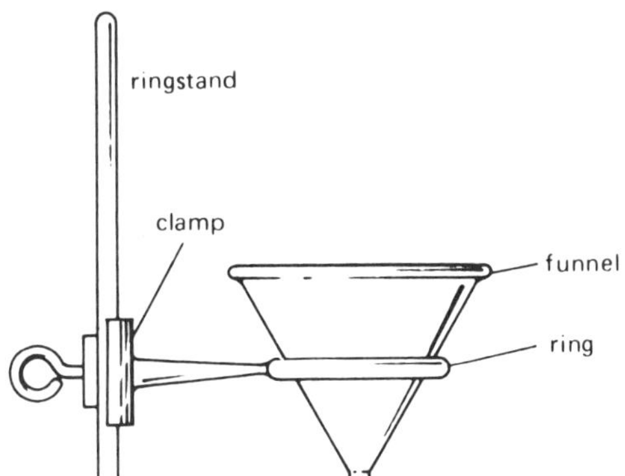

funnel and ringstand

Bunsen burner

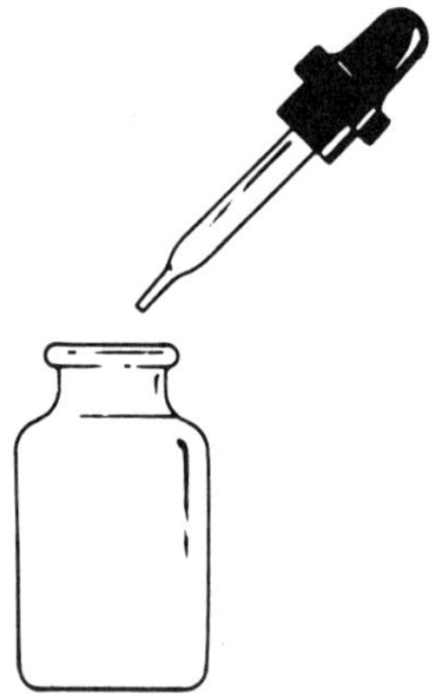
dropping bottle with dropper

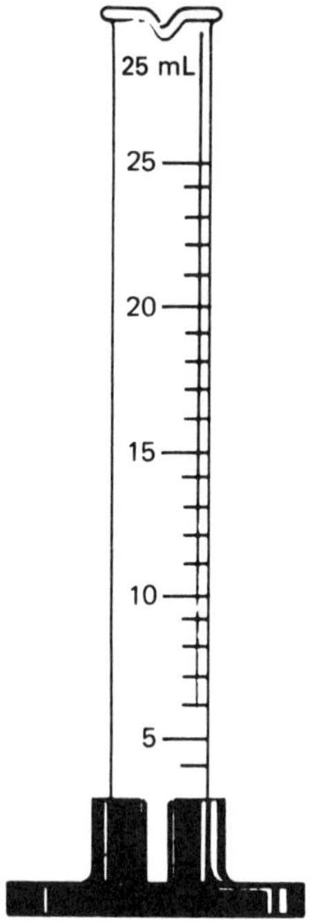

graduated cylinder

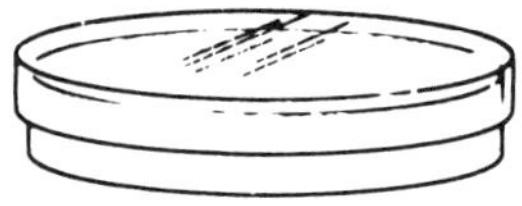
Petri plate

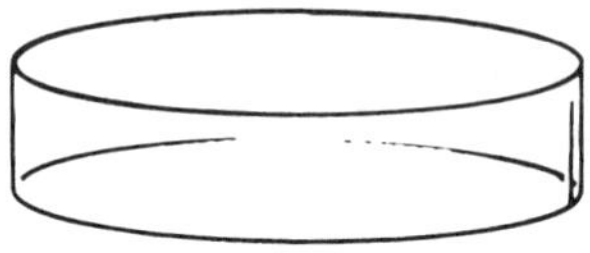
culture dish

microscope slide

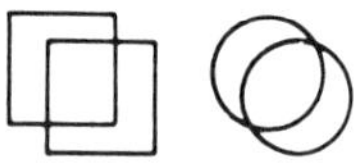
coverslips

microscope depression slide

Syracuse watch glass

graduated pipet

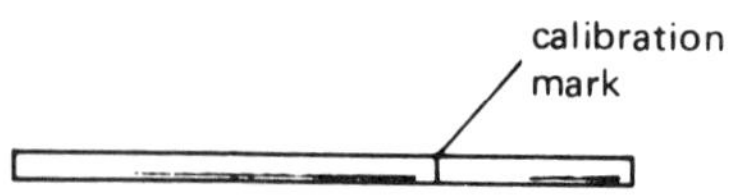

capillary pipet

watch glass

Pasteur pipet and bulb

APPENDIX D

The Metric System

INTRODUCTION

The metric system will be used throughout this course—in lecture, laboratory, and examinations. Read over this information and make sure that you can use the metric system by working the problems. The answers are provided so you can check your work.

In the tables you will find the factors for converting between English and Metric units. These are given for reference only and are rarely used. You should, however, be familiar with a few common *qualitative* equivalents so that you have some frame of reference when using the metric system. Read over the equivalents given for length, weight, volume, and temperature and learn one or two for each type of measurement.

These are the basic units of the metric system (abbreviations are in parentheses):

GRAM	METER	LITER
(g)*	(m)	(L)
mass or weight	length	volume

All other units are powers of 10 smaller or larger than the basic unit. The same prefixes are used to name these units regardless of the basic unit:

larger units ↑			larger units ↑
	kilo- (k)	10^3 × basic unit	
	BASIC UNIT		
	centi- (c)	10^{-2} × basic unit	
	milli- (m)	10^{-3} × basic unit	
	micro- (μ)	10^{-6} × basic unit	
	nano- (n)	10^{-9} × basic unit	
	pico- (p)	10^{-12} × basic unit	
↓ smaller units			↓ smaller units

CONVERSION

Within one type of measurement, such as weight, it is possible to convert between larger and smaller units by simply moving the decimal:

Example: Convert 4729 grams (g) to kilograms (kg).

Solution: Working within weight we change into a *larger* unit so our value will be *smaller*. We will *divide* to obtain a smaller value. We will use the *factor 10^3* because there are 10^3 grams/1 kilogram

$$4{,}729 \text{ g} \div 10^3 = 4.7.2.9. = 4.729 \text{ kg}.$$

Ease of conversion between units is a big advantage of the metric system. Anyone who can change dollars to cents can do it!

INTERCONVERSION

The metric system is set up so that certain units are identical *with respect to the standard reference, water:*

1g	=	1 cm^3	=	1 mL
H_2O		H_2O		H_2O

So 27 mL of H_2O will occupy a cube 3 cm on each edge with a volume of 27 cm^3 (3 cm × 3 cm × 3 cm =

* By convention, metric abbreviations never have periods or use "s" for plural: 1 gram (1 g) and 187 grams (187 g).

$27\ cm^3$), and weigh 27 g. For substances other than H_2O the density (g/mL) must be known to interconvert.

1. The first step in interconversion is to convert to one of these equivalent units:

   Weights must be converted to **grams.**
   Lengths must be converted to **cubic centimeters.**
   Volumes must be converted to **milliliters.**

2. Next, the units are interconverted (no math involved):

$$\mathbf{g = cm^3 = mL}$$

3. Finally the value is converted to the desired unit within that type of measurement:

$$\mathbf{g \times 1000\ mg/g = mg,}\ \text{etc.}$$

Example: Your queen-sized water bed measures 2.1 meters (m) × 1.5 m × 0.25 m. How much does it weigh in kg?

Solution: The volume of the bed needs to be expressed in cm^3 before converting to weight:

$210\ cm \times 150\ cm \times 25\ cm = 7.9 \times 10^5\ cm^3$
Then $7.9 \times 10^5\ cm^3$ is directly equivalent to 7.9×10^5 g
Finally $7.9 \times 10^5\ g = 790000\ g = 790\ kg$

Interconversion is a powerful tool that allows you to make "back of the envelope" calculations amazingly quickly.

PROBLEMS

Test yourself by working the following interconversion problems and checking your work with the answers below. If you can do the problems with ease, then you are in good shape. If these conversions are not very automatic for you, see a course instructor for individual help.

1. 156 centimeters = ____________ meters.
2. A cube 2 cm on each edge will hold water weighing ____________ grams.
3. 9 kilometers = ____________ meters.
4. A Pinto engine has a displacement of 1600 cm^3 or ____________ liters.
5. A wrestler sweats off 625 mL of "water." His weight loss would be ____________ kg.
6. If a cell has a diameter of 50 μm, its width is ____________ mm.
7. How many such cells could fit under a coverslip measuring 1 cm × 1 cm? (Assume "square" cells tightly packed together.) ____________
8. A grain of sand weighs 4.6 mg or ____________ μg.
9. 8 mm^3 of water will weigh ____________ mg. (Don't guess, work it out.)
10. The length of a certain caterpillar is 6.6 cm or ____________ mm.

Answers: (1) 1.56 m (2) 8 g (3) 9,000 m (4) 1.6 L (5) 0.625 kg (6) 0.050 mm (7) 40,000 cells (8) 4,600 μg (9) 8 mg (10) 66 mm

LENGTH

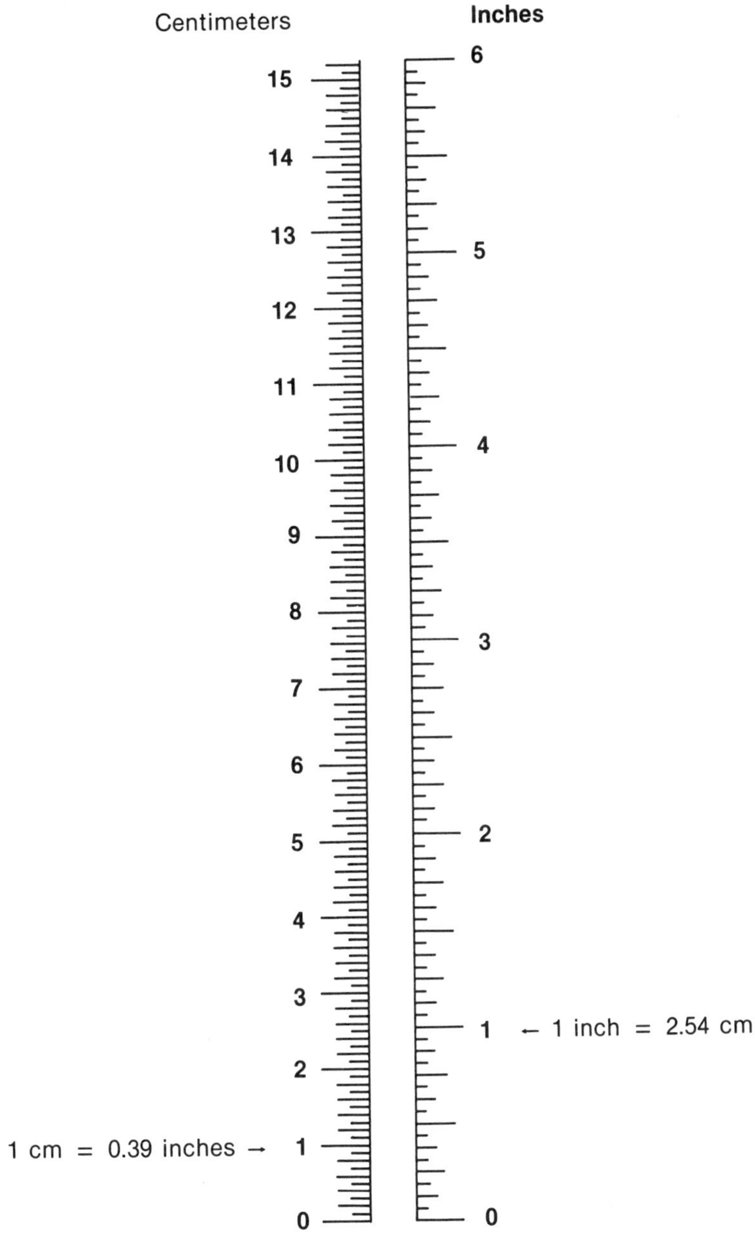

A speed limit of 85 kph is about 55 mph

A kilometer is about 0.6 miles

A meter is a little more than a yard (39 inches)

A centimeter is about the width of your index fingernail

Length Conversion

1 in. = 2.54 cm	1 mm = .04 in.
1 ft = 30.5 cm	1 cm = .39 in.
1 yd = .91 m	1 m = 39.4 in.
1 mi = 1.61 km	1 m = 1.1 yd
	1 km = .62 mi

WEIGHT

A large man weighs
120 kilograms

A small woman weighs
50 kilograms

This book weighs
about 0.8 kg

A pound of cheese
weighs about half a
kilogram

A teaspoon of sugar weighs
about 4 grams

A grain of salt weighs
about 1 milligram

VOLUME

A large auto gas tank
holds about 100 liters

One liter is a
standard "quart"
bottle of
soda, milk or wine

A cup of coffee is
about 150 mL

5 milliliters is a
standard teaspoon of
medicine

A drop of water is
about 50 microliters

Weight Conversion

1 oz. = 28 g	1 g = .035 oz.
1 lb = .45 kg	1 kg = 2.2 lb

Volume Conversion

1 tsp. = 5 mL	1 mL = .03 fl. oz.
1 tbsp. = 15 mL	1 L = 2.1 pt.
1 fl. oz. = 30 mL	1 L = 1.06 qt.
1 cup = .24 L	1 L = .26 gal.
1 pt. = .47 L	
1 qt. = .95 L	
1 gal. = 3.8 L	

TEMPERATURE

Fahrenheit °F | **Centigrade (Celsius) °C**

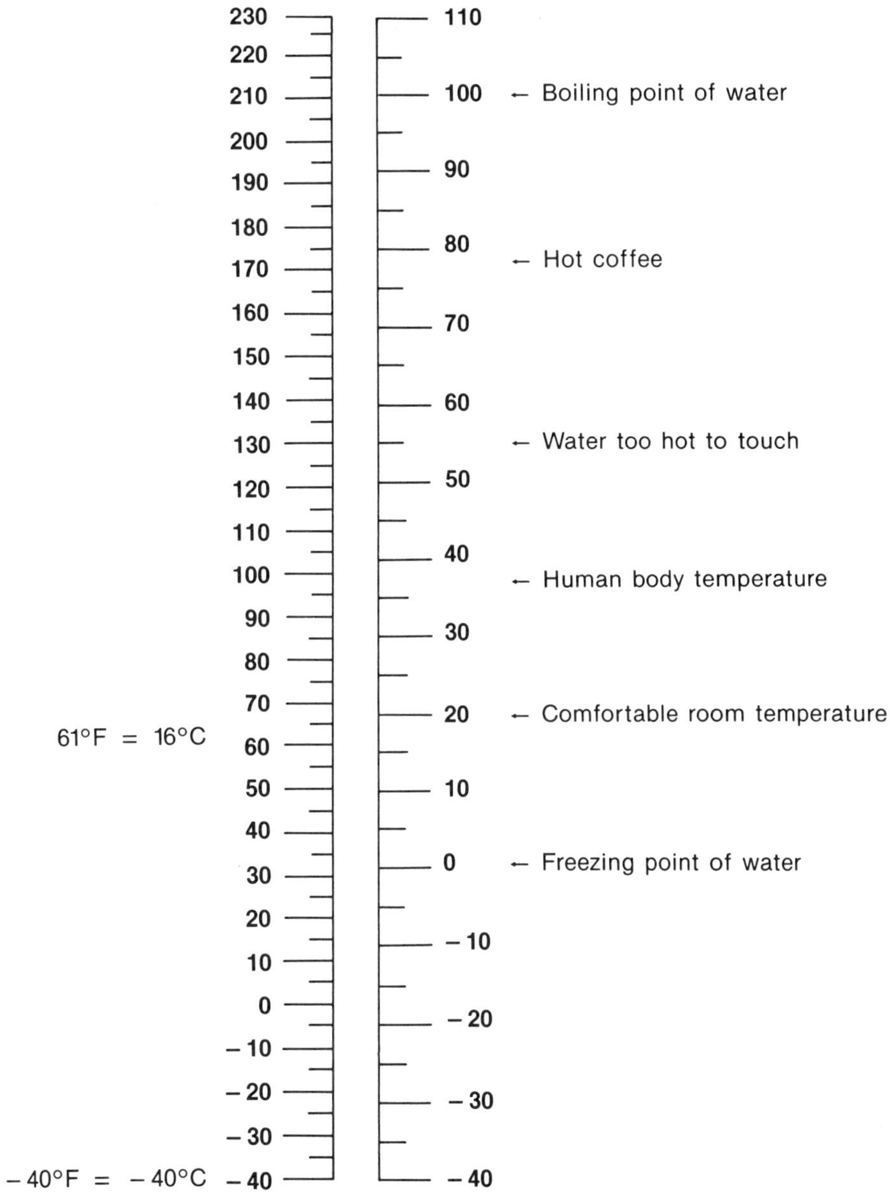

Temperature Conversion

$$°C = \frac{(°F - 32) \times 5}{9}$$

$$°F = \frac{°C \times 9}{5} + 32$$

$$1°C = 1.8°F$$

Index

NOTES

NOTES

NOTES

NOTES

NOTES

NOTES

NOTES

NOTES

NOTES

NOTES

NOTES

NOTES

NOTES

NOTES